PENGUIN BOOKS

THE BLANK SLATE

Steven Pinker is Johnstone Professor of Psychology at Harvard University. His research on visual cognition and the psychology of language has earned prizes from the National Academy of Sciences and the American Psychological Association. Pinker has also received many awards for his teaching at MIT and for his books *How the Mind Works* (which was also a finalist for the Pulitzer Prize) and *The Language Instinct*. He is an elected fellow of several scientific societies, associate editor of *Cognition*, and a member of the usage panel of the *American Heritage Dictionary*. He has written for *The New York Times, Time, The New Yorker, The New Republic, Slate*, and *Technology Review*.

∼

Praise for *The Blank Slate*

"A brilliant and forceful summary . . . A well-informed and well-written account of [human] limitations, [written with] a graceful interleaving of scientific and literary sources. . . . [This] fine book helps with a task that we all must begin to take seriously . . . Can it be that we have finally grown up?"
—Melvin Konner, *The American Prospect*

"This is a brilliant book. It is beautifully written, and addresses profound issues with courage and clarity. There is nothing else like it, and it is going to have an impact that extends well beyond the scientific academy."
—Paul Bloom, *Trends in Cognitive Sciences*

"Steven Pinker has written an extremely good book—clear, well argued, fair, learned, tough, witty, humane, stimulating. I only hope that people study it carefully before rising up ideologically against him. If they do, th   e idea of an innately flawed but wonderfully rich human nature   d, not evil."   —Colin McGinn, *T   st*

"Steven Pinker is a man of encyclopedic knowledge and an i   :-gument. His argument in *The Blank Slate* is that intellectual l   d much of our social and political policy, was increasingly d   h the twentieth century by a view of human nature that is fun   l; that this domination has been backed by something that am   c terrorism (he does not put it quite so strongly): and that   it

substantially from a more realistic view. Pinker's exposition is thoroughly readable and of enviable clarity. His explanation of such a difficult technical matter as the analysis of variance and regression in twin studies, for example, would be very hard to better. He is not afraid of using strong language . . . in addition, parts of the book are delightfully funny."

—John R. G. Turner, *The Times Literary Supplement*

"Anyone who has read Pinker's earlier books—including *How the Mind Works* and *The Language Instinct*—will rightly guess that his latest effort is similarly sweeping, erudite, sharply argued, richly footnoted and fun to read. It's also highly persuasive."

—Michael Lemonick, *Time*

"[Pinker] makes his main argument persuasively and with great verve. . . . *The Blank Slate* ought to be read by anybody who feels they have had enough of nature-nurture rows or who thinks they already know where they stand on the science wars. It could change their minds. . . . If nothing else, Mr. Pinker's book is a wonderfully readable taster of new research, much of it ingenious, designed to show that many more of our emotional biases and mental aptitudes than previously thought are hard-wired or, to use the old word, innate. . . . This is a breath of air for a topic that has been politicized for too long." —*The Economist*

"[Pinker] wades resolutely into the comforting gloom surrounding these not quite forbidden topics and calmly, lucidly marshals the facts to ground his strikingly subversive Darwinian claims—subversive not of any of the things we properly hold dear but subversive of the phony protective layers of misinformation surrounding them. . . . My reservations with Pinker's view [will be resolved] in the bright light of rational inquiry that he brings to these important topics."

—Dan Dennett, *The Times Literary Supplement*

"*The Blank Slate* brilliantly delineates the current state of play in the nature-nurture debate. Read it to understand not just the moral and aesthetic blindness of your friends, but the misguided idealism of nations. A magnificent and timely work."

—Fay Weldon, *The Daily Telegraph*

"[Pinker] points us in the direction of a more productive debate, a debate in which the implications of science are confronted forthrightly and not simply wished away by politicized scientists."

—Francis Fukuyama, *The Wall Street Journal*

"*The Blank Slate* deserves to be read carefully and with an open mind...This landmark book makes an important contribution to the argument about nature vs. nurture in humans. Whether or not most readers end up on Pinker's side of the fence, one can hope that his thoroughness and reasoning will shed light into the darker corners where research has been suppressed by taboos, and where freedom of thought and speech have been inhibited by fear of consequences for asking forbidden questions."

—Nancy Jeannette Friedlander, *The San Diego Union-Tribune*

"This book is a modern magnum opus. The scholarship alone is mind-boggling, a monument of careful research, meticulous citation, breadth of input from diverse fields, great writing and humor." —Tom Paskal, *The Montreal Gazette*

"A delightfully provocative read...A constantly dynamic, if tacit, exchange between the author and his readers." —Patrick Watson, *The Globe and Mail*

"A feast of a book. Pinker's analytical and impish mind ranges from Charles Darwin to Abigail Van Buren, from scientific studies to *Annie Hall*....It will be a rare reader who agrees with everything in this book. But it is an intelligent book that says what it means and thinks about what it is saying....Though much of the book is about human differences, the bigger idea is inherited similarity—the 'psychological unity of our species.' It is not a blank slate but a slate with a face—a face that might be called human nature. When Pinker starts describing it, the reader will surely recognize it." —Bruce Ramsey, *The Seattle Times*

THE BLANK SLATE

The Modern Denial

of Human Nature

Steven Pinker

PENGUIN BOOKS

To Don, Judy, Leda, and John

PENGUIN BOOKS
Published by the Penguin Group
Penguin Group (USA) Inc., 375 Hudson Street, New York, New York 10014, U.S.A.
Penguin Books Ltd, 80 Strand, London WC2R 0RL, England
Penguin Books Australia Ltd, 250 Camberwell Road, Camberwell, Victoria 3124, Australia
Penguin Books Canada Ltd, 10 Alcorn Avenue, Toronto, Ontario, Canada M4V 3B2
Penguin Books India (P) Ltd, 11 Community Centre, Panchsheel Park, New Delhi – 110 017, India
Penguin Books (N.Z.) Ltd, Cnr Rosedale and Airborne Roads, Albany, Auckland, New Zealand
Penguin Books (South Africa) (Pty) Ltd, 24 Sturdee Avenue, Rosebank, Johannesburg 2196, South Africa

Penguin Books Ltd, Registered Offices: 80 Strand, London WC2R 0RL, England

First published in the United States of America by Viking Penguin,
a member of Penguin Putnam Inc. 2002
Published in Penguin Books 2003

1 3 5 7 9 10 8 6 4 2

Grateful acknowledgment is made for permission to reprint the following copyrighted material. *Page 22*: Lyrics from "A Simple Desultory Philippic (or How I Was Robert McNamara'd into Submission)"; copyright © 1965, Paul Simon; used by permission of the publisher: Paul Simon Music. *Page 57*: Chart, "Percentage of Male Deaths Caused by Warfare," from *War Before Civilization* by Lawrence H. Keeley, copyright © 1996 by Oxford University Press, Inc.; used by permission of Oxford University Press, Inc. *Page 88*: Diagram of the wiring of the primate visual system from Michael Gazzaniga, *The Cognitive Neurosciences*, The MIT Press (1996). *Page 179*: Lyrics from "Gee, Officer Krupke" by Leonard Bernstein & Stephen Sondheim; © 1956, Amberson Holdings LLC and Stephen Sondheim; copyright renewed; Leonard Bernstein Music Publishing Company LLC, publisher; used by permission. *Page 199*: Diagram, "Turning the Tables," from *Mind Sights* by Roger N. Shepard, © 1990 by Roger N. Shepard; reprinted by permission of Henry Holt and Company, LLC. "Checker Shadow Illusion" © Edward Adelson, 2002; reprinted with permission. *Page 326*: Lyrics from "You Don't Mess Around with Jim," written by Jim Croce; © 1972 (renewed), Time in a Bottle/Croce Publishing (ASCAP); all rights reserved; used by permission.

THE LIBRARY OF CONGRESS HAS CATALOGUED THE HARDCOVER EDITION AS FOLLOWS:
Pinker, Steven, 1954–
The blank slate : the modern denial of human nature / Steven Pinker.
p. cm.
Includes bibliographical references (p.) and index.
ISBN 0-670-03151-8 (hc.)
ISBN 0 14 20.0334 4 (pbk.)
1. Nature and nurture. I. Title.
BF341 .P47 2002
155.2'34—dc21   2002022719

Printed in the United States of America
Set in Minion

PREFACE

"Not ANOTHER BOOK on nature and nurture! Are there really people out there who still believe that the mind is a blank slate? Isn't it obvious to anyone with more than one child, to anyone who has been in a heterosexual relationship, or to anyone who has noticed that children learn language but house pets don't, that people are born with certain talents and temperaments? Haven't we all moved beyond the simplistic dichotomy between heredity and environment and realized that all behavior comes out of an interaction between the two?"

This is the kind of reaction I got from colleagues when I explained my plans for this book. At first glance the reaction is not unreasonable. Maybe nature versus nurture *is* a dead issue. Anyone familiar with current writings on mind and behavior has seen claims to the middle ground like these:

> If the reader is now convinced that either the genetic or environmental explanation has won out to the exclusion of the other, we have not done a sufficiently good job of presenting one side or the other. It seems highly likely to us that both genes and environment have something to do with this issue. What might the mix be? We are resolutely agnostic on that issue; as far as we can determine, the evidence does not yet justify an estimate.

> This is not going to be one of those books that says everything is genetic: it isn't. The environment is just as important as the genes. The things children experience while they are growing up are just as important as the things they are born with.

> Even when a behavior is heritable, an individual's behavior is still a product of development, and thus it has a causal environmental component. . . . The modern understanding of how phenotypes are inherited through the replication of both genetic and environmental

conditions suggests that . . . cultural traditions—behaviors copied by children from their parents—are likely to be crucial.

If you think these are innocuous compromises that show that everyone has outgrown the nature-nurture debate, think again. The quotations come, in fact, from three of the most incendiary books of the last decade. The first is from *The Bell Curve* by Richard Herrnstein and Charles Murray, who argue that the difference in average IQ scores between American blacks and American whites has both genetic and environmental causes.[1] The second is from *The Nurture Assumption* by Judith Rich Harris, who argues that children's personalities are shaped by their genes as well as by their environments, so similarities between children and their parents may come from their shared genes and not just from the effects of parenting.[2] The third is from *A Natural History of Rape* by Randy Thornhill and Craig Palmer, who argue that rape is not simply a product of culture but also has roots in the nature of men's sexuality.[3] For invoking nurture *and* nature, not nurture alone, these authors have been picketed, shouted down, subjected to searing invective in the press, even denounced in Congress. Others expressing such opinions have been censored, assaulted, or threatened with criminal prosecution.[4]

The idea that nature and nurture interact to shape some part of the mind might turn out to be wrong, but it is *not* wishy-washy or unexceptionable, even in the twenty-first century, thousands of years after the issue was framed. When it comes to explaining human thought and behavior, the possibility that heredity plays any role at all still has the power to shock. To acknowledge human nature, many think, is to endorse racism, sexism, war, greed, genocide, nihilism, reactionary politics, and neglect of children and the disadvantaged. Any claim that the mind has an innate organization strikes people not as a hypothesis that might be incorrect but as a thought it is immoral to think.

This book is about the moral, emotional, and political colorings of the concept of human nature in modern life. I will retrace the history that led people to see human nature as a dangerous idea, and I will try to unsnarl the moral and political rat's nests that have entangled the idea along the way. Though no book on human nature can hope to be uncontroversial, I did not write it to be yet another "explosive" book, as dust jackets tend to say. I am not, as many people assume, countering an extreme "nurture" position with an extreme "nature" position, with the truth lying somewhere in between. In some cases, an extreme environmentalist explanation is correct: which language you speak is an obvious example, and differences among races and ethnic groups in test scores may be another. In other cases, such as certain inherited neurological disorders, an extreme hereditarian explanation is correct. In most cases the correct explanation will invoke a complex interaction between heredity and environment: culture is crucial, but culture could not exist without mental faculties that allow humans to create and learn culture to begin with. My goal in this book is not to argue that genes are everything and culture is nothing—no one believes that—but to explore why the extreme position (that culture is everything) is so often seen as moderate, and the moderate position is seen as extreme.

Nor does acknowledging human nature have the political implications so many fear. It does not, for example, require one to abandon feminism, or to accept current levels of inequality or violence, or to treat morality as a fiction. For the most part I will try not to advocate particular policies or to advance the agenda of the political left or right. I believe that controversies about policy almost always involve tradeoffs between competing values, and that science is equipped to identify the tradeoffs but not to resolve them. Many of these tradeoffs, I will show, arise from features of human nature, and by clarifying them I hope to make our collective choices, whatever they are, better informed. If I am an advocate, it is for discoveries about human nature that have been ignored or suppressed in modern discussions of human affairs.

Why is it important to sort this all out? The refusal to acknowledge human nature is like the Victorians' embarrassment about sex, only worse: it distorts our science and scholarship, our public discourse, and our day-to-day lives. Logicians tell us that a single contradiction can corrupt a set of statements and allow falsehoods to proliferate through it. The dogma that human nature does not exist, in the face of evidence from science and common sense that it does, is just such a corrupting influence.

First, the doctrine that the mind is a blank slate has distorted the study of human beings, and thus the public and private decisions that are guided by that research. Many policies on parenting, for example, are inspired by research that finds a correlation between the behavior of parents and the behavior of children. Loving parents have confident children, authoritative parents (neither too permissive nor too punitive) have well-behaved children, parents who talk to their children have children with better language skills, and so on. Everyone concludes that to grow the best children, parents must be loving, authoritative, and talkative, and if children don't turn out well it must be the parents' fault. But the conclusions depend on the belief that children are blank slates. Parents, remember, provide their children with genes, not just a home environment. The correlations between parents and children may be telling us only that the same genes that make adults loving, authoritative, and talkative make their children self-confident, well-behaved, and articulate. Until the studies are redone with adopted children (who get only their environment, not their genes, from their parents), the data are compatible with the possibility that genes make all the difference, the possibility that parenting makes all the difference, or anything in between. Yet in almost every instance, the most extreme position—that parents are everything—is the only one researchers entertain.

The taboo on human nature has not just put blinkers on researchers but turned any discussion of it into a heresy that must be stamped out. Many writers are so desperate to discredit any suggestion of an innate human constitution that they have thrown logic and civility out the window. Elementary distinctions—"some" versus "all," "probable" versus "always," "is" versus "ought"—are eagerly flouted to paint human nature as an extremist doctrine and thereby steer readers away from it. The analysis of ideas is commonly replaced by political smears and personal attacks. This poisoning of the intellectual atmosphere has left us unequipped to analyze pressing issues about human nature just as new scientific discoveries are making them acute.

The denial of human nature has spread beyond the academy and has led to a disconnect between intellectual life and common sense. I first had the idea of writing this book when I started a collection of astonishing claims from pundits and social critics about the malleability of the human psyche: that little boys quarrel and fight because they are encouraged to do so; that children enjoy sweets because their parents use them as a reward for eating vegetables; that teenagers get the idea to compete in looks and fashion from spelling bees and academic prizes; that men think the goal of sex is an orgasm because of the way they were socialized. The problem is not just that these claims are preposterous but that the writers did not acknowledge they were saying things that common sense might call into question. This is the mentality of a cult, in which fantastical beliefs are flaunted as proof of one's piety. That mentality cannot coexist with an esteem for the truth, and I believe it is responsible for some of the unfortunate trends in recent intellectual life. One trend is a stated contempt among many scholars for the concepts of truth, logic, and evidence. Another is a hypocritical divide between what intellectuals say in public and what they really believe. A third is the inevitable reaction: a culture of "politically incorrect" shock jocks who revel in anti-intellectualism and bigotry, emboldened by the knowledge that the intellectual establishment has forfeited claims to credibility in the eyes of the public.

Finally, the denial of human nature has not just corrupted the world of critics and intellectuals but has done harm to the lives of real people. The theory that parents can mold their children like clay has inflicted childrearing regimes on parents that are unnatural and sometimes cruel. It has distorted the choices faced by mothers as they try to balance their lives, and multiplied the anguish of parents whose children haven't turned out the way they hoped. The belief that human tastes are reversible cultural preferences has led social planners to write off people's enjoyment of ornament, natural light, and human scale and force millions of people to live in drab cement boxes. The romantic notion that all evil is a product of society has justified the release of dangerous psychopaths who promptly murdered innocent people. And the conviction that humanity could be reshaped by massive social engineering projects led to some of the greatest atrocities in history.

Though many of my arguments will be coolly analytical—that an acknowledgment of human nature does not, logically speaking, imply the negative outcomes so many people fear—I will not try to hide my belief that they have a positive thrust as well. "Man will become better when you show him what he is like," wrote Chekhov, and so the new sciences of human nature can help lead the way to a realistic, biologically informed humanism. They expose the psychological unity of our species beneath the superficial differences of physical appearance and parochial culture. They make us appreciate the wondrous complexity of the human mind, which we are apt to take for granted precisely because it works so well. They identify the moral intuitions that we can put to work in improving our lot. They promise a naturalness in human relationships, encouraging us to treat people in terms of how they do feel rather than how some theory says they ought to feel. They offer a touchstone by which we can identify suffering and oppression wherever they occur, unmasking the rationalizations of the powerful. They give us a way to see through the designs of self-appointed social reformers who would liberate us from our pleasures. They renew our appreciation for the achievements of democracy and of the rule of law. And they enhance the insights of artists and philosophers who have reflected on the human condition for millennia.

An honest discussion of human nature has never been more timely. Throughout the twentieth century, many intellectuals tried to rest principles of decency on fragile factual claims such as that human beings are biologically indistinguishable, harbor no ignoble motives, and are utterly free in their ability to make choices. These claims are now being called into question by discoveries in the sciences of mind, brain, genes, and evolution. If nothing else, the completion of the Human Genome Project, with its promise of an unprecedented understanding of the genetic roots of the intellect and the emotions, should serve as a wake-up call. The new scientific challenge to the denial of human nature leaves us with a challenge. If we are not to abandon values such as peace and equality, or our commitments to science and truth, then we must pry these values away from claims about our psychological makeup that are vulnerable to being proven false.

This book is for people who wonder where the taboo against human nature came from and who are willing to explore whether the challenges to the taboo are truly dangerous or just unfamiliar. It is for those who are curious about the emerging portrait of our species and curious about the legitimate criticisms of that portrait. It is for those who suspect that the taboo against human nature has left us playing without a full deck as we deal with the pressing issues confronting us. And it is for those who recognize that the sciences of mind, brain, genes, and evolution are permanently changing our view of ourselves and wonder whether the values we hold precious will wither, survive, or (as I will argue) be enhanced.

~

IT IS A pleasure to acknowledge the friends and colleagues who improved this book in innumerable ways. Helena Cronin, Judith Rich Harris, Geoffrey Miller, Orlando Patterson, and Donald Symons offered deep and insightful analyses of every aspect, and I can only hope that the final version is worthy of their wisdom. I profited as well from invaluable comments by Ned Block, David Buss, Nazli Choucri, Leda Cosmides, Denis Dutton, Michael Gazzaniga, David Geary, George Graham, Paul Gross, Marc Hauser, Owen Jones, David Kemmerer, David Lykken, Gary Marcus, Roslyn Pinker, Robert Plomin, James Rachels, Thomas Sowell, John Tooby, Margo Wilson, and William Zimmerman. My thanks also go to the colleagues who reviewed chapters in their areas of expertise: Josh Cohen, Richard Dawkins, Ronald Green, Nancy Kanwisher, Lawrence Katz, Glenn Loury, Pauline Maier, Anita Patterson, Mriganka Sur, and Milton J. Wilkinson.

I thank many others who graciously responded to requests for information or offered suggestions that found their way into the book: Mahzarin Banaji, Chris Bertram, Howard Bloom, Thomas Bouchard, Brian Boyd, Donald Brown, Jennifer Campbell, Rebecca Cann, Susan Carey, Napoleon Chagnon, Martin Daly, Irven DeVore, Dave Evans, Jonathan Freedman, Jennifer Ganger, Howard Gardner, Tamar Gendler, Adam Gopnik, Ed Hagen, David Housman, Tony Ingram, William Irons, Christopher Jencks, Henry Jenkins, Jim Johnson, Erica Jong, Douglas Kenrick, Samuel Jay Keyser, Stephen Kosslyn, Robert Kurzban, George Lakoff, Eric Lander, Loren Lomasky, Martha Nussbaum, Mary Parlee, Larry Squire, Wendy Steiner, Randy Thornhill, James Watson, Torsten Wiesel, and Robert Wright.

The themes of this book were first presented at forums whose hosts and audiences provided vital feedback. They include the Center for Bioethics at the University of Pennsylvania; the Cognition, Brain, and Art Symposium at the Getty Research Institute; the Developmental Behavior Genetics conference at the University of Pittsburgh; the Human Behavior and Evolution Society; the Humane Leadership Project at the University of Pennsylvania; the Institute on Race and Social Division at Boston University; the School of Humanities, Arts, and Social Sciences at MIT; the Neurosciences Research Program at the Neurosciences Institute; the Positive Psychology Summit; the Society for Evolutionary Analysis in Law; and the Tanner Lectures on Human Values at Yale University.

I am happy to acknowledge the superb environment for teaching and inquiry at the Massachusetts Institute of Technology, and the support of Mriganka Sur, head of the Department of Brain and Cognitive Sciences, Robert

Silbey, dean of the School of Science, Charles Vest, president of MIT, and many colleagues and students. John Bearley, the librarian of the Teuber Library, tracked down scholarly materials and answers to questions no matter how obscure. I also gratefully acknowledge the financial support of the MIT MacVicar Faculty Fellows program and the Peter de Florez chair. My research on language is supported by NIH Grant HD18381.

Wendy Wolf at Viking Penguin and Stefan McGrath at Penguin Books provided excellent advice and welcome good cheer. I thank them and my agents, John Brockman and Katinka Matson, for their efforts on behalf of the book. I am delighted that Katya Rice agreed to copy-edit this book, our fifth collaboration.

My heartfelt appreciation goes to my family, the Pinkers, Boodmans, and Subbiah-Adamses, for their love and support. Special thanks to my wife, Ilavenil Subbiah, for her wise advice and loving encouragement.

This book is dedicated to four people who have been dear friends and profound influences: Donald Symons, Judith Rich Harris, Leda Cosmides, and John Tooby.

CONTENTS

THE BLANK SLATE

PART I

THE BLANK SLATE, THE NOBLE
SAVAGE, AND THE GHOST
IN THE MACHINE

Everyone has a theory of human nature. Everyone has to anticipate the behavior of others, and that means we all need theories about what makes people tick. A tacit theory of human nature—that behavior is caused by thoughts and feelings—is embedded in the very way we think about people. We fill out this theory by introspecting on our own minds and assuming that our fellows are like ourselves, and by watching people's behavior and filing away generalizations. We absorb still other ideas from our intellectual climate: from the expertise of authorities and the conventional wisdom of the day.

Our theory of human nature is the wellspring of much in our lives. We consult it when we want to persuade or threaten, inform or deceive. It advises us on how to nurture our marriages, bring up our children, and control our own behavior. Its assumptions about learning drive our educational policy; its assumptions about motivation drive our policies on economics, law, and crime. And because it delineates what people can achieve easily, what they can achieve only with sacrifice or pain, and what they cannot achieve at all, it affects our values: what we believe we can reasonably strive for as individuals and as a society. Rival theories of human nature are entwined in different ways of life and different political systems, and have been a source of much conflict over the course of history.

For millennia, the major theories of human nature have come from religion.[1] The Judeo-Christian tradition, for example, offers explanations for much of the subject matter now studied by biology and psychology. Humans are made in the image of God and are unrelated to animals.[2] Women are derivative of men and destined to be ruled by them.[3] The mind is an immaterial substance: it has powers possessed by no purely physical structure, and can continue to exist when the body dies.[4] The mind is made up of several components, including a moral sense, an ability to love, a capacity for reason that recognizes whether an act conforms to ideals of goodness, and a decision faculty that chooses how to behave. Although the decision faculty is not bound by the laws of cause and effect, it has an innate tendency to choose sin. Our cognitive and perceptual faculties work accurately because God implanted ideals in them that correspond to reality and because he coordinates their functioning with the outside world. Mental health comes from recognizing God's purpose, choosing good and repenting sin, and loving God and one's fellow humans for God's sake.

The Judeo-Christian theory is based on events narrated in the Bible. We know that the human mind has nothing in common with the minds of animals because the Bible says that humans were created separately. We know that the design of women is based on the design of men because in the second telling of the creation of women Eve was fashioned from the rib of Adam. Human decisions cannot be the inevitable effects of some cause, we may surmise, because God held Adam and Eve responsible for eating the fruit of the tree of knowledge, implying that they could have chosen otherwise. Women are dominated by men as punishment for Eve's disobedience, and men and women inherit the sinfulness of the first couple.

The Judeo-Christian conception is still the most popular theory of human nature in the United States. According to recent polls, 76 percent of Americans believe in the biblical account of creation, 79 percent believe that the miracles in the Bible actually took place, 76 percent believe in angels, the devil, and other immaterial souls, 67 percent believe they will exist in some form after their death, and only 15 percent believe that Darwin's theory of evolution is the best explanation for the origin of human life on Earth.[5] Politicians on the right embrace the religious theory explicitly, and no mainstream politician would dare contradict it in public. But the modern sciences of cosmology, geology, biology, and archaeology have made it impossible for a scientifically literate person to believe that the biblical story of creation actually took place. As a result, the Judeo-Christian theory of human nature is no longer explicitly avowed by most academics, journalists, social analysts, and other intellectually engaged people.

Nonetheless, every society must operate with a theory of human nature, and our intellectual mainstream is committed to another one. The theory is seldom articulated or overtly embraced, but it lies at the heart of a vast number of beliefs and policies. Bertrand Russell wrote, "Every man, wherever he goes, is encompassed by a cloud of comforting convictions, which move with him like flies on a summer day." For intellectuals today, many of those convictions are about psychology and social relations. I will refer to those convictions as the Blank Slate: the idea that the human mind has no inherent structure and can be inscribed at will by society or ourselves.

That theory of human nature—namely, that it barely exists—is the topic of this book. Just as religions contain a theory of human nature, so theories of human nature take on some of the functions of religion, and the Blank Slate has become the secular religion of modern intellectual life. It is seen as a source of values, so the fact that it is based on a miracle—a complex mind arising out of nothing—is not held against it. Challenges to the doctrine from skeptics and scientists have plunged some believers into a crisis of faith and have led others to mount the kinds of bitter attacks ordinarily aimed at heretics and infidels. And just as many religious traditions eventually reconciled themselves to apparent threats from science (such as the revolutions of Copernicus and Darwin), so, I argue, will our values survive the demise of the Blank Slate.

The chapters in this part of the book (Part I) are about the ascendance of the Blank Slate in modern intellectual life, and about the new view of human nature and culture that is beginning to challenge it. In succeeding parts we will witness the anxiety evoked by this challenge (Part II) and see how the anxiety may be assuaged (Part III). Then I will show how a richer conception of human nature can provide insight into language, thought, social life, and morality (Part IV) and how it can clarify controversies on politics, violence, gender, childrearing, and the arts (Part V). Finally I will show how the passing of the Blank Slate is less disquieting, and in some ways less revolutionary, than it first appears (Part VI).

Chapter 1

The Official Theory

"Blank slate" is a loose translation of the medieval Latin term *tabula rasa*—literally, "scraped tablet." It is commonly attributed to the philosopher John Locke (1632–1704), though in fact he used a different metaphor. Here is the famous passage from *An Essay Concerning Human Understanding*:

> Let us then suppose the mind to be, as we say, white paper void of all characters, without any ideas. How comes it to be furnished? Whence comes it by that vast store which the busy and boundless fancy of man has painted on it with an almost endless variety? Whence has it all the materials of reason and knowledge? To this I answer, in one word, from EXPERIENCE.[1]

Locke was taking aim at theories of innate ideas in which people were thought to be born with mathematical ideals, eternal truths, and a notion of God. His alternative theory, empiricism, was intended both as a theory of psychology—how the mind works—and as a theory of epistemology—how we come to know the truth. Both goals helped motivate his political philosophy, often honored as the foundation of liberal democracy. Locke opposed dogmatic justifications for the political status quo, such as the authority of the church and the divine right of kings, which had been touted as self-evident truths. He argued that social arrangements should be reasoned out from scratch and agreed upon by mutual consent, based on knowledge that any person could acquire. Since ideas are grounded in experience, which varies from person to person, differences of opinion arise not because one mind is equipped to grasp the truth and another is defective, but because the two minds have had different histories. Those differences therefore ought to be tolerated rather than suppressed. Locke's notion of a blank slate also undermined a hereditary royalty and aristocracy, whose members could claim no innate wisdom or merit if their minds had started out as blank as everyone else's. It also spoke against the institution of slavery, because slaves could no longer be thought of as innately inferior or subservient.

During the past century the doctrine of the Blank Slate has set the agenda for much of the social sciences and humanities. As we shall see, psychology has sought to explain all thought, feeling, and behavior with a few simple mechanisms of learning. The social sciences have sought to explain all customs and social arrangements as a product of the socialization of children by the surrounding culture: a system of words, images, stereotypes, role models, and contingencies of reward and punishment. A long and growing list of concepts that would seem natural to the human way of thinking (emotions, kinship, the sexes, illness, nature, the world) are now said to have been "invented" or "socially constructed."[2]

The Blank Slate has also served as a sacred scripture for political and ethical beliefs. According to the doctrine, any differences we see among races, ethnic groups, sexes, and individuals come not from differences in their innate constitution but from differences in their experiences. Change the experiences—by reforming parenting, education, the media, and social rewards—and you can change the person. Underachievement, poverty, and antisocial behavior can be ameliorated; indeed, it is irresponsible not to do so. And discrimination on the basis of purportedly inborn traits of a sex or ethnic group is simply irrational.

~

THE BLANK SLATE is often accompanied by two other doctrines, which have also attained a sacred status in modern intellectual life. My label for the first of the two is commonly attributed to the philosopher Jean-Jacques Rousseau (1712–1778), though it really comes from John Dryden's *The Conquest of Granada*, published in 1670:

> I am as free as Nature first made man,
> Ere the base laws of servitude began,
> When wild in woods the noble savage ran.

The concept of the noble savage was inspired by European colonists' discovery of indigenous peoples in the Americas, Africa, and (later) Oceania. It captures the belief that humans in their natural state are selfless, peaceable, and untroubled, and that blights such as greed, anxiety, and violence are the products of civilization. In 1755 Rousseau wrote:

> So many authors have hastily concluded that man is naturally cruel, and
> requires a regular system of police to be reclaimed; whereas nothing can
> be more gentle than him in his primitive state, when placed by nature at

an equal distance from the stupidity of brutes and the pernicious good sense of civilized man. . . .

The more we reflect on this state, the more convinced we shall be that it was the least subject of any to revolutions, the best for man, and that nothing could have drawn him out of it but some fatal accident, which, for the public good, should never have happened. The example of the savages, most of whom have been found in this condition, seems to confirm that mankind was formed ever to remain in it, that this condition is the real youth of the world, and that all ulterior improvements have been so many steps, in appearance towards the perfection of individuals, but in fact towards the decrepitness of the species.[3]

First among the authors that Rousseau had in mind was Thomas Hobbes (1588–1679), who had presented a very different picture:

Hereby it is manifest, that during the time men live without a common power to keep them all in awe, they are in that condition which is called war; and such a war as is of every man against every man. . . .

In such condition there is no place for industry, because the fruit thereof is uncertain: and consequently no culture of the earth; no navigation, nor use of the commodities that may be imported by sea; no commodious building; no instruments of moving and removing such things as require much force; no knowledge of the face of the earth; no account of time; no arts; no letters; no society; and which is worst of all, continual fear, and danger of violent death; and the life of man, solitary, poor, nasty, brutish, and short.[4]

Hobbes believed that people could escape this hellish existence only by surrendering their autonomy to a sovereign person or assembly. He called it a leviathan, the Hebrew word for a monstrous sea creature subdued by Yahweh at the dawn of creation.

Much depends on which of these armchair anthropologists is correct. If people are noble savages, then a domineering leviathan is unnecessary. Indeed, by forcing people to delineate private property for the state to recognize—property they might otherwise have shared—the leviathan creates the very greed and belligerence it is designed to control. A happy society would be our birthright; all we would need to do is eliminate the institutional barriers that keep it from us. If, in contrast, people are naturally nasty, the best we can hope for is an uneasy truce enforced by police and the army. The two theories have implications for private life as well. Every child is born a savage (that is, uncivilized), so if savages are naturally gentle, childrearing is a matter of providing children with opportunities to develop their potential, and evil people are products of a society that has corrupted them. If savages are naturally nasty, then childrearing is an arena of discipline and conflict, and evil people are showing a dark side that was insufficiently tamed.

The actual writings of philosophers are always more complex than the theories they come to symbolize in the textbooks. In reality, the views of Hobbes and Rousseau are not that far apart. Rousseau, like Hobbes, believed (incorrectly) that savages were solitary, without ties of love or loyalty, and without any industry or art (and he may have out-Hobbes'd Hobbes in claiming they did not even have language). Hobbes envisioned—indeed, literally drew—his leviathan as an embodiment of the collective will, which was vested in it by a kind of social contract; Rousseau's most famous work is called *The Social Contract*, and in it he calls on people to subordinate their interests to a "general will."

Nonetheless, Hobbes and Rousseau limned contrasting pictures of the state of nature that have inspired thinkers in the centuries since. No one can fail to recognize the influence of the doctrine of the Noble Savage in contemporary consciousness. We see it in the current respect for all things natural (natural foods, natural medicines, natural childbirth) and the distrust of the man-made, the unfashionability of authoritarian styles of childrearing and education, and the understanding of social problems as repairable defects in our institutions rather than as tragedies inherent to the human condition.

～

THE OTHER SACRED doctrine that often accompanies the Blank Slate is usually attributed to the scientist, mathematician, and philosopher René Descartes (1596–1650):

> There is a great difference between mind and body, inasmuch as body is by nature always divisible, and the mind is entirely indivisible. . . . When I consider the mind, that is to say, myself inasmuch as I am only a thinking being, I cannot distinguish in myself any parts, but apprehend myself to be clearly one and entire; and though the whole mind seems to be united to the whole body, yet if a foot, or an arm, or some other part, is separated from the body, I am aware that nothing has been taken from my mind. And the faculties of willing, feeling, conceiving, etc. cannot be properly speaking said to be its parts, for it is one and the same mind which employs itself in willing and in feeling and understanding. But it is quite otherwise with corporeal or extended objects, for there is not one of them imaginable by me which my mind cannot easily divide into parts. . . . This would be sufficient to teach me that the mind or soul of man is entirely different from the body, if I had not already been apprised of it on other grounds.[5]

A memorable name for this doctrine was given three centuries later by a detractor, the philosopher Gilbert Ryle (1900–1976):

> There is a doctrine about the nature and place of minds which is so prevalent among theorists and even among laymen that it deserves to be described as the official theory. . . . The official doctrine, which hails chiefly from Descartes, is something like this. With the doubtful exception of idiots and infants in arms every human being has both a body and a mind. Some would prefer to say that every human being is both a body and a mind. His body and his mind are ordinarily harnessed together, but after the death of the body his mind may continue to exist and function. Human bodies are in space and are subject to mechanical laws which govern all other bodies in space. . . . But minds are not in space, nor are their operations subject to mechanical laws. . . .
> . . . Such in outline is the official theory. I shall often speak of it, with deliberate abusiveness, as "the dogma of the Ghost in the Machine."[6]

The Ghost in the Machine, like the Noble Savage, arose in part as a reaction to Hobbes. Hobbes had argued that life and mind could be explained in mechanical terms. Light sets our nerves and brain in motion, and that is what it means to see. The motions may persist like the wake of a ship or the vibration of a plucked string, and that is what it means to imagine. "Quantities" get added or subtracted in the brain, and that is what it means to think.

Descartes rejected the idea that the mind could operate by physical principles. He thought that behavior, especially speech, was not *caused* by anything, but freely *chosen*. He observed that our consciousness, unlike our bodies and other physical objects, does not feel as if it is divisible into parts or laid out in space. He noted that we cannot doubt the existence of our minds—indeed, we cannot doubt that we *are* our minds—because the very act of thinking presupposes that our minds exist. But we *can* doubt the existence of our bodies, because we can imagine ourselves to be immaterial spirits who merely dream or hallucinate that we are incarnate.

Descartes also found a moral bonus in his dualism (the belief that the mind is a different kind of thing from the body): "There is none which is more effectual in leading feeble spirits from the straight path of virtue, than to imagine that the soul of the brute is of the same nature as our own, and that in consequence, after this life we have nothing to fear or to hope for, any more than the flies and the ants."[7] Ryle explains Descartes's dilemma:

> When Galileo showed that his methods of scientific discovery were competent to provide a mechanical theory which should cover every

occupant of space, Descartes found in himself two conflicting motives. As a man of scientific genius he could not but endorse the claims of mechanics, yet as a religious and moral man he could not accept, as Hobbes accepted, the discouraging rider to those claims, namely that human nature differs only in degree of complexity from clockwork.[8]

It can indeed be upsetting to think of ourselves as glorified gears and springs. Machines are insensate, built to be used, and disposable; humans are sentient, possessing of dignity and rights, and infinitely precious. A machine has some workaday purpose, such as grinding grain or sharpening pencils; a human being has higher purposes, such as love, worship, good works, and the creation of knowledge and beauty. The behavior of machines is determined by the ineluctable laws of physics and chemistry; the behavior of people is freely chosen. With choice comes freedom, and therefore optimism about our possibilities for the future. With choice also comes responsibility, which allows us to hold people accountable for their actions. And of course if the mind is separate from the body, it can continue to exist when the body breaks down, and our thoughts and pleasures will not someday be snuffed out forever.

As I mentioned, most Americans continue to believe in an immortal soul, made of some nonphysical substance, which can part company with the body. But even those who do not avow that belief in so many words still imagine that somehow there must be more to us than electrical and chemical activity in the brain. Choice, dignity, and responsibility are gifts that set off human beings from everything else in the universe, and seem incompatible with the idea that we are mere collections of molecules. Attempts to explain behavior in mechanistic terms are commonly denounced as "reductionist" or "determinist." The denouncers rarely know exactly what they mean by those words, but everyone knows they refer to something bad. The dichotomy between mind and body also pervades everyday speech, as when we say "Use your head," when we refer to "out-of-body experiences," and when we speak of "John's body," or for that matter "John's brain," which presupposes an owner, John, that is somehow separate from the brain it owns. Journalists sometimes speculate about "brain transplants" when they really should be calling them "body transplants," because, as the philosopher Dan Dennett has noted, this is the one transplant operation in which it is better to be the donor than the recipient.

The doctrines of the Blank Slate, the Noble Savage, and the Ghost in the Machine—or, as philosophers call them, empiricism, romanticism, and dualism—are logically independent, but in practice they are often found together. If the slate is blank, then strictly speaking it has neither injunctions to do good nor injunctions to do evil. But good and evil are asymmetrical: there are more ways to harm people than to help them, and harmful acts can hurt them to a greater degree than virtuous acts can make them better off. So a blank slate, compared with one filled with motives, is bound to impress us more by its inability to do harm than by its inability to do good. Rousseau did not literally believe in a blank slate, but he did believe that bad behavior is a product of learning and socialization.[9] "Men are wicked," he wrote; "a sad and constant experience makes proof unnecessary."[10] But this wickedness comes from society: "There is no original perversity in the human heart. There is not a single vice to be found in it of which it cannot be said how and whence it entered."[11] If the metaphors in everyday speech are a clue, then all of us, like Rousseau, associate blankness with virtue rather than with nothingness. Think of the moral connotations of the adjectives *clean, fair, immaculate, lily-white, pure, spotless, unmarred,* and *unsullied,* and of the nouns *blemish, blot, mark, stain,* and *taint.*

The Blank Slate naturally coexists with the Ghost in the Machine, too, since a slate that is blank is a hospitable place for a ghost to haunt. If a ghost is to be at the controls, the factory can ship the device with a minimum of parts. The ghost can read the body's display panels and pull its levers, with no need for a high-tech executive program, guidance system, or CPU. The more not-clockwork there is controlling behavior, the less clockwork we need to posit. For similar reasons, the Ghost in the Machine happily accompanies the Noble Savage. If the machine behaves ignobly, we can blame the ghost, which freely chose to carry out the iniquitous acts; we need not probe for a defect in the machine's design.

~

PHILOSOPHY TODAY GETS no respect. Many scientists use the term as a synonym for effete speculation. When my colleague Ned Block told his father that he would major in the subject, his father's reply was "Luft!"—Yiddish for "air." And then there's the joke in which a young man told his mother he would become a Doctor of Philosophy and she said, "Wonderful! But what kind of disease is philosophy?"

But far from being idle or airy, the ideas of philosophers can have repercussions for centuries. The Blank Slate and its companion doctrines have infiltrated the conventional wisdom of our civilization and have repeatedly surfaced in unexpected places. William Godwin (1756–1835), one of the founders of liberal political philosophy, wrote that "children are a sort of raw material put into our hands," their minds "like a sheet of white paper."[12] More sinisterly, we find Mao Zedong justifying his radical social engineering by saying, "It is on a blank page that the most beautiful poems are written."[13] Even Walt Disney was inspired by the metaphor. "I think of a child's mind as a blank book," he wrote. "During the first years of his life, much will be written on the pages. The quality of that writing will affect his life profoundly."[14]

Locke could not have imagined that his words would someday lead to Bambi (intended by Disney to teach self-reliance); nor could Rousseau have anticipated Pocahontas, the ultimate noble savage. Indeed, the soul of Rousseau seems to have been channeled by the writer of a recent Thanksgiving op-ed piece in the *Boston Globe:*

> I would submit that the world native Americans knew was more stable, happier, and less barbaric than our society today.... there were no employment problems, community harmony was strong, substance abuse unknown, crime nearly nonexistent. What warfare there was between tribes was largely ritualistic and seldom resulted in indiscriminate or wholesale slaughter. While there were hard times, life was, for the most part, stable and predictable.... Because the native people respected what was around them, there was no loss of water or food resources because of pollution or extinction, no lack of materials for the daily essentials, such as baskets, canoes, shelter, or firewood.[15]

Not that there haven't been skeptics:

The third doctrine, too, continues to make its presence felt in modern times. In 2001 George W. Bush announced that the American government will not fund research on human embryonic stem cells if scientists have to destroy new embryos to extract them (the policy permits research on stem-cell lines that were previously extracted from embryos). He derived the policy after consulting not just with scientists but with philosophers and religious thinkers. Many of them framed the moral problem in terms of "ensoulment," the moment at which the cluster of cells that will grow into a child is endowed with a soul. Some argued that ensoulment occurs at conception, which implies that the blastocyst (the five-day-old ball of cells from which stem cells are taken) is morally equivalent to a person and that destroying it is a form of murder.[16] That argument proved decisive, which means that the

American policy on perhaps the most promising medical technology of the twenty-first century was decided by pondering the moral issue as it might have been framed centuries before: When does the ghost first enter the machine?

These are just a few of the fingerprints of the Blank Slate, the Noble Savage, and the Ghost in the Machine on modern intellectual life. In the following chapters we will see how the seemingly airy ideas of Enlightenment philosophers entrenched themselves in modern consciousness, and how recent discoveries are casting those ideas in doubt.

Chapter 2

Silly Putty

THE DANISH PHILOLOGIST Otto Jespersen (1860–1943) is one of history's most beloved linguists. His vivid books are still read today, especially *Growth and Structure of the English Language*, first published in 1905. Though Jespersen's scholarship is thoroughly modern, the opening pages remind us we are not reading a contemporary book:

> There is one expression that continually comes to my mind whenever I think of the English language and compare it with others: it seems to be positively and expressly *masculine,* it is the language of a grown-up man and has very little childish or feminine about it. . . .
>
> To bring out one of these points I select at random, by way of contrast, a passage from the language of Hawaii: "I kona hiki ana aku ilaila ua hookipa ia mai la oia me ke aloha pumehana loa." Thus it goes on, no single word ends in a consonant, and a group of two or more consonants is never found. Can any one be in doubt that even if such a language sounds pleasantly and be full of music and harmony the total impression is childlike and effeminate? You do not expect much vigor or energy in a people speaking such a language; it seems adapted only to inhabitants of sunny regions where the soil requires scarcely any labour on the part of man to yield him everything he wants, and where life therefore does not bear the stamp of a hard struggle against nature and fellow-creatures. In a lesser degree we find the same phonetic structure in such languages as Italian and Spanish; but how different are our Northern tongues.[1]

And so he continues, advertising the virility, sobriety, and logic of English— and ends the chapter: "As the language is, so also is the nation."

No modern reader can fail to be shocked by the sexism, racism, and chauvinism of the discussion: the implication that women are childlike, the stereotyping of a colonized people as indolent, the gratuitous exalting of the author's own culture. Equally surprising are the sorry standards to which the great scholar here has sunk. The suggestion that a language can be "grown-up" and "masculine" is so subjective as to be meaningless. He attributes a personality trait to an entire people without any evidence, then advances two theories—that phonology reflects personality, and that warm climates breed laziness—without invoking even correlational data, let alone proof of causation. Even on his home ground the reasoning is flimsy. Languages with a consonant-vowel syllable structure like Hawaiian call for longer words to convey the same amount of information, hardly what you would expect in a people without "vigor or energy." And the consonant-encrusted syllables of English are liable to be swallowed and misheard, hardly what you would expect from a logical, businesslike people.

But perhaps most disturbing is Jespersen's obliviousness to the possibility that he might be saying anything exceptionable. He took it for granted that his biases would be shared by his readers, whom he knew to be fellow men and speakers of "our" Northern tongues. "Can any one be in doubt?" he asked rhetorically; "you do not expect much vigor" from such a people, he asserted. The inferiority of women and other races needed neither justification nor apology.

I bring up Otto Jespersen, a man of his time, to show how standards have changed. The passage is a random sample of intellectual life a century ago; equally disturbing passages could have been taken from just about any writer of the nineteenth or early twentieth century.[2] It was a time of white men taking up the burden of leading their "new-caught sullen peoples, half-devil and half-child"; of shores teeming with huddled masses and wretched refuse; of European imperial powers looking (and sometimes throwing) daggers at one another. Imperialism, immigration, nationalism, and the legacy of slavery made differences between ethnic groups all too obvious. Some appeared educated and cultured, others ignorant and backward; some used fists and clubs to preserve their safety, others paid the police and the army to do it. It was tempting to assume that northern Europeans were an advanced race suited to rule the others. Just as convenient was the belief that women were constitutionally suited for the kitchen, church, and children, a belief supported by "research" showing that brainwork was bad for their physical and mental health.

Racial prejudice, too, had a scientific patina. Darwin's theory of evolution was commonly misinterpreted as an explanation of intellectual and moral progress rather than an explanation of how living things adapt to an ecological niche. The nonwhite races, it was easy to think, were rungs on an evolutionary ladder between the apes and the Europeans. Worse, Darwin's follower Herbert Spencer wrote that do-gooders would only interfere with the progress of evolution if they tried to improve the lot of the impoverished classes and races, who were, in Spencer's view, biologically less fit. The doctrine of Social Darwinism (or, as it ought to be called, Social Spencerism, for Darwin wanted no part of it) attracted such unsurprising spokesmen as John D. Rockefeller and Andrew Carnegie.[3] Darwin's cousin Francis Galton had suggested that human evolution should be given a helping hand by discouraging the less fit from breeding, a policy he called eugenics.[4] Within a few decades laws were passed that called for the involuntary sterilization of delinquents and the "feebleminded" in Canada, the Scandinavian countries, thirty American states, and, ominously, Germany. The Nazis' ideology of inferior races was later used to justify the murder of millions of Jews, Gypsies, and homosexuals.

We have come a long way. Though attitudes far worse than Jespersen's continue to thrive in much of the world and in parts of our society, they have been driven out of mainstream intellectual life in Western democracies. Today no respectable public figure in the United States, Britain, or Western Europe can casually insult women or sling around invidious stereotypes of other races or ethnic groups. Educated people try to be conscious of their hidden prejudices and to measure them against the facts and against the sensibilities of others. In public life we try to judge people as individuals, not as specimens of a sex or ethnic group. We try to distinguish might from right and our parochial tastes from objective merit, and therefore respect cultures that are different or poorer than ours. We realize that no mandarin is wise enough to be entrusted with directing the evolution of the species, and that it is wrong in any case for the government to interfere with such a personal decision as having a child. The very idea that the members of an ethnic group should be persecuted because of their biology fills us with revulsion.

These changes were cemented by the bitter lessons of lynchings, world wars, forced sterilizations, and the Holocaust, which showcased the grave implications of denigrating an ethnic group. But they emerged earlier in the twentieth century, the spinoff of an unplanned experiment: the massive immigration, social mobility, and diffusion of knowledge of the modern era. Most Victorian gentlemen could not have imagined that the coming century would see a nation-state forged by Jewish pioneers and soldiers, a wave of African American public intellectuals, or a software industry in Bangalore. Nor could they have anticipated that women would lead nations in wars, run huge corporations, or win Nobel Prizes in science. We now know that people of both sexes and all races are capable of attaining any station in life.

This sea change included a revolution in the treatment of human nature by scientists and scholars. Academics were swept along by the changing attitudes to race and sex, but they also helped to direct the tide by holding forth on human nature in books and magazines and by lending their expertise to government agencies. The prevailing theories of mind were refashioned to make racism and sexism as untenable as possible. The doctrine of the Blank

Slate became entrenched in intellectual life in a form that has been called the Standard Social Science Model or social constructionism.[5] The model is now second nature to people and few are aware of the history behind it.[6] Carl Degler, the foremost historian of this revolution, sums it up this way:

> What the available evidence does seem to show is that ideology or a philosophical belief that the world could be a freer and more just place played a large part in the shift from biology to culture. Science, or at least certain scientific principles or innovative scholarship also played a role in the transformation, but only a limited one. The main impetus came from the will to establish a social order in which innate and immutable forces of biology played no role in accounting for the behavior of social groups.[7]

The takeover of intellectual life by the Blank Slate followed different paths in psychology and in the other social sciences, but they were propelled by the same historical events and progressive ideology. By the second and third decades of the twentieth century, stereotypes of women and ethnic groups were starting to look silly. Waves of immigrants from southern and eastern Europe, including many Jews, were filling the cities and climbing the social ladder. African Americans had taken advantage of the new "Negro colleges," had migrated northward, and had begun the Harlem Renaissance. The graduates of flourishing women's colleges helped launch the first wave of feminism. For the first time not all professors and students were white Anglo-Saxon Protestant males. To say that this sliver of humanity was constitutionally superior had not only become offensive but went against what people could see with their own eyes. The social sciences in particular were attracting women, Jews, Asians, and African Americans, some of whom became influential thinkers.

Many of the pressing social problems of the first decades of the twentieth century concerned the less fortunate members of these groups. Should more immigrants be let in, and if so, from which countries? Once here, should they be encouraged to assimilate, and if so, how? Should women be given equal political rights and economic opportunities? Should blacks and whites be integrated? Other challenges were posed by children.[8] Education had become compulsory and a responsibility of the state. As the cities teemed and family ties loosened, troubled and troublesome children became everyone's problem, and new institutions were invented to deal with them, such as kindergartens, orphanages, reform schools, fresh-air camps, humane societies, and boys' and girls' clubs. Child development was suddenly on the front burner. These social challenges were not going to go away, and the most humane assumption was that all human beings had an equal potential to prosper if they were given the right upbringing and opportunities. Many social scientists saw it as their job to reinforce that assumption.

~

MODERN PSYCHOLOGICAL THEORY, as every introductory textbook makes clear, has roots in John Locke and other Enlightenment thinkers. For Locke the Blank Slate was a weapon against the church and tyrannical monarchs, but these threats had subsided in the English-speaking world by the nineteenth century. Locke's intellectual heir John Stuart Mill (1806–1873) was perhaps the first to apply his blank-slate psychology to political concerns we recognize today. He was an early supporter of women's suffrage, compulsory education, and the improvement of the conditions of the lower classes. This interacted with his stands in psychology and philosophy, as he explained in his autobiography:

> I have long felt that the prevailing tendency to regard all the marked distinctions of human character as innate, and in the main indelible, and to ignore the irresistible proofs that by far the greater part of those differences, whether between individuals, races, or sexes, are such as not only might but naturally would be produced by differences in circumstances, is one of the chief hindrances to the rational treatment of great social questions, and one of the greatest stumbling blocks to human improvement. . . . [This tendency is] so agreeable to human indolence, as well as to conservative interests generally, that unless attacked at the very root, it is sure to be carried to even a greater length than is really justified by the more moderate forms of intuitional philosophy.[9]

By "intuitional philosophy" Mill was referring to Continental intellectuals who maintained (among other things) that the categories of reason were innate. Mill wanted to attack their theory of psychology at the root to combat what he thought were its conservative social implications. He refined a theory of learning called associationism (previously formulated by Locke) that tried to explain human intelligence without granting it any innate organization. According to this theory, the blank slate is inscribed with sensations, which Locke called "ideas" and modern psychologists call "features." Ideas that repeatedly appear in succession (such as the redness, roundness, and sweetness of an apple) become associated, so that any one of them can call to mind the others. And similar objects in the world activate overlapping sets of ideas in the mind. For example, after many dogs present themselves to the senses, the features that they share (fur, barking, four legs, and so on) hang together to stand for the category "dog."

The associationism of Locke and Mill has been recognizable in psychology ever since. It became the core of most models of learning, especially in the approach called behaviorism, which dominated psychology from the 1920s to the 1960s. The founder of behaviorism, John B. Watson (1878–1958), wrote one of the century's most famous pronouncements of the Blank Slate:

> Give me a dozen healthy infants, well-formed, and my own specified world to bring them up in and I'll guarantee to take any one at random and train him to become any type of specialist I might select—doctor, lawyer, artist, merchant-chief, and yes, even beggar-man and thief, regardless of his talents, penchants, tendencies, abilities, vocations, and race of his ancestors.[10]

In behaviorism, an infant's talents and abilities didn't matter because there was *no such thing* as a talent or an ability. Watson had banned them from psychology, together with other contents of the mind, such as ideas, beliefs, desires, and feelings. They were subjective and unmeasurable, he said, and unfit for science, which studies only objective and measurable things. To a behaviorist, the only legitimate topic for psychology is overt behavior and how it is controlled by the present and past environment. (There is an old joke in psychology: What does a behaviorist say after making love? "It was good for you; how was it for me?")

Locke's "ideas" had been replaced by "stimuli" and "responses," but his laws of association survived as laws of conditioning. A response can be associated with a new stimulus, as when Watson presented a baby with a white rat and then clanged a hammer against an iron bar, allegedly making the baby associate fear with fur. And a response could be associated with a reward, as when a cat in a box eventually learned that pulling a string opened a door and allowed it to escape. In these cases an experimenter set up a contingency between a stimulus and another stimulus or between a response and a reward. In a natural environment, said the behaviorists, these contingencies are part of the causal texture of the world, and they inexorably shape the behavior of organisms, including humans.

Among the casualties of behaviorist minimalism was the rich psychology of William James (1842–1910). James had been inspired by Darwin's argument that perception, cognition, and emotion, like physical organs, had evolved as biological adaptations. James invoked the notion of instinct to explain the preferences of humans, not just those of animals, and he posited numerous mechanisms in his theory of mental life, including short-term and long-term memory. But with the advent of behaviorism they all joined the index of forbidden concepts. The psychologist J. R. Kantor wrote in 1923: "Brief is the answer to the question as to what is the relationship between social psychology and instincts. Plainly, there is no relationship."[11] Even sexual desire was redefined as a conditioned response. The psychologist Zing Yang Kuo wrote in 1929:

> Behavior is not a manifestation of hereditary factors, nor can it be expressed in terms of heredity. [It is] a passive and forced movement mechanically and solely determined by the structural pattern of the organism and the nature of environmental forces. . . . All our sexual appetites are the result of social stimulation. The organism possesses no ready-made reaction to the other sex, any more than it possesses innate ideas.[12]

Behaviorists believed that behavior could be understood independently of the rest of biology, without attention to the genetic makeup of the animal or the evolutionary history of the species. Psychology came to consist of the study of learning in laboratory animals. B. F. Skinner (1904–1990), the most famous psychologist in the middle decades of the twentieth century, wrote a book called *The Behavior of Organisms* in which the only organisms were rats and pigeons and the only behavior was lever pressing and key pecking. It took a trip to the circus to remind psychologists that species and their instincts mattered after all. In an article called "The Misbehavior of Organisms," Skinner's students Keller and Marian Breland reported that when they tried to use his techniques to train animals to insert poker chips into vending machines, the chickens pecked the chips, the raccoons washed them, and the pigs tried to root them with their snouts.[13] And behaviorists were as hostile to the brain as they were to genetics. As late as 1974, Skinner wrote that studying the brain was just another misguided quest to find the causes of behavior inside the organism rather than out in the world.[14]

Behaviorism not only took over psychology but infiltrated the public consciousness. Watson wrote an influential childrearing manual recommending that parents establish rigid feeding schedules for their children and give them a minimum of attention and love. If you comfort a crying child, he wrote, you will reward him for crying and thereby increase the frequency of crying behavior. (Benjamin Spock's *Baby and Child Care*, first published in 1946 and famous for recommending indulgence toward children, was in part a reaction to Watson.) Skinner wrote several bestsellers arguing that harmful behavior is neither instinctive nor freely chosen but inadvertently conditioned. If we turned society into a big Skinner box and controlled behavior deliberately rather than haphazardly, we could eliminate aggression, overpopulation, crowding, pollution, and inequality, and thereby attain utopia.[15] The noble savage became the noble pigeon.

Strict behaviorism is pretty much dead in psychology, but many of its attitudes live on. Associationism is the learning theory assumed by many mathematical models and neural network simulations of learning.[16] Many neuroscientists *equate* learning with the forming of associations, and look for an associative bond in the physiology of neurons and synapses, ignoring other kinds of computation that might implement learning in the brain.[17] (For example, storing the value of a variable in the brain, as in "$x = 3$," is a critical computational step in navigating and foraging, which are highly developed talents of animals in the wild. But this kind of learning cannot be reduced to the formation of associations, and so it has been ignored in neuroscience.) Psychologists and neuroscientists still treat organisms interchangeably, seldom asking whether a convenient laboratory animal (a rat, a cat, a monkey) is like or unlike humans in crucial ways.[18] Until recently, psychology ignored the *content* of beliefs and emotions and the possibility that the mind had evolved to treat biologically important categories in different ways.[19] Theories of memory and reasoning didn't distinguish thoughts about people from thoughts about rocks or houses. Theories of emotion didn't distinguish fear from anger, jealousy, or love.[20] Theories of social relations didn't distinguish among family, friends, enemies, and strangers.[21] Indeed, the topics in psychology that most interest laypeople—love, hate, work, play, food, sex, status, dominance, jealousy, friendship, religion, art—are almost completely absent from psychology textbooks.

One of the major documents of late twentieth-century psychology was the two-volume *Parallel Distributed Processing* by David Rumelhart, James McClelland, and their collaborators, which presented a style of neural network modeling called connectionism.[22] Rumelhart and McClelland argued that generic associationist networks, subjected to massive amounts of training, could explain all of cognition. They realized that this theory left them without a good answer to the question "Why are people smarter than rats?" Here is their answer:

> Given all of the above, the question does seem a bit puzzling. . . . People have much more cortex than rats do or even than other primates do; in particular they have very much more . . . brain structure not dedicated to input/output—and presumably, this extra cortex is strategically placed in the brain to subserve just those functions that differentiate people from rats or even apes. . . .
>
> But there must be another aspect to the difference between rats and people as well. This is that the human environment includes other people and the cultural devices that they have developed to organize their thinking processes.[23]

Humans, then, are just rats with bigger blank slates, plus something called "cultural devices." And that brings us to the other half of the twentieth-century revolution in social science.

～

He's so unhip, when you say "Dylan,"
He thinks you're talkin' about Dylan *Thomas* (whoever he was).
The man ain't got no culture.

—Simon and Garfunkel

The word *culture* used to refer to exalted genres of entertainment, such as poetry, opera, and ballet. The other familiar sense—"the totality of socially transmitted behavior patterns, arts, beliefs, institutions, and all other products of human work and thought"—is only a century old. This change in the English language is just one of the legacies of the father of modern anthropology, Franz Boas (1858–1942).

The ideas of Boas, like the ideas of the major thinkers in psychology, were rooted in the empiricist philosophers of the Enlightenment, in this case George Berkeley (1685–1753). Berkeley formulated the theory of idealism, the notion that ideas, not bodies and other hunks of matter, are the ultimate constituents of reality. After twists and turns that are too convoluted to recount here, idealism became influential among nineteenth-century German thinkers. It was embraced by the young Boas, a German Jew from a secular, liberal family.

Idealism allowed Boas to lay a new intellectual foundation for egalitarianism. The differences among human races and ethnic groups, he proposed, come not from their physical constitution but from their *culture,* a system of ideas and values spread by language and other forms of social behavior. Peoples differ because their cultures differ. Indeed, that is how we should refer to them: the Eskimo culture or the Jewish culture, not the Eskimo race or the Jewish race. The idea that minds are shaped by culture served as a bulwark against racism and was the theory one ought to prefer on moral grounds. Boas wrote, "I claim that, unless the contrary can be proved, we must assume that all complex activities are socially determined, not hereditary."[24]

Boas's case was not just a moral injunction; it was rooted in real discoveries. Boas studied native peoples, immigrants, and children in orphanages to prove that all groups of humans had equal potential. Turning Jespersen on his head, Boas showed that the languages of primitive peoples were not simpler than those of Europeans; they were just different. Eskimos' difficulty in discriminating the sounds of our language, for example, is matched by our difficulty in discriminating the sounds of theirs. True, many non-Western languages lack the means to express certain abstract concepts. They may have no words for numbers higher than three, for example, or no word for goodness in general as opposed to the goodness of a particular person. But those limitations simply reflect the daily needs of those people as they live their lives, not an infirmity in their mental abilities. As in the story of Socrates drawing abstract philosophical concepts out of a slave boy, Boas showed that he could elicit new word forms for abstract concepts like "goodness" and "pity" out of a Kwakiutl native from the Pacific Northwest. He also observed that when native peoples come into contact with civilization and acquire things that have to be counted, they quickly adopt a full-blown counting system.[25]

For all his emphasis on culture, Boas was not a relativist who believed that all cultures are equivalent, nor was he an empiricist who believed in the Blank Slate. He considered European civilization superior to tribal cultures, insisting only that all peoples were capable of achieving it. He did not deny that there might be a universal human nature, or that there might be differences among people within an ethnic group. What mattered to him was the idea that all ethnic groups are endowed with the same basic mental abilities.[26] Boas was right about this, and today it is accepted by virtually all scholars and scientists.

But Boas had created a monster. His students came to dominate American social science, and each generation outdid the previous one in its sweeping pronouncements. Boas's students insisted not just that *differences* among ethnic groups must be explained in terms of culture but that *every aspect* of human existence must be explained in terms of culture. For example, Boas had favored social explanations unless they were disproven, but his student Albert Kroeber favored them regardless of the evidence. "Heredity," he wrote, "cannot be allowed to have acted any part in history."[27] Instead, the chain of events shaping a people "involves the absolute conditioning of historical events by other historical events."[28]

Kroeber did not just deny that social behavior could be explained by innate properties of minds. He denied that it could be explained by *any* properties of minds. A culture, he wrote, is *superorganic*—it floats in its own universe, free of the flesh and blood of actual men and women: "Civilization is not mental action but a body or stream of products of mental exercise. . . . Mentality relates to the individual. The social or cultural, on the other hand, is in its essence non-individual. Civilization as such begins only where the individual ends."[29]

These two ideas—the denial of human nature, and the autonomy of culture from individual minds—were also articulated by the founder of sociology, Emile Durkheim (1858–1917), who had foreshadowed Kroeber's doctrine of the superorganic mind:

> Every time that a social phenomenon is directly explained by a psychological phenomenon, we may be sure that the explanation is false. . . . The group thinks, feels, and acts quite differently from the way in which

members would were they isolated. . . . If we begin with the individual in seeking to explain phenomena, we shall be able to understand nothing of what takes place in the group. . . . Individual natures are merely the indeterminate material that the social factor molds and transforms. Their contribution consists exclusively in very general attitudes, in vague and consequently plastic predispositions.[30]

And he laid down a law for the social sciences that would be cited often in the century to come: "The determining cause of a social fact should be sought among the social facts preceding it and not among the states of individual consciousness."[31]

Both psychology and the other social sciences, then, denied that the minds of individual people were important, but they set out in different directions from there. Psychology banished mental entities like beliefs and desires altogether and replaced them with stimuli and responses. The other social sciences located beliefs and desires in cultures and societies rather than in the heads of individual people. The different social sciences also agreed that the contents of cognition—ideas, thoughts, plans, and so on—were really phenomena of language, overt behavior that anyone could hear and write down. (Watson proposed that "thinking" really consisted of teensy movements of the mouth and throat.) But most of all they shared a dislike of instincts and evolution. Prominent social scientists repeatedly declared the slate to be blank:

Instincts do not create customs; customs create instincts, for the putative instincts of human beings are always learned and never native.
— Ellsworth Faris (1927)[32]

Cultural phenomena . . . are in no respect hereditary but are characteristically and without exception acquired.
— George Murdock (1932)[33]

Man has no nature; what he has is history.
— José Ortega y Gasset (1935)[34]

With the exception of the instinctoid reactions in infants to sudden withdrawals of support and to sudden loud noises, the human being is entirely instinctless. . . . Man is man because he has no instincts, because everything he is and has become he has learned, acquired, from his culture, from the man-made part of the environment, from other human beings.
— Ashley Montagu (1973)[35]

True, the metaphor of choice was no longer a scraped tablet or white paper. Durkheim had spoken of "indeterminate material," some kind of blob that was molded or pounded into shape by culture. Perhaps the best modern metaphor is Silly Putty, the rubbery stuff that children use both to copy printed matter (like a blank slate) and to mold into desired shapes (like indeterminate material). The malleability metaphor resurfaced in statements by two of Boas's most famous students:

> Most people are shaped to the form of their culture because of the malleability of their original endowment. . . . The great mass of individuals take quite readily the form that is presented to them.
>
> —Ruth Benedict (1934)[36]

> We are forced to conclude that human nature is almost unbelievably malleable, responding accurately and contrastingly to contrasting cultural conditions.
>
> —Margaret Mead (1935)[37]

Others likened the mind to some kind of sieve:

> Much of what is commonly called "human nature" is merely *culture* thrown against a screen of nerves, glands, sense organs, muscles, etc.
>
> —Leslie White (1949)[38]

Or to the raw materials for a factory:

> Human nature is the rawest, most undifferentiated of raw material.
>
> —Margaret Mead (1928)[39]

> Our ideas, our values, our acts, even our emotions, are, like our nervous system itself, cultural products—products manufactured, indeed, out of tendencies, capacities, and dispositions with which we were born, but manufactured nonetheless.
>
> —Clifford Geertz (1973)[40]

Or to an unprogrammed computer:

> Man is the animal most desperately dependent upon such extragenetic, outside-the-skin control mechanisms, such cultural programs, for ordering his behavior.
>
> —Clifford Geertz (1973)[41]

Or to some other amorphous entity that can have many things done to it:

> Cultural psychology is the study of the way cultural traditions and social practices regulate, express, transform, and permute the human psyche, resulting less in psychic unity for humankind than in ethnic divergences in mind, self and emotion.
>
> —Richard Shweder (1990)[42]

The superorganic or group mind also became an article of faith in social science. Robert Lowie (another Boas student) wrote, "The principles of psychology are as incapable of accounting for the phenomena of culture as is gravitation to account for architectural styles."[43] And in case you missed its full implications, the anthropologist Leslie White spelled it out:

> Instead of regarding the individual as a First Cause, as a prime mover, as the initiator and determinant of the culture process, we now see him as a component part, and a tiny and relatively insignificant part at that, of a vast, socio-cultural system that embraces innumerable individuals at any one time and extends back into their remote past as well. . . . For purposes of scientific interpretation, the culture process may be regarded as a thing *sui generis;* culture is explainable in terms of culture.[44]

In other words, we should forget about the mind of an individual person like *you,* that tiny and insignificant part of a vast sociocultural system. The mind that counts is the one belonging to the *group,* which is capable of thinking, feeling, and acting on its own.

The doctrine of the superorganism has had an impact on modern life that extends well beyond the writings of social scientists. It underlies the tendency to reify "society" as a moral agent that can be blamed for sins as if it were a person. It drives identity politics, in which civil rights and political perquisites are allocated to groups rather than to individuals. And as we shall see in later chapters, it defined some of the great divides between major political systems in the twentieth century.

~

THE BLANK SLATE was not the only part of the official theory that social scientists felt compelled to prop up. They also strove to consecrate the Noble Savage. Mead painted a Gauguinesque portrait of native peoples as peaceable, egalitarian, materially satisfied, and sexually unconflicted. Her uplifting vision of who we used to be—and therefore who we can become again—was accepted by such otherwise skeptical writers as Bertrand Russell and H. L. Mencken. Ashley Montagu (also from the Boas circle), a prominent public intellectual from the 1950s until his recent death, tirelessly invoked the doctrine

of the Noble Savage to justify the quest for brotherhood and peace and to refute anyone who might think such efforts were futile. In 1950, for example, he
drafted a manifesto for the newly formed UNESCO that declared, "Biological
studies lend support to the ethic of universal brotherhood, for man is born
with drives toward co-operation, and unless these drives are satisfied, men and
nations alike fall ill."[45] With the ashes of thirty-five million victims of World
War II still warm or radioactive, a reasonable person might wonder how "biological studies" could show anything of the kind. The draft was rejected, but
Montagu had better luck in the decades to come, when UNESCO and many
scholarly societies adopted similar resolutions.[46]

More generally, social scientists saw the malleability of humans and the
autonomy of culture as doctrines that might bring about the age-old dream of
perfecting mankind. We are not stuck with what we don't like about our current predicament, they argued. Nothing prevents us from changing it except a
lack of will and the benighted belief that we are permanently consigned to it by
biology. Many social scientists have expressed the hope of a new and improved
human nature:

> I felt (and said so early) that the environmental explanation was prefer-
> able, whenever justified by the data, because it was more optimistic,
> holding out the hope of improvement.
>
> —Otto Klineberg (1928)[47]

> Modern sociology and modern anthropology are one in saying that the
> substance of culture, or civilization, is social tradition and that this so-
> cial tradition is indefinitely modifiable by further learning on the part of
> men for happier and better ways of living together. . . . Thus the scien-
> tific study of institutions awakens faith in the possibility of remaking
> both human nature and human social life.
>
> —Charles Ellwood (1922)[48]

> Barriers in many fields of knowledge are falling below the new optimism
> which is that anybody can learn anything. . . . We have turned away
> from the concept of human ability as something fixed in the physiolog-
> ical structure, to that of a flexible and versatile mechanism subject to
> great improvement.
>
> —Robert Faris (1961)[49]

Though psychology is not as politicized as some of the other social sciences, it too is sometimes driven by a utopian vision in which changes in childrearing and education will ameliorate social pathologies and improve human
welfare. And psychological theorists sometimes try to add moral heft to arguments for connectionism or other empiricist theories with warnings about the

pessimistic implications of innatist theories. They argue, for example, that innatist theories open the door to inborn differences, which could foster racism, or that the theories imply that human traits are unchangeable, which could weaken support for social programs.[50]

~

Twentieth-century social science embraced not just the Blank Slate and the Noble Savage but the third member of the trinity, the Ghost in the Machine. The declaration that we can change what we don't like about ourselves became a watchword of social science. But that only raises the question "Who or what is the 'we'?" If the "we" doing the remaking are just other hunks of matter in the biological world, then any malleability of behavior we discover would be cold comfort, because we, the molders, would be biologically constrained and therefore might not mold people, or allow ourselves to be molded, in the most socially salutary way. A ghost in the machine is the ultimate liberator of human will—including the will to change society—from mechanical causation. The anthropologist Loren Eiseley made this clear when he wrote:

> The mind of man, by indetermination, by the power of choice and cultural communication, is on the verge of escape from the blind control of that deterministic world with which the Darwinists had unconsciously shackled man. The inborn characteristics laid upon him by the biological extremists have crumbled away. . . . Wallace saw and saw correctly, that with the rise of man the evolution of parts was to a marked degree outmoded, that mind was now the arbiter of human destiny.[51]

The "Wallace" that Eiseley is referring to is Alfred Russel Wallace (1823–1913), the co-discoverer with Darwin of natural selection. Wallace parted company from Darwin by claiming that the human mind could not be explained by evolution and must have been designed by a superior intelligence. He certainly did believe that the mind of man could escape "the blind control of a deterministic world." Wallace became a spiritualist and spent the later years of his career searching for a way to communicate with the souls of the dead.

The social scientists who believed in an absolute separation of culture from biology may not have literally believed in a spook haunting the brain. Some used the analogy of the difference between living and nonliving matter. Kroeber wrote: "The dawn of the social \ . . is not a link in any chain, not a step in a path, but a leap to another plane. . . . [It is like] the first occurrence of life in the hitherto lifeless universe. . . . From this moment on there should be two worlds in place of one."[52] And Lowie insisted that it was "not mysticism, but sound scientific method" to say that culture was *"sui generis"* and could be explained only by culture, because everyone knows that in biology a living cell can come only from another living cell.[53]

At the time that Kroeber and Lowie wrote, they had biology on their side. Many biologists still thought that living things were animated by a special essence, an *élan vital,* and could not be reduced to inanimate matter. A 1931 history of biology, referring to genetics as it was then understood, said, "Thus the last of the biological theories leaves us where we first started, in the presence of a power called life or psyche which is not only of its own kind but unique in each and all of its exhibitions."[54] In the next chapter we will see that the analogy between the autonomy of culture and the autonomy of life would prove to be more telling than these social scientists realized.

Chapter 3

The Last Wall to Fall

In 1755 Samuel Johnson wrote that his dictionary should not be expected to "change sublunary nature, and clear the world at once from folly, vanity, and affectation." Few people today are familiar with the lovely word *sublunary*, literally "below the moon." It alludes to the ancient belief in a strict division between the pristine, lawful, unchanging cosmos above and our grubby, chaotic, fickle Earth below. The division was already obsolete when Johnson used the word: Newton had shown that the same force that pulled an apple toward the ground kept the moon in its celestial orbit.

Newton's theory that a single set of laws governed the motions of all objects in the universe was the first event in one of the great developments in human understanding: the unification of knowledge, which the biologist E. O. Wilson has termed consilience.[1] Newton's breaching of the wall between the terrestrial and the celestial was followed by a collapse of the once equally firm (and now equally forgotten) wall between the creative past and the static present. That happened when Charles Lyell showed that the Earth was sculpted in the past by forces we see today (such as earthquakes and erosion) acting over immense spans of time.

The living and nonliving, too, no longer occupy different realms. In 1628 William Harvey showed that the human body is a machine that runs by hydraulics and other mechanical principles. In 1828 Friedrich Wöhler showed that the stuff of life is not a magical, pulsating gel but ordinary compounds following the laws of chemistry. Charles Darwin showed how the astonishing diversity of life and its ubiquitous signs of design could arise from the physical process of natural selection among replicators. Gregor Mendel, and then James Watson and Francis Crick, showed how replication itself could be understood in physical terms.

The unification of our understanding of life with our understanding of matter and energy was the greatest scientific achievement of the second half of the twentieth century. One of its many consequences was to pull the rug out from under social scientists like Kroeber and Lowie who had invoked the "sound scientific method" of placing the living and nonliving in parallel universes. We now know that cells did not always come from other cells and that the emergence of life did not create a second world where before there was just one. Cells evolved from simpler replicating molecules, a nonliving part of the physical world, and may be understood as collections of molecular machinery—fantastically complicated machinery, of course, but machinery nonetheless.

This leaves one wall standing in the landscape of knowledge, the one that twentieth-century social scientists guarded so jealously. It divides matter from mind, the material from the spiritual, the physical from the mental, biology from culture, nature from society, and the sciences from the social sciences, humanities, and arts. The division was built into each of the doctrines of the official theory: the blank slate given by biology versus the contents inscribed by experience and culture, the nobility of the savage in the state of nature versus the corruption of social institutions, the machine following inescapable laws versus the ghost that is free to choose and to improve the human condition.

But this wall, too, is falling. New ideas from four frontiers of knowledge—the sciences of mind, brain, genes, and evolution—are breaching the wall with a new understanding of human nature. In this chapter I will show how they are filling in the blank slate, declassing the noble savage, and exorcising the ghost in the machine. In the following chapter I will show that this new conception of human nature, connected to biology from below, can in turn be connected to the humanities and social sciences above. That new conception can give the phenomena of culture their due without segregating them into a parallel universe.

~

THE FIRST BRIDGE between biology and culture is the science of mind, cognitive science.[2] The concept of mind has been perplexing for as long as people have reflected on their thoughts and feelings. The very idea has spawned paradoxes, superstitions, and bizarre theories in every period and culture. One can almost sympathize with the behaviorists and social constructionists of the first half of the twentieth century, who looked on minds as enigmas or conceptual traps that were best avoided in favor of overt behavior or the traits of a culture.

But beginning in the 1950s with the cognitive revolution, all that changed. It is now possible to make sense of mental processes and even to study them in the lab. And with a firmer grasp on the concept of mind, we can see that many tenets of the Blank Slate that once seemed appealing are now unnecessary or even incoherent. Here are five ideas from the cognitive revolution that have revamped how we think and talk about minds.

The first idea: *The mental world can be grounded in the physical world by the concepts of information, computation, and feedback.* A great divide between mind and matter has always seemed natural because behavior appears to have a different kind of trigger than other physical events. Ordinary events have *causes,* it seems, but human behavior has *reasons.* I once participated in a BBC television debate on whether "science can explain human behavior." Arguing against the resolution was a philosopher who asked how we might explain why someone was put in jail. Say it was for inciting racial hatred. The intention, the hatred, and even the prison, she said, cannot be described in the language of physics. There is simply no way to define "hatred" or "jail" in terms of the movements of particles. Explanations of behavior are like *narratives,* she argued, couched in the intentions of actors—a plane completely separate from natural science. Or take a simpler example. How might we explain why Rex just walked over to the phone? We would not say that phone-shaped stimuli caused Rex's limbs to swing in certain arcs. Rather, we might say that he wanted to speak to his friend Cecile and knew that Cecile was home. No explanation has as much predictive power as that one. If Rex was no longer on speaking terms with Cecile, or if he remembered that Cecile was out bowling that night, his body would not have risen off the couch.

For millennia the gap between physical events, on the one hand, and meaning, content, ideas, reasons, and intentions, on the other, seemed to cleave the universe in two. How can something as ethereal as "inciting hatred" or "wanting to speak to Cecile" actually cause matter to move in space? But the cognitive revolution unified the world of ideas with the world of matter using a powerful new theory: that mental life can be explained in terms of information, computation, and feedback. Beliefs and memories are collections of information—like facts in a database, but residing in patterns of activity and structure in the brain. Thinking and planning are systematic transformations of these patterns, like the operation of a computer program. Wanting and trying are feedback loops, like the principle behind a thermostat: they receive information about the discrepancy between a goal and the current state of the world, and then they execute operations that tend to reduce the difference. The mind is connected to the world by the sense organs, which transduce physical energy into data structures in the brain, and by motor programs, by which the brain controls the muscles.

This general idea may be called the computational theory of mind. It is not the same as the "computer metaphor" of the mind, the suggestion that the mind literally works like a human-made database, computer program, or thermostat. It says only that we can explain minds and human-made information processors using some of the same principles. It is just like other cases in which the natural world and human engineering overlap. A physiologist might invoke the same laws of optics to explain how the eye works and how a camera works without implying that the eye is like a camera in every detail.

The computational theory of mind does more than explain the existence of knowing, thinking, and trying without invoking a ghost in the machine (though that would be enough of a feat). It also explains how those processes can be *intelligent*—how rationality can emerge from a mindless physical process. If a sequence of transformations of information stored in a hunk of matter (such as brain tissue or silicon) mirrors a sequence of deductions that obey the laws of logic, probability, or cause and effect in the world, they will generate correct predictions about the world. And making correct predictions in pursuit of a goal is a pretty good definition of "intelligence."[3]

Of course there is no new thing under the sun, and the computational theory of mind was foreshadowed by Hobbes when he described mental activity as tiny motions and wrote that "reasoning is but reckoning." Three and a half centuries later, science has caught up to his vision. Perception, memory, imagery, reasoning, decision making, language, and motor control are being studied in the lab and successfully modeled as computational paraphernalia such as rules, strings, matrices, pointers, lists, files, trees, arrays, loops, propositions, and networks. For example, cognitive psychologists are studying the graphics system in the head and thereby explaining how people "see" the solution to a problem in a mental image. They are studying the web of concepts in long-term memory and explaining why some facts are easier to recall than others. They are studying the processor and memory used by the language system to learn why some sentences are a pleasure to read and others a difficult slog.

And if the proof is in the computing, then the sister field of artificial intelligence is confirming that ordinary matter can perform feats that were supposedly performable by mental stuff alone. In the 1950s computers were already being called "electronic brains" because they could calculate sums, organize data, and prove theorems. Soon they could correct spelling, set type, solve equations, and simulate experts on restricted topics such as picking stocks and diagnosing diseases. For decades we psychologists preserved human bragging rights by telling our classes that no computer could read text, decipher speech, or recognize faces, but these boasts are obsolete. Today software that can recognize printed letters and spoken words comes packaged with home computers. Rudimentary programs that understand or translate sentences are available in many search engines and Help programs, and they are steadily improving. Face-recognition systems have advanced to the point that civil libertarians are concerned about possible abuse when they are used with security cameras in public places.

Human chauvinists can still write off these low-level feats. Sure, they say, the input and output processing can be fobbed off onto computational modules, but you still need a human user with the capacity for judgment, reflection, and creativity. But according to the computational theory of mind, these capacities are themselves forms of information processing and can be implemented in a computational system. In 1997 an IBM computer called Deep

Blue defeated the world chess champion Garry Kasparov, and unlike its predecessors, it did not just evaluate trillions of moves by brute force but was fitted with strategies that intelligently responded to patterns in the game. *Newsweek* called the match "The Brain's Last Stand." Kasparov called the outcome "the end of mankind."

You might still object that chess is an artificial world with discrete moves and a clear winner, perfectly suited to the rule-crunching of a computer. People, on the other hand, live in a messy world offering unlimited moves and nebulous goals. Surely this requires human creativity and intuition—which is why everyone knows that computers will never compose a symphony, write a story, or paint a picture. But everyone may be wrong. Recent artificial intelligence systems have written credible short stories,[4] composed convincing Mozart-like symphonies,[5] drawn appealing pictures of people and landscapes,[6] and conceived clever ideas for advertisements.[7]

None of this is to say that the brain works like a digital computer, that artificial intelligence will ever duplicate the human mind, or that computers are conscious in the sense of having first-person subjective experience. But it does suggest that reasoning, intelligence, imagination, and creativity are forms of information processing, a well-understood physical process. Cognitive science, with the help of the computational theory of mind, has exorcised at least one ghost from the machine.

A second idea: *The mind cannot be a blank slate, because blank slates don't do anything.* As long as people had only the haziest concept of what a mind was or how it might work, the metaphor of a blank slate inscribed by the environment did not seem too outrageous. But as soon as one starts to think seriously about what kind of computation enables a system to see, think, speak, and plan, the problem with blank slates becomes all too obvious: they don't do anything. The inscriptions will sit there forever unless something notices patterns in them, combines them with patterns learned at other times, uses the combinations to scribble new thoughts onto the slate, and reads the results to guide behavior toward goals. Locke recognized this problem and alluded to something called "the understanding," which looked at the inscriptions on the white paper and carried out the recognizing, reflecting, and associating. But of course explaining how the mind understands by invoking something called "the understanding" is circular.

This argument against the Blank Slate was stated pithily by Gottfried Wilhelm Leibniz (1646–1716) in a reply to Locke. Leibniz repeated the empiricist motto "There is nothing in the intellect that was not first in the senses," then added, "except the intellect itself."[8] *Something* in the mind must be innate, if it is only the mechanisms that do the learning. Something has to see a world of objects rather than a kaleidoscope of shimmering pixels. Something has to infer the content of a sentence rather than parrot back the exact wording.

Something has to interpret other people's behavior as their attempts to achieve goals rather than as trajectories of jerking arms and legs.

In the spirit of Locke, one could attribute these feats to an abstract noun—perhaps not to "the understanding" but to "learning," "intelligence," "plasticity," or "adaptiveness." But as Leibniz remarked, to do so is to "[save appearances] by fabricating faculties or occult qualities, . . . and fancying them to be like little demons or imps which can without ado perform whatever is wanted, as though pocket watches told the time by a certain horological faculty without needing wheels, or as though mills crushed grain by a fractive faculty without needing anything in the way of millstones."[9] Leibniz, like Hobbes (who had influenced him), was ahead of his time in recognizing that intelligence is a form of information processing and needs complex machinery to carry it out. As we now know, computers don't understand speech or recognize text as they roll off the assembly line; someone has to install the right software first. The same is likely to be true of the far more demanding performance of the human being. Cognitive modelers have found that mundane challenges like walking around furniture, understanding a sentence, recalling a fact, or guessing someone's intentions are formidable engineering problems that are at or beyond the frontiers of artificial intelligence. The suggestion that they can be solved by a lump of Silly Putty that is passively molded by something called "culture" just doesn't cut the mustard.

This is not to say that cognitive scientists have put the nature-nurture debate completely behind them; they are still spread out along a continuum of opinion on how much standard equipment comes with the human mind. At one end are the philosopher Jerry Fodor, who has suggested that all concepts might be innate (even "doorknob" and "tweezers"), and the linguist Noam Chomsky, who believes that the word "learning" is misleading and we should say that children "grow" language instead.[10] At the other end are the connectionists, including Rumelhart, McClelland, Jeffrey Elman, and Elizabeth Bates, who build relatively simple computer models and train the living daylights out of them.[11] Fans locate the first extreme, which originated at the Massachusetts Institute of Technology, at the East Pole, the mythical place from which all directions are west. They locate the second extreme, which originated at the University of California, San Diego, at the West Pole, the mythical place from which all directions are east. (The names were suggested by Fodor during an MIT seminar at which he was fulminating against a "West Coast theorist" and someone pointed out that the theorist worked at Yale, which is, technically, on the East Coast.)[12]

But here is why the East Pole–West Pole debate is different from the ones that preoccupied philosophers for millennia: neither side believes in the Blank Slate. Everyone acknowledges that there can be no learning without innate circuitry to do the learning. In their West Pole manifesto *Rethinking Innateness*,

Bates and Elman and their coauthors cheerfully concede this point: "No learning rule can be entirely devoid of theoretical content nor can the *tabula* ever be completely *rasa*."[13] They explain:

> There is a widespread belief that connectionist models (and modelers) are committed to an extreme form of empiricism; and that any form of innate knowledge is to be avoided like the plague.... We obviously do not subscribe to this point of view.... There are good reasons to believe that some kinds of prior constraints [on learning models] are necessary. In fact, all connectionist models necessarily make some assumptions which must be regarded as constituting innate constraints.[14]

The disagreements between the two poles, though significant, are over the details: how many innate learning networks there are, and how specifically engineered they are for particular jobs. (We will explore some of these disagreements in Chapter 5.)

A third idea: *An infinite range of behavior can be generated by finite combinatorial programs in the mind.* Cognitive science has undermined the Blank Slate and the Ghost in the Machine in another way. People can be forgiven for scoffing at the suggestion that human behavior is "in the genes" or "a product of evolution" in the senses familiar from the animal world. Human acts are not selected from a repertoire of knee-jerk reactions like a fish attacking a red spot or a hen sitting on eggs. Instead, people may worship goddesses, auction kitsch on the Internet, play air guitar, fast to atone for past sins, build forts out of lawn chairs, and so on, seemingly without limit. A glance at *National Geographic* shows that even the strangest acts in our own culture do not exhaust what our species is capable of. If anything goes, one might think, then perhaps we are Silly Putty, or unconstrained agents, after all.

But that impression has been made obsolete by the computational approach to the mind, which was barely conceivable in the era in which the Blank Slate arose. The clearest example is the Chomskyan revolution in language.[15] Language is the epitome of creative and variable behavior. Most utterances are brand-new combinations of words, never before uttered in the history of humankind. We are nothing like Tickle Me Elmo dolls who have a fixed list of verbal responses hard-wired in. But, Chomsky pointed out, for all its open-endedness language is not a free-for-all; it obeys rules and patterns. An English speaker can utter unprecedented strings of words such as *Every day new universes come into existence,* or *He likes his toast with cream cheese and ketchup,* or *My car has been eaten by wolverines.* But no one would say *Car my been eaten has wolverines by* or most of the other possible orderings of English words. Something in the head must be capable of generating not just any combinations of words but highly systematic ones.

That something is a kind of software, a generative grammar that can crank out new arrangements of words. A battery of rules such as "An English sentence contains a subject and a predicate," "A predicate contains a verb, an object, and a complement," and "The subject of *eat* is the eater" can explain the boundless creativity of a human talker. With a few thousand nouns that can fill the subject slot and a few thousand verbs that can fill the predicate slot, one already has several million ways to open a sentence. The possible combinations quickly multiply out to unimaginably large numbers. Indeed, the repertoire of sentences is theoretically infinite, because the rules of language use a trick called recursion. A recursive rule allows a phrase to contain an example of itself, as in *She thinks that he thinks that they think that he knows* and so on, ad infinitum. And if the number of sentences is infinite, the number of possible thoughts and intentions is infinite too, because virtually every sentence expresses a different thought or intention. The combinatorial grammar for language meshes with other combinatorial programs in the head for thoughts and intentions. A fixed collection of machinery in the mind can generate an infinite range of behavior by the muscles.[16]

Once one starts to think about mental software instead of physical behavior, the radical differences among human cultures become far smaller, and that leads to a fourth new idea: *Universal mental mechanisms can underlie superficial variation across cultures.* Again, we can use language as a paradigm case of the open-endedness of behavior. Humans speak some six thousand mutually unintelligible languages. Nonetheless, the grammatical programs in their minds differ far less than the actual speech coming out of their mouths. We have known for a long time that all human languages can convey the same kinds of ideas. The Bible has been translated into hundreds of non-Western languages, and during World War II the U.S. Marine Corps conveyed secret messages across the Pacific by having Navajo Indians translate them to and from their native language. The fact that any language can be used to convey any proposition, from theological parables to military directives, suggests that all languages are cut from the same cloth.

Chomsky proposed that the generative grammars of individual languages are variations on a single pattern, which he called Universal Grammar. For example, in English the verb comes before the object *(drink beer)* and the preposition comes before the noun phrase *(from the bottle)*. In Japanese the object comes before the verb *(beer drink)* and the noun phrase comes before the preposition, or, more accurately, the postposition *(the bottle from)*. But it is a significant discovery that both languages have verbs, objects, and pre- or postpositions to start with, as opposed to having the countless other conceivable kinds of apparatus that could power a communication system. And it is even more significant that unrelated languages build their phrases by assembling a head (such as a verb or preposition) and a complement (such as a noun phrase) and assigning a consistent order to the two. In English the head comes first; in Japanese the head comes last. But everything else about the structure of phrases in the two languages is pretty much the same. And so it goes with phrase after phrase and language after language. The common kinds of heads and complements can be ordered in 128 logically possible ways, but 95 percent of the world's languages use one of two: either the English ordering or its mirror image the Japanese ordering.[17] A simple way to capture this uniformity is to say that all languages have the same grammar except for a parameter or switch that can be flipped to either the "head-first" or "head-last" setting. The linguist Mark Baker has recently summarized about a dozen of these parameters, which succinctly capture most of the known variation among the languages of the world.[18]

Distilling the variation from the universal patterns is not just a way to tidy up a set of messy data. It can also provide clues about the innate circuitry that makes learning possible. If the universal part of a rule is embodied in the neural circuitry that guides babies when they first learn language, it could explain how children learn language so easily and uniformly and without the benefit of instruction. Rather than treating the sound coming out of Mom's mouth as just an interesting noise to mimic verbatim or to slice and dice in arbitrary ways, the baby listens for heads and complements, pays attention to how they are ordered, and builds a grammatical system consistent with that ordering.

This idea can make sense of other kinds of variability across cultures. Many anthropologists sympathetic to social constructionism have claimed that emotions familiar to us, like anger, are absent from some cultures.[19] (A few anthropologists say there are cultures with no emotions at all!)[20] For example, Catherine Lutz wrote that the Ifaluk (a Micronesian people) do not experience our "anger" but instead undergo an experience they call *song*. *Song* is a state of dudgeon triggered by a moral infraction such as breaking a taboo or acting in a cocky manner. It licenses one to shun, frown at, threaten, or gossip about the offender, though not to attack him physically. The target of *song* experiences another emotion allegedly unknown to Westerners: *metagu*, a state of dread that impels him to appease the *song*-ful one by apologizing, paying a fine, or offering a gift.

The philosophers Ron Mallon and Stephen Stich, inspired by Chomsky and other cognitive scientists, point out that the issue of whether to call Ifaluk *song* and Western *anger* the same emotion or different emotions is a quibble about the meaning of emotion words: whether they should be defined in terms of surface behavior or underlying mental computation.[21] If an emotion is defined by behavior, then emotions certainly do differ across cultures. The Ifaluk react emotionally to a woman working in the taro gardens while menstruating or to a man entering a birthing house, and we do not. We react emotionally to someone shouting a racial epithet or raising the middle finger, but as far as we know, the Ifaluk do not. But if an emotion is defined by mental mechanisms—what psychologists like Paul Ekman and Richard Lazarus call "affect programs" or "if-then formulas" (note the computational vocabulary)—we and the Ifaluk are not so different after all.[22] We might all be equipped with a program that responds to an affront to our interests or our dignity with an unpleasant burning feeling that motivates us to punish or to exact compensation. But what counts as an affront, whether we feel it is permissible to glower in a particular setting, and what kinds of retribution we think we are entitled to, depend on our culture. The stimuli and responses may differ, but the mental states are the same, whether or not they are perfectly labeled by words in our language.

And as in the case of language, without *some* innate mechanism for mental computation, there would be no way to learn the parts of a culture that do have to be learned. It is no coincidence that the situations that provoke *song* among the Ifaluk include violating a taboo, being lazy or disrespectful, and refusing to share, but do not include respecting a taboo, being kind and deferential, and standing on one's head. The Ifaluk construe the first three as similar because they evoke the same affect program—they are perceived as affronts. That makes it easier to learn that they call for the same reaction and makes it more likely that those three would be lumped together as the acceptable triggers for a single emotion.

The moral, then, is that familiar categories of behavior—marriage customs, food taboos, folk superstitions, and so on—certainly do vary across cultures and have to be learned, but the deeper mechanisms of mental computation that generate them may be universal and innate. People may dress differently, but they may all strive to flaunt their status via their appearance. They may respect the rights of the members of their clan exclusively or they may extend that respect to everyone in their tribe, nation-state, or species, but all divide the world into an in-group and an out-group. They may differ in which outcomes they attribute to the intentions of conscious beings, some allowing only that artifacts are deliberately crafted, others believing that illnesses come from magical spells cast by enemies, still others believing that the entire world was brought into being by a creator. But all of them explain certain events by invoking the existence of entities with minds that strive to bring about goals. The behaviorists got it backwards: it is the mind, not behavior, that is lawful.

A fifth idea: *The mind is a complex system composed of many interacting parts.* The psychologists who study emotions in different cultures have made another important discovery. Candid facial expressions appear to be the same everywhere, but people in some cultures learn to keep a poker face in polite company.[23] A simple explanation is that the affect programs fire up facial expressions in the same way in all people, but a separate system of "display rules" governs when they can be shown.

The difference between these two mechanisms underscores another insight of the cognitive revolution. Before the revolution, commentators invoked enormous black boxes such as "the intellect" or "the understanding," and they made sweeping pronouncements about human nature, such as that we are essentially noble or essentially nasty. But we now know that the mind is not a homogeneous orb invested with unitary powers or across-the-board traits. The mind is modular, with many parts cooperating to generate a train of thought or an organized action. It has distinct information-processing systems for filtering out distractions, learning skills, controlling the body, remembering facts, holding information temporarily, and storing and executing rules. Cutting across these data-processing systems are mental faculties (sometimes called multiple intelligences) dedicated to different kinds of content, such as language, number, space, tools, and living things. Cognitive scientists at the East Pole suspect that the content-based modules are differentiated largely by the genes;[24] those at the West Pole suspect they begin as small innate biases in attention and then coagulate out of statistical patterns in the sensory input.[25] But those at both poles agree that the brain is not a uniform meatloaf. Still another layer of information-processing systems can be found in the affect programs, that is, the systems for motivation and emotion.

The upshot is that an urge or habit coming out of one module can be translated into behavior in different ways—or suppressed altogether—by some other module. To take a simple example, cognitive psychologists believe that a module called the "habit system" underlies our tendency to produce certain responses habitually, such as responding to a printed word by pronouncing it silently. But another module, called the "supervisory attention system," can override it and focus on the information relevant to a stated problem, such as naming the color of the ink the word is printed in, or thinking up an action that goes with the word.[26] More generally, the interplay of mental systems can explain how people can entertain revenge fantasies that they never act on, or can commit adultery only in their hearts. In this way the theory of human nature coming out of the cognitive revolution has more in common with the Judeo-Christian theory of human nature, and with the psychoanalytic theory proposed by Sigmund Freud, than with behaviorism, social constructionism, and other versions of the Blank Slate. Behavior is not just emitted or elicited, nor does it come directly out of culture or society. It comes from an internal struggle among mental modules with differing agendas and goals.

The idea from the cognitive revolution that the mind is a system of universal, generative computational modules obliterates the way that debates on human nature have been framed for centuries. It is now simply misguided to ask whether humans are flexible or programmed, whether behavior is universal or varies across cultures, whether acts are learned or innate, whether we are essentially good or essentially evil. Humans behave flexibly *because* they are programmed: their minds are packed with combinatorial software that can generate an unlimited set of thoughts and behavior. Behavior may vary across cultures, but the design of the mental programs that generate it need not vary. Intelligent behavior is learned successfully because we have innate systems that do the learning. And all people may have good and evil motives, but not everyone may translate them into behavior in the same way.

~

THE SECOND BRIDGE between mind and matter is neuroscience, especially cognitive neuroscience, the study of how cognition and emotion are implemented in the brain.[27] Francis Crick wrote a book about the brain called *The Astonishing Hypothesis,* alluding to the idea that all our thoughts and feelings, joys and aches, dreams and wishes consist in the physiological activity of the brain.[28] Jaded neuroscientists, who take the idea for granted, snickered at the title, but Crick was right: the hypothesis *is* astonishing to most people the first time they stop to ponder it. Who cannot sympathize with the imprisoned Dmitri Karamazov as he tries to make sense of what he has just learned from a visiting academic?

> Imagine: inside, in the nerves, in the head—that is, these nerves are there in the brain . . . (damn them!) there are sort of little tails, the little tails of those nerves, and as soon as they begin quivering . . . that is, you see, I look at something with my eyes and then they begin quivering, those little tails . . . and when they quiver, then an image appears . . . it doesn't appear at once, but an instant, a second, passes . . . and then something like a moment appears; that is, not a moment—devil take the moment!—but an image; that is, an object, or an action, damn it! That's why I see and then think, because of those tails, not at all because I've got a soul, and that I am some sort of image and likeness. All that is nonsense! Rakitin explained it all to me yesterday, brother, and it simply bowled me over. It's magnificent, Alyosha, this science! A new man's arising—that I understand. . . . And yet I am sorry to lose God![29]

Dostoevsky's prescience is itself astonishing, because in 1880 only the rudiments of neural functioning were understood, and a reasonable person could have doubted that all experience arises from quivering nerve tails. But no longer. One can say that the information-processing activity of the brain *causes* the mind, or one can say that it *is* the mind, but in either case the evidence is overwhelming that every aspect of our mental lives depends entirely on physiological events in the tissues of the brain.

When a surgeon sends an electrical current into the brain, the person can have a vivid, lifelike experience. When chemicals seep into the brain, they can alter the person's perception, mood, personality, and reasoning. When a patch of brain tissue dies, a part of the mind can disappear: a neurological patient may lose the ability to name tools, recognize faces, anticipate the outcome of his behavior, empathize with others, or keep in mind a region of space or of his own body. (Descartes was thus wrong when he said that "the mind is entirely indivisible" and concluded that it must be completely different from the body.) Every emotion and thought gives off physical signals, and the new technologies for detecting them are so accurate that they can literally read a person's mind and tell a cognitive neuroscientist whether the person is imagining a face or a place. Neuroscientists can knock a gene out of a mouse (a gene also found in humans) and prevent the mouse from learning, or insert extra copies and make the mouse learn faster. Under the microscope, brain tissue shows a staggering complexity—a hundred billion neurons connected by a hundred trillion synapses—that is commensurate with the staggering complexity of human thought and experience. Neural network modelers have begun to show how the building blocks of mental computation, such as storing and retrieving a pattern, can be implemented in neural circuitry. And when the brain dies, the person goes out of existence. Despite concerted efforts by Alfred Russel Wallace and other Victorian scientists, it is apparently not possible to communicate with the dead.

Educated people, of course, know that perception, cognition, language, and emotion are rooted in the brain. But it is still tempting to think of the brain as it was shown in old educational cartoons, as a control panel with gauges and levers operated by a user—the self, the soul, the ghost, the person, the "me." But cognitive neuroscience is showing that the self, too, is just another network of brain systems.

The first hint came from Phineas Gage, the nineteenth-century railroad worker familiar to generations of psychology students. Gage was using a yard-long spike to tamp explosive powder into a hole in a rock when a spark ignited the powder and sent the spike into his cheekbone, through his brain, and out the top of his skull. Phineas survived with his perception, memory, language, and motor functions intact. But in the famous understatement of a co-worker, "Gage was no longer Gage." A piece of iron had literally turned him into a different person, from courteous, responsible, and ambitious to rude, unreliable, and shiftless. It did this by impaling his ventromedial prefrontal cortex, the region of the brain above the eyes now known to be involved in reasoning about other people. Together with other areas of the prefrontal lobes and the limbic system (the seat of the emotions), it anticipates the consequences of one's actions and selects behavior consonant with one's goals.[30]

Cognitive neuroscientists have not only exorcised the ghost but have shown that the brain does not even have a part that does exactly what the ghost is supposed to do: review all the facts and make a decision for the rest of the brain to carry out.[31] Each of us *feels* that there is a single "I" in control. But that is an illusion that the brain works hard to produce, like the impression that our visual fields are rich in detail from edge to edge. (In fact, we are blind to detail outside the fixation point. We quickly move our eyes to whatever looks interesting, and that fools us into thinking that the detail was there all along.) The brain does have supervisory systems in the prefrontal lobes and anterior cingulate cortex, which can push the buttons of behavior and override habits and urges. But those systems are gadgets with specific quirks and limitations; they are not implementations of the rational free agent traditionally identified with the soul or the self.

One of the most dramatic demonstrations of the illusion of the unified self comes from the neuroscientists Michael Gazzaniga and Roger Sperry, who showed that when surgeons cut the corpus callosum joining the cerebral hemispheres, they literally cut the self in two, and each hemisphere can exercise free will without the other one's advice or consent. Even more disconcertingly, the left hemisphere constantly weaves a coherent but false account of the behavior chosen without its knowledge by the right. For example, if an experimenter flashes the command "WALK" to the right hemisphere (by keeping it in the part of the visual field that only the right hemisphere can see), the person will comply with the request and begin to walk out of the room. But when the person (specifically, the person's left hemisphere) is asked why he just got up, he will say, in all sincerity, "To get a Coke"—rather than "I don't really know" or "The urge just came over me" or "You've been testing me for years since I had the surgery, and sometimes you get me to do things but I don't know exactly what you asked me to do." Similarly, if the patient's left hemisphere is shown a chicken and his right hemisphere is shown a snowfall, and both hemispheres have to select a picture that goes with what they see (each using a different hand), the left hemisphere picks a claw (correctly) and the right picks a shovel (also correctly). But when the left hemisphere is asked why the whole person made those choices, it blithely says, "Oh, that's simple. The chicken claw goes with the chicken, and you need a shovel to clean out the chicken shed."[32]

The spooky part is that we have no reason to think that the baloney-generator in the patient's left hemisphere is behaving any differently from *ours* as *we* make sense of the inclinations emanating from the rest of *our* brains. The conscious mind—the self or soul—is a spin doctor, not the commander in chief. Sigmund Freud immodestly wrote that "humanity has in the course of time had to endure from the hands of science three great outrages upon its naïve self-love": the discovery that our world is not the center of the celestial spheres but rather a speck in a vast universe, the discovery that we were not specially created but instead descended from animals, and the discovery that often our conscious minds do not control how we act but merely tell us a story about our actions. He was right about the cumulative impact, but it was cognitive neuroscience rather than psychoanalysis that conclusively delivered the third blow.

Cognitive neuroscience is undermining not just the Ghost in the Machine but also the Noble Savage. Damage to the frontal lobes does not only dull the person or subtract from his behavioral repertoire but can unleash aggressive attacks.[33] That happens because the damaged lobes no longer serve as inhibitory brakes on parts of the limbic system, particularly a circuit that links the amygdala to the hypothalamus via a pathway called the stria terminalis. Connections between the frontal lobe in each hemisphere and the limbic system provide a lever by which a person's knowledge and goals can override other mechanisms, and among those mechanisms appears to be one designed to generate behavior that harms other people.[34]

Nor is the physical structure of the brain a blank slate. In the mid-nineteenth century the neurologist Paul Broca discovered that the folds and wrinkles of the cerebral cortex do not squiggle randomly like fingerprints but have a recognizable geometry. Indeed, the arrangement is so consistent from brain to brain that each fold and wrinkle can be given a name. Since that time neuroscientists have discovered that the gross anatomy of the brain—the sizes, shapes, and connectivity of its lobes and nuclei, and the basic plan of the cerebral cortex—is largely shaped by the genes in normal prenatal development.[35] So is the quantity of gray matter in the different regions of the brains of different people, including the regions that underlie language and reasoning.[36]

This innate geometry and cabling can have real consequences for thinking, feeling, and behavior. As we shall see in a later chapter, babies who suffer damage to particular areas of the brain often grow up with permanent deficits in particular mental faculties. And people born with variations on the typical plan have variations in the way their minds work. According to a recent study of the brains of identical and fraternal twins, differences in the amount of gray matter in the frontal lobes are not only genetically influenced but are significantly correlated with differences in intelligence.[37] A study of Albert Einstein's brain revealed that he had large, unusually shaped inferior parietal lobules, which participate in spatial reasoning and intuitions about number.[38] Gay men are likely to have a smaller third interstitial nucleus in the anterior hypothalamus, a nucleus known to have a role in sex differences.[39] And convicted murderers and other violent, antisocial people are likely to have a smaller and less active prefrontal cortex, the part of the brain that governs decision making and inhibits impulses.[40] These gross features of the brain are almost certainly not sculpted by information coming in from the senses, which implies that differences in intelligence, scientific genius, sexual orientation, and impulsive violence are not entirely learned.

Indeed, until recently the innateness of brain structure was an embarrassment for neuroscience. The brain could not possibly be wired by the genes down to the last synapse, because there isn't nearly enough information in the genome to do so. And we know that people learn throughout their lives, and the products of that learning have to be stored in the brain somehow. Unless you believe in a ghost in the machine, everything a person learns has to affect some part of the brain; more accurately, learning *is* a change in some part of the brain. But it was difficult to find the features of the brain that reflected those changes amid all that innate structure. Becoming stronger in math or motor coordination or visual discrimination does not bulk up the brain the way becoming stronger at weightlifting bulks up the muscles.

Now, at last, neuroscience is beginning to catch up with psychology by discovering changes in the brain that underlie learning. As we shall see, the boundaries between swatches of cortex devoted to different body parts, talents, and even physical senses can be adjusted by learning and practice. Some neuroscientists are so excited by these discoveries that they are trying to push the pendulum in the other direction, emphasizing the plasticity of the cerebral cortex. But for reasons that I will review in Chapter 5, most neuroscientists believe that these changes take place within a matrix of genetically organized structure. There is much we don't understand about how the brain is laid out in development, but we know that it is not indefinitely malleable by experience.

THE THIRD BRIDGE between the biological and the mental is behavioral genetics, the study of how genes affect behavior.[41] All the potential for thinking, learning, and feeling that distinguishes humans from other animals lies in the information contained in the DNA of the fertilized ovum. This is most obvious when we compare species. Chimpanzees brought up in a human home do not speak, think, or act like people, and that is because of the information in the ten megabytes of DNA that differ between us. Even the two species of chimpanzees, common chimps and bonobos, which differ in just a few tenths of one percent of their genomes, part company in their behavior, as zookeepers first discovered when they inadvertently mixed the two. Common chimps are among the most aggressive mammals known to zoology, bonobos among the most peaceable; in common chimps the males dominate the females, in bonobos the females have the upper hand; common chimps have sex for procreation, bonobos for recreation. Small differences in the genes can lead to large differences in behavior. They can affect the size and shape of the different parts of the brain, their wiring, and the nanotechnology that releases, binds, and recycles hormones and neurotransmitters.

The importance of genes in organizing the normal brain is underscored by the many ways in which nonstandard genes can give rise to nonstandard minds. When I was an undergraduate an exam question in Abnormal Psychology asked, "What is the best predictor that a person will become schizophrenic?"

The answer was, "Having an identical twin who is schizophrenic." At the time it was a trick question, because the reigning theories of schizophrenia pointed to societal stress, "schizophrenogenic mothers," double binds, and other life experiences (none of which turned out to have much, if any, importance); hardly anyone thought about genes as a possible cause. But even then the evidence was there: schizophrenia is highly concordant within pairs of identical twins, who share all their DNA and most of their environment, but far less concordant within pairs of fraternal twins, who share only half their DNA (of the DNA that varies in the population) and most of their environment. The trick question could be asked—and would have the same answer—for virtually every cognitive and emotional disorder or difference ever observed. Autism, dyslexia, language delay, language impairment, learning disability, left-handedness, major depressions, bipolar illness, obsessive-compulsive disorder, sexual orientation, and many other conditions run in families, are more concordant in identical than in fraternal twins, are better predicted by people's biological relatives than by their adoptive relatives, and are poorly predicted by any measurable feature of the environment.[42]

Genes not only push us toward exceptional conditions of mental functioning but scatter us within the normal range, producing much of the variation in ability and temperament that we notice in the people around us. The famous Chas Addams cartoon from *The New Yorker* is only a slight exaggeration:

Separated at birth, the Mallifert twins meet accidentally.

Identical twins think and feel in such similar ways that they sometimes suspect they are linked by telepathy. When separated at birth and reunited as adults, they say they feel they have known each other all their lives. Testing confirms that identical twins, whether separated at birth or not, are eerily alike (though far from identical) in just about any trait one can measure. They are similar in verbal, mathematical, and general intelligence, in their degree of life satisfaction, and in personality traits such as introversion, agreeableness, neuroticism, conscientiousness, and openness to experience. They have similar attitudes toward controversial issues such as the death penalty, religion, and modern music. They resemble each other not just in paper-and-pencil tests but in consequential behavior such as gambling, divorcing, committing crimes, getting into accidents, and watching television. And they boast dozens of shared idiosyncrasies such as giggling incessantly, giving interminable answers to simple questions, dipping buttered toast in coffee, and—in the case of Abigail van Buren and Ann Landers—writing indistinguishable syndicated advice columns. The crags and valleys of their electroencephalograms (brainwaves) are as alike as those of a single person recorded on two occasions, and the wrinkles of their brains and distribution of gray matter across cortical areas are also similar.[43]

The effects of differences in genes on differences in minds can be measured, and the same rough estimate—substantially greater than zero, but substantially less than 100 percent—pops out of the data no matter what measuring stick is used. Identical twins are far more similar than fraternal twins, whether they are raised apart or together; identical twins raised apart are highly similar; biological siblings, whether raised together or apart, are far more similar than adoptive siblings. Many of these conclusions come from massive studies in Scandinavian countries where governments keep huge databases on their citizens, and they employ the best-validated measuring instruments known to psychology. Skeptics have offered alternative explanations that try to push the effects of the genes to zero—they suggest that identical twins separated at birth may have been placed in similar adoptive homes, that they may have contacted each other before being tested, that they look alike and hence may have been treated alike, and that they shared a womb in addition to their genes. But as we shall see in the chapter on children, these explanations have all been tested and rejected. Recently a new kind of evidence may be piled on the heap. "Virtual twins" are the mirror image of identical twins raised apart: they are unrelated siblings, one or both adopted, who are raised together from infancy. Though they are the same age and are growing up in the same family, the psychologist Nancy Segal found that their IQ scores are barely correlated.[44] One father in the study said that despite efforts to treat them alike, the virtual twins are "like night and day."

Twinning and adoption are natural experiments that offer strong indirect evidence that differences in minds can come from differences in genes. Recently geneticists have pinpointed some of the genes that can cause the differences. A single wayward nucleotide in a gene called *FOXP2* causes a hereditary disorder in speech and language.[45] A gene on the same chromosome, *LIM-kinase1,* produces a protein found in growing neurons that helps install the faculty of spatial cognition: when the gene is deleted, the person has normal intelligence but cannot assemble objects, arrange blocks, or copy shapes.[46] One version of the gene *IGF2R* is associated with high general intelligence, accounting for as many as four IQ points and two percent of the variation in intelligence among normal individuals.[47] If you have a longer than average version of the *D4DR* dopamine receptor gene, you are more likely to be a thrill seeker, the kind of person who jumps out of airplanes, clambers up frozen waterfalls, or has sex with strangers.[48] If you have a shorter version of a stretch of DNA that inhibits the serotonin transporter gene on chromosome 17, you are more likely to be neurotic and anxious, the kind of person who can barely function at social gatherings for fear of offending someone or acting like a fool.[49]

Single genes with large consequences are the most dramatic examples of the effects of genes on the mind, but they are not the most representative examples. Most psychological traits are the product of many genes with small effects that are modulated by the presence of other genes, rather than the product of a single gene with a large effect that shows up come what may. That is why studies of identical twins (two people who share *all* their genes) consistently show powerful genetic effects on a trait even when the search for a *single* gene for that trait is unsuccessful.

In 2001 the complete sequence of the human genome was published, and with it came a powerful new ability to identify genes and their products, including those that are active in the brain. In the coming decade, geneticists will identify genes that differentiate us from chimpanzees, infer which of them were subject to natural selection during the millions of years our ancestors evolved into humans, identify which combinations are associated with normal, abnormal, and exceptional mental abilities, and begin to trace the chain of causation in fetal development by which genes shape the brain systems that let us learn, feel, and act.

People sometimes fear that if the genes affect the mind at all they must determine it in every detail. That is wrong, for two reasons. The first is that most effects of genes are probabilistic. If one identical twin has a trait, there is usually no more than an even chance that the other will have it, despite their having a complete genome in common. Behavioral geneticists estimate that only about half of the variation in most psychological traits within a given environment correlates with the genes. In the chapter on children, we will explore what this means and where the other half of the variation comes from.

The second reason that genes aren't everything is that their effects can vary depending on the environment. A simple example may be found in any genetics textbook. While different strains of corn grown in a single field will vary in height because of their genes, a single strain of corn grown in different fields—one arid, the other irrigated—will vary in height because of the environment. A human example comes from Woody Allen. Though his fame, fortune, and ability to attract beautiful women may depend on having genes that enhance a sense of humor, in *Stardust Memories* he explains to an envious childhood friend that there is a crucial environmental factor as well: "We live in a society that puts a big value on jokes. . . . If I had been an Apache Indian, those guys didn't need comedians, so I'd be out of work."

The meaning of findings in behavioral genetics for our understanding of human nature has to be worked out for each case. An aberrant gene that causes a disorder shows that the standard version of the gene is necessary to have a normal human mind. But what the standard version does is not immediately obvious. If a gear with a broken tooth goes *clunk* on every turn, we do not conclude that the tooth in its intact form was a clunk-suppressor. And so a gene that disrupts a mental ability need not be a defective version of a gene that is "for" that ability. It may produce a toxin that interferes with normal brain development, or it may leave a chink in the immune system that allows a pathogen to infect the brain, or it may make the person look stupid or sinister and thereby affect how other people react to him. In the past, geneticists couldn't rule out the boring possibilities (the ones that don't involve brain function directly), and skeptics intimated that *all* genetic effects might be boring, merely warping or defacing a blank slate rather than being an ineffective version of a gene that helps to give structure to a complex brain. But increasingly researchers are able to tie genes to the brain.

A promising example is the *FOXP2* gene, associated with a speech and language disorder in a large family.[50] The aberrant nucleotide has been found in every impaired member of the family (and in one unrelated person with the same syndrome), but it was not found in any of the unimpaired members, nor was it found in 364 chromosomes from unrelated normal people. The gene belongs to a family of genes for transcription factors—proteins that turn on other genes—that are known to play important roles in embryogenesis. The mutation disrupts the part of the protein that latches onto a particular region of DNA, the key step in turning on the right gene at the right time. The gene appears to be strongly active in fetal brain tissue, and a closely related version found in mice is active in the developing cerebral cortex. These are signs, according to the authors of the study, that the normal version of the gene triggers a cascade of events that help organize a part of the developing brain.

The meaning of genetic variation among normal individuals (as opposed to genetic defects that cause a disorder) also has to be thought through with care. An innate *difference* among people is not the same thing as an innate human nature that is *universal* across the species. Documenting the ways that people vary will not directly reveal the workings of human nature, any more than documenting the ways that automobiles vary will directly reveal how car engines work. Nonetheless, genetic variation certainly has implications for human nature. If there are many ways for a mind to vary genetically, the mind must have many genetically influenced parts and attributes that make the variation possible. Also, any modern conception of human nature that is rooted in biology (as opposed to traditional conceptions of human nature that are rooted in philosophy, religion, or common sense) must predict that the faculties making up human nature show quantitative variation, even if their fundamental design (how they work) is universal. Natural selection depends on genetic variation, and though it reduces that variation as it shapes organisms over the generations, it never uses it up completely.[51]

Whatever their exact interpretation turns out to be, the findings of behavioral genetics are highly damaging to the Blank Slate and its companion doctrines. The slate cannot be blank if different genes can make it more or less smart, articulate, adventurous, shy, happy, conscientious, neurotic, open, introverted, giggly, spatially challenged, or likely to dip buttered toast in coffee. For genes to affect the mind in all these ways, the mind must have many parts and features for the genes to affect. Similarly, if the mutation or deletion of a gene can target a cognitive ability as specific as spatial construction or a personality trait as specific as sensation-seeking, that trait may be a distinct component of a complex psyche.

Moreover, many of the traits affected by genes are far from noble. Psychologists have discovered that our personalities differ in five major ways: we are to varying degrees introverted or extroverted, neurotic or stable, incurious or open to experience, agreeable or antagonistic, and conscientious or undirected. Most of the 18,000 adjectives for personality traits in an unabridged dictionary can be tied to one of these five dimensions, including such sins and flaws as being *aimless, careless, conforming, impatient, narrow, rude, self-pitying, selfish, suspicious, uncooperative,* and *undependable.* All five of the major personality dimensions are heritable, with perhaps 40 to 50 percent of the variation in a typical population tied to differences in their genes. The unfortunate wretch who is introverted, neurotic, narrow, selfish, and undependable is probably that way in part because of his genes, and so, most likely, are the rest of us who have tendencies in any of those directions as compared with our fellows.

It's not just unpleasant temperaments that are partly heritable, but actual behavior with real consequences. Study after study has shown that a willingness to commit antisocial acts, including lying, stealing, starting fights, and destroying property, is partly heritable (though like all heritable traits it is exercised more in some environments than in others).[52] People who commit truly heinous acts, such as bilking elderly people out of their life savings, raping a succession of women, or shooting convenience store clerks lying on the floor during a robbery, are often diagnosed with "psychopathy" or "antisocial personality disorder."[53] Most psychopaths showed signs of malice from the time they were children. They bullied smaller children, tortured animals, lied habitually, and were incapable of empathy or remorse, often despite normal family backgrounds and the best efforts of their distraught parents. Most experts on psychopathy believe that it comes from a genetic predisposition, though in some cases it may come from early brain damage.[54] In either case genetics and neuroscience are showing that a heart of darkness cannot always be blamed on parents or society.

And the genes, even if they by no means seal our fate, don't sit easily with the intuition that we are ghosts in machines either. Imagine that you are agonizing over a choice—which career to pursue, whether to get married, how to vote, what to wear that day. You have finally staggered to a decision when the phone rings. It is the identical twin you never knew you had. During the joyous conversation it comes out that she has just chosen a similar career, has decided to get married at around the same time, plans to cast her vote for the same presidential candidate, and is wearing a shirt of the same color—just as the behavioral geneticists who tracked you down would have bet. How much discretion did the "you" making the choices actually have if the outcome could have been predicted in advance, at least probabilistically, based on events that took place in your mother's Fallopian tubes decades ago?

THE FOURTH BRIDGE from biology to culture is evolutionary psychology, the study of the phylogenetic history and adaptive functions of the mind.[55] It holds out the hope of understanding the *design* or *purpose* of the mind—not in some mystical or teleological sense, but in the sense of the simulacrum of engineering that pervades the natural world. We see these signs of engineering everywhere: in eyes that seem designed to form images, in hearts that seem designed to pump blood, in wings that seem designed to lift birds in flight.

Darwin showed, of course, that the illusion of design in the natural world can be explained by natural selection. Certainly an eye is too well engineered to have arisen by chance. No wart or tumor or product of a big mutation could be lucky enough to have a lens, an iris, a retina, tear ducts, and so on, all perfectly arranged to form an image. Nor is the eye a masterpiece of engineering literally fashioned by a cosmic designer who created humans in his own image. The human eye is uncannily similar to the eyes of other organisms and has quirky vestiges of extinct ancestors, such as a retina that appears to have been installed backwards.[56] Today's organs are replicas of organs in our ancestors whose design worked better than the alternatives, thereby enabling them to *become* our ancestors.[57] Natural selection is the only physical process we know of that can simulate engineering, because it is the only process in which how well something works can play a causal role in how it came to be.

Evolution is central to the understanding of life, including human life. Like all living things, we are outcomes of natural selection; we got here because we inherited traits that allowed our ancestors to survive, find mates, and reproduce. This momentous fact explains our deepest strivings: why having a thankless child is sharper than a serpent's tooth, why it is a truth universally acknowledged that a single man in possession of a good fortune must be in want of a wife, why we do not go gentle into that good night but rage, rage against the dying of the light.

Evolution is central to understanding ourselves because signs of design in human beings do not stop at the heart or the eye. For all its exquisite engineering, an eye is useless without a brain. Its output is not the meaningless patterns of a screen saver, but raw material for circuitry that computes a representation of the external world. That representation feeds other circuits that make sense of the world by imputing causes to events and placing them in categories that allow useful predictions. And that sense-making, in turn, works in the service of motives such as hunger, fear, love, curiosity, and the pursuit of status and esteem. As I mentioned, abilities that seem effortless to us—categorizing events, deducing cause and effect, and pursuing conflicting goals—are major challenges in designing an intelligent system, ones that robot designers strive, still unsuccessfully, to duplicate.

So signs of engineering in the human mind go all the way up, and that is why psychology has always been evolutionary. Cognitive and emotional faculties have always been recognized as nonrandom, complex, and useful, and that means they must be products either of divine design or of natural selection. But until recently evolution was seldom explicitly invoked within psychology, because with many topics, folk intuitions about what is adaptive are good enough to make headway. You don't need an evolutionary biologist to tell you that depth perception keeps an animal from falling off cliffs and bumping into trees, that thirst keeps it from drying out, or that it's better to remember what works and what doesn't than to be an amnesiac.

But with other aspects of our mental life, particularly in the social realm, the function of a faculty is not so easy to guess. Natural selection favors organisms that are good at reproducing in some environment. When the environment consists of rocks, grass, and snakes, it's fairly obvious which strategies work and which ones don't. But when the relevant environment consists of other members of the species evolving their own strategies, it is not so obvious. In the game of evolution, is it better to be monogamous or polygamous? Gentle or aggressive? Cooperative or selfish? Indulgent with children or stern with them? Optimistic, pragmatic, or pessimistic?

For questions like these, hunches are unhelpful, and that is why evolutionary biology has increasingly been brought into psychology. Evolutionary biologists tell us that it is a mistake to think of anything conducive to people's well-being—group cohesion, the avoidance of violence, monogamous pair bonding, aesthetic pleasure, self-esteem—as an "adaptation." What is "adaptive" in everyday life is not necessarily an "adaptation" in the technical sense of being a trait that was favored by natural selection in a species' evolutionary history. Natural selection is the morally indifferent process in which the most effective replicators outreproduce the alternatives and come to prevail in a population. The selected genes will therefore be the "selfish" ones, in Richard Dawkins's metaphor—more accurately, the megalomaniacal ones, those that make the most copies of themselves.[58] An adaptation is anything brought about by the genes that helps them fulfill this metaphorical obsession, whether or not it also fulfills human aspirations. And this is a strikingly different conception from our everyday intuitions about what our faculties were designed for.

The megalomania of the genes does not mean that benevolence and co-operation cannot evolve, any more than the law of gravity proves that flight cannot evolve. It means only that benevolence, like flight, is a special state of affairs in need of an explanation, not something that just happens. It can evolve only in particular circumstances and has to be supported by a suite of cognitive and emotional faculties. Thus benevolence (and other social motives) must be dragged into the spotlight rather than treated as part of the furniture. In the sociobiological revolution of the 1970s, evolutionary biologists replaced the fuzzy feeling that organisms evolve to serve the greater good with deductions of what kinds of motives are likely to evolve when organisms interact with offspring, mates, siblings, friends, strangers, and adversaries.

When the predictions were combined with some basic facts about the hunter-gatherer lifestyle in which humans evolved, parts of the psyche that were previously inscrutable turned out to have a rationale as legible as those for depth perception and the regulation of thirst. An eye for beauty, for example, locks onto faces that show signs of health and fertility—just as one would predict if it had evolved to help the beholder find the fittest mate.[59] The emotions of sympathy, gratitude, guilt, and anger allow people to benefit from co-operation without being exploited by liars and cheats.[60] A reputation for toughness and a thirst for revenge were the best defense against aggression in a world in which one could not call 911 to summon the police.[61] Children acquire spoken language instinctively but written language only by the sweat of their brow, because spoken language has been a feature of human life for tens or hundreds of millennia whereas written language is a recent and slow-spreading invention.[62]

None of this means that people literally strive to replicate their genes. If that's how the mind worked, men would line up outside sperm banks and women would pay to have their eggs harvested and given away to infertile couples. It means only that inherited systems for learning, thinking, and feeling have a design that would have led, on average, to enhanced survival and reproduction in the environment in which our ancestors evolved. People enjoy eating, and in a world without junk food, that led them to nourish themselves, even if the nutritional content of the food never entered their minds. People love sex and love children, and in a world without contraception, that was enough for the genes to take care of themselves.

The difference between the mechanisms that impel organisms to behave in real time and the mechanisms that shaped the design of the organism over evolutionary time is important enough to merit some jargon. A *proximate* cause of behavior is the mechanism that pushes behavior buttons in real time, such as the hunger and lust that impel people to eat and have sex. An *ultimate* cause is the adaptive rationale that led the proximate cause to evolve, such as the need for nutrition and reproduction that gave us the drives of hunger and lust. The distinction between proximate and ultimate causation is indispensable in understanding ourselves because it determines the answer to every question of the form "Why did that person act as he did?" To take a simple example, ultimately people crave sex in order to reproduce (because the ultimate cause of sex is reproduction), but proximately they may do everything they can not to reproduce (because the proximate cause of sex is pleasure).

The difference between proximate and ultimate goals is another kind of proof that we are not blank slates. Whenever people strive for obvious rewards like health and happiness, which make sense both proximately and ultimately, one could plausibly suppose that the mind is equipped only with a desire to be happy and healthy and a cause-and-effect calculus that helps them get what they want. But people often have desires that subvert their proximate well-being, desires that they cannot articulate and that they (and their society) may try unsuccessfully to extirpate. They may covet their neighbor's spouse, eat themselves into an early grave, explode over minor slights, fail to love their stepchildren, rev up their bodies in response to a stressor that they cannot fight or flee, exhaust themselves keeping up with the Joneses or climbing the corporate ladder, and prefer a sexy and dangerous partner to a plain but dependable one. These personally puzzling drives have a transparent evolutionary rationale, and they suggest that the mind is packed with cravings shaped by natural selection, not with a generic desire for personal well-being.

Evolutionary psychology also explains *why* the slate is not blank. The mind was forged in Darwinian competition, and an inert medium would have been outperformed by rivals outfitted with high technology—with acute perceptual systems, savvy problem-solvers, cunning strategists, and sensitive feedback circuits. Worse still, if our minds were truly malleable they would be easily manipulated by our rivals, who could mold or condition us into serving their needs rather than our own. A malleable mind would quickly be selected out.

Researchers in the human sciences have begun to flesh out the hypothesis that the mind evolved with a universal complex design. Some anthropologists have returned to an ethnographic record that used to trumpet differences among cultures and have found an astonishingly detailed set of aptitudes and tastes that all cultures have in common. This shared way of thinking, feeling, and living makes us look like a single tribe, which the anthropologist Donald Brown has called the Universal People, after Chomsky's Universal Grammar.[63] Hundreds of traits, from fear of snakes to logical operators, from romantic love to humorous insults, from poetry to food taboos, from exchange of goods to mourning the dead, can be found in every society ever documented. It's not that every universal behavior directly reflects a universal component of human nature—many arise from an interplay between universal properties of the mind, universal properties of the body, and universal properties of the world. Nonetheless, the sheer richness and detail in the rendering of the Universal People comes as a shock to any intuition that the mind is a blank slate or that cultures can vary without limit, and there is something on the list to refute almost any theory growing out of those intuitions. Nothing can substitute for seeing Brown's list in full; it is reproduced, with his permission, as an appendix (see p. 435).

The idea that natural selection has endowed humans with a universal complex mind has received support from other quarters. Child psychologists no longer believe that the world of an infant is a blooming, buzzing confusion, because they have found signs of the basic categories of mind (such as those for objects, people, and tools) in young babies.[64] Archaeologists and paleontologists have found that prehistoric humans were not brutish troglodytes but exercised their minds with art, ritual, trade, violence, cooperation, technology, and symbols.[65] And primatologists have shown that our hairy relatives are not like lab rats waiting to be conditioned but are outfitted with many complex faculties that used to be considered uniquely human, including concepts, a spatial sense, tool use, jealousy, parental love, reciprocity, peacemaking, and differences between the sexes.[66] With so many mental abilities appearing in all human cultures, in children before they have acquired culture, and in creatures that have little or no culture, the mind no longer looks like a formless lump pounded into shape by culture.

But it is the doctrine of the Noble Savage that has been most mercilessly debunked by the new evolutionary thinking. A thoroughly noble *anything* is an unlikely product of natural selection, because in the competition among genes for representation in the next generation, noble guys tend to finish last. Conflicts of interest are ubiquitous among living things, since two animals cannot both eat the same fish or monopolize the same mate. To the extent that social motives are adaptations that maximize copies of the genes that produced them, they should be designed to prevail in such conflicts, and one way to prevail is to neutralize the competition. As William James put it, just a bit too flamboyantly, "We, the lineal representatives of the successful enactors of one scene of slaughter after another, must, whatever more pacific virtues we may also possess, still carry about with us, ready at any moment to burst into flame, the smoldering and sinister traits of character by means of which they lived through so many massacres, harming others, but themselves unharmed."[67]

From Rousseau to the Thanksgiving editorialist of Chapter 1, many intellectuals have embraced the image of peaceable, egalitarian, and ecology-loving natives. But in the past two decades anthropologists have gathered data on life and death in pre-state societies rather than accepting the warm and fuzzy stereotypes. What did they find? In a nutshell: Hobbes was right, Rousseau was wrong.

To begin with, the stories of tribes out there somewhere who have never heard of violence turn out to be urban legends. Margaret Mead's descriptions of peace-loving New Guineans and sexually nonchalant Samoans were based on perfunctory research and turned out to be almost perversely wrong. As the anthropologist Derek Freeman later documented, Samoans may beat or kill their daughters if they are not virgins on their wedding night, a young man who cannot woo a virgin may rape one to extort her into eloping, and the family of a cuckolded husband may attack and kill the adulterer.[68] The !Kung San of the Kalahari Desert had been described by Elizabeth Marshall Thomas as "the harmless people" in a book with that title. But as soon as anthropologists camped out long enough to accumulate data, they discovered that the !Kung San have a murder rate higher than that of American inner cities. They learned as well that a group of the San had recently avenged a murder by sneaking into the killer's group and executing every man, woman, and child as they slept.[69] But at least the !Kung San exist. In the early 1970s the *New York Times Magazine* reported the discovery of the "gentle Tasaday" of the Philippine rainforest, a people with no words for conflict, violence, or weapons. The Tasaday turned out to be local farmers dressed in leaves for a photo opportunity so that cronies of Ferdinand Marcos could set aside their "homeland" as a preserve and enjoy exclusive mineral and logging rights.[70]

Anthropologists and historians have also been counting bodies. Many intellectuals tout the small numbers of battlefield casualties in pre-state societies as evidence that primitive warfare is largely ritualistic. They do not notice that two deaths in a band of fifty people is the equivalent of ten million deaths in a country the size of the United States. The archaeologist Lawrence Keeley has summarized the proportion of male deaths caused by war in a number of societies for which data are available:[71]

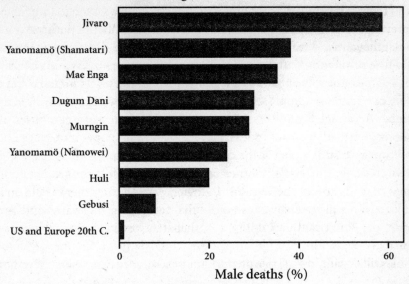

Percentage of male deaths caused by warfare

Jivaro	
Yanomamö (Shamatari)	
Mae Enga	
Dugum Dani	
Murngin	
Yanomamö (Namowei)	
Huli	
Gebusi	
US and Europe 20th C.	

Male deaths (%)

The first eight bars, which range from almost 10 percent to almost 60 percent, come from indigenous peoples in South America and New Guinea. The nearly invisible bar at the bottom represents the United States and Europe in the twentieth century and includes the statistics from two world wars. Moreover, Keeley and others have noted that native peoples are dead serious when they carry out warfare. Many of them make weapons as damaging as their technology permits, exterminate their enemies when they can get away with it, and enhance the experience by torturing captives, cutting off trophies, and feasting on enemy flesh.[72]

Counting societies instead of bodies leads to equally grim figures. In 1978 the anthropologist Carol Ember calculated that 90 percent of hunter-gatherer societies are known to engage in warfare, and 64 percent wage war at least once every two years.[73] Even the 90 percent figure may be an underestimate, because anthropologists often cannot study a tribe long enough to measure outbreaks that occur every decade or so (imagine an anthropologist studying the peaceful Europeans between 1918 and 1938). In 1972 another anthropologist, W. T. Divale, investigated 99 groups of hunter-gatherers from 37 cultures, and found that 68 were at war at the time, 20 had been at war five to twenty-five years before, and all the others reported warfare in the more distant past.[74] Based on these and other ethnographic surveys, Donald Brown includes conflict, rape, revenge, jealousy, dominance, and male coalitional violence as human universals.[75]

It is, of course, understandable that people are squeamish about acknowledging the violence of pre-state societies. For centuries the stereotype of the savage savage was used as a pretext to wipe out indigenous peoples and steal their lands. But surely it is unnecessary to paint a false picture of a people as peaceable and ecologically conscientious in order to condemn the great crimes against them, as if genocide were wrong only when the victims are nice guys.

The prevalence of violence in the kinds of environments in which we evolved does not mean that our species has a death wish, an innate thirst for blood, or a territorial imperative. There are good evolutionary reasons for the members of an intelligent species to try to live in peace. Many computer simulations and mathematical models have shown that cooperation pays off in evolutionary terms as long as the cooperators have brains with the right combination of cognitive and emotional faculties.[76] Thus while conflict is a human universal, so is conflict resolution. Together with all their nasty and brutish motives, all peoples display a host of kinder, gentler ones: a sense of morality, justice, and community, an ability to anticipate consequences when choosing how to act, and a love of children, spouses, and friends.[77] Whether a group of people will engage in violence or work for peace depends on which set of motives is engaged, a topic I will pursue at length in later chapters.

Not everyone will be comforted by such reassurances, though, because they eat away at the third cherished assumption of modern intellectual life. Love, will, and conscience are in the traditional job description for the soul and have always been placed in opposition to mere "biological" functions. If those faculties are "biological" too—that is, evolutionary adaptations implemented in the circuitry of the brain—then the ghost is left with even less to do and might as well be pensioned off for good.

Chapter 4

Culture Vultures

Like all men of Babylon, I have been proconsul; like all, I have been a slave. Look here—my right hand has no index finger. Look here—through this gash in my cape you can see on my stomach a crimson tattoo—it is the second letter, *Beth*. On nights when the moon is full, this symbol gives me power over men with the mark of Gimel, but it subjects me to those with the Aleph, who on nights when there is no moon owe obedience to those marked with the Gimel. In the half-light of dawn, in a cellar, standing before a black altar, I have slit the throats of sacred bulls. Once, for an entire lunar year, I was declared invisible—I would cry out and no one would heed my call, I would steal bread and not be beheaded. . . .

I owe that almost monstrous variety to an institution—the Lottery—which is unknown in other nations, or at work in them imperfectly or secretly.[1]

JORGE LUIS BORGES'S story "The Lottery in Babylon" is perhaps the best depiction of the idea that culture is a set of roles and symbols that mysteriously descend on passive individuals. His lottery began as the familiar game in which a winning ticket was rewarded by a jackpot. But to enhance the suspense the operators added a few numbers that presented the ticket holder with a fine rather than a reward. They then imposed prison sentences on those who did not pay the fines, and the system expanded into a variety of nonmonetary punishments and rewards. The lottery became free, compulsory, omnipotent, and increasingly mysterious. People began to speculate on how it worked and whether it even continued to exist.

At first glance human cultures do appear to have the monstrous variety of a Borgesian lottery. Members of *Homo sapiens* ingest everything from maggots and worms to cow urine and human flesh. They bind, cut, scar, and stretch body parts in ways that would make the most perforated Western teenager wince. They sanction kinky sexual practices like teenagers receiving daily fellatio from younger boys and parents arranging marriages between their five-year-olds. The apparent caprice of cultural variation leads naturally to the doctrine that culture lives in a separate universe from brains, genes, and evolution. And this separation depends in turn on the concept of a slate that is left blank by biology and written upon by culture. Now that I have tried to convince you that the slate is not blank, it is time to put culture back into the picture. That will complete the consilience that runs from the life sciences through the sciences of human nature to the social sciences, humanities, and arts.

In this chapter I will lay out an alternative to the belief that culture is like a lottery. Culture can be seen instead as a part of the human phenotype: the distinctive design that allows us to survive, prosper, and perpetuate our lineages. Humans are a knowledge-using, cooperative species, and culture emerges naturally from that lifestyle. To preview: The phenomena we call "culture" arise as people pool and accumulate their discoveries, and as they institute conventions to coordinate their labors and adjudicate their conflicts. When groups of people separated by time and geography accumulate different discoveries and conventions, we use the plural and call them cultures. Different cultures, then, don't come from different kinds of genes—Boas and his heirs were right about that—but they don't live in a separate world or stamp a shape onto formless minds either.

~

THE FIRST STEP in connecting culture to the sciences of human nature is to recognize that culture, for all its importance, is not some miasma that seeps into people through their skin. Culture relies on neural circuitry that accomplishes the feat we call learning. Those circuits do not make us indiscriminate mimics but have to work in surprisingly subtle ways to make the transmission of culture possible. That is why a focus on innate faculties of mind is not an alternative to a focus on learning, culture, and socialization, but rather an attempt to explain how they work.

Take the case of a person's mother tongue, which is a learned cultural skill par excellence. A parrot and a child both learn *something* when exposed to speech, but only the child has a mental algorithm that extracts words and rules from the sound wave and uses them to utter and understand an unlimited number of new sentences. The innate endowment for language is in fact an innate mechanism for *learning* language.[2] In the same way, for children to learn about culture they cannot be mere video cameras that passively record sights and sounds. They must be equipped with mental machinery that can extract the beliefs and values underlying other people's behavior so that the children themselves can become competent members of the culture.[3]

Even the humblest act of cultural learning—imitating the behavior of a parent or a peer—is more complicated than it looks. To appreciate what goes on in our minds when we effortlessly learn from other people, we have to imagine what it would be like to have some *other* kind of mind. Fortunately, cognitive scientists have imagined it for us by plumbing the minds of robots, animals, and people whose minds are impaired.

The artificial intelligence researcher Rodney Brooks, who wants to build a robot capable of learning by imitation, immediately faced this problem when he considered using techniques for learning that are common in computer science:

> The robot is observing a person opening a glass jar. The person approaches the robot and places the jar on a table near the robot. The person rubs his hands together and then sets himself to removing the lid from the jar. He grasps the glass jar in one hand and the lid in the other and begins to unscrew the lid by turning it counter-clockwise. While he is opening the jar, he pauses to wipe his brow, and glances at the robot to see what it is doing. He then resumes opening the jar. The robot then attempts to imitate the action. [But] which parts of the action to be imitated are important (such as turning the lid counter-clockwise), and which aren't (such as wiping your brow)? . . . How can the robot abstract the knowledge gained from this experience and apply it to a similar situation?[4]

The answer is that the robot has to be equipped with an ability to see into the mind of the person being imitated, so that it can infer the person's goals and pick out the aspects of behavior that the person intended to achieve the goal. Cognitive scientists call this ability intuitive psychology, folk psychology, or a theory of mind. (The "theory" here refers to the tacit beliefs held by a person, animal, or robot, not to the explicit beliefs of scientists.) No existing robot comes close to having this ability.

Another mind that finds it difficult to infer others' goals is the chimpanzee's. The psychologist Laura Petitto was the principal sign language trainer for the animal known as Nim Chimpsky and lived with him for a year in a university mansion. At first glance Nim seemed to "imitate" her washing the dishes, but with an important difference. A dish was not necessarily any cleaner after Nim rubbed it with a sponge than before, and if he was given a spotless dish, Nim would "wash" it just as if it were dirty. Nim didn't get the concept of "washing," namely using liquid to make something clean. He just mimicked her rubbing motion while enjoying the sensation of warm water over his fingers. Many laboratory experiments have shown something similar. Though chimpanzees and other primates have a reputation as imitators ("Monkey see, monkey do"), their ability to imitate in the way people do—

replicating another person's *intent* rather than going through the motions—is rudimentary, because their intuitive psychology is rudimentary.[5]

A mind unequipped to discern other people's beliefs and intentions, even if it can learn in other ways, is incapable of the kind of learning that perpetuates culture. People with autism suffer from an impairment of this kind. They can grasp physical representations like maps and diagrams but cannot grasp *mental* representations—that is, they cannot read other people's minds.[6] Though they certainly imitate, they do it in bizarre ways. Some are prone to echolalia, repeating other people's utterances verbatim rather than extracting the grammatical patterns that would allow them to compose their own sentences. Autistics who do learn to speak on their own often use the word *you* as if it were their own name, because other people refer to them as *you* and it never occurs to them that the word is defined relative to who is addressing it to whom. If a parent knocks over a glass and says, "Oh, damn!" an autistic child might use *oh damn* as the word for a glass—disproving the empiricist theory that normal children can learn words merely by associating sounds and events that overlap in time. None of this is a consequence of low intelligence. Autistic children can be competent (or even savants) when solving other problems, and retarded children without autism don't show the same foibles with language and imitation. Autism is an innate neurological condition with strong genetic roots.[7] Together with robots and chimpanzees, people with autism remind us that cultural learning is possible only because neurologically normal people have innate equipment to accomplish it.

Scientists often interpret the long childhood of members of *Homo sapiens* as an adaptation that allows children to acquire the vast store of information from their culture before striking out on their own as adults. If cultural learning depends on special psychological equipment, we should see the equipment up and running early in childhood. And indeed we do.

Experiments show that one-and-a-half-year-old babies are not associationists who connect overlapping events indiscriminately. They are intuitive psychologists who psych out other people's intentions before copying what they do. When an adult first exposes a baby to a word, as in "That's a *toma*," the baby will remember it as the name of the toy the *adult* was looking at at the time, not as the name of the toy the baby herself was looking at.[8] If an adult fiddles with a gadget but indicates that the action was an accident (by saying "Whoops!"), a baby will not even bother trying to imitate him. But if the adult does the same thing but indicates that he intended the action, the baby *will* imitate him.[9] And when an adult tries and fails to accomplish something (like trying to press the button on a buzzer, or trying to string a loop around a peg), the baby will imitate what the adult tried to do, not what he did do.[10] As someone who studies language acquisition in children, I have continually been amazed at how early they "get" the logic of language, availing themselves of most of the spoken vernacular by the age of three.[11] That, too, may be an attempt by the genome to get our culture-acquiring apparatus online as early in life as the growing brain can handle it.

~

OUR MINDS, THEN, are fitted with mechanisms designed to read the goals of other people so we can copy their intended acts. But why would we want to? Though we take it for granted that acquiring culture is a good thing, the act of acquiring it is often spoken of with scorn. The longshoreman and philosopher Eric Hoffer wrote, "When people are free to do as they please, they usually imitate each other." And we have a menagerie of metaphors that equate this quintessentially human ability with the behavior of animals: along with *monkey see, monkey do,* we have *aping, parroting, sheep, lemmings, copycats,* and *a herd mentality.*

Social psychologists have amply documented that people have a powerful urge to do as their neighbors do. When unwitting subjects are surrounded by confederates of the experimenter who have been paid to do something odd, many or most will go along. They will defy their own eyes and call a long line "short" or vice versa, nonchalantly fill out a questionnaire as smoke pours out of a heating vent, or (in a *Candid Camera* sketch) suddenly strip down to their underwear for no apparent reason.[12] But the social psychologists point out that human conformity, no matter how hilarious it looks in contrived experiments, has a genuine rationale in social life—indeed, two rationales.[13]

The first is *informational,* the desire to benefit from other people's knowledge and judgment. Weary veterans of committees say that the IQ of a group is the lowest IQ of any member of the group divided by the number of people in the group, but that is too pessimistic. In a species equipped with language, an intuitive psychology, and a willingness to cooperate, a group can pool the hard-won discoveries of members present and past and end up far smarter than a race of hermits. Hunter-gatherers accumulate the know-how to make tools, control fire, outsmart prey, and detoxify plants, and can live by this collective ingenuity even if no member could re-create it all from scratch. Also, by coordinating their behavior (say, in driving game or taking turns watching children while others forage), they can act like a big multi-headed, multi-limbed beast and accomplish feats that a die-hard individualist could not. And an array of interconnected eyes, ears, and heads is more robust than a single set with all its shortcomings and idiosyncrasies. There is a Yiddish expression offered as a reality check to malcontents and conspiracy theorists: The whole world isn't crazy.

Much of what we call culture is simply accumulated local wisdom: ways of fashioning artifacts, selecting food, dividing up windfalls, and so on. Some anthropologists, like Marvin Harris, argue that even practices that seem as arbitrary as a lottery may in fact be solutions to ecological problems.[14] Cows really should be sacred in India, he points out; they supply food (milk and butter), fuel (dung), and power (by pulling plows), so the customs protecting them thwart the temptation to kill the goose that laid the golden egg. Other cultural differences may have a rationale in reproduction.[15] In some societies, men live with their paternal families and support their wives and children; in others, they live with their maternal families and support their sisters and nieces and nephews. The second arrangement tends to be found in societies where men have to spend long periods of time away from home and adultery is relatively common, so they cannot be sure that their wives' children are theirs. Since the children of a man's mother's daughter have to be his biological kin regardless of who has been sleeping with whom, a matrilocal family allows men to invest in children who are guaranteed to carry some of their genes.

Of course, only Procrustes could argue that all cultural practices have a direct economic or genetic payoff. The second motive for conformity is *normative*, the desire to follow the norms of a community, whatever they are. But this, too, is not as stupidly lemminglike as it first appears. Many cultural practices are arbitrary in their specific form but not in their reason for being. There is no good reason for people to drive on the right side of the road as opposed to the left side, or vice versa, but there is every reason for people to drive on the *same* side. So an arbitrary choice of which side to drive on, and a widespread conformity with that choice, make a great deal of sense. Other examples of arbitrary but coordinated choices, which economists called "cooperative equilibria," include money, designated days of rest, and the pairings of sound and meaning that make up the words in a language.

Shared arbitrary practices also help people cope with the fact that while many things in life are arranged along a continuum, decisions must often be binary.[16] Children do not become adults instantaneously, nor do dating couples become monogamous partners. Rites of passage and their modern equivalent, pieces of paper like ID cards and marriage licenses, allow third parties to decide how to treat ambiguous cases—as a child or as an adult, as committed or as available—without endless haggling over differences of opinion.

And the fuzziest categories of all are other people's intentions. Is he a loyal member of the coalition (one that I would want to have in my foxhole) or a quisling who will bail out when times get tough? Does his heart lie with his father's clan or with his father-in-law's? Is she a suspiciously merry widow or just getting on with her life? Is he dissing me or just in a hurry? Initiation rites, tribal badges, prescribed periods of mourning, and ritualized forms of address may not answer these questions definitively, but they can remove clouds of suspicion that would otherwise hang over people's heads.

When conventions are widely enough entrenched, they can become a kind of reality even though they exist only in people's minds. In his book *The Construction of Social Reality* (not to be confused with the social construction of reality), the philosopher John Searle points out that certain facts are objectively true just because people act as if they are true.[17] For example, it is a matter of fact, not opinion, that George W. Bush is the forty-third president of the United States, that O. J. Simpson was found not guilty of murder, that the Boston Celtics won the NBA World Championship in 1986, and that a Big Mac (at the time of this writing) costs $2.62. But though these are objective facts, they are not facts about the physical world, like the atomic number of cadmium or the classification of a whale as a mammal. They consist in a shared understanding in the minds of most members in a community, usually agreements to grant (or deny) power or status to certain other people.

Life in complex societies is built on social realities, the most obvious examples being money and the rule of law. But a social fact depends entirely on the willingness of people to treat it as a fact. It is specific to a community, as we see when people refuse to honor a foreign currency or fail to recognize the sovereignty of a self-proclaimed leader. And it can dissolve with changes in the collective psychology, as when a currency becomes worthless through hyperinflation or a regime collapses because people defy the police and army en masse. (Searle points out that Mao was only half right when he said that "political power grows out of the barrel of a gun." Since no regime can keep a gun trained on every last citizen, political power grows out of a regime's ability to command the fear of enough people at the same time.) Social reality exists only within a group of people, but it depends on a cognitive ability present in each individual: the ability to understand a public agreement to confer power or status, and to honor it as long as other people do.

How does a psychological event—an invention, an affectation, a decision to treat a certain kind of person in a certain way—turn into a sociocultural fact—a tradition, a custom, an ethos, a way of life? We should understand culture, according to the cognitive anthropologist Dan Sperber, as the *epidemiology* of mental representations: the spread of ideas and practices from person to person.[18] Many scientists now use the mathematical tools of epidemiology (how diseases spread) or of population biology (how genes and organisms spread) to model the evolution of culture.[19] They have shown how a tendency of people to adopt the innovations of other people can lead to effects that we understand using metaphors like epidemics, wildfire, snowballs, and tipping points. Individual psychology turns into collective culture.

⁓

CULTURE, THEN, IS a pool of technological and social innovations that people accumulate to help them live their lives, not a collection of arbitrary roles and symbols that happen to befall them. This idea helps explain what makes cultures different and similar. When a splinter group leaves the tribe and is cut off by an ocean, a mountain range, or a demilitarized zone, an innovation on one side of the barrier has no way of diffusing to the other side. As each group modifies its own collection of discoveries and conventions, the collections will diverge and the groups will have different cultures. Even when two groups stay within shouting distance, if their relationship has an edge of hostility they may adopt behavioral identity badges that advertise which side someone is on, further exaggerating any differences. This branching and differentiation is easily visible in the evolution of languages, perhaps the clearest example of cultural evolution. And as Darwin pointed out, it has a close parallel in the origin of species, which often arise when a population splits in two and the groups of descendants evolve in different directions.[20] As with languages and species, cultures that split apart more recently tend to be more similar. The traditional cultures of Italy and France, for example, are more similar to each other than either is to the cultures of the Maoris and Hawaiians.

The psychological roots of culture also help explain why some bits of culture change and others stay put. Some collective practices have enormous inertia because they impose a high cost on the first individual who would try to change them. A switch from driving on the left to driving on the right could not begin with a daring nonconformist or a grass-roots movement but would have to be imposed from the top down (which is what happened in Sweden at 5 A.M., Sunday, September 3, 1967). Other examples are laying down your weapons when hostile neighbors are armed to the teeth, abandoning the QWERTY keyboard layout, and pointing out that the emperor is not wearing any clothes.

But traditional cultures can change, too, and more dramatically than most people realize. Preserving cultural diversity is considered a supreme virtue today, but the members of the diverse cultures don't always see it that way. People have wants and needs, and when cultures rub shoulders, people in one culture are bound to notice when their neighbors are satisfying those desires better than they are. When they do notice, history tells us, they shamelessly borrow whatever works best. Far from being self-preserving monoliths, cultures are porous and constantly in flux. Language, once again, is a clear example. Notwithstanding the perennial lamentations of purists and the sanctions of language academies, no language is ever spoken the way it was centuries before. Just compare contemporary English with the language of Shakespeare, or the language of Shakespeare with the language of Chaucer. Many other "traditional" practices are surprisingly recent. The ancestors of the Hasidic Jews did not wear black coats and fur-lined hats in Levantine deserts, nor did the Plains Indians ride horses before the arrival of the Europeans. National cuisines, too, have shallow roots. Potatoes in Ireland, paprika in Hungary, tomatoes in Italy, hot chile peppers in India and China, and cassava in Africa come from New World plants, and were brought to their "traditional" homes in the centuries after the arrival of Columbus in the Americas.[21]

The idea that a culture is a tool for living can even explain the fact that first led Boas to argue the opposite, that a culture is an autonomous system of ideas. The most obvious cultural difference on the planet is that some cultures are materially more successful than others. In past centuries, cultures from Europe and Asia decimated the cultures of Africa, the Americas, Australia, and the Pacific. Even within Europe and Asia the fortunes of cultures have varied widely, some developing expansive civilizations rich in art, science, and technology, others stuck in poverty and helpless to resist conquest. What allowed small groups of Spaniards to cross the Atlantic and defeat the great empires of the Incas and Aztecs, rather than the other way around? Why didn't African tribes colonize Europe instead of vice versa? The immediate answer is that the wealthy conquerors had better technology and a more complex political and economic organization. But that simply pushes back the question of why some cultures develop more complex ways of life than others.

Boas helped overthrow the bad racial science of the nineteenth century that attributed these disparities to differences in how far each race had biologically evolved. In its place his successors stipulated that behavior is determined by culture and that culture is autonomous from biology.[22] Unfortunately, that left the dramatic differences among cultures unexplained, as if they were random outcomes of the lottery in Babylon. Indeed, the differences were not just unexplained but unmentionable, out of a fear that people would misinterpret the observation that some cultures were more technologically sophisticated than others as some kind of moral judgment that advanced societies were better than primitive ones. But no one can fail to notice that some cultures can accomplish things that all people want (like health and comfort) better than others. The dogma that cultures vary capriciously is a feeble refutation of any private opinion that some races have what it takes to develop science, technology, and government and others don't.

But recently two scholars, working independently, have decisively shown that there is no need to invoke race to explain differences among cultures. Both arrived at that conclusion by eschewing the Standard Social Science Model, in which cultures are arbitrary symbol systems that exist apart from the minds of individual people. In his trilogy *Race and Culture, Migrations and Cultures,* and *Conquests and Cultures,* the economist Thomas Sowell explained his starting point for an analysis of cultural differences:

> A culture is not a symbolic pattern, preserved like a butterfly in amber. Its place is not in a museum but in the practical activities of daily life, where it evolves under the stress of competing goals and other competing cultures. Cultures do not exist as simply static "differences" to be celebrated but compete with one another as better and worse ways of getting things done—better and worse, not from the standpoint of some observer, but from the standpoint of the peoples themselves, as they cope and aspire amid the gritty realities of life.[23]

The physiologist Jared Diamond is a proponent of ideas in evolutionary psychology and of consilience between the sciences and the humanities, particularly history.[24] In *Guns, Germs, and Steel* he rejected the standard assumption that history is just one damn thing after another and tried to explain the sweep of human history over tens of thousands of years in the context of human evolution and ecology.[25] Sowell and Diamond have made an authoritative case that the fates of human societies come neither from chance nor from race but from the human drive to adopt the innovations of others, combined with the vicissitudes of geography and ecology.

Diamond begins at the beginning. For most of human evolutionary history we lived as hunter-gatherers. The trappings of civilization—sedentary living, cities, a division of labor, government, professional armies, writing, metallurgy—sprang from a recent development, farming, about ten thousand years ago. Farming depends on plants and animals that can be tamed and exploited, and only a few species are suited to it. They happened to be concentrated in a few parts of the world, including the Fertile Crescent, China, and Central and South America. The first civilizations arose in those regions.

From then on, geography was destiny. Diamond and Sowell point out that Eurasia, the world's largest landmass, is an enormous catchment area for local innovations. Traders, sojourners, and conquerors can collect them and spread them, and people living at the crossroads can concentrate them into a high-tech package. Also, Eurasia runs in an east-west direction, whereas Africa and the Americas run north-south. Crops and animals that are domesticated in one region can easily be spread to others along lines of latitude, which are also lines of similar climate. But they cannot be spread as easily along lines of longitude, where a few hundred miles can spell the difference between temperate and tropical climates. Horses domesticated in the Asian steppes, for example, could make their way westward to Europe and eastward to China, but llamas and alpacas domesticated in the Andes never made it northward to Mexico, so the Mayan and Aztec civilizations were left without pack animals. And until recently the transportation of heavy goods over long distances (and with them traders and their ideas) was possible only by water. Europe and parts of Asia are blessed by a notchy, furrowed geography with many natural harbors and navigable rivers. Africa and Australia are not.

So Eurasia conquered the world not because Eurasians are smarter but because they could best take advantage of the principle that many heads are better than one. The "culture" of any of the conquering nations of Europe, such as Britain, is in fact a greatest-hits collection of inventions assembled across thousands of miles and years. The collection is made up of cereal crops and alphabetic writing from the Middle East, gunpowder and paper from China, domesticated horses from Ukraine, and many others. But the necessarily insular cultures of Australia, Africa, and the Americas had to make do with a few homegrown technologies, and as a result they were no match for their pluralistic conquerors. Even within Eurasia and (later) the Americas, cultures that were isolated by mountainous geography—for example, in the Appalachians, the Balkans, and the Scottish highlands—remained backward for centuries in comparison with the vast network of people around them.

The extreme case, Diamond points out, is Tasmania. The Tasmanians, who were nearly exterminated by Europeans in the nineteenth century, were the most technologically primitive people in recorded history. Unlike the Aborigines on the Australian mainland, the Tasmanians had no way of making fire, no boomerangs or spear throwers, no specialized stone tools, no axes with handles, no canoes, no sewing needles, and no ability to fish. Amazingly, the archaeological record shows that their ancestors from the Australian mainland had arrived with these technologies ten thousand years before. But then the land bridge connecting Tasmania to the mainland was submerged and the island was cut off from the rest of the world. Diamond speculates that any technology can be lost from a culture at some point in its history. Perhaps a raw material came to be in short supply and people stopped making the products that depended on it. Perhaps all the skilled artisans in a generation were killed by a freak storm. Perhaps some prehistoric Luddite or ayatollah imposed a taboo on the practice for one inane reason or another. Whenever this happens in a culture that rubs up against other ones, the lost technology can eventually be reacquired as the people clamor for the higher standard of living enjoyed by their neighbors. But in lonely Tasmania, people would have had to reinvent the proverbial wheel every time it was lost, and so their standard of living ratcheted downward.

The ultimate irony of the Standard Social Science Model is that it failed to accomplish the very goal that brought it into being: explaining the different fortunes of human societies without invoking race. The best explanation today is thoroughly cultural, but it depends on seeing a culture as a product of human desires rather than as a shaper of them.

~

HISTORY AND CULTURE, then, can be grounded in psychology, which can be grounded in computation, neuroscience, genetics, and evolution. But this kind of talk sets off alarms in the minds of many nonscientists. They fear that consilience is a smokescreen for a hostile takeover of the humanities, arts, and social sciences by philistines in white coats. The richness of their subject matter would be dumbed down into a generic palaver about neurons, genes, and evolutionary urges. This scenario is often called "reductionism," and I will conclude the chapter by showing why consilience does not call for it.

Reductionism, like cholesterol, comes in good and bad forms. Bad reductionism—also called "greedy reductionism" or "destructive reductionism"—consists of trying to explain a phenomenon in terms of its smallest or simplest constituents. Greedy reductionism is not a straw man. I know several scientists who believe (or at least say to granting agencies) that we will make breakthroughs in education, conflict resolution, and other social concerns by studying the biophysics of neural membranes or the molecular structure of the synapse. But greedy reductionism is far from the majority view, and it is easy to show why it is wrong. As the philosopher Hilary Putnam has pointed out, even the simple fact that a square peg won't fit into a round hole cannot be explained in terms of molecules and atoms but only at a higher level of analysis involving rigidity (regardless of what makes the peg rigid) and geometry.[26] And if anyone really thought that sociology or literature or history could be replaced by biology, why stop there? Biology could in turn be ground up into chemistry, and chemistry into physics, leaving one struggling to explain the causes of World War I in terms of electrons and quarks. Even if World War I consisted of nothing but a very, very large number of quarks in a very, very complicated pattern of motion, no insight is gained by describing it that way.

Good reductionism (also called hierarchical reductionism) consists not of *replacing* one field of knowledge with another but of *connecting* or *unifying* them. The building blocks used by one field are put under a microscope by another. The black boxes get opened; the promissory notes get cashed. A geographer might explain why the coastline of Africa fits into the coastline of the Americas by saying that the landmasses were once adjacent but sat on different plates, which drifted apart. The question of why the plates move gets passed on to the geologists, who appeal to an upwelling of magma that pushes them apart. As for how the magma got so hot, they call in the physicists to explain the reactions in the Earth's core and mantle. None of the scientists is dispensable. An isolated geographer would have to invoke magic to move the continents, and an isolated physicist could not have predicted the shape of South America.

So, too, for the bridge between biology and culture. The big thinkers in the sciences of human nature have been adamant that mental life has to be understood at several levels of analysis, not just the lowest one. The linguist Noam Chomsky, the computational neuroscientist David Marr, and the ethologist Niko Tinbergen have independently marked out a set of levels of analysis for understanding a faculty of the mind. These levels include its function (what it accomplishes in an ultimate, evolutionary sense); its real-time operation (how it works proximately, from moment to moment); how it is implemented in neural tissue; how it develops in the individual; and how it evolved in the species.[27] For example, language is based on a combinatorial grammar designed to communicate an unlimited number of thoughts. It is utilized by people in real time via an interplay of memory lookup and rule application. It is implemented in a network of regions in the center of the left cerebral hemisphere that must coordinate memory, planning, word meaning, and grammar.

It develops in the first three years of life in a sequence from babbling to words to word combinations, including errors in which rules may be overapplied. It evolved through modifications of a vocal tract and brain circuitry that had other uses in earlier primates, because the modifications allowed our ancestors to prosper in a socially interconnected, knowledge-rich lifestyle. None of these levels can be replaced by any of the others, but none can be fully understood in isolation from the others.

Chomsky distinguishes all of these from yet another level of analysis (one that he himself has little use for but that other language scholars invoke). The vantage points I just mentioned treat language as an internal, individual entity, such as the knowledge of Canadian English that I possess in my head. But language can also be understood as an external entity: the "English language" as a whole, with its fifteen-hundred-year history, its countless dialects and hybrids spanning the globe, its half a million words in the *Oxford English Dictionary*. An external language is an abstraction that pools the internal languages of hundreds of millions of people living in different places and times. It could not exist without the internal languages in the minds of real humans conversing with one another, but it cannot be reduced to what any of them knows either. For example, the statement "English has a larger vocabulary than Japanese" could be true even if no English speaker has a larger vocabulary than any Japanese speaker.

The English language was shaped by broad historical events that did not take place inside a single head. They include the Scandinavian and Norman invasions in medieval times, which infected it with non-Anglo-Saxon words; the Great Vowel Shift of the fifteenth century, which scrambled the pronunciation of the long vowels and left its spelling system an irregular mess; the expansion of the British Empire, which budded off a variety of Englishes (American, Australian, Singaporean); and the development of global electronic media, which may rehomogenize the language as we all read the same web pages and watch the same television shows.

At the same time, none of these forces can be understood without taking into account the thought processes of flesh-and-blood people. They include the Britons who reanalyzed French words when they absorbed them into English, the children who failed to remember irregular past-tense forms like *writhe-wrothe* and *crow-crew* and converted them into regular verbs, the aristocrats who affected fussy pronunciations to differentiate themselves from the rabble, the mumblers who swallowed consonants to leave us *made* and *had* (originally *maked* and *haved*), and the clever speakers who first converted *I had the house built* to *I had built the house* and inadvertently gave English its perfect tense. Language is re-created every generation as it passes through the minds of the humans who speak it.[28]

External language is, of course, a fine example of culture, the province of social scientists and scholars in the humanities. The way that language can be understood at some half-dozen connected levels of analysis, from the brain and evolution to the cognitive processes of individuals to vast cultural systems, shows how culture and biology may be connected. The possibilities for connections in other spheres of human knowledge are plentiful, and we will encounter them throughout the book. The moral sense can illuminate legal and ethical codes. The psychology of kinship helps us understand sociopolitical arrangements. The mentality of aggression helps to make sense of war and conflict resolution. Sex differences are relevant to gender politics. Human aesthetics and emotion can enlighten our understanding of the arts.

What is the payoff for connecting the social and cultural levels of analysis to the psychological and biological ones? It is the thrill of discoveries that could never be made within the boundaries of a single discipline, such as universals of beauty, the logic of language, and the components of the moral sense. And it is the uniquely satisfying understanding we have enjoyed from the unification of the other sciences—the explanation of muscles as tiny magnetic ratchets, of flowers as lures for insects, of the rainbow as a splaying of wavelengths that ordinarily blend into white. It is the difference between stamp collecting and detective work, between slinging around jargon and offering insight, between saying that something just is and explaining why it had to be that way as opposed to some other way it could have been. In a talk-show parody in *Monty Python's Flying Circus,* an expert on dinosaurs trumpets her new theory of the brontosaurus: "All brontosauruses are thin at one end; much, much thicker in the middle; and then thin again at the far end." We laugh because she has not explained her subject in terms of deeper principles—she has not "reduced" it, in the good sense. Even the word *understand*—literally, "stand under"—alludes to descending to a deeper level of analysis.

Our understanding of life has only been enriched by the discovery that living flesh is composed of molecular clockwork rather than quivering protoplasm, or that birds soar by exploiting the laws of physics rather than defying them. In the same way, our understanding of ourselves and our cultures can only be enriched by the discovery that our minds are composed of intricate neural circuits for thinking, feeling, and learning rather than blank slates, amorphous blobs, or inscrutable ghosts.

Chapter 5

The Slate's Last Stand

HUMAN NATURE IS a scientific topic, and as new facts come in, our conception of it will change. Sometimes the facts may show that a theory grants our minds too much innate structure. For example, perhaps our language faculties are equipped not with nouns, verbs, adjectives, and prepositions but only with a distinction between more nounlike and more verblike parts of speech. At other times a theory may turn out to have granted our minds too little innate structure. No current theory of personality can explain why both members of a pair of identical twins reared apart liked to keep rubber bands around their wrists and pretend to sneeze in crowded elevators.

Also up for grabs is exactly how our minds use the information coming in from the senses. Once our faculties for language and social interaction are up and running, some kinds of learning may consist of simply recording information for future use, like the name of a person or the content of a new piece of legislation. Others may be more like setting a dial, flipping a switch, or computing an average, where the apparatus is in place but a parameter is left open so the mind can track variation in the local environment. Still others may use the information provided by all normal environments, such as the presence of gravity or the statistics of colors and lines in the visual field, to tune up our sensorimotor systems. There are yet other ways that nature and nurture might interact, and many will blur the distinction between the two.

This book is based on the estimation that whatever the exact picture turns out to be, a universal complex human nature will be part of it. I think we have reason to believe that the mind is equipped with a battery of emotions, drives, and faculties for reasoning and communicating, and that they have a common logic across cultures, are difficult to erase or redesign from scratch, were shaped by natural selection acting over the course of human evolution, and owe some of their basic design (and some of their variation) to information in the genome. This general picture is meant to embrace a variety of theories, present and future, and a range of foreseeable scientific discoveries.

But the picture does not embrace just *any* theory or discovery. Conceivably scientists might discover that there is insufficient information in the genome to specify any innate circuitry, or no known mechanism by which it could be wired into the brain. Or perhaps they will discover that brains are made out of general-purpose stuff that can soak up just about any pattern in the sensory input and organize itself to accomplish just about any goal. The former discovery would make innate organization impossible; the latter would make it unnecessary. Those discoveries would call into question the very concept of human nature. Unlike the moral and political objections to the concept of human nature (objections that I discuss in the rest of this book), these would be *scientific* objections. If such discoveries are on the horizon, I had better look at them carefully.

This chapter is about three scientific developments that are sometimes interpreted as undermining the possibility of a complex human nature. The first comes from the Human Genome Project. When the sequence of the human genome was published in 2001, geneticists were surprised that the number of genes was lower than they had predicted. The estimates hovered around 34,000 genes, which lies well outside the earlier range of 50,000 to 100,000.[1] Some editorialists concluded that the smaller gene count refuted any claim about innate talents or tendencies, because the slate is too small to contain much writing. Some even saw it as vindicating the concept of free will: the smaller the machine, the more room for a ghost.

The second challenge comes from the use of computer models of neural networks to explain cognitive processes. These artificial neural networks can often be quite good at learning statistical patterns in their input. Some modelers from the school of cognitive science called connectionism suggest that generic neural networks can account for all of human cognition, with little or no innate tailoring for particular faculties such as social reasoning or language. In Chapter 2 we met the founders of connectionism, David Rumelhart and James McClelland, who suggested that people are smarter than rats only because they have more associative cortex and because their environment contains a culture to organize it.

The third comes from the study of neural plasticity, which examines how the brain develops in the womb and early childhood and how it records experience as the animal learns. Neuroscientists have recently shown how the brain changes in response to learning, practice, and input from the senses. One spin on these discoveries may be called extreme plasticity. According to this slant, the cerebral cortex—the convoluted gray matter responsible for perception, thinking, language, and memory—is a protean substance that can be shaped almost limitlessly by the structure and demands of the environment. The blank slate becomes the plastic slate.

Connectionism and extreme plasticity are popular among cognitive scientists at the West Pole, who reject a completely blank slate but want to restrict innate organization to simple biases in attention and memory. Extreme plasticity also appeals to neuroscientists who wish to boost the importance of their field for education and social policy, and to entrepreneurs selling products to speed up infant development, cure learning disabilities, or slow down aging. Outside the sciences, all three developments have been welcomed by some scholars in the humanities who want to beat back the encroachments of biology.[2] The lean genome, connectionism, and extreme plasticity are the Blank Slate's last stand.

The point of this chapter is that these claims are not vindications of the doctrine of the Blank Slate but *products* of the Blank Slate. Many people (including a few scientists) have selectively read the evidence, sometimes in bizarre ways, to fit with a prior belief that the mind cannot possibly have any innate structure, or with simplistic notions of how innate structure, if it did exist, would be encoded in the genes and develop in the brain.

I should say at the outset that I find these latest-and-best blank-slate theories highly implausible—indeed, barely coherent. Nothing comes out of nothing, and the complexity of the brain has to come from somewhere. It cannot come from the environment alone, because the whole point of having a brain is to accomplish certain goals, and the environment has no idea what those goals are. A given environment can accommodate organisms that build dams, migrate by the stars, trill and twitter to impress the females, scent-mark trees, write sonnets, and so on. To one species, a snatch of human speech is a warning to flee; to another, it is an interesting new sound to incorporate into its own vocal repertoire; to a third, it is grist for grammatical analysis. Information in the world doesn't tell you what to do with it.

Also, brain tissue is not some genie that can grant its owner any power that would come in handy. It is a physical mechanism, an arrangement of matter that converts inputs to outputs in particular ways. The idea that a single generic substance can see in depth, control the hands, attract a mate, bring up children, elude predators, outsmart prey, and so on, without *some* degree of specialization, is not credible. Saying that the brain solves these problems because of its "plasticity" is not much better than saying it solves them by magic.

Still, in this chapter I will examine the latest scientific objections to human nature carefully. Each of the discoveries is important on its own terms, even if it does not support the extravagant conclusions that have been drawn. And once the last supports for the Blank Slate have been evaluated, I can properly sum up the scientific case for the alternative.

~

THE HUMAN GENOME is often seen as the essence of our species, so it is not surprising that when its sequence was announced in 2001 commentators rushed to give it the correct interpretation for human affairs. Craig Venter, whose company had competed with a public consortium in the race to sequence the genome, said at a press conference that the smaller-than-expected gene count shows that "we simply do not have enough genes for this idea of biological determinism to be right. The wonderful diversity of the human species is not hard-wired in our genetic code. Our environments are critical." In the United Kingdom, *The Guardian* headlined its story, "Revealed: The Secret of Human Behaviour. Environment, Not Genes, Key to Our Acts."[3] An editorial in another British newspaper concluded that "we are more free, it seems, than we had realized." Moreover, the finding "offers comfort for the left, with its belief in the potential of all, however deprived their background. But it is damning for the right, with its fondness for ruling classes and original sin."[4]

All this from the number 34,000! Which leads to the question, What number of genes would have proven that the diversity of our species was wired into our genetic code, or that we are less free than we had realized, or that the political right is right and the left is wrong? 50,000? 150,000? Conversely, if it turned out that we had only 20,000 genes, would that have made us even freer, or the environment even more important, or the political left even more comfortable? The fact is that no one knows what these numbers mean. No one has the slightest idea how many genes it would take to build a system of hard-wired modules, or a general-purpose learning program, or anything in between—to say nothing of original sin or the superiority of the ruling class. In our current state of ignorance of how the genes build a brain, the number of genes in the human genome is just a number.

If you don't believe this, consider the roundworm *Caenorhabditis elegans,* which has about 18,000 genes. By the logic of the genome editorialists, it should be twice as free, be twice as diverse, and have twice as much potential as a human being. In fact, it is a microscopic worm composed of 959 cells grown by a rigid genetic program, with a nervous system consisting of exactly 302 neurons in a fixed wiring diagram. As far as behavior is concerned, it eats, mates, approaches and avoids certain smells, and that's about it. This alone should make it obvious that our freedom and diversity of behavior come from having a complex biological makeup, not a simple one.

Now, it is a genuine puzzle why humans, with their hundred trillion cells and hundred billion neurons, need only twice as many genes as a humble little worm. Many biologists believe that the human genes have been undercounted. The number of genes in a genome can only be estimated; right now they cannot literally be totted up. Gene-estimating programs look for sequences in the DNA that are similar to known genes and that are active enough to be caught in the act of building a protein.[5] Genes that are unique to humans or active only in the developing brain of the fetus—the genes most relevant to human nature—and other inconspicuous genes could evade the software and get left out of the estimates. Alternative estimates of 57,000, 75,000, and even 120,000 human genes are currently being bruited about.[6] Still, even if humans had six times as many genes as a roundworm rather than just twice as many, the puzzle would remain.

Most biologists who are pondering the puzzle don't conclude that humans are less complex than we thought. Instead they conclude that the number of genes in a genome has little to do with the complexity of the organism.[7] A single gene does not correspond to a single component in such a way that an organism with 20,000 genes has 20,000 components, an organism with 30,000 genes has 30,000 components, and so on. Genes specify proteins, and some of the proteins do become the meat and juices of an organism. But other proteins turn genes on or off, speed up or slow down their activity, or cut and splice other proteins into new combinations. James Watson points out that we should recalibrate our intuitions about what a given number of genes can do: "Imagine watching a play with thirty thousand actors. You'd get pretty confused."

Depending on how the genes interact, the assembly process can be much more intricate for one organism than for another with the same number of genes. In a simple organism, many of the genes simply build a protein and dump it into the stew. In a complex organism, one gene may turn on a second one, which speeds up the activity of a third one (but only if a fourth one is active), which then turns off the original gene (but only if a fifth one is inactive), and so on. This defines a kind of recipe that can build a more complex organism out of the same number of genes. The complexity of an organism thus depends not just on its gene count but on the intricacy of the box-and-arrow diagram that captures how each gene impinges on the activity of the other genes.[8] And because adding a gene doesn't just add an ingredient but can multiply the number of ways that the genes can interact with one another, the complexity of organisms depends on the number of possible *combinations* of active and inactive genes in their genomes. The geneticist Jean-Michel Claverie suggests that it might be estimated by the number two (active versus inactive) raised to the power of the number of genes. By that measure, a human genome is not twice as complex as a roundworm genome but $2^{16,000}$ (a one followed by 4,800 zeroes) times as complex.[9]

There are two other reasons why the complexity of the genome is not reflected in the number of genes it contains. One is that a given gene can produce not just one protein but several. A gene is typically broken into stretches of DNA that code for fragments of protein (exons) separated by stretches of DNA that don't (introns), a bit like a magazine article interrupted by ads. The segments of a gene can then be spliced together in multiple ways. A gene composed of exons A, B, C, and D might give rise to proteins corresponding to

ABC, ABD, ACD, and so on—as many as ten different proteins per gene. This happens to a greater degree in complex organisms than in simple ones.[10]

Second, the 34,000 genes take up only about 3 percent of the human genome. The rest consists of DNA that does not code for protein and that used to be dismissed as "junk." But as one biologist recently put it, "The term 'junk DNA' is a reflection of our ignorance."[11] The size, placement, and content of the noncoding DNA can have dramatic effects on the way that nearby genes are activated to make proteins. Information in the billions of bases in the noncoding regions of the genome is part of the specification of a human being, above and beyond the information contained in the 34,000 genes.

The human genome, then, is fully capable of building a complex brain, in spite of the bizarre proclamations of how wonderful it is that people are almost as simple as worms. *Of course* "the wonderful diversity of the human species is not hard-wired in our genetic code," but we didn't need to count genes to figure that out—we already know it from the fact that a child growing up in Japan speaks Japanese but the same child growing up in England would speak English. It is an example of a syndrome we will meet elsewhere in this book: scientific findings spin-doctored beyond recognition to make a moral point that could have been made more easily on other grounds.

～

THE SECOND SCIENTIFIC defense of the Blank Slate comes from connectionism, the theory that the brain is like the artificial neural networks simulated on computers to learn statistical patterns.[12]

Cognitive scientists agree that the elementary processes that make up the instruction set of the brain—storing and retrieving an association, sequencing elements, focusing attention—are implemented in the brain as networks of densely interconnected neurons (brain cells). The question is whether a generic kind of network, after being shaped by the environment, can explain all of human psychology, or whether the genome tailors different networks to the demands of particular domains: language, vision, morality, fear, lust, intuitive psychology, and so on. The connectionists, of course, do not believe in a blank slate, but they do believe in the closest mechanistic equivalent, a general-purpose learning device.

What is a neural network? Connectionists use the term to refer not to real neural circuitry in the brain but to a kind of computer program based on the metaphor of neurons and neural circuits. In the most common approach, a "neuron" carries information by being more or less active. The activity level indicates the presence or absence (or intensity or degree of confidence) of a simple feature of the world. The feature may be a color, a line with a certain slant, a letter of the alphabet, or a property of an animal such as having four legs.

A *network* of neurons can represent different concepts, depending on which ones are active. If neurons for "yellow," "flies," and "sings" are active, the network is thinking about a canary; if neurons for "silver," "flies," and "roars" are active, it is thinking about an airplane. An artificial neural network computes in the following manner. Neurons are linked to other neurons by connections that work something like synapses. Each neuron counts up the inputs from other neurons and changes its activity level in response. The network learns by allowing the input to change the *strengths* of the connections. The strength of a connection determines the likelihood that the input neuron will excite or inhibit the output neuron.

Depending on what the neurons stand for, how they are innately wired, and how the connections change with training, a connectionist network can learn to compute various things. If everything is connected to everything else, a network can soak up the correlations among features in a set of objects. For example, after exposure to descriptions of many birds it can predict that feathered singing things tend to fly or that feathered flying things tend to sing or that singing flying things tend to have feathers. If a network has an input layer connected to an output layer, it can learn associations between ideas, such as that small soft flying things are animals but large metallic flying things are vehicles. If its output layer feeds back to earlier layers, it can crank out ordered sequences, such as the sounds making up a word.

The appeal of neural networks is that they automatically generalize their training to similar new items. If a network has been trained that tigers eat Frosted Flakes, it will tend to generalize that lions eat Frosted Flakes, because "eating Frosted Flakes" has been associated not with "tigers" but with simpler features like "roars" and "has whiskers," which make up part of the representation of lions, too. The school of connectionism, like the school of associationism championed by Locke, Hume, and Mill, asserts that these generalizations are the crux of intelligence. If so, highly trained but otherwise generic neural networks can explain intelligence.

Computer modelers often set their models on simplified toy problems to prove that they can work in principle. The question then becomes whether the models can "scale up" to more realistic problems, or whether, as skeptics say, the modeler "is climbing trees to get to the moon." Here we have the problem with connectionism. Simple connectionist networks can manage impressive displays of memory and generalization in circumscribed problems like reading a list of words or learning stereotypes of animals. But they are simply too underpowered to duplicate more realistic feats of human intelligence like understanding a sentence or reasoning about living things.

Humans don't just loosely associate things that resemble each other, or things that tend to occur together. They have combinatorial minds that entertain propositions about what is true of what, and about who did what to whom, when and where and why. And that requires a computational architecture that is more sophisticated than the uniform tangle of neurons used in generic connectionist networks. It requires an architecture equipped with logical apparatus like rules, variables, propositions, goal states, and different kinds of data structures, organized into larger systems. Many cognitive scientists have made this point, including Gary Marcus, Marvin Minsky, Seymour Papert, Jerry Fodor, Zenon Pylyshyn, John Anderson, Tom Bever, and Robert Hadley, and it is acknowledged as well by neural network modelers who are not in the connectionist school, such as John Hummel, Lokendra Shastri, and Paul Smolensky.[13] I have written at length on the limits of connectionism, both in scholarly papers and in popular books; here is a summary of my own case.[14]

In a section called "Connectoplasm" in *How the Mind Works,* I laid out some simple logical relationships that underlie our understanding of a complete thought (such as the meaning of a sentence) but that are difficult to represent in generic networks.[15] One is the distinction between a kind and an individual: between ducks in general and this duck in particular. Both have the same features (swims, quacks, has feathers, and so on), and both are thus represented by the same set of active units in a standard connectionist model. But people know the difference.

A second talent is compositionality: the ability to entertain a new, complex thought that is not just the sum of the simple thoughts composing it but depends on their relationships. The thought that cats chase mice, for example, cannot be captured by activating a unit each for "cats," "mice," and "chase," because that pattern could just as easily stand for mice chasing cats.

A third logical talent is quantification (or the binding of variables): the difference between fooling some of the people all of the time and fooling all of the people some of the time. Without the computational equivalent of x's, y's, parentheses, and statements like "For all x," a model cannot tell the difference.

A fourth is recursion: the ability to embed one thought inside another, so that we can entertain not only the thought that Elvis lives, but the thought that the *National Enquirer* reported that Elvis lives, that some people believe the *National Enquirer* report that Elvis lives, that it is amazing that some people believe the *National Enquirer* report that Elvis lives, and so on. Connectionist networks would superimpose these propositions and thereby confuse their various subjects and predicates.

A final elusive talent is our ability to engage in categorical, as opposed to fuzzy, reasoning: to understand that Bob Dylan is a grandfather, even though he is not very grandfatherly, or that shrews are not rodents, though they look just like mice. With nothing but a soup of neurons to stand for an object's properties, and no provision for rules, variables, and definitions, the networks fall back on stereotypes and are bamboozled by atypical examples.

In *Words and Rules* I aimed a microscope on a single phenomenon of language that has served as a test case for the ability of generic associative networks to account for the essence of language: assembling words, or pieces of words, into new combinations. People don't just memorize snatches of language but create new ones. A simple example is the English past tense. Given a neologism like *to spam* or *to snarf,* people don't have to run to the dictionary to look up their past-tense forms; they instinctively know that they are *spammed* and *snarfed.* The talent for assembling new combinations appears as early as age two, when children overapply the past-tense suffix to irregular verbs, as in *We holded the baby rabbits* and *Horton heared a Who.*[16]

The obvious way to explain this talent is to appeal to two kinds of computational operations in the mind. Irregular forms like *held* and *heard* are stored in and retrieved from memory, just like any other word. Regular forms like *walk-walked* can be generated by a mental version of the grammatical rule "Add *-ed* to the verb." The rule can apply whenever memory fails. It may be used when a word is unfamiliar and no past-tense form had been stored in memory, as in *to spam,* and it may be used by children when they cannot recall an irregular form like *heard* and need some way of marking its tense. Combining a suffix with a verb is a small example of an important human talent: combining words and phrases to create new sentences and thereby express new thoughts. It is one of the new ideas of the cognitive revolution introduced in Chapter 3, and one of the logical challenges for connectionism I listed in the preceding discussion.

Connectionists have used the past tense as a proving ground to see if they could duplicate this textbook example of human creativity without using a rule and without dividing the labor between a system for memory and a system for grammatical combination. A series of computer models have tried to generate past-tense forms using simple pattern associator networks. The networks typically connect the sounds in verbs with the sounds in the past-tense form: *-am* with *-ammed, -ing* with *-ung,* and so on. The models can then generate new forms by analogy, just like the generalization from tigers to lions: trained on *crammed,* a model can guess *spammed;* trained on *folded,* it tends to say *holded.*

But human speakers do far more than associate sounds with sounds, and the models thus fail to do them justice. The failures come from the absence of machinery to handle logical relationships. Most of the models are baffled by new words that sound different from familiar words and hence cannot be generalized by analogy. Given the novel verb *to frilg,* for example, they come up not with *frilged,* as people do, but with an odd mishmash like *freezled.* That is because they lack the device of a variable, like *x* in algebra or "verb" in grammar, which can apply to any member of a category, regardless of how familiar its properties are. (This is the gadget that allows people to engage in categorical rather than fuzzy reasoning.) The networks can only associate bits of sound with bits of sound, so when confronted with a new verb that does not sound like anything they were trained on, they assemble a pastiche of the most similar sounds they can find in their network.

The models also cannot properly distinguish among verbs that have the same sounds but different past-tense forms, such as *ring the bell–rang the bell* and *ring the city–ringed the city*. That is because the standard models represent only sound and are blind to the grammatical differences among verbs that call for different conjugations. The key difference here is between simple roots like *ring* in the sense of "resonate" (past tense *rang*) and complex verbs derived from nouns like *ring* in the sense of "form a ring around" (past tense *ringed*). To register that difference, a language-using system has to be equipped with compositional data structures (such as "a verb made from the noun *ring*") and not just a beanbag of units.

Yet another problem is that connectionist networks track the statistics of the input closely: how many verbs of each sound pattern they have encountered. That leaves them unable to account for the epiphany in which young children discover the *-ed* rule and start making errors like *holded* and *heared*. Connectionist modelers can induce these errors only by bombarding the network with regular verbs (so as to burn in the *-ed*) in a way that is unlike anything real children experience. Finally, a mass of evidence from cognitive neuroscience shows that grammatical combination (including regular verbs) and lexical lookup (including irregular verbs) are handled by different systems in the brain rather than by a single associative network.

It's not that neural networks are incapable of handling the meanings of sentences or the task of grammatical conjugation. (They had better not be, since the very idea that thinking is a form of neural computation requires that *some* kind of neural network duplicate whatever the mind can do.) The problem lies in the credo that one can do everything with a generic model as long as it is sufficiently trained. Many modelers have beefed up, retrofitted, or combined networks into more complicated and powerful systems. They have dedicated hunks of neural hardware to abstract symbols like "verb phrase" and "proposition" and have implemented additional mechanisms (such as synchronized firing patterns) to bind them together in the equivalent of compositional, recursive symbol structures. They have installed banks of neurons for words, or for English suffixes, or for key grammatical distinctions. They have built hybrid systems, with one network that retrieves irregular forms from memory and another that combines a verb with a suffix.[17]

A system assembled out of beefed-up subnetworks could escape all the criticisms. But then we would no longer be talking about a generic neural network! We would be talking about a complex system innately tailored to compute a task that people are good at. In the children's story called "Stone Soup," a hobo borrows the use of a woman's kitchen ostensibly to make soup from a stone. But he gradually asks for more and more ingredients to balance the flavor until he has prepared a rich and hearty stew at her expense. Connectionist modelers who claim to build intelligence out of generic neural networks without requiring anything innate are engaged in a similar business. The design choices that make a neural network system smart—what each of the neurons represents, how they are wired together, what kinds of networks are assembled into a bigger system, in which way—embody the innate organization of the part of the mind being modeled. They are typically hand-picked by the modeler, like an inventor rummaging through a box of transistors and diodes, but in a real brain they would have evolved by natural selection (indeed, in some networks, the architecture of the model does evolve by a simulation of natural selection).[18] The only alternative is that some previous episode of learning left the networks in a state ready for the current learning, but of course the buck has to stop at *some* innate specification of the first networks that kick off the learning process.

So the rumor that neural networks can replace mental structure with statistical learning is not true. Simple, generic networks are not up to the demands of ordinary human thinking and speaking; complex, specialized networks are a stone soup in which much of the interesting work has been done in setting up the innate wiring of the network. Once this is recognized, neural network modeling becomes an indispensable complement to the theory of a complex human nature rather than a replacement for it.[19] It bridges the gap between the elementary steps of cognition and the physiological activity of the brain and thus serves as an important link in the long chain of explanation between biology and culture.

~

FOR MOST OF its history, neuroscience was faced with an embarrassment: the brain looked as if it were innately specified in every detail. When it comes to the body, we can see many of the effects of a person's life experience: it may be tanned or pale, callused or soft, scrawny or plump or chiseled. But no such marks could be found in the brain. Now, something has to be wrong with this picture. People learn, and learn massively: they learn their language, their culture, their know-how, their database of facts. Also, the hundred trillion connections in the brain cannot possibly be specified individually by a 750-megabyte genome. The brain somehow must change in response to its input; the only question is how.

We are finally beginning to understand how. The study of neural plasticity is hot. Almost every week sees a discovery about how the brain gets wired in the womb and tuned outside it. After all those decades in which no one could find *anything* that changed in the brain, it is not surprising that the discovery of plasticity has given the nature-nurture pendulum a push. Some people describe plasticity as a harbinger of an expansion of human potential in which the powers of the brain will be harnessed to revolutionize childrearing, education, therapy, and aging. And several manifestos have proclaimed that plasticity proves that the brain cannot have any significant innate organization.[20] In *Rethinking Innateness,* Jeffrey Elman and a team of West Pole connectionists write that predispositions to think about different things in different ways (language, people, objects, and so on) may be implemented in the brain only as "attention-grabbers" that ensure that the organism will receive "massive experience of certain inputs prior to subsequent learning."[21] In a "constructivist manifesto," the theoretical neuroscientists Stephen Quartz and Terrence Sejnowski write that "although the cortex is not a tabula rasa . . . it is largely equipotential at early stages," and therefore that innatist theories "appear implausible."[22]

Neural development and plasticity unquestionably make up one of the great frontiers of human knowledge. How a linear string of DNA can direct the assembly of an intricate three-dimensional organ that lets us think, feel, and learn is a problem to stagger the imagination, to keep neuroscientists engaged for decades, and to belie any suggestion that we are approaching "the end of science."

And the discoveries themselves are fascinating and provocative. The cerebral cortex (outer gray matter) of the brain has long been known to be divided into areas with different functions. Some represent particular body parts; others represent the visual field or the world of sound; still others concentrate on aspects of language or thinking. We now know that with learning and practice some of their boundaries can move around. (This does not mean that the brain tissue literally grows or shrinks, only that if the cortex is probed with electrodes or monitored with a scanner, the boundary where one ability leaves off and the next one begins can shift.) Violinists, for example, have an expanded region of cortex representing the fingers of the left hand.[23] If a person or a monkey is trained on a simple task like recognizing shapes or attending to a location in space, neuroscientists can watch as parts of the cortex, or even individual neurons, take on the job.[24]

The reallocation of brain tissue to new tasks is especially dramatic when people lose the use of a sense or body part. Congenitally blind people use their visual cortex to read Braille.[25] Congenitally deaf people use part of their auditory cortex to process sign language.[26] Amputees use the part of the cortex formerly serving the missing limb to represent other parts of their bodies.[27] Young children can grow up relatively normal after traumas to the brain that would turn adults into basket cases—even removal of the entire left hemisphere, which in adults underlies language and logical reasoning.[28] All this suggests that the allocation of brain tissue to perceptual and cognitive processes is not done permanently and on the basis of the exact location of the tissue in the skull, but depends on how the brain itself processes information.

This dynamic allocation of tissue can also be seen as the brain puts itself together in the womb. Unlike a computer that gets assembled in a factory and is turned on for the first time when complete, the brain is active *while* it is being assembled, and that activity may take part in the assembly process. Experiments on cats and other mammals have shown that if a brain is chemically silenced during fetal development it may end up with significant abnormalities.[29] And patches of cortex develop differently depending on the kind of input they receive. In an experimental tour de force, the neuroscientist Mriganka Sur literally *rewired* the brains of ferrets so that signals from their eyes fed into the primary auditory cortex, the part of the brain that ordinarily receives signals from the ears.[30] When he then probed the auditory cortex with electrodes, he found that it acted in many ways like the visual cortex. Locations in the visual field were laid out like a map, and individual neurons responded to lines and stripes at a particular orientation and direction of movement, similar to the neurons in an ordinary visual cortex. The ferrets could even use their rewired brains to move toward objects that were detectable by sight alone. The input to the sensory cortex must help to organize it: visual input makes the auditory cortex work something like the visual cortex.

What do these discoveries mean? Do they show that the brain is "able to be shaped, molded, modeled, or sculpted," as the dictionary definition of *plastic* would suggest? In the rest of this chapter I will show you that the answer is no.[31] Discoveries of how the brain changes with experience do not show that learning is more powerful than we thought, that the brain can be dramatically reshaped by its input, or that the genes do not shape the brain. Indeed, demonstrations of the plasticity of the brain are less radical than they first appear: the supposedly plastic regions of cortex are doing pretty much the same thing they would have been doing if they had never been altered. And the most recent discoveries on brain development have refuted the idea that the brain is largely plastic. Let me go over these points in turn.

～

THE FACT THAT the brain changes when we learn is not, as some have claimed, a radical discovery with profound implications for nature and nurture or human potential. Dmitri Karamazov could have deduced it in his nineteenth-century prison cell as he mulled over the fact that thinking comes from quivering nerve tails rather than an immaterial soul. If thought and action are products of the physical activity of the brain, and if thought and action can be affected by experience, then experience has to leave a trace in the physical structure of the brain.

So there is no scientific question as to whether experience, learning, and practice affect the brain; they surely do if we are even vaguely on the right track. It is not surprising that people who can play the violin have different brains from those who cannot, or that masters of sign language or of Braille have different brains from people who speak and read. Your brain changes when you are introduced to a new person, when you hear a bit of gossip, when you watch the Oscars, when you polish your golf stroke—in short, whenever an experience leaves a trace in the mind. The only question is *how* learning affects the brain. Are memories stored in protein sequences, in new neurons or synapses, or in changes in the strength of existing synapses? When someone learns a new skill, is it stored only in organs dedicated to learning skills (like the cerebellum and the basal ganglia), or does it also adjust the cortex? Does an increase in dexterity depend on using more square centimeters of cortex or on using a greater concentration of synapses in the same number of square centimeters? These are important scientific problems, but they say nothing about whether people can learn, or how much. We already knew trained violinists play better than beginners or we would never have put their heads in the scanner to begin with. Neural plasticity is just another name for learning and development, described at a different level of analysis.

All this should be obvious, but nowadays any banality about learning can be dressed up in neurospeak and treated like a great revelation of science. According to a *New York Times* headline, "Talk therapy, a psychiatrist maintains, can alter the structure of the patient's brain."[32] I should hope so, or else the psychiatrist would be defrauding her clients. "Environmental manipulation can change the way [a child's] brain develops," the pediatric neurologist Harry Chugani told the *Boston Globe*. "A child surrounded by aggression, violence, or inadequate stimulation will reflect these connections in the brain and behavior."[33] Well, yes; if the environment affects the child at all, it would do so by changing connections in the brain. A special issue of the journal *Educational Technology and Society* was intended "to examine the position that learning takes place in the brain of the learner, and that pedagogies and technologies should be designed and evaluated on the basis of the effect they have on student brains." The guest editor (a biologist) did not say whether the alternative was that learning takes place in some other organ of the body like the pancreas or that it takes place in an immaterial soul. Even professors of neuroscience sometimes proclaim "discoveries" that would be news only to believers in a ghost in the machine: "Scientists have found that the brain is capable of altering its connections. . . . You have the ability to change the synaptic connections within the brain."[34] Good thing, because otherwise we would be permanent amnesiacs.

This neuroscientist is an executive at a company that "uses brain research and technology to develop products intended to enhance human learning and performance," one of many new companies with that aspiration. "The human being has unlimited creativity if focused and nurtured properly," says a consultant who teaches clients to draw diagrams that "map their neural patterns." "The older you get, the more connections and associations your brain should be making," said a satisfied customer; "Therefore you should have more information stored in your brain. You just need to tap into it."[35] Many people have been convinced by the public pronouncements of neuroscience advocates— on the basis of no evidence whatsoever—that varying the route you take when driving home can stave off the effects of aging.[36] And then there is the marketing genius who realized that blocks, balls, and other toys "provide visual and tactile stimulation" and "encourage movement and tracking," part of a larger movement of "brain-based" childrearing and education that we will meet again in the chapter on children.[37]

These companies tap into people's belief in a ghost in the machine by implying that any form of learning that affects the brain (as opposed, presumably, to the kinds of learning that don't affect the brain) is unexpectedly real or deep or powerful. But this is mistaken. All learning affects the brain. It is undeniably exciting when scientists make a discovery about *how* learning affects the brain, but that does not make the learning itself any more pervasive or profound.

～

A SECOND MISINTERPRETATION of neural plasticity can be traced to the belief that there is nothing in the mind that was not first in the senses. The most highly publicized discoveries about cortical plasticity concern primary sensory cortex, the patches of gray matter that first receive signals from the senses (via the thalamus and other subcortical organs). Writers who use plasticity to prop up the Blank Slate assume that if primary sensory cortex is plastic, the rest of the brain must be even *more* plastic, because the mind is built out of sensory experience. For example, one neuroscientist was quoted as saying that Sur's rewiring experiments "challenge the recent emphasis on the power of the genes" and "will push people back toward more consideration of environmental factors in creating normal brain organization."[38]

But if the brain is a complex organ with many parts, the moral does not follow. Primary sensory cortex is not the bedrock of the mind but a gadget, one of many in the brain, that happens to be specialized for certain kinds of signal processing in the first stages of sensory analysis. Let's suppose that primary sensory cortex really were formless, getting all its structure from the input. Would that mean that the entire brain is formless and gets all of its structure from the input? Not at all. For one thing, even primary sensory cortex is just one part of a huge, intricate system. To put things in perspective, here is a recent diagram of the wiring of the primate visual system:[39]

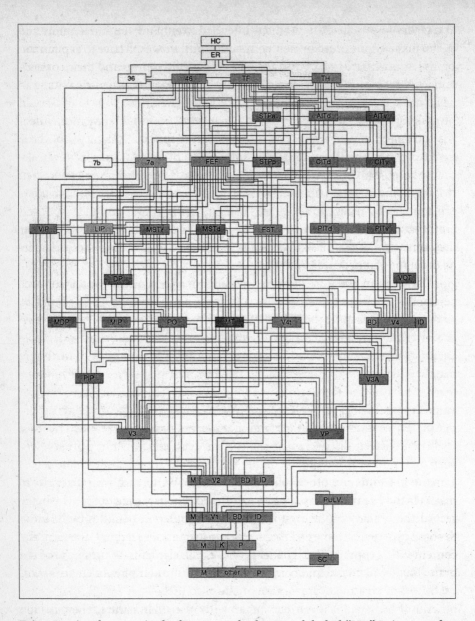

Primary visual cortex is the box near the bottom labeled "V1." It is one of at least fifty distinct brain areas devoted to visual processing, and they are interconnected in precise ways. (Despite the spaghetti-like appearance, not everything is connected to everything else. Only about a third of the logically possible connections between components are actually present in the brain.) Primary visual cortex, by itself, is not enough to see with. Indeed, it is so deeply buried in the visual system that Francis Crick and the neuroscientist Christof Koch have argued that we are not conscious of anything that goes on in it.[40] What we see—familiar colored objects arranged in a scene or moving in particular ways—is a product of the entire contraption. So even if the innards of the V1 box were completely specified by its input, we would have to explain the architecture of the rest of the visual system—the fifty boxes and their connections. I don't mean to imply that the entire block diagram is genetically specified, but much of it almost certainly is.[41]

And of course the visual system itself must be put into perspective, because it is just one part of the brain. The visual system dominates some half-dozen of the more than fifty major areas of the cortex that can be distinguished by their anatomy and connections. Many of the others underlie other functions such as language, reasoning, planning, and social skills. Though no one knows to what extent they are genetically prepared for their computational roles, there are hints that the genetic influence is substantial.[42] The divisions are established in the womb, even if the cortex is cut off from sensory input during development. As development proceeds, different sets of genes are activated in different regions. The brain has a well-stocked toolbox of mechanisms to interconnect neurons, including molecules that attract or repel axons (the output fibers of neurons) to guide them to their targets, and molecules that glue them in place or ward them away. The number, size, and connectivity of cortical areas differ among species of mammals, and they differ between humans and other primates. This diversity is caused by genetic changes in the course of evolution that are beginning to be understood.[43] Geneticists recently discovered, for example, that different sets of genes are activated in the developing brain of humans and the developing brains of chimpanzees.[44]

The possibility that cortical areas are specialized for different tasks has been obscured by the fact that different parts of the cortex look similar under a microscope. But because the brain is an information-processing system, that means little. The microscopic pits on a CD look the same regardless of what is recorded on it, and the strings of characters in different books look the same to someone who cannot read them. In an information-carrying medium, the content lies in combinatorial *patterns* among the elements—in the case of the brain, the details of the microcircuitry—and not in their physical appearance.

And the cortex itself is not the entire brain. Tucked beneath the cortex are other brain organs that drive important parts of human nature. They include the hippocampus, which consolidates memory and supports mental maps, the amygdala, which colors experience with certain emotions, and the hypothalamus, which originates sexual desire and other appetites. Many neuroscientists, even when they are impressed by the plasticity of the cortex, acknowledge that subcortical structures are far less plastic.[45] This is not a minor cavil about anatomy. Some commentators have singled out evolutionary psychology as a casualty of neural plasticity, saying that the changeability of the cortex proves that the brain cannot support evolutionary specializations.[46] But most proposals in evolutionary psychology are about drives like fear, sex, love, and aggression, which reside largely in subcortical circuitry. More generally, on anyone's theory an innately shaped human ability would have to be implemented in a *network* of cortical and subcortical areas, not in a single patch of sensory cortex.

 ∽

ANOTHER BASIC POINT about the brain has been lost in the recent enthusiasm for plasticity. A discovery that neural activity is crucial for brain development does not show either that learning is crucial in shaping the brain or that genes fail to shape the brain.

The study of neural development is often framed in terms of nature and nurture, but it is more fruitful to think of it as a problem in developmental biology—how a ball of identical cells differentiates into a functioning organ. Doing so stands the conventional assumptions of associationism on their head. Primary sensory cortex, rather than being the firmest part of the brain on top of which successive stories can only be even more plastic, may be the part of the brain that is *most* dependent on the input for proper development.

In assembling a brain, a complete genetic blueprint is out of the question for two reasons. One is that a gene cannot anticipate every detail of the environment, including the environment consisting of the other genes in the genome. It has to specify an adaptive developmental program that ensures that the organism as a whole functions properly across variations in nutrition, other genes, growth rates over the lifespan, random perturbations, and the physical and social environment. And that requires feedback from the way the rest of the organism is developing.

Take the development of the body. The genes that build a femur cannot specify the exact shape of the ball on top, because the ball has to articulate with the socket in the pelvis, which is shaped by other genes, nutrition, age, and chance. So the ball and the socket adjust their shapes as they rotate against each other while the baby kicks in the womb. (We know this because experimental animals that are paralyzed while they develop end up with grossly deformed joints.) Similarly, the genes shaping the lens of the growing eye cannot know how far back the retina is going to be or vice versa. So the brain of the baby is equipped with a feedback loop that uses signals about the sharpness of the image on the retina to slow down or speed up the physical growth of the eyeball. These are good examples of "plasticity," but the metaphor of plastic material is misleading. The mechanisms are not designed to allow variable environments to shape variable organs. They do the opposite: they ensure that despite variable environments, a *constant* organ develops, one that is capable of doing its job.

Like the body, the brain must use feedback circuits to shape itself into a working system. This is especially true in the sensory areas, which have to cope with growing sense organs. For that reason alone we would expect the activity of the brain to play a role in its own development, even if its end state, like those of the femur and the eyeball, is in some sense genetically specified. How this happens is still largely a mystery, but we know that patterns of neural stimulation can trigger the expression of a gene and that one gene can trigger many others.[47] Since every brain cell contains a complete genetic program, the machinery exists, in principle, for neural activity to trigger the development of an innately organized neural circuitry in any of several different regions. If so, brain activity would not be sculpting the brain; it would merely be telling the genome where in the brain a certain neural circuit should go.

So even an extreme innatist need not believe that the brain differentiates itself by the equivalent of GPS coordinates in the skull, following rules like "If you are between the left temple and the left ear, become a language circuit" (or a fear circuit, or a circuit for recognizing faces). A developmental program may be triggered in a part of the developing brain by some combination of the source of the stimulation, the firing pattern, the chemical environment, and other signals. The end result may be a faculty that is seated in different parts of the brain in different people. After all, the brain is the organ of computation, and the same computation can happen in different places as long as the pattern of information flow is the same. In your computer, a file or program may sit in different parts of memory or be fragmented across different sectors of the disk and work the same way in every case. It would not be surprising if the growing brain were at least that dynamic in allocating neural resources to computational demands.

The other reason that brains can't rely on a complete genetic blueprint is that the genome is a limited resource. Genes are constantly mutating over evolutionary time, and natural selection can weed out the bad ones only slowly. Most evolutionary biologists believe that natural selection can support a genome that is only so big. That means that the genetic plans for a complex brain have to be compressed to the minimum size that is consistent with the brain's developing and working properly. Though more than half the genome is put to work primarily or exclusively in the brain, that is not nearly enough to specify the brain's connection diagram.

The development program for the brain has to be resourceful. Take the problem of getting every axon (output fiber) from the eyes to connect to the brain in an orderly way. Neighboring points in the eye must connect to neighboring points in the brain (an arrangement called topographic mapping), and corresponding locations in the two eyes should end up near each other in the brain but not get mixed up with each other.

Rather than give each axon a genetically specified address, the mammalian brain may organize the connections in a cleverer way. In her studies of brain development in cats, the neuroscientist Carla Shatz has discovered that waves of activity flow across each retina, first in one direction, then in some other direction.[48] That means that neurons that are next to each other in a single eye will tend to fire at around the same time, because they are often hit by the same wavefront. But axons from different eyes, or from distant locations in the same eye, will be uncorrelated in their activity, because a wave passing over one will miss the other. Just as you could reconstruct the seating diagram of a stadium if the fans were doing "the wave" along various directions and you knew only who stood up at which time (since people who stood up at the same time had to be seated near each other), the brain could reconstruct the spatial layout of the two eyes by listening for which sets of input neurons were firing at the same time. One of the rules of learning in neural networks, first outlined by the psychologist D. O. Hebb, is that "neurons that fire together wire together; neurons out of synch fail to link." As the waves crisscross the retina for days and weeks, the visual thalamus downstream could organize itself into layers, each from a single eye, with adjacent neurons responding to adjacent parts of the retina. The cortex, in theory, could organize its wiring in a similar way.[49]

Which parts of the brain actually use this auto-installation technique is another matter. The visual system does not appear to need the technique to grow topographically organized wiring; a rough topographic map develops under the direct control of the genes. Some neuroscientists believe that the fire-together-wire-together technique may still be used to make the maps more precise or to segregate the inputs from the two eyes.[50] That, too, has been challenged, but let us assume it is correct and see what it means.

The fire-together-wire-together process could, in theory, be set in motion by letting the eyeballs gaze at the world. The world has lines and edges that stimulate neighboring parts of the retina at the same time, and that provides the information the brain needs to set up or fine-tune an orderly map. But in the case of Shatz's cats, it works *without any environmental input at all*. The visual system develops in the pitch-dark womb, before the animal's eyes are open and before its rods and cones are even hooked up and functioning. The retinal waves are generated endogenously by the tissues of the retina during the period in which the visual brain has to wire itself up. In other words, the eye generates a *test pattern,* and the brain uses it to complete its own assembly. Ordinarily, axons from the eye carry information about things in the world, but the developmental program co-opted those axons to carry information about which neurons come from the same eye or the same place in the eye. A rough analogy occurred to me when I watched the cable TV installer figure out which cable in the basement led to a particular room upstairs. He attached a tone generator called a "screamer" to the end in the bedroom and then ran downstairs to listen for the signal on each cable in the bouquet coming out of the wall. Though the cables were designed to carry a television signal upstairs, not a test tone downstairs, they lent themselves to this other use during the installation process because an information conduit is useful for both purposes.

The moral is that a discovery that brain development depends on brain activity may say nothing about learning or experience, only that the brain takes advantage of its own information-transmission abilities while wiring itself up.

Fire-together-wire-together is a trick that solves a particular kind of wiring problem: connecting a surface of receptors to a maplike representation in the cortex. The problem is found not just in the visual system but in other spatial senses such as touch. That is because the problem of tiling a patch of primary visual cortex, which receives information from the 2-D surface of the retina, is similar to the problem of tiling a patch of primary somatosensory cortex, which receives information from the 2-D surface of the skin. Even the auditory system may use the trick, because the inputs representing different sound frequencies (roughly, pitches) originate in a 1-D membrane in the inner ear, and the brain treats pitch in audition the way it treats space in vision and touch.

But the trick may be useless elsewhere in the brain. The olfactory (smell) system, for example, wires itself by a completely different technique. Unlike sights, sounds, and touches, which are arranged by location when they arrive at the sensory cortex, smells arrive all mixed together, and they are analyzed in terms of the chemical compounds making them up, each detected by a different receptor in the nose. Each receptor connects to a neuron that carries its signal into the brain, and in this case the genome really does use a different gene for each axon when wiring them into their respective places in the brain, a thousand genes in all. It economizes on genes in a remarkable way. The protein produced by each gene is used twice: once in the nose, as a receptor to detect an airborne chemical, and a second time in the brain, as a probe at the end of the corresponding axon to direct it to its proper spot in the olfactory bulb.[51]

The wiring problems are different again for other parts of the brain, such as the medulla, which generates the swallowing reflex and other fixed action patterns; the amygdala, which handles fear and other emotions; and the ventromedial frontal cortex, which is involved in social reasoning. The fire-together-wire-together technique may be an ideal method for sensory maps and other structures that simply have to reproduce redundancies in the world or in other parts of the brain, such as primary sensory cortex for seeing, touching, and hearing. But other regions evolved with different functions, such as smelling or swallowing or avoiding danger or winning friends, and they have to be wired by more complicated techniques. This is simply a corollary of the general point with which I began the chapter: the environment cannot tell the various parts of an organism what their goals are.

The doctrine of extreme plasticity has used the plasticity discovered in primary sensory cortex as a *metaphor* for what happens elsewhere in the brain. The upshot of these two sections is that it is not a very good metaphor. If the plasticity of sensory cortex symbolized the plasticity of mental life as a whole, it should be easy to change what we don't like about ourselves or other people.

Take a case very different from vision, sexual orientation. Most gay men feel stirrings of attraction to other males around the time of the first hormonal changes that presage puberty. No one knows why some boys become gay—genes, prenatal hormones, other biological causes, and chance may all play a role—but my point is not so much about becoming gay as about becoming straight. In the less tolerant past, unhappy gay men sometimes approached psychiatrists (and sometimes were coerced into approaching them) for help in changing their sexual orientation. Even today, some religious groups pressure their gay members to "choose" heterosexuality. Many techniques have been foisted on them: psychoanalysis, guilt mongering, and conditioning techniques that use impeccable fire-together-wire-together logic (for example, having them look at *Playboy* centerfolds while sexually aroused). The techniques are all failures.[52] With a few dubious exceptions (which are probably instances of conscious self-control rather than a change in desire), the sexual orientation of most gay men cannot be reversed by experience. Some parts of the mind just aren't plastic, and no discoveries about how sensory cortex gets wired will change that fact.

~

WHAT IS THE brain actually doing when it undergoes the changes we call plasticity? One commentator called it "the brain equivalent of Christ turning water into wine" and thus a disproof of any theory that parts of the brain have been specialized for their jobs by evolution.[53] Those who don't believe in miracles are skeptical. Neural tissue is not a magical substance that can assume any form demanded of it but a mechanism that obeys the laws of cause and effect. When we take a closer look at the prominent examples of plasticity, we discover that the changes are not miracles after all. In every case, the altered cortex is not doing anything very different from what it ordinarily does.

Most demonstrations of plasticity involve remappings within primary sensory cortex. A brain area for an amputated or immobilized finger may be taken over by an adjacent finger, or a brain area for a stimulated finger expands its borders at the expense of a neighbor. The brain's ability to reweight its inputs is indeed remarkable, but the kind of information processing done by the taken-over cortex has not fundamentally changed: the cortex is still processing information about the surface of the skin and the angles of the joints. And the representation of a digit or part of the visual field cannot grow indefinitely, no matter how much it is stimulated; the intrinsic wiring of the brain would prevent it.[54]

What about the takeover of the visual cortex by Braille in blind people? At first glance it looks like real transubstantiation. But maybe not. We are not witnessing just any talent taking over just any vacant lot in the cortex. Braille reading may use the anatomy of the visual cortex in the same way that seeing does.

Neuroanatomists have long known that there are as many fibers bringing information down into the visual cortex from other brain areas as there are bringing information up from the eyes.[55] These top-down connections could have several uses. They may aim a spotlight of attention on portions of the visual field, or coordinate vision with the other senses, or group pixels into regions, or implement mental imagery, the ability to visualize things in the mind's eye.[56] Blind people may simply be using these prewired top-down connections to read Braille. They may be "imagining" the rows of dots as they feel them, much as a blindfolded person can imagine objects placed in his hand, though of course far more rapidly. (Previous research has established that blind people have mental images—perhaps even visual images—containing spatial information.)[57] The visual cortex is well suited to the kind of computation needed for Braille. In sighted people the eyes scan around a scene, bringing fine detail into the fovea, the high-resolution center of the retina. This is similar to moving the hands over a line of Braille, bringing fine detail under the high-resolution skin of the fingertips. So the visual system may be functioning in blind people much as it does in sighted ones, despite the lack of input from the eyes. Years of practice at imagining the tactile world and attending to the details of Braille have led the visual cortex to make maximal use of the innate inputs from other parts of the brain.

With deafness, too, one of the senses is taking over the controls of suitable circuitry, rather than just moving into any old unoccupied territory. Laura Petitto and her colleagues found that deaf people use the superior gyrus of the temporal lobe (a region near the primary auditory cortex) to recognize the elements of signs in sign languages, just as hearing people use it to process speech sounds in spoken languages. They also found that the deaf use the lateral prefrontal cortex to retrieve signs from memory, just as hearing people use it to retrieve words from memory.[58] This should come as no surprise. As linguists have long known, sign languages are organized much like spoken languages. They use words, a grammar, and even phonological rules that combine meaningless gestures into meaningful signs, just as phonological rules in spoken languages combine meaningless sounds into meaningful words.[59] Spoken languages, moreover, are partly modular: the representations for words and rules can be distinguished from the input-output systems that connect them to the ears and the mouth. The simplest interpretation, endorsed by Petitto and her colleagues, is that the cortical areas recruited in signers are specialized for *language* (words and rules), not for speech per se. What the areas are doing in deaf people is the same as what they are doing in hearing people.

Let me turn to the most amazing plasticity of all: the rewired ferrets whose eyes fed their auditory thalamus and cortex and made those areas work like a visual thalamus and cortex. Even here, water is not being turned into wine. Sur and his colleagues noted the redirected input did not change the actual wiring of the auditory brain, only the pattern of synaptic strengths. As a result they found many differences between the co-opted auditory brain and a normal visual brain.[60] The representation of the visual field in the auditory brain was fuzzier and more disorganized, because the tissue is optimized for auditory, not visual, analysis. The map of the visual field, for instance, was far more precise in the left-right direction than in the up-down direction. That is because the left-right direction was mapped onto an axis of the auditory cortex that in normal animals represents different sound frequencies and thus gets inputs from the inner ear that are precisely arranged in order of frequency. But the up-down direction was mapped onto the perpendicular axis of the auditory cortex, which ordinarily gets a mass of inputs of the same frequency. Sur also notes that the connections between the primary auditory cortex and other brain areas for hearing (the equivalent of the wiring diagram for the visual system on page 88) were unchanged by the new input.

So patterns in the input can tune a patch of sensory cortex to mesh with that input, but only within the limits of the wiring already present. Sur suggests that the reason the auditory cortex in the rewired ferrets can process visual information at all is that certain kinds of signal processing may be useful to perform on raw sensory input, whether it is visual, auditory, or tactile:

> On this view, one function of sensory thalamus or cortex is to perform certain stereotypical operations on input regardless of modality [vision, hearing, or touch]; the specific type of sensory input of course provides the substrate information that is transmitted and processed. . . . If the normal organization of central auditory structures is not altered, or at least not altered significantly, by visual input, then we might expect some operations similar to those we observe on visual inputs in operated ferrets to be carried out as well in the auditory pathway in normal ferrets. In other words, the animals with visual inputs induced into the auditory pathway provide a different window on some of the same operations that should occur normally in auditory thalamus and cortex.[61]

The suggestion that the auditory cortex is inherently suited to analyze visual input is not far-fetched. I mentioned that frequency (pitch) in hearing behaves a lot like space in vision. The mind treats soundmakers with different pitches as if they were objects at different locations, and it treats jumps in pitch like motions in space.[62] This means that some of the analyses performed on sights may be the same as the analyses performed on sounds, and could be computed, at least in part, by similar kinds of circuitry. Inputs from an ear represent different frequencies; inputs from an eye represent spots at different locations. Neurons in the sensory cortex (both visual and auditory) receive information from a neighborhood of input fibers and extract simple patterns from them. Therefore neurons in the auditory cortex that ordinarily detect rising or falling glides, rich or pure tones, and sounds that come from specific places may, in the rewired ferrets, automatically be capable of detecting lines of specific slants, places, and directions of movement.

This is not to say that the primary auditory cortex can handle visual input right out of the box. The cortex still must tune its synaptic connections in response to the patterns in the input. The rewired ferrets are a remarkable demonstration of how the developing sensory cortex organizes itself into a well-functioning system. But as in the other examples of plasticity, they do not show that input from the senses can transform an amorphous brain into doing whatever would come in handy. The cortex has an intrinsic structure that allows it to perform certain kinds of computation. Many examples of "plasticity" may consist of making the input mesh with that structure.

~

ANYONE WHO HAS watched the Discovery Channel has seen footage of baby wildebeests or zebras falling out of the birth canal, wobbling on shaky legs for a minute or two, and then prancing around their mothers with their senses, drives, and motor control fully operational. It happens far too quickly for patterned experience to have organized their brains, so there must be genetic mechanisms capable of shaping the brain before birth. Neuroscientists were aware of this before plasticity came into vogue. The first studies of the development of the visual system by David Hubel and Torsten Wiesel showed that the microcircuitry of monkeys is pretty much complete at birth.[63] Even their famous demonstrations that the visual systems of cats can be altered by experience during a critical period of development (by being reared in the dark, in striped cylinders, or with one eye sewn shut) show only that experience is necessary to *maintain* the visual system and to retune it as the animal grows. They do not show that experience is necessary to wire up the brain to start with.

We know in a general way how the brain assembles itself under the guidance of the genes.[64] Even before the cortex has been formed, the neurons destined to make up different areas are organized into a "proto-map." Each area in the proto-map is composed of neurons with different properties, molecular mechanisms that attract different input fibers, and different patterns of responses to the input. Axons are attracted and repelled by many kinds of molecules dissolved in the surrounding fluid or attached to the membranes of neighboring cells. And different sets of genes are expressed in different parts of the growing cortex. The neuroscientist Lawrence Katz has lamented that fire-together-wire-together has become a "dogma" keeping neuroscientists from exploring the full reach of these genetic mechanisms.[65]

But the tide is beginning to turn, and recent discoveries are showing how parts of the brain can organize themselves without any information from the senses. In experiments that the journal *Science* called "heretical," Katz's team removed one or both eyes from a developing ferret, depriving the visual cortex of all its input. Nonetheless, the visual cortex developed with the standard arrangement of connections from the two eyes.[66]

Genetically engineered mice have provided especially important clues, because knocking out a single gene can be more precise than the conventional techniques of poisoning neurons or slicing up the brain. One team invented a mouse whose synapses were completely shut down, preventing neurons from signaling to one another. Its brain developed fairly normally, complete with layered structures, fiber pathways, and synapses in the right places.[67] (The brain degenerated quickly after birth, showing again that neural activity may be more important in maintaining the brain than in wiring it.) Another team designed a mouse with a useless thalamus, depriving the entire cortex of its input. But the cortex differentiated into the normal layers and regions, each with a different set of turned-on genes.[68] A third study did the opposite, inventing mice that were missing one of the genes that lay down gradients of molecules that help organize the brain by triggering other genes in particular places. The missing gene made a big difference: the boundaries among cortical areas were badly warped.[69] The studies with knockout mice, then, suggest that genes may be more important than neural activity in organizing the cortex. Neural activity undoubtedly plays a role, which depends on the species, the stage of development, and the part of the brain, but it is just one capability of the brain rather than the source of its structure.

What about our own species? Recall that a recent study of twins showed that differences in the anatomy of the cortex, particularly the amount of gray matter in different cortical regions, are under genetic control, paralleling differences in intelligence and other psychological traits.[70] And demonstrations of the plasticity of the human brain do not rule out substantial genetic organization. One of the most commonly cited examples of plasticity in both humans and monkeys is that the cortex dedicated to an amputated or numbed body part may get reallocated to some other body part. But the fact that the input can change the brain once it is built does not mean that the input molded the brain in the first place. Most amputees experience phantom limbs: vivid, detailed hallucinations of the missing body part. Amazingly, a substantial proportion of people who were *born* with a limb missing experience these apparitions as well.[71] They can describe the anatomy of their phantom limb (for example, how many toes they feel in a nonexistent foot) and may even feel that they are gesturing with their phantom hands during conversation. One girl solved arithmetic problems by counting on her phantom fingers! The psychologist Ronald Melzack, who documented many of these cases, proposed that the brain contains an innate "neuromatrix," distributed across several cortical and subcortical regions, dedicated to representing the body.

The impression that human brains are limitlessly plastic has also come from demonstrations that children can sometimes recover from early brain damage. But the existence of cerebral palsy—lifelong difficulties with motor control and speech caused by malformations or early damage in the brain—shows that even the plasticity of a child's brain has severe limits. The most famous evidence for extreme plasticity in humans had been the ability of some children to grow up relatively normal even with an entire hemisphere surgically removed in infancy.[72] But that may be a special case, which arises from the fact that the primate brain is fundamentally a symmetrical organ. The typically human asymmetries—language more on the left, spatial attention and some emotions more on the right—are superimposed on that mostly symmetrical design. It would not be surprising if the hemispheres were genetically programmed with pretty much the same abilities, together with small biases that lead each hemisphere to specialize in some talents while letting others wither. With one hemisphere gone, the remaining one has to put all its capabilities to full use.

What happens when a child loses a part of the cortex in *both* hemispheres, so neither hemisphere can take over the job of the missing part in the other? If cortical regions are interchangeable, plastic, and organized by the input, then an intact part of the brain should take over the function of the missing parts. The child may be a bit slower because he is working with less brain tissue, but he should develop a full complement of human faculties. But that is not what seems to happen. Several decades ago, neurologists studied a boy who suffered a temporary loss of oxygen to the brain and lost both the standard language areas in the left hemisphere and their mirror images on the right. Though he was just ten days old when he sustained the damage, he grew into a child with permanent difficulties in speaking and understanding.[73]

That case study, like many in pediatric neurology, is not scientifically pure, but recent studies on two other mental faculties echo the point that babies' brains may be less plastic than many people think. The psychologist Martha Farah and her collaborators recently reported the case of a sixteen-year-old boy who contracted meningitis when he was one day old and suffered damage to the visual cortex and to the bottom of the temporal lobes on both sides of his brain.[74] When adults sustain such damage, they lose the ability to recognize faces and also have some trouble recognizing animals, though they often can recognize words, tools, furniture, and other shapes. The boy had exactly this syndrome. Though he grew up with normal verbal intelligence, he was utterly incapable of recognizing faces. He could not even recognize pictures of the cast of his favorite television show, *Baywatch,* which he had seen for an hour a day for the preceding year and a half. Without the appropriate strips of brain, sixteen years of seeing faces and plenty of available cortex were not enough to give him the basic human ability to recognize other people by sight.

The neuroscientists Steven Anderson, Hannah and Antonio Damasio, and their colleagues recently tested two young adults who had sustained damage to their ventromedial and orbital prefrontal cortex when they were young children.[75] These are the parts of the brain that sit above the eyes and are important for empathy, social skills, and self-management (as we know from Phineas Gage, the railroad worker whose brain was impaled by a tamping iron). Both children recovered from their injuries and grew up with average IQs in stable homes with normal siblings and college-educated parents. If the brain were really homogeneous and plastic, the healthy parts should have been shaped by the normal social environment and taken over the functions of the damaged parts. But that is not what happened with either of the children. One, who had been run over by a car when she was fifteen months old, grew into an intractable child who ignored punishment and lied compulsively. As a teenager she shoplifted, stole from her parents, failed to win friends, showed no empathy or remorse, and was dangerously uninterested in her own baby. The other patient was a young man who had lost similar parts of his brain to a tumor when he was three months old. He too grew up friendless, shiftless, thieving, and hotheaded. Along with their bad behavior, both had trouble thinking through simple moral problems, despite having IQs in the normal range. They could not, for example, say what two people should do if they disagreed on which TV channel to watch, or decide whether a man ought to steal a drug to save his dying wife.

These cases do more than refute the doctrine of extreme plasticity. They set a challenge for the genetics and neuroscience of the twenty-first century. How does the genome tell a developing brain to differentiate into neural networks that are prepared for such abstract computational problems as recognizing a face or thinking about the interests of other people?

THE BLANK SLATE has made its last stand, but, as we have seen, its latest scientific fortifications are illusory. The human genome may have a smaller number of genes than biologists had previously estimated, but that only shows that the number of genes in a genome has little to do with the complexity of the organism. Connectionist networks may explain some of the building blocks of cognition, but they are too underpowered to account for thought and language on their own; they must be innately engineered and assembled for the tasks. Neural plasticity is not a magical protean power of the brain but a set of tools that help turn megabytes of genome into terabytes of brain, that make sensory cortex dovetail with its input, and that implement the process called learning.

Therefore genomics, neural networks, and neural plasticity fit into the picture that has emerged in recent decades of a complex human nature. It is not, of course, a nature that is rigidly programmed, impervious to the input, free of culture, or endowed with the minutiae of every concept and feeling. But it is a nature that is rich enough to take on the demands of seeing, moving, planning, talking, staying alive, making sense of the environment, and negotiating the world of other people.

The aftermath of the Blank Slate's last stand is a good time to take stock of the case for the alternative. Here is my summary of the evidence for a complex human nature, some of it reiterating arguments from previous chapters, some of it anticipating arguments in chapters to come.

Simple logic says there can be no learning without innate mechanisms to do the learning. Those mechanisms must be powerful enough to account for all the kinds of learning that humans accomplish. Learnability theory—the mathematical analysis of how learning can work in principle—tells us there are always an infinite number of generalizations that a learner can draw from a finite set of inputs.[76] The sentences heard by a child, for example, can be grounds for repeating them back verbatim, producing any combination of words with the same proportion of nouns to verbs, or analyzing the underlying grammar and producing sentences that conform to it. The sight of someone washing dishes can, with equal logical justification, prompt a learner to try to get dishes clean or to let warm water run over his fingers. A successful learner, then, must be constrained to draw some conclusions from the input and not others. Artificial intelligence reinforces this point. Computers and robots programmed to do humanlike feats are invariably endowed with many complex modules.[77]

Evolutionary biology has shown that complex adaptations are ubiquitous in the living world, and that natural selection is capable of evolving them, including complex cognitive and behavioral adaptations.[78] The study of the behavior of animals in their natural habitat shows that species differ innately from one another in their drives and abilities, some of them (like celestial navigation and food caching) requiring complicated and specialized neural systems.[79] The study of humans from an evolutionary perspective has shown that many psychological faculties (such as our hunger for fatty food, for social status, and for risky sexual liaisons) are better adapted to the evolutionary demands of our ancestral environment than to the actual demands of the current environment.[80] Anthropological surveys have shown that hundreds of universals, pertaining to every aspect of experience, cut across the world's cultures.[81]

Cognitive scientists have discovered that distinct kinds of representations and processes are used in different domains of knowledge, such as words and rules for language, the concept of an enduring object for understanding the physical world, and a theory of mind for understanding other people.[82] Developmental psychology has shown that these distinct modes of interpreting experience come on line early in life: infants have a basic grasp of objects, numbers, faces, tools, language, and other domains of human cognition.[83]

The human genome contains an enormous amount of information, both in the genes and in the noncoding regions, to guide the construction of a complex organism. In a growing number of cases, particular genes can be tied to aspects of cognition, language, and personality.[84] When psychological traits vary, much of the variation comes from differences in genes: identical twins are more similar than fraternal twins, and biological siblings are more similar than adoptive siblings, whether reared together or apart.[85] A person's temperament and personality emerge early in life and remain fairly constant throughout the lifespan.[86] And both personality and intelligence show few or no effects of children's particular home environments within their culture: children reared in the same family are similar mainly because of their shared genes.[87]

Finally, neuroscience is showing that the brain's basic architecture develops under genetic control. The importance of learning and plasticity notwithstanding, brain systems show signs of innate specialization and cannot arbitrarily substitute for one another.[88]

In these three chapters I have given you a summary of the current scientific case for a complex human nature. The rest of the book is about its implications.

PART II

FEAR AND LOATHING

B y the middle of the second half of the twentieth century, the ideals of
the social scientists of the first half had enjoyed a well-deserved victory.
Eugenics, Social Darwinism, colonial conquest, Dickensian policies toward children, overt expressions of racism and sexism among the educated,
and official discrimination against women and minorities had been eradicated, or at least were rapidly fading, from mainstream Western life.

At the same time, the doctrine of the Blank Slate, which had been blurred
with ideals of equality and progress for much of the century, was beginning to
show cracks. As the new sciences of human nature began to flourish, it was becoming clear that thinking is a physical process, that people are not psychological clones, that the sexes differ above the neck as well as below it, that the
human brain was not exempt from the process of evolution, and that people in
all cultures share mental traits that might be illuminated by new ideas in evolutionary biology.

These developments presented intellectuals with a choice. Cooler heads
could have explained that the discoveries were irrelevant to the political ideals
of equal opportunity and equal rights, which are moral doctrines on how we
ought to treat people rather than scientific hypotheses about what people are
like. Certainly it is wrong to enslave, oppress, discriminate against, or kill people regardless of any foreseeable datum or theory that a sane scientist would
offer.

But it was not a time for cool heads. Rather than detach the moral doctrines from the scientific ones, which would ensure that the clock would not be
turned back no matter what came out of the lab and field, many intellectuals,
including some of the world's most famous scientists, made every effort to
connect the two. The discoveries about human nature were greeted with fear
and loathing because they were thought to threaten progressive ideals. All this
could be relegated to the history books were it not for the fact that these intellectuals, who once called themselves radicals, are now the establishment, and

the dread they sowed about human nature has taken root in modern intellectual life.

This part of the book is about the politically motivated reactions to the new sciences of human nature. Though the opposition was originally a brainchild of the left, it is becoming common on the right, whose spokespeople are fired up by some of the same moral objections. In Chapter 6 I recount the shenanigans that erupted as a reaction to the new ideas about human nature. In Chapter 7 I show how these reactions came from a moral imperative to uphold the Blank Slate, the Noble Savage, and the Ghost in the Machine.

Chapter 6

Political Scientists

THE FIRST LECTURE I attended as a graduate student at Harvard in 1976 was by the famous computer scientist Joseph Weizenbaum. He was an early contributor to artificial intelligence (AI) and is best remembered for the program Eliza, which fooled people into thinking that the computer was conversing though it was just spouting canned repartee. Weizenbaum had just published *Computer Power and Human Reason*, a critique of artificial intelligence and computer models of cognition, praised as "the most important computer book of the past decade." I had misgivings about the book, which was short on argument and long on sanctimony. (For example, he wrote that certain ideas in artificial intelligence, such as a science-fiction proposal for a hybrid of nervous systems and computers, were "simply obscene. These are [applications] whose very contemplation ought to give rise to feelings of disgust in every civilized person. . . . One must wonder what must have happened to the proposers' perception of life, hence to their perceptions of themselves as part of the continuum of life, that they can even think of such a thing.")[1] Still, nothing could have prepared me for the performance in store at the Science Center that afternoon.

Weizenbaum discussed an AI program by the computer scientists Alan Newell and Herbert Simon that relied on analogy: if it knew the solution to one problem, it applied the solution to other problems with a similar logical structure. This, Weizenbaum told us, was really designed to help the Pentagon come up with counterinsurgency strategies in Vietnam. The Vietcong had been said to "move in the jungle as fish move in water." If the program were fed this information, he said, it could deduce that just as you can drain a pond to expose the fish, you can denude the jungle to expose the Vietcong. Turning to research on speech recognition by computer, he said that the only conceivable reason to study speech perception was to allow the CIA to monitor millions of telephone conversations simultaneously, and he urged the students in the audience to boycott the topic. But, he added, it didn't really matter if we ignored his advice because he was completely certain—there was not the slightest doubt in his mind—that by the year 2000 we would all be dead. And with that inspiring charge to the younger generation he ended the talk.

The rumors of our death turned out to be greatly exaggerated, and the other prophecies of the afternoon fared no better. The use of analogy in reasoning, far from being the work of the devil, is today a major research topic in cognitive science and is widely considered a key to what makes us smart. Speech-recognition software is routinely used in telephone information services and comes packaged with home computers, where it has been a godsend for the disabled and for people with repetitive strain injuries. And Weizenbaum's accusations stand as a reminder of the political paranoia and moral exhibitionism that characterized university life in the 1970s, the era in which the current opposition to the sciences of human nature took shape.

It was not how I imagined that scholarly discourse would be conducted in the Athens of America, but perhaps I should not have been surprised. Throughout history, battles of opinion have been waged by noisy moralizing, demonizing, hyperbole, and worse. Science was supposed to be a beachhead in which ideas rather than people are attacked and in which verifiable facts are separated from political opinions. But when science began to edge toward the topic of human nature, onlookers reacted differently from how they would to discoveries about, say, the origin of comets or the classification of lizards, and scientists reverted to the moralistic mindset that comes so naturally to our species.

Research on human nature would be controversial in any era, but the new sciences picked a particularly bad decade in which to attract the spotlight. In the 1970s many intellectuals had become political radicals. Marxism was correct, liberalism was for wimps, and Marx had pronounced that "the ruling ideas of each age have ever been the ideas of its ruling class." The traditional misgivings about human nature were folded into a hard-left ideology, and scientists who examined the human mind in a biological context were now considered tools of a reactionary establishment. The critics announced they were part of a "radical science movement," giving us a convenient label for the group.[2]

Weizenbaum was repelled by the attempt within artificial intelligence and cognitive science to unify mind and mechanism, but the other sciences of human nature evoked acrimony as well. In 1971 the psychologist Richard Herrnstein published an article called "IQ" in the *Atlantic Monthly*.[3] Herrnstein's argument, he was the first to point out, should have been banal. He wrote that as social status becomes less strongly determined by arbitrary legacies such as race, parentage, and inherited wealth, it will become more strongly determined by talent, especially (in a modern economy) intelligence. Since differences in intelligence are partly inherited, and since intelligent people tend to marry other intelligent people, when a society becomes more just it will also become more stratified along genetic lines. Smarter people will tend to float into the higher strata, and their children will tend to stay there. The basic argument should be banal because it is based on a mathematical necessity: as the proportion of variance in social status caused by nongenetic factors goes down, the proportion caused by genetic factors has to go up. It could be completely false only if there were no variation in social status based on intellectual talent (which would require that people not preferentially hire and trade with the talented) or if there were no genetic variation in intelligence (which would require that people be either blank slates or clones).

Herrnstein's argument does not imply that any differences in average intelligence between races are innate (a distinct hypothesis that had been broached by the psychologist Arthur Jensen two years earlier),[4] and he explicitly denied that he was making such a claim. School desegregation was less than a generation old, civil rights legislation less than a decade, so the differences that had been documented in average IQ scores of blacks and whites could easily be explained by differences in opportunity. Indeed, to say that Herrnstein's syllogism implied that black people would end up at the bottom of a genetically stratified society was to add the gratuitous assumption that blacks were on average genetically less intelligent, which Herrnstein took pains to avoid.

Nonetheless, the influential psychiatrist Alvin Poussaint wrote that Herrnstein "has become the enemy of black people and his pronouncements are a threat to the survival of every black person in America." He asked rhetorically, "Shall we carry banners for Herrnstein proclaiming his right to freedom of speech?" Leaflets were handed out at Boston-area universities urging students to "Fight Harvard Prof's Fascist Lies," and Harvard Square was plastered with his photograph above the caption WANTED FOR RACISM and five misquotations purportedly from his article. Herrnstein received a death threat and found that he could no longer speak about his research specialty, learning in pigeons, because wherever he went the lecture halls were filled with chanting mobs. At Princeton, for example, students declared they would block the doors of the auditorium to force him to answer questions on the IQ controversy. Several lectures were canceled when the hosting universities said they could not guarantee his safety.[5]

The topic of innate differences among people has obvious political implications, which I will examine in later chapters. But some scholars were incensed by the seemingly warm-and-fuzzy claim that people have innate *commonalities*. In the late 1960s the psychologist Paul Ekman discovered that smiles, frowns, sneers, grimaces, and other facial expressions were displayed and understood worldwide, even among foraging peoples with no prior contact with the West. These findings, he argued, vindicated two claims that Darwin had made in his 1872 book *The Expression of the Emotions in Man and*

Animals. One was that humans had been endowed with emotional expressions by the process of evolution; the other, radical in Darwin's time, was that all races had recently diverged from a common ancestor.[6] Despite these uplifting messages, Margaret Mead called Ekman's research "outrageous," "appalling," and "a disgrace"—and these were some of the milder responses.[7] At the annual meeting of the American Anthropological Association, Alan Lomax Jr. rose from the audience shouting that Ekman should not be allowed to speak because his ideas were fascist. On another occasion an African American activist accused him of racism for claiming that black facial expressions were no different from white ones. (Sometimes you can't win.) And it was not just claims about innate faculties in the human species that drew the radicals' ire, but claims about innate faculties in any species. When the neuroscientist Torsten Wiesel published his historic work with David Hubel showing that the visual system of cats is largely complete at birth, another neuroscientist angrily called him a fascist and vowed to prove him wrong.

SOME OF THESE protests were signs of the times and faded with the decline of radical chic. But the reaction to two books on evolution continued for decades and became part of the intellectual mainstream.

The first was E. O. Wilson's *Sociobiology,* published in 1975.[8] *Sociobiology* synthesized a vast literature on animal behavior using new ideas on natural selection from George Williams, William Hamilton, John Maynard Smith, and Robert Trivers. It reviewed principles on the evolution of communication, altruism, aggression, sex, and parenting, and applied them to the major taxa of social animals such as insects, fishes, and birds. The twenty-seventh chapter did the same for *Homo sapiens,* treating our species like another branch of the animal kingdom. It included a review of the literature on universals and variation among societies, a discussion of language and its effects on culture, and the hypothesis that some universals (including the moral sense) may come from a human nature shaped by natural selection. Wilson expressed the hope that this idea might connect biology to the social sciences and philosophy, a forerunner of the argument in his later book *Consilience.*

The first attack on *Sociobiology* zeroed in on its main heresy. In a book-length critique, the anthropologist Marshall Sahlins defined "vulgar sociobiology" as the challenge to Durkheim's and Kroeber's doctrine of the superorganism: the belief that culture and society lived in a separate realm from individual people and their thoughts and feelings. "Vulgar sociobiology," Sahlins wrote, "consists in the explication of human social behavior as the expression of the needs and drives of the human organism, such propensities having been constructed in human nature by biological evolution."[9] Acknowledging fear of an incursion into his academic turf, he added, "The central intellectual problem does come down to the autonomy of culture and of the study of culture. *Sociobiology* challenges the integrity of culture as a thing-in-itself, as a distinctive and symbolic human creation."[10]

Sahlins's book was called *The Use and Abuse of Biology*. An example of the alleged abuse was the idea that Hamilton's theory of inclusive fitness could help explain the importance of family ties in human life. Hamilton had shown how a tendency to make sacrifices for relatives could have evolved. Relatives share genes, so any gene that nudges an organism to help a relative would be indirectly helping a copy of itself. The gene will proliferate if the cost incurred by the favor is less than the benefit conferred to the relative, discounted by the degree of relatedness (one-half for a full sibling or offspring, one-eighth for a first cousin, and so on). That can't be true, Sahlins wrote, because people in most cultures don't have words for fractions. This leaves them unable to figure out the coefficients of relatedness that would tell them which relatives to favor and by how much. His objection is a textbook confusion of a proximate cause with an ultimate cause. It is like saying that people can't possibly see in depth, because most cultures haven't worked out the trigonometry that underlies stereoscopic vision.

In any case, "vulgar" wasn't the half of it. Following a favorable review in the *New York Review of Books* by the distinguished biologist C. H. Waddington, the "Sociobiology Study Group" (including two of Wilson's colleagues, the paleontologist Stephen Jay Gould and the geneticist Richard Lewontin) published a widely circulated philippic called "Against 'Sociobiology.'" After lumping Wilson with proponents of eugenics, Social Darwinism, and Jensen's hypothesis of innate racial differences in intelligence, the signatories wrote:

> The reason for the survival of these recurrent determinist theories is that they consistently tend to provide a genetic justification of the *status quo* and of existing privileges for certain groups according to class, race, or sex. . . . These theories provided an important basis for the enactment of sterilization laws and restrictive immigration laws by the United States between 1910 and 1930 and also for the eugenics policies which led to the establishment of gas chambers in Nazi Germany.
>
> . . . What Wilson's book illustrates to us is the enormous difficulty in separating out not only the effects of environment (e.g., cultural transmission) but also the personal and social class prejudices of the researcher. Wilson joins the long parade of biological determinists whose work has served to buttress the institutions of their society by exonerating them from responsibility for social problems.[11]

They also accused Wilson of discussing "the salutary advantages of genocide" and of making "institutions such as slavery . . . seem natural in human societies because of their 'universal' existence in the biological kingdom." In case the connection wasn't clear enough, one of the signatories wrote elsewhere that "in the last analysis it was sociobiological scholarship . . . that provided the conceptual framework by which eugenic theory was transformed into genocidal practice" in Nazi Germany.[12]

One can certainly find things to criticize in the final chapter of *Sociobiology*. We now know that some of Wilson's universals are inaccurate or too coarsely stated, and his claim that moral reasoning will someday be superseded by evolutionary biology is surely wrong. But the criticisms in "Against 'Sociobiology'" were demonstrably false. Wilson was called a "determinist," someone who believes that human societies conform to a rigid genetic formula. But this is what he had written:

> The first and most easily verifiable diagnostic trait [about human societies] is statistical in nature. The parameters of social organization . . . vary far more among human populations than among those of any other primate species. . . . Why are human societies this flexible?[13]

Similarly, Wilson was accused of believing that people are locked into castes determined by their race, class, sex, and individual genome. But in fact he had written that "there is little evidence of any hereditary solidification of status"[14] and that "human populations are not very different from one another genetically."[15] Moreover:

> Human societies have effloresced to levels of extreme complexity because their members have the intelligence and flexibility to play roles of virtually any degree of specification, and to switch them as the occasion demands. Modern man is an actor of many parts who may well be stretched to his limit by the constantly shifting demands of the environment.[16]

As for the inevitability of aggression—another dangerous idea he was accused of holding—what Wilson had written was that in the course of human evolution "aggressiveness was constrained and the old forms of primate dominance replaced by complex social skills."[17] The accusation that Wilson (a lifelong liberal Democrat) was led by personal prejudice to defend racism, sexism, inequality, slavery, and genocide was especially unfair—and irresponsible, because Wilson became a target of vilification and harassment by people who read the manifesto but not the book.[18]

At Harvard there were leaflets and teach-ins, a protester with a bullhorn calling for Wilson's dismissal, and invasions of his classroom by slogan-shouting students. When he spoke at other universities, posters called him the "Right-Wing Prophet of Patriarchy" and urged people to bring noisemakers to his lectures.[19] Wilson was about to speak at a 1978 meeting of the American Association for the Advancement of Science when a group of people carrying placards (one with a swastika) rushed onto the stage chanting, "Racist Wilson, you can't hide, we charge you with genocide." One protester grabbed the microphone and harangued the audience while another doused Wilson with a pitcher of water.

As the notoriety of *Sociobiology* grew in the ensuing years, Hamilton and Trivers, who had thought up many of the ideas, also became targets of picketers, as did the anthropologists Irven DeVore and Lionel Tiger when they tried to teach the ideas. The insinuation that Trivers was a tool of racism and right-wing oppression was particularly galling because Trivers was himself a political radical, a supporter of the Black Panthers, and a scholarly collaborator of Huey Newton's.[20] Trivers had argued that sociobiology is, if anything, a force for political progress. It is rooted in the insight that organisms did not evolve to benefit their family, group, or species, because the individuals making up those groups have genetic conflicts of interest with one another and would be selected to defend those interests. This immediately subverts the comfortable belief that those in power rule for the good of all, and it throws a spotlight on hidden actors in the social world, such as females and the younger generation. Also, by finding an evolutionary basis for altruism, sociobiology shows that a sense of justice has a deep foundation in people's minds and need not run against our organic nature. And by showing that self-deception is likely to evolve (because the best liar is the one who believes his own lies), sociobiology encourages self-scrutiny and helps undermine hypocrisy and corruption.[21] (I will return to the political beliefs of Trivers and other "Darwinian leftists" in the chapter on politics.)

Trivers later wrote of the attacks on sociobiology, "Although some of the attackers were prominent biologists, the attack seemed intellectually feeble and lazy. Gross errors in logic were permitted as long as they appeared to give some tactical advantage in the political struggle. . . . Because we were hirelings of the dominant interests, said these fellow hirelings of the same interests, we were their mouthpieces, employed to deepen the [deceptions] with which the ruling elite retained their unjust advantage. Although it follows from evolutionary reasoning that individuals will tend to argue in ways that are ultimately (sometimes unconsciously) self-serving, it seemed a priori unlikely that evil should reside so completely in one set of hirelings and virtue in the other."[22]

The "prominent biologists" that Trivers had in mind were Gould and Lewontin, and together with the British neuroscientist Steven Rose they became the intellectual vanguard of the radical science movement. For twenty-five years they have indefatigably fought a rearguard battle against behavioral genetics, sociobiology (and later evolutionary psychology), and the neuroscience of politically sensitive topics such as sex differences and mental illness.[23] Other than Wilson, the major target of their attacks has been Richard Dawkins. In his 1976 book *The Selfish Gene,* Dawkins covered many of the same ideas as Wilson but concentrated on the logic of the new evolutionary theories rather than the zoological details. He said almost nothing about humans.

The radical scientists' case against Wilson and Dawkins can be summed up in two words: "determinism" and "reductionism."[24] Their writings are peppered with these words, used not in any technical sense but as vague terms of abuse. For example, here are two representative passages in a book by Lewontin, Rose, and the psychologist Leon Kamin with the defiantly Blank Slate title *Not in Our Genes:*

> Sociobiology is a reductionist, biological determinist explanation of human existence. Its adherents claim . . . that the details of present and past social arrangements are the inevitable manifestations of the specific action of genes.[25]

> [Reductionists] argue that the properties of a human society are . . . no more than the sums of the individual behaviors and tendencies of the individual humans of which that society is composed. Societies are "aggressive" because the individuals who compose them are "aggressive," for instance.[26]

The quotations from Wilson we saw earlier in the chapter show that he never expressed anything close to these ridiculous beliefs, and neither, of course, did Dawkins. For example, after discussing the tendency in mammals for males to seek a greater number of sexual partners than females do, Dawkins devoted a paragraph to human societies in which he wrote:

> What this astonishing variety suggests is that man's way of life is largely determined by culture rather than by genes. However, it is still possible that human males in general have a tendency towards promiscuity, and females a tendency to monogamy, as we would predict on evolutionary grounds. Which of these tendencies wins in particular societies depends on details of cultural circumstance, just as in different animal species it depends on ecological details.[27]

What exactly do "determinism" and "reductionism" mean? In the precise sense in which mathematicians use the word, a "deterministic" system is one whose states are caused by prior states with absolute certainty, rather than probabilistically. Neither Dawkins nor any other sane biologist would ever dream of proposing that human behavior is deterministic, as if people *must* commit acts of promiscuity, aggression, or selfishness at every opportunity. Among the radical scientists and the many intellectuals they have influenced, "determinism" has taken on a meaning that is diametrically opposed to its true meaning. The word is now used to refer to any claim that people have a *tendency* to act in certain ways in certain circumstances. It is a sign of the tenacity of the Blank Slate that a probability greater than zero is equated with a probability of 100 percent. Zero innateness is the only acceptable belief, and all departures from it are treated as equivalent.

So much for genetic determinism. What about "reductionism" (a concept we examined in Chapter 4) and the claim that Dawkins is "the most reductionist of sociobiologists," one who believes that every trait has its own gene? Lewontin, Rose, and Kamin try to educate their readers on how living things really work according to their alternative to reductionism, which they call "dialectical biology":

> Think, for example, of the baking of a cake: the taste of the product is the result of a complex interaction of components—such as butter, sugar, and flour—exposed for various periods to elevated temperatures; it is not dissociable into such-or-such a percent of flour, such-or-such of butter, etc., although each and every component . . . has its contribution to make to the final product.[28]

I will let Dawkins comment:

> When put like that, this dialectical biology seems to make a lot of sense. Perhaps even *I* can be a dialectical biologist. Come to think of it, isn't there something familiar about that cake? Yes, here it is, in a 1981 publication by the most reductionist of sociobiologists:
> ". . . If we follow a particular recipe, word for word, in a cookery book, what finally emerges from the oven is a cake. We cannot now break the cake into its component crumbs and say: this crumb corresponds to the first word in the recipe; this crumb corresponds to the second word in the recipe, etc. With minor exceptions such as the cherry on top, there is no one-to-one mapping from words of recipe to 'bits' of cake. The whole recipe maps onto the whole cake."
> I am not, of course, interested in claiming priority for the cake. . . . But what I do hope is that this little coincidence may at least give Rose and Lewontin pause. Could it be that their targets are not quite the naïvely atomistic reductionists they would desperately like them to be?[29]

Indeed, the accusation of reductionism is topsy-turvy because Lewontin and Rose, in their own research, are card-carrying reductionist biologists who explain phenomena at the level of genes and molecules. Dawkins, in contrast, was trained as an ethologist and writes about the behavior of animals in their natural habitat. Wilson, for his part, is a pioneer of research in ecology and a passionate defender of the endangered field that molecular biologists dismissively refer to as "birdsy-woodsy" biology.

All else having failed, Lewontin, Rose, and Kamin finally pinned a damning quotation on Dawkins: "They [the genes] control us, body and mind."[30] That does sound pretty deterministic. But what the man wrote was, "They *created* us, body and mind," which is very different.[31] Lewontin has used the doctored quotation in five different places.[32]

Is there any charitable explanation of these "gross errors," as Trivers called them? One possibility may be Dawkins's and Wilson's use of the expression "a gene for X" in discussing the evolution of social behavior like altruism, monogamy, and aggression. Lewontin, Rose, and Gould repeatedly pounce on this language, which refers, they think, to a gene that *always* causes the behavior and that is the *only* cause of the behavior. But Dawkins made it clear that the phrase refers to a gene *that increases the probability* of a behavior compared with alternative genes at that locus. And that probability is an average computed over the other genes that have accompanied it over evolutionary time, and over the environments that the organisms possessing the gene have lived in. This nonreductionist, nondeterminist use of the phrase "a gene for X" is routine among geneticists and evolutionary biologists because it is indispensable to what they do. *Some* behavior must be affected by *some* genes, or we could never explain why lions act differently from lambs, why hens sit on their eggs rather than eat them, why stags butt heads but gerbils don't, and so on. The point of evolutionary biology is to explain how these animals ended up with those genes, as opposed to genes with different effects. Now, a given gene may not have the *same* effect in all environments, nor the same effect in all genomes, but it has to have an *average* effect. That average is what natural selection selects (all things being equal), and that is all that the "for" means in "a gene for X." It is hard to believe that Gould and Lewontin, who are evolutionary biologists, could literally have been confused by this usage, but if they were, it would explain twenty-five years of pointless attacks.

How low can one go? Ridiculing an opponent's sex life would seem to come right out of a bad satirical novel on academic life. But Lewontin, Rose, and Kamin bring up a suggestion by the sociologist Steven Goldberg that women are skilled at manipulating others' emotions, and they comment, "What a touching picture of Goldberg's vulnerability to seduction is thus revealed!"[33] Later they mention a chapter in Donald Symons's groundbreaking book *The Evolution of Human Sexuality* which shows that in all societies, sex is typically conceived of as a female service or favor. "In reading sociobiology," they comment, "one has the constant feeling of being a voyeur, peeping into the autobiographical memoirs of its proponents."[34] Rose was so pleased with this joke that he repeated it fourteen years later in his book *Lifelines: Biology Beyond Determinism*.[35]

~

ANY HOPE THAT these tactics are a thing of the past was dashed by events in the year 2000. Anthropologists have long been hostile to anyone who discusses human aggression in a biological context. In 1976 the American Anthropological Association nearly passed a motion censuring *Sociobiology* and banning two symposia on the topic, and in 1983 they did pass one decreeing that Derek Freeman's *Margaret Mead and Samoa* was "poorly written, unscientific, irresponsible, and misleading."[36] But that was mild compared with what was to come.

In September 2000, the anthropologists Terence Turner and Leslie Sponsel sent the executives of the association a letter (which quickly proliferated throughout cyberspace) warning of a scandal for anthropology that was soon to be divulged in a book by the journalist Patrick Tierney.[37] The alleged perpetrators were the geneticist James Neel, a founder of the modern science of human genetics, and the anthropologist Napoleon Chagnon, famous for his thirty-year study of the Yanomamö people of the Amazon rainforest. Turner and Sponsel wrote:

> This nightmarish story—a real anthropological heart of darkness beyond the imagining of even a Josef Conrad (though not, perhaps, a Josef Mengele)—will be seen (rightly in our view) by the public, as well as most anthropologists, as putting the whole discipline on trial. As another reader of the galleys put it, This book should shake anthropology to its very foundations. It should cause the field to understand how the corrupt and depraved protagonists could have spread their poison for so long while they were accorded great respect throughout the Western World and generations of undergraduates received their lies as the introductory substance of anthropology. This should never be allowed to happen again.

The accusations were truly shocking. Turner and Sponsel charged Neel and Chagnon with deliberately infecting the Yanomamö with measles (which is often fatal among indigenous peoples) and then withholding medical care in order to test Neel's "eugenically slanted genetic theories." According to Turner and Sponsel's rendition of these theories, polygynous headmen in foraging societies were biologically fitter than coddled Westerners because they possessed "dominant genes" for "innate ability" that were selected when the headmen engaged in violent competition for wives. Neel believed, said Turner and Sponsel, that "democracy, with its free breeding for the masses and its sentimental supports for the weak," is a mistake. They reasoned, "The political implication of this fascistic eugenics is clearly that society should be reorganized into small breeding isolates in which genetically superior males could emerge into dominance, eliminating or subordinating the male losers in the competition for leadership and women, and amassing harems of brood females."

The accusations against Chagnon were just as lurid. In his books and papers on the Yanomamö, Chagnon had documented their frequent warfare and raiding, and had presented data suggesting that men who had participated in a killing had more wives and offspring than those who had not.[38] (The finding is provocative because if that payoff was typical of the pre-state societies in which humans evolved, the strategic use of violence would have been selected over evolutionary time.) Turner and Sponsel accused him of fabricating his data, of *causing* the violence among the Yanomamö (by sending them into a frenzy over the pots and knives with which he paid his informants), and of staging lethal fights for documentary films. Chagnon's portrayal of the Yanomamö, they charged, had been used to justify an invasion of gold miners into their territory, abetted by Chagnon's collusion with "sinister" Venezuelan politicians. The Yanomamö have unquestionably been decimated by disease and by the depredations of the miners, so to lay these tragedies and crimes at Chagnon's feet is literally to accuse him of genocide. For good measure, Turner and Sponsel added that Tierney's book contained "passing references to Chagnon . . . demanding that villagers bring him girls for sex."

Headlines such as "Scientist 'Killed Amazon Indians to Test Race Theory' " soon appeared around the world, followed by an excerpt of Tierney's book in *The New Yorker* and then the book itself, titled *Darkness in El Dorado: How Scientists and Journalists Devastated the Amazon.*[39] Under pressure from the publisher's libel lawyers, some of the more sensational accusations in the book had been excised, watered down, or put in the mouths of Venezuelan journalists or untraceable informants. But the substance of the charges remained.[40]

Turner and Sponsel admitted that their charge against Neel "remains only an inference in the present state of our knowledge: there is no 'smoking gun' in the form of a written text or recorded speech by Neel." That turned out to be an understatement. Within days, scholars with direct knowledge of the events—historians, epidemiologists, anthropologists, and filmmakers—demolished the charges point by point.[41]

Far from being a depraved eugenicist, James Neel (who died shortly before the accusations came out) was an honored and beloved scientist who had consistently *attacked* eugenics. Indeed, he is often credited with purging human genetics of old eugenic theories and thereby making it a respectable science. The cockamamie theory that Turner and Sponsel attributed to him was incoherent on the face of it and scientifically illiterate (for example, they confused a "dominant gene" with a gene for dominance). In any case there is not the slightest evidence that Neel held any belief close to it. Records show that Neel and Chagnon were surprised by the measles epidemic already in progress and made heroic efforts to contain it. The vaccine they administered, which Tierney had charged was the source of the epidemic, has never caused contagious transmission of measles in the hundreds of millions of people all over the world who have received it, and in all probability the efforts of Neel and Chagnon saved hundreds of Yanomamö lives.[42] Confronted with public statements from epidemiologists refuting his claims, Tierney lamely said, "Experts I spoke to then had very different opinions than the ones they are expressing in public now."[43]

Though no one can prove that Neel and Chagnon did not inadvertently introduce the disease in other places by their very presence, the odds are strongly against it. The Yanomamö, who are spread out over tens of thousands of square miles, had many more contacts with other Europeans than they did with Chagnon or Neel, because thousands of missionaries, traders, miners, and adventurers move through the area. Indeed, Chagnon himself had documented that a Catholic Salesian missionary was the likely source of an earlier outbreak. Together with Chagnon's criticism of the mission for providing the Yanomamö with shotguns, this earned him the missionaries' undying enmity. Not coincidentally, most of Tierney's Yanomamö informants were associated with the mission.

The specific accusations against Chagnon crumbled as quickly as those against Neel. Chagnon, contrary to Tierney's charges, had not exaggerated Yanomamö violence or ignored the rest of their lifestyle; in fact, he had meticulously described their techniques for conflict resolution.[44] The suggestion that Chagnon *introduced* them to violence is simply incredible. Raiding and warfare among the Yanomamö have been described since the mid-1800s and were documented throughout the first half of the twentieth century, long before Chagnon set foot in the Amazon. (One revealing account was a first-person narrative called *Yanoáma: The Story of Helena Valero, a Girl Kidnapped by Amazonian Indians*.)[45] And Chagnon's main empirical claims have met the gold standard of science: independent replication. In surveys of rates of death by warfare in pre-state societies, Chagnon's estimates for the Yanomamö fall well within the range, as we saw in the graph in Chapter 3.[46] Even his most controversial claim, that killers had more wives and offspring, has been replicated in other groups, though there is controversy over the interpretation. It is instructive to compare Tierney's summary of a book supposedly refuting Chagnon with the author's own words. Tierney reports:

> Among the Jivaro, head-hunting was a ritual obligation of all males and
> a required male initiation for teenagers. There, too, most men died in

war. Among the Jivaro leaders, however, those who captured the most heads had the fewest wives, and those who had the most wives captured the fewest heads.[47]

The author, the anthropologist Elsa Redmond, had actually written:

> Yanomamo men who have killed tend to have more wives, which they have acquired either by abducting them from raiding villages, or by the usual marriage alliances in which they are considered more attractive as mates. The same is true of Jivaro war leaders, who might have four to six wives; as a matter of fact, a great war leader on the Upano River in the 1930s by the name of Tuki or José Grande had eleven wives. Distinguished warriors also have more offspring, due mainly to their greater marital success.[48]

Turner and Sponsel had long been among Chagnon's most vehement critics (and, not coincidentally, major sources for Tierney's book, despite their professed shock at learning of its contents). They are open about their ideological agenda, which is to defend the doctrine of the Noble Savage. Sponsel wrote that he is committed to "the anthropology of peace" in order to promote a "more nonviolent and peaceful world," which he believes is "latent in human nature."[49] He is opposed to a "Darwinian emphasis on violence and competition" and recently pronounced that "nonviolence and peace were likely the norm throughout most of human prehistory and that intrahuman killing was probably rare."[50] He even admits that much of his criticism of Chagnon comes from "an almost automatic reaction against any biological explanation of human behavior, the possibility of biological reductionism, and the associated political implications."[51]

Also familiar from the radical science days is an irredentist leftism that considers even moderate and liberal positions reactionary. According to Tierney, Neel "was convinced that democracy, with its free breeding for the masses and its sentimental support for the weak, violated natural selection"[52] and was thus "a eugenic mistake." But in fact Neel was a political liberal who had protested the diversion of money from poor children to research on aging that he thought would benefit the affluent. He also advocated increasing investment in prenatal care, medical care for children and adolescents, and universal quality education.[53] As for Chagnon, Tierney calls him "a militant anti-Communist and free-market advocate." His evidence? A quotation from Turner (!) stating that Chagnon is "a kind of right-wing character who has a paranoid attitude on people he considers lefty." To explain how he came by these right-wing leanings, Tierney informs readers that Chagnon grew up in a part of rural Michigan "where differences were not welcomed, where xenophobia, linked to anti-Communist feeling, ran high, and where Senator Joseph McCarthy enjoyed strong support." Unaware of the irony, Tierney concludes that Chagnon is an "offspring" of McCarthy who had "received a full portion of [McCarthy's] spirit." Chagnon, in fact, is a political moderate who had always voted for Democrats.[54]

An autobiographical comment in Tierney's preface is revealing: "I gradually changed from being an observer to being an advocate. . . . traditional, objective journalism was no longer an option for me."[55] Tierney believes that accounts of Yanomamö violence might be used by invaders to depict them as primitive savages who should be removed or assimilated for their own good. Defaming messengers like Chagnon is, in this view, an ennobling form of social action and a step for the cultural survival of indigenous peoples (despite the fact that Chagnon himself has repeatedly acted to protect the interests of the Yanomamö).

The decimation of native Americans by European disease and genocide over five hundred years is indeed one of the great crimes of history. But it is bizarre to blame the crime on a handful of contemporary scientists struggling to document their lifestyle before it vanishes forever under the pressures of assimilation. And it is a dangerous tactic. Surely indigenous peoples have a right to survive in their lands whether or not they—like all human societies—are prone to violence and warfare. Self-appointed "advocates" who link the survival of native peoples to the doctrine of the Noble Savage paint themselves into a terrible corner. When the facts show otherwise they either have inadvertently weakened the case for native rights or must engage in any means necessary to suppress the facts.

~

No one should be surprised that claims about human nature are controversial. Obviously any such claim should be scrutinized and any logical and empirical flaws pointed out, just as with any scientific hypothesis. But the criticism of the new sciences of human nature went well beyond ordinary scholarly debate. It turned into harassment, slurs, misrepresentation, doctored quotations, and, most recently, blood libel. I think there are two reasons for this illiberal behavior.

One is that in the twentieth century the Blank Slate became a sacred doctrine that, in the minds of its defenders, had to be either avowed with a perfect faith or renounced in every aspect. Only such black-and-white thinking could lead people to convert the idea that *some* aspects of behavior are innate into the idea that *all* aspects of behavior are innate, or convert the proposal that genetic traits *influence* human affairs into the idea that they *determine* human affairs. Only if it is theologically necessary for 100 percent of the differences in intelligence to be caused by the environment could anyone be incensed over the mathematical banality that as the proportion of variance due to nongenetic causes goes down, the proportion due to genetic causes must go up. Only if the mind is required to be a scraped tablet could anyone be outraged by the claim that human nature makes us smile, rather than scowl, when we are pleased.

A second reason is that "radical" thinkers got trapped by their own moralizing. Once they staked themselves to the lazy argument that racism, sexism, war, and political inequality were factually incorrect because there is no such thing as human nature (as opposed to being morally despicable regardless of the details of human nature), every discovery about human nature was, by their own reasoning, tantamount to saying that those scourges were not so bad after all. That made it all the more pressing to discredit the heretics making the discoveries. If ordinary standards of scientific argumentation were not doing the trick, other tactics had to be brought in, because a greater good was at stake.

Chapter 7

The Holy Trinity

BEHAVIORAL SCIENCE IS not for sissies. Researchers may wake up to discover that they are despised public figures because of some area they have chosen to explore or some datum they have stumbled upon. Findings on certain topics—daycare, sexual behavior, childhood memories, the treatment of substance abuse—may bring on vilification, harassment, intervention by politicians, and physical assault.[1] Even a topic as innocuous as left-handedness turns out to be booby-trapped. In 1991 the psychologists Stanley Coren and Diane Halpern published statistics in a medical journal showing that lefties on average had more prenatal and perinatal complications, are victims of more accidents, and die younger than righties. They were soon showered with abuse—including the threat of a lawsuit, numerous death threats, and a ban on the topic in a scholarly journal—from enraged left-handers and their advocates.[2]

Are the dirty tricks of the preceding chapter just another example of people taking offense at claims about behavior that make them uncomfortable? Or, as I have hinted, are they part of a systematic intellectual current: the attempt to safeguard the Blank Slate, the Noble Savage, and the Ghost in the Machine as a source of meaning and morality? The leading theoreticians of the radical science movement deny that they believe in a blank slate, and it is only fair that their positions be examined carefully. In addition, I will look at the attacks on the sciences of human nature that have come from their political opposites, the contemporary right.

~

COULD THE RADICAL scientists really believe in the Blank Slate? The doctrine might seem plausible to some of the scholars who live in a world of disembodied ideas. But could hardheaded boffins who live in a mechanistic world of neurons and genes really think that the psyche soaks into the brain from the surrounding culture? They deny it in the abstract, but when it comes to specifics their position is plainly in the tradition of the tabula rasa social science of the early twentieth century. Stephen Jay Gould, Richard Lewontin, and the other signatories of the "Against 'Sociobiology' " manifesto wrote:

> We are not denying that there are genetic components to human behavior. But we suspect that human biological universals are to be discovered more in the generalities of eating, excreting, and sleeping than in such specific and highly variable habits as warfare, sexual exploitation of women and the use of money as a medium of exchange.[3]

Note the tricky framing of the issue. The notion that money is a genetically coded universal is so ridiculous (and not, incidentally, something Wilson ever proposed) that *any* alternative has to be seen as more plausible than that. But if we take the alternative on its own terms, rather than as one prong in a false dichotomy, Gould and Lewontin seem to be saying that the genetic components of human behavior will be discovered primarily in the "generalities of eating, excreting, and sleeping." The rest of the slate, presumably, is blank.

This debating tactic—first deny the Blank Slate, then make it look plausible by pitting it against a straw man—can be found elsewhere in the writings of the radical scientists. Gould, for instance, writes:

> Thus, my criticism of Wilson does not invoke a non-biological "environmentalism"; it merely pits the concept of biological potentiality, with a brain capable of a full range of human behaviors and predisposed to none, against the idea of biological determinism, with specific genes for specific behavioral traits.[4]

The idea of "biological determinism"—that genes cause behavior with 100 percent certainty—and the idea that every behavioral trait has its own gene, are obviously daft (never mind that Wilson never embraced them). So Gould's dichotomy would seem to leave "biological potentiality" as the only reasonable choice. But what does that mean? The claim that the brain is "capable of a full range of human behaviors" is almost a tautology: how could the brain *not* be capable of a full range of human behaviors? And the claim that the brain is not predisposed to *any* human behavior is just a version of the Blank Slate. "Predisposed to none" literally means that all human behaviors have identical probabilities of occurring. So if any person anywhere on the planet has ever committed some act in some circumstance—abjuring food or sex, impaling himself with spikes, killing her child—then the brain has no predisposition to avoid that act as compared with the alternatives, such as enjoying food and sex, protecting one's body, or cherishing one's child.

Lewontin, Rose, and Kamin also deny that they are saying that humans are blank slates.[5] But they grant only two concessions to human nature. The first comes not from an appeal to evidence or logic but from their politics: "If [a blank slate] were the case, there could be no social evolution." Their support for this "argument" consists of an appeal to the authority of Marx, whom they quote as saying, "The materialist doctrine that men are the products of circumstances and upbringing, and that, therefore, changed men are products of other circumstances and changed upbringing, forgets that it is men that change circumstances and that the educator himself needs educating."[6] Their own view is that "the only sensible thing to say about human nature is that it is 'in' that nature to construct its own history."[7] The implication is that any other statement about the psychological makeup of our species—about our capacity for language, our love of family, our sexual emotions, our typical fears, and so on—is not "sensible."

Lewontin, Rose, and Kamin do make one concession to biology—not to the organization of the mind and brain but to the size of the body. "Were human beings only six inches tall there could be no human culture at all as we understand it," they note, because a Lilliputian could not control fire, break rocks with a pick-axe, or carry a brain big enough to support language. It is their only acknowledgment of the possibility that human biology affects human social life.

Eight years later Lewontin reiterated this theory of what is innate in humans: "The most important fact about human genes is that they help to make us as big as we are and to have a central nervous system with as many connections as it has."[8] Once again, the rhetoric has to be unpacked with care. If we take the sentence literally, Lewontin is referring only to "the most important fact" about human genes. Then again, if we take it literally, the sentence is meaningless. How could one ever rank-order the thousands of effects of the genes, all necessary to our existence, and point to one or two at the top of the list? Is our stature more important than the fact that we have a heart, or lungs, or eyes? Is our synapse number more important than our sodium pumps, without which our neurons would fill up with positive ions and shut down? So taking the sentence literally is pointless. The only sensible reading, and the one that fits in the context, is that these are the *only* important facts about human genes for the human mind. The tens of thousands of genes that are expressed primarily or exclusively in the brain do nothing important but give it lots of connections; the *pattern* of connections and the *organization* of the brain (into structures like the hippocampus, amygdala, hypothalamus, and a cerebral cortex divided into areas) are random, or might as well be. The genes do not give the brain multiple memory systems, complicated visual and motor tracts, an ability to learn a language, or a repertoire of emotions (or else the genes do provide these faculties, but they are not "important").

In an update of John Watson's claim that he could turn any infant into a "doctor, lawyer, artist, merchant-chief, and yes, even beggar-man and thief, regardless of his talents, penchants, tendencies, abilities, vocations, and race of his ancestors," Lewontin wrote a book whose jacket précis claims that "our genetic endowments confer a plasticity of psychic and physical development, so that in the course of our lives, from conception to death, each of us, irrespective of race, class, or sex, can develop virtually any identity that lies within the human ambit."[9] Watson admitted he was "going beyond my facts," which was forgivable because at the time he wrote there were no facts. But the declaration on Lewontin's book that any individual can assume any identity (even granting the equivalence of races, sexes, and classes), in defiance of six decades of research in behavioral genetics, is an avowal of faith of uncommon purity. And in a passage that re-erects Durkheim's wall between the biological and the cultural, Lewontin concludes a 1992 book by writing that the genes "have been replaced by an entirely new level of causation, that of social interaction with its own laws and its own nature that can be understood and explored only through that unique form of experience, social action."[10]

So while Gould, Lewontin, and Rose deny that they believe in a blank slate, their concessions to evolution and genetics—that they let us eat, sleep, urinate, defecate, grow bigger than a squirrel, and bring about social change—reveal them to be empiricists more extreme than Locke himself, who at least recognized the need for an innate faculty of "understanding."

~

THE NOBLE SAVAGE, too, is a cherished doctrine among critics of the sciences of human nature. In *Sociobiology*, Wilson mentioned that tribal warfare was common in human prehistory. The against-sociobiologists declared that this had been "strongly rebutted both on the basis of historical and anthropological studies." I looked up these "studies," which were collected in Ashley Montagu's *Man and Aggression*. In fact they were just hostile reviews of books by the ethologist Konrad Lorenz, the playwright Robert Ardrey, and the novelist William Golding (author of *Lord of the Flies*).[11] Some of the criticisms were, to be sure, deserved: Ardrey and Lorenz believed in archaic theories such as that aggression was like the discharge of a hydraulic pressure and that evolution acted for the good of the species. But far stronger criticisms of Ardrey and Lorenz had been made by the sociobiologists themselves. (On the second page of *The Selfish Gene*, for example, Dawkins wrote, "The trouble with these books is that the authors got it totally and utterly wrong.") In any case, the reviews contained virtually no data about tribal warfare. Nor did Montagu's summary essay, which simply rehashed attacks on the concept of "instinct" from decades of behaviorists. One of the only chapters with data "refuted" Lorenz's claims about warfare and raiding in the Ute Indians by saying they didn't do it any more than other native groups!

Twenty years later, Gould wrote that "*Homo sapiens* is not an evil or destructive species." His new argument comes from what he calls the Great

Asymmetry. It is "an essential truth," he writes, that "good and kind people outnumber all others by thousands to one."[12] Moreover, "we perform 10,000 acts of small and unrecorded kindness for each surpassingly rare, but sadly balancing, moment of cruelty."[13] The statistics making up this "essential truth" are pulled out of the air and are certainly wrong: psychopaths, who are definitely not "good and kind people," make up about three or four percent of the male population, not several hundredths of a percent.[14] But even if we accept the figures, the argument assumes that for a species to count as "evil and destructive," it would have to be evil and destructive all the time, like a deranged postal worker on a permanent rampage. It is precisely because one act *can* balance ten thousand kind ones that we call it "evil." Also, does it make sense to judge our entire species, as if we were standing en masse at the pearly gates? The issue is not whether our *species* is "evil and destructive" but whether we house evil and destructive *motives*, together with the beneficent and constructive ones. If we do, one can try to understand what they are and how they work.

Gould has objected to any attempt to understand the motives for war in the context of human evolution, because "each case of genocide can be matched with numerous incidents of social beneficence; each murderous band can be paired with a pacific clan."[15] Once again a ratio has been conjured out of the blue; the data reviewed in Chapter 3 show that "pacific clans" either do not exist or are considerably outnumbered by the "murderous bands."[16] But for Gould, such facts are beside the point, because he finds it necessary to believe in the pacific clans on moral grounds. Only if humans lack any predisposition for good or evil or anything else, he suggests, do we have grounds for opposing genocide. Here is how he imagines the position of the evolutionary psychologists he disagrees with:

Perhaps the most popular of all explanations for our genocidal capacity cites evolutionary biology as an unfortunate source—and as an ultimate escape from full moral responsibility. . . . A group devoid of xenophobia and unschooled in murder might invariably succumb to others replete with genes to encode a propensity for such categorization and destruction. Chimpanzees, our closest relatives, will band together and systematically kill the members of adjacent groups. Perhaps we are programmed to act in such a manner as well. These grisly propensities once promoted the survival of groups armed with nothing more destructive than teeth and stones. In a world of nuclear bombs, such unchanged (and perhaps unchangeable) inheritances may now spell our undoing (or at least propagate our tragedies)—but we cannot be blamed for these moral failings. Our accursed genes have made us creatures of the night.[17]

In this passage Gould presents a more-or-less reasonable summary of why scientists might think that human violence can be illuminated by evolution. But then he casually slips in some outrageous non sequiturs ("an ultimate escape from full moral responsibility," "we cannot be blamed"), as if the scientists had no choice but to believe those, too. He concludes his essay:

> In 1525, thousands of German peasants were slaughtered . . . , and Michelangelo worked on the Medici Chapel. . . . Both sides of this dichotomy represent our common, evolved humanity. Which, ultimately, shall we choose? As to the potential path of genocide and destruction, let us take this stand. It need not be. We can do otherwise.[18]

The implication is that anyone who believes that the causes of genocide might be illuminated by an understanding of the evolved makeup of human beings is in fact taking a stand *in favor* of genocide!

WHAT ABOUT THE third member of the trinity, the Ghost in the Machine? The radical scientists are thoroughgoing materialists and could hardly believe in an immaterial soul. But they are equally uncomfortable with any clearly stated alternative, because it would cramp their political belief that we can collectively implement any social arrangement we choose. To update Ryle's description of Descartes's dilemma: as men of scientific acumen they cannot but endorse the claims of biology, yet as political men they cannot accept the discouraging rider to those claims, namely that human nature differs only in degree of complexity from clockwork.

Ordinarily it is not cricket to bring up the political beliefs of scholars in discussing their scholarly arguments, but it is Lewontin and Rose who insist that their scientific beliefs are inseparable from their political ones. Lewontin wrote a book with the biologist Richard Levins called *The Dialectical Biologist,* which they dedicated to Friedrich Engels ("who got it wrong a lot of the time but got it right where it counted"). In it they wrote, "As working scientists in the field of evolutionary genetics and ecology, we have been attempting with some success to guide our research by a conscious application of Marxist philosophy."[19] In *Not in Our Genes,* Lewontin, Rose, and Kamin declared that they "share a commitment to the prospect of a more socially just—a socialist—society" and see their "critical science as an integral part of the struggle to create that society."[20] At one point they frame their disagreement with "reductionism" as follows:

> Against this economic reduction as the explanatory principle underlying all human behavior, we could counterpose the . . . revolutionary practitioners and theorists like Mao Tse-tung on the power of human consciousness in both interpreting and changing the world, a power

based on an understanding of the essential dialectical unity of the biological and the social, not as two distinct spheres, or separable components of action, but as ontologically coterminous.[21]

Lewontin and Rose's commitment to the "dialectical" approach of Marx, Engels, and Mao explains why they deny human nature and also deny that they deny it. The very idea of a durable human nature that can be discussed separately from its ever-changing interaction with the environment is, in their view, a dull-witted mistake. The mistake lies not just in ignoring interactions with the environment—Lewontin and Rose already knocked over the straw men who do that. The deeper mistake, as they see it, lies in trying to analyze behavior as an interaction between human nature and the human environment (including society) in the first place.[22] The very act of separating them in one's mind, even for the purpose of figuring out how the two interact, "supposes the alienation of the organism and the environment." That contradicts the principles of dialectical understanding, which says that the two are "ontologically coterminous"—not just in the trivial sense that no organism lives in a vacuum, but in the sense that they are inseparable in every aspect of their being.

Since the dialectic between organism and environment constantly changes over historical time, with neither one directly causing the other, organisms can alter that dialectic. Thus Rose repeatedly counters the "determinists" with the declaration "We have the ability to construct our own futures, albeit not in circumstances of our own choosing"[23]—presumably echoing Marx's statement that "men make their own history, but they do not make it just as they please; they make it under circumstances directly encountered, given and transmitted from the past." But Rose never explains who the "we" is, if not highly structured neural circuits, which must get that structure in part from genes and evolution. We can call this doctrine the Pronoun in the Machine.

Gould is not a doctrinarian like Rose and Lewontin, but he too uses the first-person plural pronoun as if it somehow disproved the relevance of genes and evolution to human affairs: "Which . . . shall we choose? . . . Let us take this stand. . . . We can do otherwise." And he too cites Marx's "wonderful aphorism" about making our own history and believes that Marx vindicated the concept of free will:

Marx himself had a much more subtle view than most of his contemporaries of the differences between human and natural history. He understood that the evolution of consciousness, and the consequent development of social and economic organization, introduced elements of difference and volition that we usually label as "free will."[24]

Subtle indeed is the argument that explains free will in terms of its synonym "volition" (with or without "elements of difference," whatever that means) and attributes it to the equally mysterious "evolution of consciousness." Basically, Rose and Gould are struggling to make sense of the dichotomy they invented between a naturally selected, genetically organized brain on one side and a desire for peace, justice, and equality on the other. In Part III we will see that the dichotomy is a false one.

The doctrine of the Pronoun in the Machine is not a casual oversight in the radical scientists' world view. It is consistent with their desire for radical political change and their hostility to "bourgeois" democracy. (Lewontin repeatedly uses "bourgeois" as an epithet.) If the "we" is truly unfettered by biology, then once "we" see the light we can carry out the vision of radical change that we deem correct. But if the "we" is an imperfect product of evolution—limited in knowledge and wisdom, tempted by status and power, and blinded by self-deception and delusions of moral superiority—then "we" had better think twice before constructing all that history. As the chapter on politics will explain, constitutional democracy is based on a jaundiced theory of human nature in which "we" are eternally vulnerable to arrogance and corruption. The checks and balances of democratic institutions were explicitly designed to stalemate the often dangerous ambitions of imperfect humans.

~

THE GHOST IN the Machine, of course, is far dearer to the political right than to the political left. In his book *The New Know-Nothings: The Political Foes of the Scientific Study of Human Nature,* the psychologist Morton Hunt has shown that the foes include people on the left, people on the right, and a motley collection of single-issue fanatics in between.[25] So far I have discussed the far-left outrage because it has been deployed in the battlefield of ideas in the universities and the mainstream press. Those on the far right have also been outraged, though until recently they have aimed at different targets and have fought in different arenas.

The longest-standing right-wing opposition to the sciences of human nature comes from the religious sectors of the coalition, especially Christian fundamentalism. Anyone who doesn't believe in evolution is certainly not going to believe in the evolution of the mind, and anyone who believes in an immaterial soul is certainly not going to believe that thought and feeling consist of information processing in the tissues of the brain.

The religious opposition to evolution is fueled by several moral fears. Most obviously, the fact of evolution challenges the literal truth of the creation story in the Bible and thus the authority that religion draws from it. As one creationist minister put it, "If the Bible gets it wrong in biology, then why should I trust the Bible when it talks about morality and salvation?"[26]

But the opposition to evolution goes beyond a desire to defend biblical literalism. Modern religious people may not believe in the literal truth of every miracle narrated in the Bible, but they do believe that humans were designed in God's image and placed on earth for a larger purpose—namely, to live a moral life by following God's commandments. If humans are accidental products of the mutation and selection of chemical replicators, they worry, morality would have no foundation and we would be left mindlessly obeying biological urges. One creationist, testifying to this danger in front of the U.S. House Judiciary Committee, cited the lyrics of a rock song: "You and me baby ain't nothin' but mammals / So let's do it like they do it on the Discovery Channel."[27] After the 1999 lethal rampage by two teenagers at Columbine High School in Colorado, Tom Delay, the Republican Majority Whip in the House of Representatives, said that such violence is inevitable as long as "our school systems teach children that they are nothing but glorified apes, evolutionized out of some primordial soup of mud."[28]

The most damaging effect of the right-wing opposition to evolution is the corruption of American science education by activists in the creationist movement. Until a Supreme Court decision in 1968, states were allowed to ban the teaching of evolution outright. Since then, creationists have tried to hobble it in ways that they hope will pass constitutional muster. These include removing evolution from science proficiency standards, demanding disclaimers that it is "only a theory," watering down the curriculum, and opposing textbooks with good coverage of evolution or imposing ones with coverage of creationism. In recent years the National Center for Science Education has learned of new instances of these tactics at a rate of about one a week, coming from forty states.[29]

The religious right is discomfited not just by evolution but by neuroscience. By exorcising the ghost in the machine, brain science is undermining two moral doctrines that depend on it. One is that every person has a soul, which finds value, exercises free will, and is responsible for its choices. If behavior is controlled instead by circuits in the brain that follow the laws of chemistry, choice and value would be myths and the possibility of moral responsibility would evaporate. As the creationist advocate John West put it, "If human beings (and their beliefs) really are the mindless products of their material existence, then everything that gives meaning to human life—religion, morality, beauty—is revealed to be without objective basis."[30]

The other moral doctrine (which is found in some, but not all, Christian denominations) is that the soul enters the body at conception and leaves it at death, thereby defining who is a person with a right to life. The doctrine makes abortion, euthanasia, and the harvesting of stem cells from blastocysts equivalent to murder. It makes humans fundamentally different from animals. And it makes human cloning a violation of the divine order. All this would seem to be threatened by neuroscientists, who say that the self or the soul inheres in neural activity that develops gradually in the brain of an embryo, that can be seen in the brains of animals, and that can break down piecemeal with aging and disease. (We will return to this issue in Chapter 13.)

But the right-wing opposition to the sciences of human nature can no longer be associated only with Bible-thumpers and televangelists. Today evolution is being challenged by some of the most cerebral theorists in the formerly secular neoconservative movement. They are embracing a hypothesis called Intelligent Design, originated by the biochemist Michael Behe.[31] The molecular machinery of cells cannot function in a simpler form, Behe argues, and therefore it could not have evolved piecemeal by natural selection. Instead it must have been conceived as a working invention by an intelligent designer. The designer could, in theory, have been an advanced alien from outer space, but everyone knows that the subtext of the theory is that it must have been God.

Biologists reject Behe's argument for a number of reasons.[32] His specific claims about the "irreducible complexity" of biochemistry are unproven or just wrong. He takes every phenomenon whose evolutionary history has not yet been figured out and chalks it up to design by default. When it comes to the intelligent designer, Behe suddenly jettisons all scientific scruples and does not question where the designer came from or how the designer works. And he ignores the overwhelming evidence that the process of evolution, far from being intelligent and purposeful, is wasteful and cruel.

Nonetheless, Intelligent Design has been embraced by leading neoconservatives, including Irving Kristol, Robert Bork, Roger Kimball, and Gertrude Himmelfarb. Other conservative intellectuals have also sympathized with creationism for moral reasons, such as the law professor Philip Johnson, the writer William F. Buckley, the columnist Tom Bethell, and, disconcertingly, the bioethicist Leon Kass—chair of George W. Bush's new Council on Bioethics and thus a shaper of the nation's policies on biology and medicine.[33] A story entitled "The Deniable Darwin" appeared, astonishingly, on the cover of *Commentary,* which means that a magazine that was once a leading forum for secular Jewish intellectuals is now more skeptical of evolution than is the Pope![34]

It is not clear whether these worldly thinkers are really convinced that Darwinism is false or whether they think it is important for other people to believe it is false. In a scene from *Inherit the Wind,* the play about the Scopes Monkey Trial, the prosecutor and defense attorney (based on William Jennings Bryan and Clarence Darrow) are relaxing together after a day in court. The prosecutor says of the Tennessee locals:

They're simple people, Henry; poor people. They work hard and they need to believe in something, something beautiful. Why do you want to take it away from them? It's all they have.

That is not far from the attitude of the neocons. Kristol has written:

If there is one indisputable fact about the human condition it is that no community can survive if it is persuaded—or even if it suspects—that its members are leading meaningless lives in a meaningless universe.[35]

He spells out the moral corollary:

There are different kinds of truths for different kinds of people. There are truths appropriate for children; truths that are appropriate for students; truths that are appropriate for educated adults; and truths that are appropriate for highly educated adults, and the notion that there should be one set of truths available to everyone is a modern democratic fallacy. It doesn't work.[36]

As the science writer Ronald Bailey observes, "Ironically, today many modern conservatives fervently agree with Karl Marx that religion is 'the opium of the people'; they add a heartfelt, 'Thank God!' "[37]

Many conservative intellectuals join fundamentalist Christians in deploring neuroscience and evolutionary psychology, which they see as explaining away the soul, eternal values, and free choice. Kass writes:

With science, the leading wing of modern rationalism, has come the progressive demystification of the world. Falling in love, should it still occur, is for the modern temper to be explained not by demonic possession (Eros) born of the soul-smiting sight of the beautiful (Aphrodite) but by a rise in the concentration of some still-to-be-identified polypeptide hormone in the hypothalamus. The power of religious sensibilities and understandings fades too. Even if it is true that the great majority of Americans still profess a belief in God, He is for few of us a God before whom one trembles in fear of judgment.[38]

Similarly, the journalist Andrew Ferguson warns his readers that evolutionary psychology "is sure to give you the creeps," because "whether behavior is moral, whether it signifies virtue, is a judgment that the new science, and materialism in general, cannot make."[39] The new sciences, he writes, claim that people are nothing but "meat puppets," a frightening shift from the traditional Judeo-Christian view in which "human beings [are] persons from the start, endowed with a soul, created by God, and infinitely precious."[40]

Even the left-baiting author Tom Wolfe, who admires neuroscience and evolutionary psychology, worries about their moral implications. In his essay "Sorry, but Your Soul Just Died," he writes that when science has finally killed the soul ("that last refuge of values"), "the lurid carnival that will ensue may make [Nietzsche's] phrase 'the total eclipse of all values' seem tame":

Meanwhile, the notion of a self—a self who exercises self-discipline, postpones gratification, curbs the sexual appetite, stops short of aggression and criminal behavior—a self who can become more intelligent and lift itself to the very peaks of life by its own bootstraps through study, practice, perseverance, and refusal to give up in the face of great odds—this old-fashioned notion (what's a *boot*strap, for God's sake?) of success through enterprise and true grit is already slipping away, slipping away . . . slipping away . . . [41]

"Where does that leave self-control?" he asks. "Where, indeed, if people believe this ghostly self does not even exist, and brain imaging proves it, once and for all?"[42]

An irony in the modern denial of human nature is that partisans at opposite extremes of the political spectrum, who ordinarily can't stand the sight of each other, find themselves strange bedfellows. Recall how the signatories of "Against 'Sociobiology' " wrote that theories like Wilson's "provided an important basis for . . . the eugenics policies which led to the establishment of gas chambers in Nazi Germany." In May 2001 the Education Committee of the Louisiana House of Representatives resolved that "Adolf Hitler and others have exploited the racist views of Darwin and those he influenced . . . to justify the annihilation of millions of purportedly racially inferior individuals."[43] The sponsor of the resolution (which was eventually defeated) cited in its defense a passage by Gould, which is not the first time that he has been cited approvingly in creationist propaganda.[44] Though Gould has been a tireless opponent of creationism, he has been an equally tireless opponent of the idea that evolution can explain mind and morality, and that is the implication of Darwinism that creationists fear most.

The left and the right also agree that the new sciences of human nature threaten the concept of moral responsibility. When Wilson suggested that in humans, as in many other mammals, males have a greater desire for multiple sexual partners than do females, Rose accused him of really saying:

Don't blame your mates for sleeping around, ladies, it's not their fault they are genetically programmed.[45]

Compare Tom Wolfe, tongue only partly in cheek:

The male of the human species is genetically hardwired to be polygamous, i.e., unfaithful to his legal mate. Any magazine-reading male gets the picture soon enough. (Three million years of evolution made me do it!)[46]

On one wing we have Gould asking the rhetorical question:

> Why do we want to fob off responsibility for our violence and sexism upon our genes?[47]

And on the other wing we find Ferguson raising the same point:

> The "scientific belief" would . . . appear to be corrosive of any notion of free will, personal responsibility, or universal morality.[48]

For Rose and Gould the ghost in the machine is a "we" that can construct history and change the world at will. For Kass, Wolfe, and Ferguson it is a "soul" that makes moral judgments according to religious precepts. But all of them see genetics, neuroscience, and evolution as threats to this irreducible locus of free choice.

~

WHERE DOES THIS leave intellectual life today? The hostility to the sciences of human nature from the religious right is likely to increase, but the influence of the right will be felt more in direct appeals to politicians than from changes in the intellectual climate. Any inroads of the religious right into mainstream intellectual life will be limited by their opposition to the theory of evolution itself. Whether it is known as creationism or by the euphemism Intelligent Design, a denial of the theory of natural selection will founder under the weight of the mass of evidence that the theory is correct. How much additional damage the denial will do to science education and biomedical research before it sinks is unknown.

The hostility from the radical left, on the other hand, has left a substantial mark on modern intellectual life, because the so-called radical scientists are now the establishment. I have met many social and cognitive scientists who proudly say they have learned all their biology from Gould and Lewontin.[49] Many intellectuals defer to Lewontin as the infallible pontiff of evolution and genetics, and many philosophers of biology spent time as his apprentice. A sneering review by Rose of every new book on human evolution or genetics has become a fixture of British journalism. As for Gould, Isaac Asimov probably did not intend the irony when he wrote in a book blurb that "Gould can do no wrong," but that is precisely the attitude of many journalists and social scientists. A recent article in *New York* magazine on the journalist Robert Wright called him a "stalker" and a "young punk" with "penis envy" because he had the temerity to criticize Gould on his logic and facts.[50]

In part the respect awarded to the radical scientists has been earned. Quite aside from their scientific accomplishments, Lewontin is an incisive analyst on many scientific and social issues, Gould has written hundreds of superb essays on natural history, and Rose wrote a fine book on the neuroscience of memory. But they have also positioned themselves shrewdly on the intellectual landscape. As the biologist John Alcock explains, "Stephen Jay Gould abhors violence, he speaks out against sexism, he despises Nazis, he finds genocide horrific, he is unfailingly on the side of the angels. Who can argue with such a person?"[51] This immunity from argument allowed the radical scientists' unfair attacks on others to become part of the conventional wisdom.

Many writers today casually equate behavioral genetics with eugenics, as if studying the genetic correlates of behavior were the same as coercing people in their decisions about having children. Many equate evolutionary psychology with Social Darwinism, as if studying our evolutionary roots were the same as justifying the station of the poor. The confusions do not come only from the scientifically illiterate but may be found in prestigious publications such as *Scientific American* and *Science*.[52] After Wilson argued in *Consilience* that divisions between fields of human knowledge were becoming obsolete, the historian Tzvetan Todorov wrote sarcastically, "I have a proposal for Wilson's next book . . . [an] analysis of Social Darwinism, the doctrine that was adopted by Hitler, and of the ways it differs from sociobiology."[53] When the Human Genome Project was completed in 2001, its leaders made a ritual denunciation of "genetic determinism," the belief—held by no one—that "all characteristics of the person are 'hard-wired' into our genome."[54]

Even many scientists are perfectly content with the radicals' social constructionism, not so much because they agree with it but because they are preoccupied in their labs and need picketers outside their window like they need another hole in the head. As the anthropologist John Tooby and the psychologist Leda Cosmides note, the dogma that biology is intrinsically disconnected from the human social order offers scientists "safe conduct across the politicized minefield of modern academic life."[55] As we shall see, even today people who challenge the Blank Slate or the Noble Savage are still sometimes silenced by demonstrators or denounced as Nazis. Even when such attacks are sporadic, they create an atmosphere of intimidation that distorts scholarship far and wide.

But the intellectual climate is showing signs of change. Ideas about human nature, while still anathema to some academics and pundits, are beginning to get a hearing. Scientists, artists, scholars in the humanities, legal theorists, and thoughtful laypeople have expressed a thirst for the new insights about the mind that have been coming out of the biological and cognitive sciences. And the radical science movement, for all its rhetorical success, has turned out to be an empirical wasteland. Twenty-five years of data have not been kind to its predictions. Chimpanzees are not peaceful vegetarians, as Montagu claimed, nor is the heritability of intelligence indistinguishable from zero, IQ a "reification" unrelated to the brain, personality and social behavior without any genetic basis, gender differences a product only of "psychocultural expectations," or the number of murderous clans equal to the number of pacific bands.[56] Today the idea of guiding scientific research by "a conscious application of Marxist philosophy" is just embarrassing, and as the evolutionary psychologist Martin Daly pointed out, "Sufficient research to fill a first issue of *Dialectical Biology* has yet to materialize."[57]

In contrast, sociobiology did not, as Sahlins had predicted, turn out to be a passing fad. The title of Alcock's 2001 book *The Triumph of Sociobiology* says it all: in the study of animal behavior, no one even talks about "sociobiology" or "selfish genes" anymore, because the ideas are part and parcel of the science.[58] In the study of humans, there are major spheres of human experience—beauty, motherhood, kinship, morality, cooperation, sexuality, violence—in which evolutionary psychology provides the only coherent theory and has spawned vibrant new areas of empirical research.[59] Behavioral genetics has revivified the study of personality and will only expand with the application of knowledge from the Human Genome Project.[60] Cognitive neuroscience will not shrink from applying its new tools to every aspect of mind and behavior, including the emotionally and politically charged ones.

The question is not whether human nature will increasingly be explained by the sciences of mind, brain, genes, and evolution, but what we are going to do with the knowledge. What in fact are the implications for our ideals of equality, progress, responsibility, and the worth of the person? The opponents of the sciences of human nature from the left and the right are correct about one thing: these are vital questions. But that is all the more reason that they be confronted not with fear and loathing but with reason. That is the goal of the next part of the book.

PART III

HUMAN NATURE WITH
A HUMAN FACE

When Galileo attracted the unwanted attention of the Inquisition in 1633, more was at stake than issues in astronomy. By stating that the Earth revolved around the sun rather than vice versa, Galileo was contradicting the literal truth of the Bible, such as the passage in which Joshua issued the successful command "Sun, stand thou still." Worse, he was challenging a theory of the moral order of the universe.

According to the theory, developed in medieval times, the sphere of the moon divided the universe into an unchanging perfection in the heavens above and a corrupt degeneration in the Earth below (hence Samuel Johnson's disclaimer that he could not "change sublunary nature"). Surrounding the moon were spheres for the inner planets, the sun, the outer planets, and the fixed stars, each cranked by a higher angel. And surrounding them all were the heavens, home to God. Contained within the sphere of the moon, and thus a little lower than the angels, were human souls, and then, in descending order, human bodies, animals (in the order beasts, birds, fish, insects), then plants, minerals, the inanimate elements, nine layers of devils, and finally, at the center of the Earth, Lucifer in hell. The universe was thus arranged in a hierarchy, a Great Chain of Being.

The Great Chain was thick with moral implications. Our home, it was thought, lay at the center of the universe, reflecting the importance of our existence and behavior. People lived their lives in their proper station (king, duke, or peasant), and after death their souls rose to a higher place or sank to a lower one. Everyone had to be mindful that the human abode was a humble place in the scheme of things and that they must look up to catch a glimpse of heavenly perfection. And in a world that seemed always to teeter on the brink of famine and barbarism, the Great Chain offered the comfort of knowing that the nature of things was orderly. If the planets wandered from their spheres, chaos would break out, because everything was connected in the cosmic order.

As Alexander Pope wrote, "From Nature's chain whatever link you strike, / Tenth, or ten thousandth, breaks the chain alike."[1]

None of this escaped Galileo as he was pounding away at his link. He knew that he could not simply argue on empirical grounds that the division between a corrupt Earth and the unchanging heavens was falsified by sunspots, novas, and moons drifting across Jupiter. He also argued that the moral trappings of the geocentric theory were as dubious as its empirical claims, so if the theory turned out to be false, no one would be the worse. Here is Galileo's alter ego in *Dialogue Concerning the Two Chief World Systems,* wondering what is so great about being invariant and inalterable:

> For my part I consider the earth very noble and admirable precisely be-
> cause of the diverse alterations, changes, generations, etc. that occur in it
> incessantly. If, not being subject to any changes, it were a vast desert of
> sand or mountain of jasper, or if at the time of the flood the waters
> which covered it had frozen, and it had remained an enormous globe of
> ice where nothing was ever born or ever altered or changed, I should
> deem it a useless lump in the universe, devoid of activity and, in a word,
> superfluous and essentially nonexistent. This is exactly the difference
> between a living animal and a dead one; and I say the same of the moon,
> of Jupiter, and of all other world globes.
>
> . . . Those who so greatly exalt incorruptibility, inalterability, et
> cetera, are reduced to talking this way, I believe, by their great desire to
> go on living, and by the terror they have of death. They do not reflect
> that if men were immortal, they themselves would never have come into
> the world. Such men really deserve to encounter a Medusa's head which
> would transmute them into statues of jasper or diamond, and thus make
> them more perfect than they are.[2]

Today we see things Galileo's way. It's hard for us to imagine why the three-dimensional arrangement of rock and gas in space should have anything to do with right and wrong or with the meaning and purpose of our lives. The moral sensibilities of Galileo's time eventually adjusted to the astronomical facts, not just because they had to give a nod to reality but because the very idea that morality has something to do with a Great Chain of Being was daffy to begin with.

We are now living, I think, through a similar transition. The Blank Slate is today's Great Chain of Being: a doctrine that is widely embraced as a rationale for meaning and morality and that is under assault from the sciences of the day. As in the century following Galileo, our moral sensibilities will adjust to the biological facts, not only because facts are facts but because the moral credentials of the Blank Slate are just as spurious.

This part of the book will show why a renewed conception of meaning and morality will survive the demise of the Blank Slate. I am not, to say the least, proposing a novel philosophy of life like the spiritual leader of some new cult. The arguments I will lay out have been around for centuries and have been advanced by some of history's greatest thinkers. My goal is to put them down in one place and connect them to the apparent moral challenges from the sciences of human nature, to serve as a reminder of why the sciences will not lead to a Nietzschean total eclipse of all values.

The anxiety about human nature can be boiled down to four fears:

- If people are innately different, oppression and discrimination would be justified.
- If people are innately immoral, hopes to improve the human condition would be futile.
- If people are products of biology, free will would be a myth and we could no longer hold people responsible for their actions.
- If people are products of biology, life would have no higher meaning and purpose.

Each will get a chapter. I will first explain the basis of the fear: which claims about human nature are at stake, and why they are thought to have treacherous implications. I will then show that in each case the logic is faulty; the implications simply do not follow. But I will go farther than that. It's not just that claims about human nature are less dangerous than many people think. It's that the *denial* of human nature can be *more* dangerous than people think. This makes it imperative to examine claims about human nature objectively, without putting a moral thumb on either side of the scale, and to figure out how we can live with the claims should they turn out to be true.

Chapter 8

The Fear of Inequality

THE GREATEST MORAL appeal of the doctrine of the Blank Slate comes from a simple mathematical fact: zero equals zero. This allows the Blank Slate to serve as a guarantor of political equality. Blank is blank, so if we are all blank slates, the reasoning goes, we must all be equal. But if the slate of a newborn is not blank, different babies could have different things inscribed on their slates. Individuals, sexes, classes, and races might differ innately in their talents, abilities, interests, and inclinations. And that, it is thought, could lead to three evils.

The first is prejudice: if groups of people are biologically different, it could be rational to discriminate against the members of some of the groups. The second is Social Darwinism: if differences among groups in their station in life—their income, status, and crime rate, for example—come from their innate constitutions, the differences cannot be blamed on discrimination, and that makes it easy to blame the victim and tolerate inequality. The third is eugenics: if people differ biologically in ways that other people value or dislike, it would invite them to try to improve society by intervening biologically—by encouraging or discouraging people's decisions to have children, by taking that decision out of their hands, or by killing them outright. The Nazis carried out the "final solution" because they thought Jews and other ethnic groups were biologically inferior. The fear of the terrible consequences that might arise from a discovery of innate differences has thus led many intellectuals to insist that such differences do not exist—or even that human nature does not exist, because if it did, innate differences would be possible.

I hope that once this line of reasoning is laid out, it will immediately set off alarm bells. We should not concede that *any* foreseeable discovery about humans could have such horrible implications. The problem is not with the possibility that people might differ from one another, which is a factual question that could turn out one way or the other. The problem is with the line of reasoning that says that if people do turn out to be different, then discrimination, oppression, or genocide would be OK after all. Fundamental values (such as equality and human rights) should not be held hostage to some factual conjecture about blank slates that might be refuted tomorrow. In this chapter we will see how these values might be put on a more secure foundation.

~

WHAT KINDS OF differences are there to worry about? The chapters on gender and children will review the current evidence on differences between sexes and individuals, together with their implications and non-implications. The goal of this part of the chapter is more general: to lay out the kinds of differences that research *could* turn up over the long term, based on our understanding of human evolution and genetics, and to lay out the moral issues they raise.

This book is primarily about human nature—an endowment of cognitive and emotional faculties that is universal to healthy members of *Homo sapiens*. Samuel Johnson wrote, "We are all prompted by the same motives, all deceived by the same fallacies, all animated by hope, obstructed by danger, entangled by desire, and seduced by pleasure."[1] The abundant evidence that we share a human nature does not mean that the *differences* among individuals, races, or sexes are also in our nature. Confucius could have been right when he wrote, "Men's natures are alike; it is their habits that carry them far apart."[2]

Modern biology tells us that the forces that make people alike are not the same as the forces that make people different.[3] (Indeed, they tend to be studied by different scientists: the similarities by evolutionary psychologists, the differences by behavioral geneticists.) Natural selection works to homogenize a species into a standard overall design by concentrating the effective genes—the ones that build well-functioning organs—and winnowing out the ineffective ones. When it comes to an explanation of what makes us tick, we are thus birds of a feather. Just as we all have the same physical organs (two eyes, a liver, a four-chambered heart), we have the same mental organs. This is most obvious in the case of language, where every neurologically intact child is equipped to acquire any human language, but it is true of other parts of the mind as well. Discarding the Blank Slate has thrown far more light on the psychological unity of humankind than on any differences.[4]

We are all pretty much alike, but we are not, of course, clones. Except in the case of identical twins, each person is genetically unique. That is because random mutations infiltrate the genome and take time to be eliminated, and they are shuffled together in new combinations when individuals sexually reproduce. Natural selection tends to preserve some degree of genetic heterogeneity at the microscopic level in the form of small, random variations among proteins. That variation twiddles the combinations of an organism's molecular locks and keeps its descendants one step ahead of the microscopic germs that are constantly evolving to crack those locks.

All species harbor genetic variability, but *Homo sapiens* is among the less variable ones. Geneticists call us a "small" species, which sounds like a bad joke given that we have infested the planet like roaches. What they mean is that the amount of genetic variation found among humans is what a biologist would expect in a species with a small number of members.[5] There are more genetic differences among chimpanzees, for instance, than there are among humans, even though we dwarf them in number. The reason is that our ancestors passed through a population bottleneck fairly recently in our evolutionary history (less than a hundred thousand years ago) and dwindled to a small number of individuals with a correspondingly small amount of genetic variation. The species survived and rebounded, and then underwent a population explosion after the invention of agriculture about ten thousand years ago. That explosion bred many copies of the genes that were around when we were sparse in number; there has not been much time to accumulate many new versions of the genes.

At various points after the bottleneck, differences between races emerged. But the differences in skin and hair that are so obvious when we look at people of other races are really a trick played on our intuitions. Racial differences are largely adaptations to climate. Skin pigment was a sunscreen for the tropics, eyelid folds were goggles for the tundra. The parts of the body that face the elements are also the parts that face the eyes of other people, which fools them into thinking that racial differences run deeper than they really do.[6] Working in opposition to the adaptation to local climates, which makes groups different on the skin, is an evolutionary force that makes neighboring groups similar inside. Rare genes can offer immunity to endemic diseases, so they get sucked into one group from a neighboring group like ink on a blotter, even if members of one group mate with members of the other infrequently.[7] That is why Jews, for example, tend to be genetically similar to their non-Jewish neighbors all over the world, even though until recently they tended to marry other Jews. As little as one conversion, affair, or rape involving a gentile in every generation can be enough to blur genetic boundaries over time.[8]

Taking all these processes into account, we get the following picture. People are qualitatively the same but may differ quantitatively. The quantitative differences are small in biological terms, and they are found to a far greater extent *among* the individual members of an ethnic group or race than *between* ethnic groups or races. These are reassuring findings. Any racist ideology that holds that the members of an ethnic group are all alike, or that one ethnic group differs fundamentally from another, is based on false assumptions about our biology.

But biology does not let us off the hook entirely. Individuals are not genetically identical, and it is unlikely that the differences affect every part of the body except the brain. And though genetic differences between races and ethnic groups are much smaller than those among individuals, they are not nonexistent (as we see in their ability to give rise to physical differences and to different susceptibilities to genetic diseases such as Tay-Sachs and sickle cell anemia). Nowadays it is popular to say that races do not exist but are purely social constructions. Though that is certainly true of bureaucratic pigeon-holes such as "colored," "Hispanic," "Asian/Pacific Islander," and the one-drop rule for being "black," it is an overstatement when it comes to human differences in general. The biological anthropologist Vincent Sarich points out that a race is just a very large and partly inbred family. Some racial distinctions thus may have a degree of biological reality, even though they are not exact boundaries between fixed categories. Humans, having recently evolved from a single founder population, are all related, but Europeans, having mostly bred with other Europeans for millennia, are on average more closely related to other Europeans than they are to Africans or Asians, and vice versa. Because oceans, deserts, and mountain ranges have prevented people from choosing mates at random in the past, the large inbred families we call races are still discernible, each with a somewhat different distribution of gene frequencies. In theory, some of the varying genes could affect personality or intelligence (though any such differences would at most apply to averages, with vast overlap between the group members). This is not to say that such genetic differences are expected or that we have evidence for them, only that they are biologically possible.

(My own view, incidentally, is that in the case of the most discussed racial difference—the black-white IQ gap in the United States—the current evidence does not call for a genetic explanation. Thomas Sowell has documented that in most of the twentieth century and throughout the world, ethnic differences in IQ were the rule, not the exception.[9] Members of minority groups who were out of the cultural mainstream commonly had average IQs that fell below that of the majority, including immigrants to the United States from southern and eastern Europe, the children of white mountaineers in the United States, children who grew up on canal boats in Britain, and Gaelic-speaking children in the Hebrides. The differences were at least as large as the current black-white gap but disappeared within a few generations. For many reasons, the experience of African Americans in the United States under slavery and segregation is not comparable to those of immigrants or rural isolates, and their transition to mainstream cultural patterns could easily take longer.)[10]

And then there are the sexes. Unlike ethnic groups and races, in which any differences are biologically minor and haphazard, the two sexes differ in at least one way that is major and systematic: they have different reproductive organs. On evolutionary grounds one might expect men and women to differ somewhat in the neural systems that control how they use those organs—in their sexuality, parental instincts, and mating tactics. By the same logic, one would expect them not to differ as much in the neural systems that deal with the challenges both sexes face, such as those for general intelligence (as we will see in the chapter on gender).

～

So could discoveries in biology turn out to justify racism and sexism? Absolutely not! The case against bigotry is not a factual claim that humans are biologically indistinguishable. It is a moral stance that condemns judging an *individual* according to the average traits of certain *groups* to which the individual belongs. Enlightened societies choose to ignore race, sex, and ethnicity in hiring, promotion, salary, school admissions, and the criminal justice system because the alternative is morally repugnant. Discriminating against people on the basis of race, sex, or ethnicity would be unfair, penalizing them for traits over which they have no control. It would perpetuate the injustices of the past, in which African Americans, women, and other groups were enslaved or oppressed. It would rend society into hostile factions and could escalate into horrific persecution. But none of these arguments against discrimination depends on whether groups of people are or are not genetically indistinguishable.

Far from being conducive to discrimination, a conception of human nature is the reason we oppose it. Here is where the distinction between innate variation and innate universals is crucial. Regardless of IQ or physical strength or any other trait that can vary, all humans can be assumed to have certain traits in common. No one likes being enslaved. No one likes being humiliated. No one likes being treated unfairly, that is, according to traits that the person cannot control. The revulsion we feel toward discrimination and slavery comes from a conviction that however much people vary on some traits, they do not vary on these. This conviction contrasts, by the way, with the supposedly progressive doctrine that people have no inherent concerns, which implies that they could be conditioned to enjoy servitude or degradation.

The idea that political equality is a moral stance, not an empirical hypothesis, has been expressed by some of history's most famous exponents of equality. The Declaration of Independence proclaims, "We hold these truths to be self-evident; that all men are created equal." The author, Thomas Jefferson, made it clear that he was referring to an equality of rights, not a biological sameness. For example, in an 1813 letter to John Adams he wrote: "I agree with you that there is a natural aristocracy among men. The grounds of this are virtue and talents. . . . For experience proves, that the moral and physical qualities of man, whether good or evil, are transmissible in a certain degree from father to son."[11] (The fact that the Declaration originally was applied only to white men, and that Jefferson was far from an egalitarian in the conduct of his own life, does not change the argument. Jefferson defended political equality among white men—a novel idea in his time—even as he acknowledged innate differences among white men.) Similarly, Abraham Lincoln thought that the signers of the Declaration "did not mean to say all were equal in color, size, intellect, moral development or social capacity," but only in respect to "certain inalienable rights."[12]

Some of the most influential contemporary thinkers about biology and human nature have drawn the same distinction. Ernst Mayr, one of the founders of the modern theory of evolution, wisely anticipated nearly four decades of debate when he wrote in 1963:

> Equality in spite of evident nonidentity is a somewhat sophisticated concept and requires a moral stature of which many individuals seem to be incapable. They rather deny human variability and equate equality with identity. Or they claim that the human species is exceptional in the organic world in that only morphological characters are controlled by genes and all other traits of the mind or character are due to "conditioning" or other nongenetic factors. Such authors conveniently ignore the results of twin studies and of the genetic analysis of nonmorphological traits in animals. An ideology based on such obviously wrong premises can only lead to disaster. Its championship of human equality is based on a claim of identity. As soon as it is proved that the latter does not exist, the support of equality is likewise lost.[13]

Noam Chomsky made the same point in an article entitled "Psychology and Ideology." Though he disagreed with Herrnstein's argument about IQ (discussed in Chapter 6), he denied the popular charge that Herrnstein was a racist and distanced himself from fellow radical scientists who were denouncing the facts as dangerous:

> A correlation between race and IQ (were this shown to exist) entails no social consequences except in a racist society in which each individual is assigned to a racial category and dealt with not as an individual in his own right, but as a representative of this category. Herrnstein mentions a possible correlation between height and IQ. Of what social importance is that? None of course, since our society does not suffer under discrimination by height. We do not insist on assigning each adult to the category "below six feet in height" or "above six feet in height" when we ask what sort of education he should receive or where he should live or what work he should do. Rather, he is what he is, quite independent of the mean IQ of people of his height category. In a nonracist society, the category of race would be of no greater significance. The mean IQ of individuals of a certain racial background is irrelevant to the situation of a particular individual who is what he is. . . .
> It is, incidentally, surprising to me that so many commentators should find it disturbing that IQ might be heritable, perhaps largely so.

Would it also be disturbing to discover that relative height or musical talent or rank in running the one-hundred-yard dash is in part genetically determined? Why should one have preconceptions one way or another about these questions, and how do the answers to them, whatever they may be, relate either to serious scientific issues (in the present state of our knowledge) or to social practice in a decent society?[14]

Some readers may not be reassured by this lofty stance. If all ethnic groups and both sexes were identical in all talents, then discrimination would simply be self-defeating, and people would abandon it as soon as the facts were known. But if they are not identical, it would be *rational* to take those differences into account. After all, according to Bayes' theorem a decision maker who needs to make a prediction (such as whether a person will succeed in a profession) *should* factor in the prior probability, such as the base rate of success for people in that group. If races or sexes are different on average, racial profiling or gender stereotyping would be actuarially sound, and it would be naïve to expect information about race and sex not to be used for prejudicial ends. So a policy to treat people as individuals seems like a thin reed on which to hang any hope of reducing discrimination.

An immediate reply to this worry is that the danger arises whether the differences between groups are genetic or environmental in origin. An average is an average, and an actuarial decision maker should care only about what it is, not what caused it.

Moreover, the fact that discrimination can be economically rational would be truly dangerous only if our policies favored ruthless economic optimization regardless of all other costs. But in fact we have many policies that allow moral principles to trump economic efficiency. For example, it is illegal to sell your vote, sell your organs, or sell your children, even though an economist could argue that any voluntary exchange leaves both parties better off. These decisions come naturally in modern democracies, and we can just as resolutely choose public policies and private mores that disallow race and gender prejudice.[15]

Moral and legal proscriptions are not the only way to reduce discrimination in the face of possible group differences. The more information we have about the qualifications of an individual, the less impact a race-wide or sex-wide average would have in any statistical decision concerning that person. The best cure for discrimination, then, is more accurate and more extensive testing of mental abilities, because it would provide so much predictive information about an individual that no one would be tempted to factor in race or gender. (This, however, is an idea with no political future.)

Discrimination—in the sense of using a statistically predictive trait of an individual's group to make a decision about the individual—is not always immoral, or at least we don't always treat it as immoral. To predict someone's behavior perfectly we would need an X-ray machine for the soul. Even predicting someone's behavior with the tools we do have—such as tests, interviews, background checks, and recommendations—would require unlimited resources if we were to use them to the fullest. Decisions that have to be made with finite time and resources, and which have high costs for certain kinds of errors, must use *some* trait as a basis for judging a person. And that necessarily judges the person according to a stereotype.

In some cases the overlap between two groups is so small that we feel comfortable discriminating against one of the groups absolutely. For example, no one objects to keeping chimpanzees out of our schools, even though it is conceivable that if we tested every chimp on the planet we might find one that could learn to read and write. We apply a speciesist stereotype that chimps cannot profit from a human education, figuring that the odds of finding an exception do not outweigh the costs of examining every last one.

In more realistic circumstances we have to decide on a case-by-case basis whether the discrimination is justifiable. Denying driving and voting rights to young teenagers is a form of age discrimination that is unfair to responsible teens. But we are not willing to pay either the financial costs of developing a test for psychological maturity or the moral costs of classification errors, such as teens wrapping their cars around trees. Almost everyone is appalled by racial profiling—pulling over motorists for "driving while black." But after the 2001 terrorist attacks on the World Trade Center and the Pentagon, about half of Americans polled said they were not opposed to ethnic profiling—scrutinizing passengers for "flying while Arab."[16] People who distinguish the two must reason that the benefits of catching a marijuana dealer do not outweigh the harm done to innocent black drivers, but the benefits of stopping a suicide hijacker do outweigh the harm done to innocent Arab passengers. Cost-benefit analyses are also sometimes used to justify racial preferences: the benefits of racially diverse workplaces and campuses are thought to outweigh the costs of discriminating against whites.

The possibility that men and women are not the same in all respects also presents policymakers with choices. It would be reprehensible for a bank to hire a man over a woman as a manager for the reason that he is less likely to quit after having a child. Would it also be reprehensible for a couple to hire a woman over a man as a nanny for their daughter because she is less likely to sexually abuse the child? Most people believe that the punishment for a given crime should be the same regardless of who commits it. But knowing the typical sexual emotions of the two sexes, should we apply the same punishment to a man who seduces a sixteen-year-old girl and to a woman who seduces a sixteen-year-old boy?

These are some of the issues that face the people of a democracy in deciding what to do about discrimination. The point is not that group differences may *never* be used as a basis for discrimination. The point is that they do not *have* to be used that way, and sometimes we can decide on moral grounds that they must not be used that way.

~

THE BLANK SLATE, then, is not necessary to combat racism and sexism. Nor is it necessary to combat Social Darwinism, the belief that the rich and the poor deserve their status and so we must abandon any principle of economic justice in favor of extreme laissez-faire policies.

Because of a fear of Social Darwinism, the idea that class has anything to do with genes is treated by modern intellectuals like plutonium, even though it is hard to imagine how it could not be true in part. To adapt an example from the philosopher Robert Nozick, suppose a million people are willing to pay ten dollars to hear Pavarotti sing and are unwilling to pay ten dollars to hear me sing, in part because of genetic differences between us. Pavarotti will be ten million dollars richer and will live in an economic stratum that my genes keep me out of, even in a society that is totally fair.[17] It is a brute fact that greater rewards will go to people with greater inborn talent if other people are willing to pay more for the fruits of those talents. The only way that cannot happen is if people are locked into arbitrary castes, if all economic transactions are controlled by the state, or if there is no such thing as inborn talent because we are blank slates.

A surprising number of intellectuals, particularly on the left, do deny that there is such a thing as inborn talent, especially intelligence. Stephen Jay Gould's 1981 bestseller *The Mismeasure of Man* was written to debunk "the abstraction of intelligence as a single entity, its location within the brain, its quantification as one number for each individual, and the use of these numbers to rank people in a single series of worthiness, invariably to find that oppressed and disadvantaged groups—races, classes, or sexes—are innately inferior and deserve their status."[18] The philosopher Hilary Putnam argued that the concept of intelligence is part of a social theory called "elitism" that is specific to capitalist societies:

> Under a less competitive form of social organization, the theory of elitism might well be replaced by a different theory—the theory of egalitarianism. This theory might say that ordinary people can do anything that is in their interest and do it well when (1) they are highly motivated, and (2) they work collectively.[19]

In other words, any of us could become a Richard Feynman or a Tiger Woods if only we were highly enough motivated and worked collectively.

I find it truly surreal to read academics denying the existence of intelligence. Academics are *obsessed* with intelligence. They discuss it endlessly in considering student admissions, in hiring faculty and staff, and especially in their gossip about one another. Nor can citizens or policymakers ignore the concept, regardless of their politics. People who say that IQ is meaningless will quickly invoke it when the discussion turns to executing a murderer with an IQ of 64, removing lead paint that lowers a child's IQ by five points, or the presidential qualifications of George W. Bush. In any case, there is now ample evidence that intelligence is a stable property of an individual, that it can be linked to features of the brain (including overall size, amount of gray matter in the frontal lobes, speed of neural conduction, and metabolism of cerebral glucose), that it is partly heritable among individuals, and that it predicts some of the variation in life outcomes such as income and social status.[20]

The existence of inborn talents, however, does *not* call for Social Darwinism. The anxiety that one must lead to the other is based on two fallacies. The first is an all-or-none mentality that often infects discussions of the social implications of genetics. The likelihood that inborn differences are *one* contributor to social status does not mean that it is the *only* contributor. The other ones include sheer luck, inherited wealth, race and class prejudice, unequal opportunity (such as in schooling and connections), and cultural capital: habits and values that promote economic success. Acknowledging that talent matters doesn't mean that prejudice and unequal opportunity do not matter.

But more important, even if inherited talents can lead to socioeconomic success, it doesn't mean that the success is *deserved* in a moral sense. Social Darwinism is based on Spencer's assumption that we can look to evolution to discover what is right—that "good" can be boiled down to "evolutionarily successful." This lives in infamy as a reference case for the "naturalistic fallacy": the belief that what happens in nature is good. (Spencer also confused people's social success—their wealth, power, and status—with their evolutionary success, the number of their viable descendants.) The naturalistic fallacy was named by the moral philosopher G. E. Moore in his 1903 *Principia Ethica*, the book that killed Spencer's ethics.[21] Moore applied "Hume's Guillotine," the argument that no matter how convincingly you show that something is true, it never follows logically that it *ought* to be true. Moore noted that it is sensible to ask, "This conduct is more evolutionarily successful, but is it good?" The mere fact that the question makes sense shows that evolutionary success and goodness are not the same thing.

Can one really reconcile biological differences with a concept of social justice? Absolutely. In his famous theory of justice, the philosopher John Rawls asks us to imagine a social contract drawn up by self-interested agents negotiating under a veil of ignorance, unaware of the talents or status they will inherit at birth—ghosts ignorant of the machines they will haunt. He argues that a just society is one that these disembodied souls would agree to be born into, knowing that they might be dealt a lousy social or genetic hand.[22] If you agree that this is a reasonable conception of justice, and that the agents would insist on a broad social safety net and redistributive taxation (short of eliminating incentives that make everyone better off), then you can justify compensatory social policies *even if you think differences in social status are 100 percent genetic*. The policies would be, quite literally, a matter of justice, not a consequence of the indistinguishability of individuals.

Indeed, the existence of innate differences in ability makes Rawls's conception of social justice especially acute and eternally relevant. If we were blank slates, and if a society ever did eliminate discrimination, the poorest could be said to deserve their station because they must have chosen to do less with their standard-issue talents. But if people differ in talents, people might find themselves in poverty in a nonprejudiced society even if they applied themselves to the fullest. That is an injustice that, a Rawlsian would argue, ought to be rectified, and it would be overlooked if we didn't recognize that people differ in their abilities.

~

SOME PEOPLE HAVE suggested to me that these grandiloquent arguments are just too fancy for the dangerous world we live in. Granted, there is evidence that people are different, but since data in the social sciences are never perfect, and since a conclusion of inequality might be used to the worst ends by bigots or Social Darwinists, shouldn't we err on the side of caution and stick with the null hypothesis that people are identical? Some believe that even if we were *certain* that people differ genetically, we might still want to promulgate the fiction that they are the same, because it is less open to abuse.

This argument is based on the fallacy that the Blank Slate has nothing but good moral implications and a theory of human nature nothing but bad ones. In the case of human differences, as in the case of human universals, the dangers go both ways. If people in different stations are mistakenly thought to differ in their inherent ability, we might overlook discrimination and unequal opportunity. In Darwin's words, "If the misery of the poor be caused not by the laws of nature, but by our institutions, great is our sin." But if people in different stations are mistakenly thought to be the same, then we might envy them the rewards they've earned fair and square and might implement coercive policies to hammer down the nails that stick up. The economist Friedrich Hayek wrote, "It is just not true that humans are born equal; . . . if we treat them equally, the result must be inequality in their actual position; . . . [thus] the only way to place them in an equal position would be to treat them differently. Equality before the law and material equality are, therefore, not only different but in conflict with each other."[23] The philosophers Isaiah Berlin, Karl Popper, and Robert Nozick have made similar points.

Unequal treatment in the name of equality can take many forms. Some forms have both defenders and detractors, such as soak-the-rich taxation, heavy estate taxes, streaming by age rather than ability in schools, quotas and preferences that favor certain races or regions, and prohibitions against private medical care or other voluntary transactions. But some can be downright dangerous. If people are assumed to start out identical but some end up wealthier than others, observers may conclude that the wealthier ones must be more rapacious. And as the diagnosis slides from talent to sin, the remedy can shift from redistribution to vengeance. Many atrocities of the twentieth century were committed in the name of egalitarianism, targeting people whose success was taken as evidence of their criminality. The kulaks ("bourgeois peasants") were exterminated by Lenin and Stalin in the Soviet Union; teachers, former landlords, and "rich peasants" were humiliated, tortured, and murdered during China's Cultural Revolution; city dwellers and literate professionals were worked to death or executed during the reign of the Khmer Rouge in Cambodia.[24] Educated and entrepreneurial minorities who have prospered in their adopted regions, such as the Indians in East Africa and Oceania, the Ibos in Nigeria, the Armenians in Turkey, the Chinese in Indonesia and Malaysia, and the Jews almost everywhere, have been expelled from their homes or killed in pogroms because their visibly successful members were seen as parasites and exploiters.[25]

A nonblank slate means that a tradeoff between freedom and material equality is inherent to all political systems. The major political philosophies can be defined by how they deal with the tradeoff. The Social Darwinist right places no value on equality; the totalitarian left places no value on freedom. The Rawlsian left sacrifices some freedom for equality; the libertarian right sacrifices some equality for freedom. While reasonable people may disagree about the best tradeoff, it is unreasonable to pretend there *is* no tradeoff. And that in turn means that any discovery of innate differences among individuals is not forbidden knowledge to be suppressed but information that might help us decide on these tradeoffs in an intelligent and humane manner.

~

THE SPECTER OF eugenics can be disposed of as easily as the specters of discrimination and Social Darwinism. Once again, the key is to distinguish biological facts from human values.

If people differ genetically in intelligence and character, could we selectively breed for smarter and nicer people? Possibly, though the intricacies of genetics and development would make it far harder than the fans of eugenics imagined. Selective breeding is straightforward for genes with additive effects—that is, genes that have the same impact regardless of the other genes in the genome. But some traits, such as scientific genius, athletic virtuosity, and musical giftedness, are what behavioral geneticists call emergenic: they materialize only with certain *combinations* of genes and therefore don't "breed true."[26] Moreover, a given gene can lead to different behavior in different environments. When the biochemist (and radical scientist) George Wald was solicited for a semen sample by William Shockley's sperm bank for Nobel Prize–winning scientists, he replied, "If you want sperm that produces Nobel Prize winners you should be contacting people like my father, a poor immigrant tailor. What have my sperm given the world? Two guitarists!"[27]

Whether or not we *can* breed for certain traits, *should* we do it? It would require a government wise enough to know which traits to select, knowledgeable enough to know how to implement the breeding, and intrusive enough to encourage or coerce people's most intimate decisions. Few people in a democracy would grant their government that kind of power even if it did promise a better society in the future. The costs in freedom to individuals and in possible abuse by authorities are unacceptable.

Contrary to the belief spread by the radical scientists, eugenics for much of the twentieth century was a favorite cause of the left, not the right.[28] It was championed by many progressives, liberals, and socialists, including Theodore Roosevelt, H. G. Wells, Emma Goldman, George Bernard Shaw, Harold Laski, John Maynard Keynes, Sidney and Beatrice Webb, Margaret Sanger, and the Marxist biologists J. B. S. Haldane and Hermann Muller. It's not hard to see why the sides lined up this way. Conservative Catholics and Bible Belt Protestants hated eugenics because it was an attempt by intellectual and scientific elites to play God. Progressives loved eugenics because it was on the side of reform rather than the status quo, activism rather than laissez-faire, and social responsibility rather than selfishness. Moreover, they were comfortable expanding state intervention in order to bring about a social goal. Most abandoned eugenics only when they saw how it led to forced sterilizations in the United States and Western Europe and, later, to the policies of Nazi Germany. The history of eugenics is one of many cases in which the moral problems posed by human nature cannot be folded into familiar left-right debates but have to be analyzed afresh in terms of the conflicting values at stake.

⁓

THE MOST SICKENING associations of a biological conception of human nature are the ones to Nazism. Though the opposition to the idea of a human nature began decades earlier, historians agree that bitter memories of the Holocaust were the main reason that human nature became taboo in intellectual life after World War II.

Hitler was undeniably influenced by the bastardized versions of Darwinism and genetics that were popular in the early decades of the twentieth century, and he specifically cited natural selection and the survival of the fittest in laying out his poisonous doctrine. He believed in an extreme Social Darwinism in which groups were the unit of selection and a struggle among groups was necessary for national strength and vigor. He believed that the groups were constitutionally distinct races, that their members shared a distinctive biological makeup, and that they differed from one another in strength, courage, honesty, intelligence, and civic-mindedness. He wrote that the extinction of inferior races was part of the wisdom of nature, that the superior races owed their vitality and virtue to their genetic purity, and that the superior races were in danger of being degraded by interbreeding with the inferior ones. He used these beliefs to justify his war of conquest and his genocide of Jews, Gypsies, Slavs, and homosexuals.[29]

The misuse of biology by the Nazis is a reminder that perverted ideas can have horrifying consequences and that intellectuals have a responsibility to take reasonable care that their ideas not be misused for evil ends. But part of that responsibility is not to trivialize the horror of Nazism by exploiting it for rhetorical clout in academic catfights. Linking the people you disagree with to Nazism does nothing for the memory of Hitler's victims or for the effort to prevent other genocides. It is precisely because these events are so grave that we have a special responsibility to identify their causes precisely.

An idea is not false or evil because the Nazis misused it. As the historian Robert Richards wrote of an alleged connection between Nazism and evolutionary biology, "If such vague similarities suffice here, we should all be hustled to the gallows."[30] Indeed, if we censored ideas that the Nazis abused, we would have to give up far more than the application of evolution and genetics to human behavior. We would have to censor the study of evolution and genetics, period. And we would have to suppress many other ideas that Hitler twisted into the foundations of Nazism:

- *The germ theory of disease:* The Nazis repeatedly cited Pasteur and Koch to argue that the Jews were like an infectious bacillus that had to be eradicated to control a contagious disease.
- *Romanticism, environmentalism, and the love of nature:* The Nazis amplified a Romantic strain in German culture that believed the *Volk* were a people of destiny with a mystical bond to nature and the land. The Jews and other minorities, in contrast, took root in the degenerate cities.
- *Philology and linguistics:* The concept of the Aryan race was based on a prehistoric tribe posited by linguists, the Indo-Europeans, who were thought to have spilled out of an ancient homeland thousands of years ago and to have conquered much of Europe and Asia.
- *Religious belief:* Though Hitler disliked Christianity, he was not an atheist, and was emboldened by the conviction that he was carrying out a divinely ordained plan.[31]

The danger that we might distort our own science as a reaction to the Nazis' distortions is not hypothetical. The historian of science Robert Proctor has shown that American public health officials were slow to acknowledge that smoking causes cancer because it was the Nazis who had originally established the link.[32] And some German scientists argue that biomedical research has been crippled in their country because of vague lingering associations to Nazism.[33]

Hitler was evil because he caused the deaths of thirty million people and inconceivable suffering to countless others, not because his beliefs made reference to biology (or linguistics or nature or smoking or God). Smearing the guilt from his actions to every conceivable aspect of his factual beliefs can only backfire. Ideas are connected to other ideas, and should any of Hitler's turn out to have some grain of truth—if races, for example, turn out to have any biological reality, or if the Indo-Europeans really were a conquering tribe—we would not want to concede that Nazism wasn't so wrong after all.

The Nazi Holocaust was a singular event that changed attitudes toward countless political and scientific topics. But it was not the only ideologically inspired holocaust in the twentieth century, and intellectuals are only beginning to assimilate the lessons of the others: the mass killings in the Soviet Union, China, Cambodia, and other totalitarian states carried out in the name of Marxism. The opening of Soviet archives and the release of data and memoirs on the Chinese and Cambodian revolutions are forcing a reevaluation of the consequences of ideology as wrenching as that following World War II. Historians are currently debating whether the Communists' mass executions, forced marches, slave labor, and man-made famines led to one hundred million deaths or "only" twenty-five million. They are debating whether these atrocities are morally worse than the Nazi Holocaust or "only" the equivalent.[34]

And here is the remarkable fact: though both Nazi and Marxist ideologies led to industrial-scale killing, *their biological and psychological theories were opposites.* Marxists had no use for the concept of race, were averse to the notion of genetic inheritance, and were hostile to the very idea of a human nature rooted in biology.[35] Marx and Engels did not explicitly embrace the doctrine of the Blank Slate in their writings, but they were adamant that human nature has no enduring properties. It consists only in the interactions of groups of people with their material environments in a historical period, and constantly changes as people change their environment and are simultaneously changed by it.[36] The mind therefore has no innate structure but emerges from the dialectical processes of history and social interaction. As Marx put it:

All history is nothing but a continuous transformation of human nature.[37]

Circumstances make men just as much as men make circumstances.[38]

The mode of production of material life conditions the social, political, and intellectual life processes in general. It is not the consciousness of men that determines their being, but, on the contrary, their social being that determines their consciousness.[39]

In a foreshadowing of Durkheim's and Kroeber's insistence that individual human minds are not worthy of attention, Marx wrote:

Man is not an abstract being, squatting outside the world. Man is the *world of men,* the State and Society. The essence of man is not an abstraction inherent in each particular individual. The real nature of man is the totality of social relations.[40]

Individuals are dealt with only in so far as they are the personifications of economic categories, embodiments of particular class-relations and class interests.[41]

[Death] seems to be a harsh victory of the species over the *particular* individual and to contradict their unity. But the particular individual is only a *particular species-being,* and as such mortal.[42]

Marx's twentieth-century followers did embrace the Blank Slate, or at least the related metaphor of malleable material. Lenin endorsed Nikolai Bukharin's ideal of "the manufacturing of Communist man out of the human material of the capitalist age."[43] Lenin's admirer Maxim Gorky wrote, "The working classes are to Lenin what minerals are to the metallurgist"[44] and "Human raw material is immeasurably more difficult to work with than wood" (the latter while admiring a canal built by slave labor).[45] We come across the metaphor of the blank slate in the writings of a man who may have been responsible for sixty-five million deaths:

A blank sheet of paper has no blotches, and so the newest and most beautiful words can be written on it, the newest and most beautiful pictures can be painted on it.

—Mao Zedong[46]

And we find it in a saying of a political movement that killed a quarter of its countrymen:

Only the newborn baby is spotless.

—Khmer Rouge slogan[47]

The new realization that government-sponsored mass murder can come from an anti-innatist belief system as easily as from an innatist one upends the postwar understanding that biological approaches to behavior are uniquely sinister. An accurate appraisal of the cause of state genocides must look for beliefs common to Nazism and Marxism that launched them on their parallel trajectories, and for the beliefs specific to Marxism that led to the unique atrocities committed in its name. A new wave of historians and philosophers is doing exactly that.[48]

Nazism and Marxism shared a desire to reshape humanity. "The alteration of men on a mass scale is necessary," wrote Marx; "the will to create mankind anew" is the core of National Socialism, wrote Hitler.[49] They also shared a revolutionary idealism and a tyrannical certainty in pursuit of this dream, with no patience for incremental reform or adjustments guided by the human consequences of their policies. This alone was a recipe for disaster. As Aleksandr Solzhenitsyn wrote in *The Gulag Archipelago*, "Macbeth's self-justifications were feeble—and his conscience devoured him. Yes, even Iago was a little lamb too. The imagination and the spiritual strength of Shakespeare's evildoers stopped short at a dozen corpses. Because they had no *ideology*."

The ideological connection between Marxist socialism and National Socialism is not fanciful.[50] Hitler read Marx carefully while living in Munich in 1913, and may have picked up from him a fateful postulate that the two ideologies would share.[51] It is the belief that history is a preordained succession of conflicts between groups of people and that improvement in the human condition can come only from the victory of one group over the others. For the Nazis the groups were races; for the Marxists they were classes. For the Nazis the conflict was Social Darwinism; for the Marxists, it was class struggle. For the Nazis the destined victors were the Aryans; for the Marxists, they were the proletariat. The ideologies, once implemented, led to atrocities in a few steps: struggle (often a euphemism for violence) is inevitable and beneficial; certain groups of people (the non-Aryan races or the bourgeoisie) are morally inferior; improvements in human welfare depend on their subjugation or elimination. Aside from supplying a direct justification for violent conflict, the ideology of intergroup struggle ignites a nasty feature of human social psychology: the tendency to divide people into in-groups and out-groups and to treat the out-groups as less than human. It doesn't matter whether the groups are thought to be defined by their biology or by their history. Psychologists have found that they can create instant intergroup hostility by sorting people on just about any pretext, including the flip of a coin.[52]

The ideology of group-against-group struggle explains the similar outcomes of Marxism and Nazism. The ideology of the Blank Slate helps explain some of the features that were unique to the Marxist states:

- If people do not differ in psychological traits like talent or drive, then anyone who is better off must be avaricious or larcenous (as I mentioned earlier). Massive killing of kulaks and "rich" or "bourgeois" peasants was a feature of Lenin's and Stalin's Soviet Union, Mao's China, and Pol Pot's Cambodia.

- If the mind is structureless at birth and shaped by its experience, a society that wants the right kind of minds must control the experience ("It is on a blank page that the most beautiful poems are written").[53] Twentieth-century Marxist states were not just dictatorships but *totalitarian* dictatorships. They tried to control every aspect of life: childrearing, education, clothing, entertainment, architecture, the arts, even food and sex. Authors in the Soviet Union were enjoined to become "engineers of human souls." In China and Cambodia, mandatory communal dining halls, same-sex adult dormitories, and the separation of children from parents were recurring (and detested) experiments.

- If people are shaped by their social environments, then growing up bourgeois can leave a permanent psychological stain ("Only the newborn baby is spotless"). The descendants of landlords and "rich peasants" in postrevolutionary regimes bore a permanent stigma and were persecuted as readily as if bourgeois parentage were a genetic trait. Worse, since parentage is invisible but discoverable by third parties, the practice of outing people with a "bad background" became a weapon of social competition. That led to the atmosphere of denunciation and paranoia that made life in these regimes an Orwellian nightmare.

- If there is no human nature leading people to favor the interests of their families over "society," then people who produce more crops on their own plots than on communal farms whose crops are confiscated by the state must be greedy or lazy and punished accordingly. Fear rather than self-interest becomes the incentive to work.

- Most generally, if individual minds are interchangeable components of a superorganic entity called society, then the society, not the individual, is the natural unit of health and well-being and the proper beneficiary of human striving. The rights of the individual person have no place.

None of this is meant to impugn the Blank Slate as an evil doctrine, any more than a belief in human nature is an evil doctrine. Both are separated by a great many steps from the wicked acts committed under their banners, and they must be evaluated on factual grounds. But it *is* meant to overturn the simplistic linkage of the sciences of human nature with the moral catastrophes of the twentieth century. That glib association stands in the way of our desire to understand ourselves, and it stands in the way of the imperative to understand the causes of those catastrophes. All the more so if the causes have something to do with a side of ourselves we do not fully understand.

Chapter 9

The Fear of Imperfectibility

But Nature then was sovereign in my mind,
And mighty forms, seizing a youthful fancy,
Had given a charter to irregular hopes.
In any age of uneventful calm
Among the nations, surely would my heart
Have been possessed by similar desire;
But Europe at that time was thrilled with joy,
France standing on the top of golden hours,
And human nature seeming born again.
—William Wordsworth[1]

In Wordsworth's reminiscence we find the second fear raised by an innate psyche. The Romantic poet is exhilarated by the thought that human nature can be born again, and could only be depressed by the possibility that we are permanently saddled with our fatal flaws and deadly sins. Romantic political thinkers have the same reaction, because an unchanging human nature would seem to subvert all hope for reform. Why try to make the world a better place if people are rotten to the core and will just foul it up no matter what you do? It is no coincidence that the writings of Rousseau inspired both the Romantic movement in literature and the French Revolution in history, or that the 1960s would see a resurfacing of romanticism and radical politics in tandem. The philosopher John Passmore has shown that a yearning for a better world through a new and improved human nature is a recurring motif in Western thought, which he summarizes in a remark by D. H. Lawrence: "The Perfectibility of Man! Ah, heaven, what a dreary theme!"[2]

The dread of a permanently wicked human nature takes two forms. One is a practical fear: that social reform is a waste of time because human nature is unchangeable. The other is a deeper concern, which grows out of the Romantic belief that what is natural is good. According to the worry, if scientists suggest it is "natural"—part of human nature—to be adulterous, violent, ethnocentric, and selfish, they would be implying that these traits are *good,* not just unavoidable.

As with the other convictions surrounding the Blank Slate, the fear of imperfectibility makes some sense in the context of twentieth-century history. A revulsion to the idea that people are naturally bellicose or xenophobic is an understandable reaction to an ideology that glorified war. One of the most memorable images I came across as a graduate student was a painting of a dead soldier in a muddy field. A uniformed ghost floated up from his corpse, one arm around a cloaked and faceless man, the other around a bare-breasted blond valkyrie. The caption read, "Happy those who with a glowing faith in one embrace clasped death and victory." Was it a kitschy poster recruiting cannon fodder for an imperial exploit? A jingoistic monument in the castle of a Prussian military aristocrat? No, *Death and Victory* was painted in 1922 by the great American artist John Singer Sargent and hangs prominently in one of the world's most famous scholarly libraries, the Widener at Harvard University.

That a piece of pro-death iconography should decorate these hallowed halls of learning is a testament to the warmongering mentality of decades past. War was thought to be invigorating, ennobling, the natural aspiration of men and nations. This belief led world leaders to sleepwalk into World War I and millions of men to enlist eagerly, oblivious to the carnage that lay ahead. Beginning with the disillusionment following that war and culminating in the widespread opposition to the war in Vietnam, Western sensibilities have steadily recoiled from the glorification of combat. Even recent works meant to honor the courage of fighting men, such as the movie *Saving Private Ryan,* show war as a hell that brave men endured at terrible cost to eliminate an identified evil, not something they could possibly feel "happy" about. Real wars today are waged with remote-control gadgetry to minimize casualties, sometimes at the cost of downgrading the war's objectives. In this climate any suggestion that war is "natural" will be met with indignant declarations to the contrary, such as the recurring Statements on Violence by social scientists averring that it is "scientifically incorrect" to say that humans have tendencies toward aggression.[3]

A hostility to the idea that selfish sexual urges might be rooted in our nature comes from feminism. For millennia women have suffered under a double standard based on assumptions about differences between the sexes. Laws and customs punished the philandering of women more harshly than the philandering of men. Fathers and husbands stripped women of control over their sexuality by constraining their appearance and movement. Legal systems exonerated rapists or mitigated their punishment if the victim was thought to have aroused an irresistible urge by her dress or behavior. Authorities brushed off victims of harassment, stalking, and battering by assuming that these crimes were normal features of courtship or marriage. Because of a fear of accepting any idea that would seem to make these outrages "natural" or unavoidable, some schools of feminism have rejected any suggestion that men are born with greater sexual desire or jealousy. We saw in Chapter 7 that the claim that men want casual sex more than women do has been denounced by both the right and the left. Even heavier bipartisan fire has recently been aimed at Randy Thornhill and Craig Palmer for suggesting in their book *A Natural History of Rape* that rape is a consequence of men's sexuality. A spokesperson from the Feminist Majority Foundation called the book "scary" and "regressive" because it "almost validates the crime and blames the victim."[4] A spokesperson for the Discovery Institute, a creationist organization, testified at a U.S. congressional hearing that the book threatened the moral fabric upon which America is founded.[5]

A third vice with political implications is selfishness. If people, like other animals, are driven by selfish genes, selfishness might seem inevitable or even a virtue. The argument is fallacious from the start because selfish genes do not necessarily grow selfish organisms. Still, let us consider the possibility that people might have some tendency to value their own interests and those of their family and friends above the interests of the tribe, society, or species. The political implications are spelled out in the two major philosophies of how societies should be organized, which make opposite assumptions about innate human selfishness:

> It is not from the benevolence of the butcher, the brewer, or the baker that we expect our dinner, but from their regard to their own interest. We address ourselves, not to their humanity but to their self-love.
>
> —Adam Smith

> From each according to his abilities, to each according to his needs.
>
> —Karl Marx

Smith the explainer of capitalism assumes that people will selfishly give their labor according to their needs and will be paid according to their abilities (because the payers are selfish, too). Marx the architect of communism and socialism assumes that in a socialist society of the future the butcher, the brewer, and the baker will provide us with dinner out of benevolence or self-actualization—for why else would they cheerfully exert themselves according to their abilities and not according to their needs?

Those who believe that communism or socialism is the most rational form of social organization are aghast at the suggestion that they run against our selfish natures. For that matter, everyone, regardless of politics, has to be appalled at people who impose costs on society in pursuit of their individual interests—hunting endangered species, polluting rivers, destroying historic sites to build shopping malls, spraying graffiti on public monuments, inventing weapons that elude metal detectors. Equally disturbing are the outcomes of actions that make sense to the individual choosing them but are costly to society when everyone chooses them. Examples include overfishing a harbor, overgrazing a commons, commuting on a bumper-to-bumper freeway, or buying a sport utility vehicle to protect oneself in a collision because everyone else is driving a sport utility vehicle. Many people dislike the suggestion that humans are inclined to selfishness because it would seem to imply that these self-defeating patterns of behavior are inevitable, or at least reducible only through permanent coercive measures.

~

THE FEAR OF imperfectibility and the resultant embrace of the Blank Slate are rooted in a pair of fallacies. We have already met the naturalistic fallacy, the belief that whatever happens in nature is good. One might think that the belief was irreversibly tainted by Social Darwinism, but it was revived by the romanticism of the 1960s and 1970s. The environmentalist movement, in particular, often appeals to the goodness of nature to promote conservation of natural environments, despite their ubiquitous gore. For example, predators such as wolves, bears, and sharks have been given an image makeover as euthanists of the old and the lame, and thus worthy of preservation or reintroduction. It would seem to follow that anything we have inherited from this Eden is healthy and proper, so a claim that aggression or rape is "natural," in the sense of having been favored by evolution, is tantamount to saying that it is good.

The naturalistic fallacy leads quickly to its converse, the moralistic fallacy: that if a trait is moral, it must be found in nature. That is, not only does "is" imply "ought," but "ought" implies "is." Nature, including human nature, is stipulated to have only virtuous traits (no needless killings, no rapacity, no exploitation), or no traits at all, because the alternative is too horrible to accept. That is why the naturalistic and moralistic fallacies are so often associated with the Noble Savage and the Blank Slate.

Defenders of the naturalistic and moralistic fallacies are not made of straw but include prominent scholars and writers. For example, in response to Thornhill's earlier writings on rape, the feminist scholar Susan Brownmiller wrote, "It seems quite clear that the biologicization of rape and the dismissal of social or 'moral' factors will . . . tend to legitimate rape. . . . It is reductive and reactionary to isolate rape from other forms of violent antisocial behavior and dignify it with adaptive significance."[6] Note the fallacy: if something is explained with biology, it has been "legitimated"; if something is shown to be adaptive, it has been "dignified." Similarly, Stephen Jay Gould wrote of another discussion of rape in animals, "By falsely describing an inherited behavior in birds with an old name for a deviant human action, we subtly suggest that true rape—our own kind—might be a natural behavior with Darwinian advantages to certain people as well."[7] The implicit rebuke is that to describe an act as "natural" or as having "Darwinian advantages" is somehow to condone it.

The moralistic fallacy, like the naturalistic fallacy, is, well, a fallacy, as we learn from this *Arlo and Janis* cartoon:

Arlo & Janis reprinted by permission of Newspaper Enterprise Association, Inc.

The boy has biology on his side.[8] George Williams, the revered evolutionary biologist, describes the natural world as "grossly immoral."[9] Having no foresight or compassion, natural selection "can honestly be described as a process for maximizing short-sighted selfishness." On top of all the miseries inflicted by predators and parasites, the members of a species show no pity to their own kind. Infanticide, siblicide, and rape can be observed in many kinds of animals; infidelity is common even in so-called pair-bonded species; cannibalism can be expected in all species that are not strict vegetarians; death from fighting is more common in most animal species than it is in the most violent American cities.[10] Commenting on how biologists used to describe the killing of starving deer by mountain lions as an act of mercy, Williams wrote:

> The simple facts are that both predation and starvation are painful prospects for deer, and that the lion's lot is no more enviable. Perhaps biology would have been able to mature more rapidly in a culture not dominated by Judeo-Christian theology and the Romantic tradition. It might have been well served by the First Holy Truth from [Buddha's] Sermon at Benares: "Birth is painful, old age is painful, sickness is painful, death is painful . . ."[11]

As soon as we recognize that there is nothing morally commendable about the products of evolution, we can describe human psychology honestly, without the fear that identifying a "natural" trait is the same as condoning it. As Katharine Hepburn says to Humphrey Bogart in *The African Queen,* "Nature, Mr. Allnut, is what we are put in this world to rise above."

Crucially, this cuts both ways. Many commentators from the religious and cultural right believe that any behavior that strikes them as biologically atypical, such as homosexuality, voluntary childlessness, and women who assume traditional male roles or vice versa, should be condemned because it is "unnatural." For example, the popular talk-show host Laura Schlesinger has declared, "I am getting people to stop doing wrong and start doing right." As part of this crusade she has called on gay people to submit to therapy to change their sexual orientation, because homosexuality is a "biological error." This kind of moral reasoning can come only from people who know nothing about biology. Most activities that moral people extol—being faithful to one's spouse, turning the other cheek, treating every child as precious, loving thy neighbor as thyself—are "biological errors" and are utterly unnatural in the rest of the living world.

Acknowledging the naturalistic fallacy does not mean that facts about human nature are *irrelevant* to our choices.[12] The political scientist Roger Masters, noting that the naturalistic fallacy can be invoked too glibly to deny the relevance of biology to human affairs, points out, "When the physician says a patient ought to have an operation because the facts show appendicitis, the patient is unlikely to complain about a fallacious logical deduction."[13] Acknowledging the naturalistic fallacy implies only that discoveries about human nature do not, by themselves, *dictate* our choices. The facts must be combined with a statement of values and a method of resolving conflicts among them. Given the fact of appendicitis, the value that health is desirable, and the conviction that the pain and expense of the operation are outweighed by the resulting gain in health, one ought to have the operation.

Suppose rape is rooted in a feature of human nature, such as that men want sex across a wider range of circumstances than women do. It is *also* a feature of human nature, just as deeply rooted in our evolution, that women want control over when and with whom they have sex. It is inherent to our value system that the interests of women should not be subordinated to those of men, and that control over one's body is a fundamental right that trumps other people's desires. So rape is not tolerated, regardless of any possible connection to the nature of men's sexuality. Note how this calculus requires a "deterministic" and "essentialist" claim about human nature: that women abhor being raped. Without that claim we would have no way to choose between trying to deter rape and trying to socialize women to accept it, which would be perfectly compatible with the supposedly progressive doctrine that we are malleable raw material.

In other cases, the best way to resolve a conflict is not as obvious. The psychologists Martin Daly and Margo Wilson have documented that stepparents are far more likely to abuse a child than are biological parents. The discovery was by no means banal: many parenting experts insist that the abusive stepparent is a myth originating in Cinderella stories and that parenting is a "role" that anyone can take on. Daly and Wilson had originally examined the abuse statistics to test a prediction from evolutionary psychology.[14] Parental love is selected over evolutionary time because it compels parents to protect and nurture their children, who are likely to carry the genes giving rise to parental love. In any species in which someone else's offspring are likely to enter the family circle, selection will favor a tendency to prefer one's own, because in the cold reckoning of natural selection an investment in the unrelated children would go to waste. A parent's patience will tend to run out with stepchildren more quickly than with biological children, and in extreme cases this can lead to abuse.

Does all this mean that social service agencies should monitor stepparents more closely than biological parents? Not so fast. The vast majority of both kinds of parents never commit abuse, so putting stepparents under a cloud of suspicion would be unfair to millions of innocent people. As the legal scholar Owen Jones points out, the evolutionary analysis of stepparenting—or of anything else—has no automatic policy implications. Rather, it delineates a tradeoff and forces us to choose an optimum along it. In this case, the tradeoff is between minimizing child abuse while stigmatizing stepparents, on one hand, and being maximally fair to stepparents while tolerating an increase in child abuse, on the other.[15] If we did not know that people are predisposed to lose patience with stepchildren faster than with biological children, we would implicitly choose one end of this tradeoff—ignoring stepparenting as a risk factor altogether, and tolerating the extra cases of child abuse—without even realizing it.

An understanding of human nature with all its weaknesses can enrich not just our policies but our personal lives. Families with stepchildren tend to be less happy and more fragile than families with biological children, largely because of tensions over how much time, patience, and money the stepparents should expend. Many stepparents, nonetheless, *are* kind and generous to a spouse's children, in part out of love for the spouse. Still, there is a difference between the instinctive love that parents automatically lavish on their own children and the deliberate kindness and generosity that wise stepparents extend to their stepchildren. Understanding this difference, Daly and Wilson suggest, could enhance a marriage.[16] Though a marriage based on strict tit-for-tat reciprocity is generally miserable, a good marriage finds each spouse appreciating the sacrifices that the other has made over the long haul. Acknowledging a partner's conscious benevolence toward one's children may ultimately breed less resentment and misunderstanding than *demanding* such benevolence as a matter of course and begrudging any ambivalence the partner may feel. It is one of many ways in which a realism about the imperfect emotions we actually have may bring more happiness than an illusion about the ideal emotions we wish we had.

～

SO IF WE are put in this world to rise above nature, how do we do it? Where in the causal chain of evolved genes building a neural computer do we find a chink into which we can fit the seemingly unmechanical event of "choosing values"? By allowing for choice, are we just inviting a ghost back into the machine?

The question is itself a symptom of the Blank Slate. If one starts off thinking the slate is blank, then when someone proposes an innate desire one will mentally plunk it onto the barren surface in one's imagination and conclude that it must be an ineluctable urge, because there is nothing else on the slate to counteract it. Selfish thoughts translate into selfish behavior, aggressive urges beget natural-born killers, a taste for multiple sexual partners means that men just can't help fooling around. For example, when the primatologist Michael Ghiglieri appeared on the National Public Radio program *Science Friday* to talk about his book on violence, the interviewer asked, "You explain rape and murder and war and all the bad things that men do as something—if I would just boil it down—something they can't help because of its—it's locked up in their evolutionary genes there?"[17]

If, however, the mind is a system with many parts, then an innate desire is just one component among others. Some faculties may endow us with greed or lust or malice, but others may endow us with sympathy, foresight, self-respect, a desire for respect from others, and an ability to learn from our own experiences and those of our neighbors. These are physical circuits residing in the prefrontal cortex and other parts of the brain, not occult powers of a poltergeist, and they have a genetic basis and an evolutionary history no less than the primal urges. It is only the Blank Slate and the Ghost in the Machine that make people think that drives are "biological" but that thinking and decision making are something else.

The faculties underlying empathy, foresight, and self-respect are information-processing systems that accept input and commandeer other parts of the brain and body. They are combinatorial systems, like the mental grammar underlying language, capable of cranking out an unlimited number of ideas and courses of action. Personal and social change can come about when people exchange information that affects those mechanisms—even if we *are* nothing but meat puppets, glorified clockwork, or lumbering robots created by selfish genes.

Not only is acknowledging human nature compatible with social and moral progress, but it can help explain the obvious progress that has taken place over millennia. Customs that were common throughout history and prehistory—slavery, punishment by mutilation, execution by torture, genocide for convenience, endless blood feuds, the summary killing of strangers, rape as the spoils of war, infanticide as a form of birth control, and the legal ownership of women—have vanished from large parts of the world.

The philosopher Peter Singer has shown how continuous moral progress can emerge from a fixed moral sense.[18] Suppose we are endowed with a conscience that treats other persons as targets of sympathy and inhibits us from harming or exploiting them. Suppose, too, that we have a mechanism for assessing whether a living thing gets to be classified as a person. (After all, we don't want to classify plants as persons and starve before we would eat them.) Singer explains moral improvement in the title of his book: *The Expanding Circle*. People have steadily expanded the mental dotted line that embraces the entities considered worthy of moral consideration. The circle been poked outward from the family and village to the clan, the tribe, the nation, the race, and most recently (as in the Universal Declaration of Human Rights) to all of humanity. It has been slackened from royalty, aristocracy, and property holders to all men. It has grown from including only men to including women, children, and newborns. It has crept outward to embrace criminals, prisoners of war, enemy civilians, the dying, and the mentally handicapped.

Nor are the possibilities for moral progress over. Today some people want to enlarge the circle to include great apes, warm-blooded creatures, or animals with central nervous systems. Some want to count in zygotes, blastocysts, fetuses, and the brain-dead. Still others want to embrace species, ecosystems, or the entire planet. This sweeping change in sensibilities, the driving force in the moral history of our species, did not require a blank slate or a ghost in the machine. It could have arisen from a moral gadget containing a single knob or slider that adjusts the size of the circle embracing the entities whose interests we treat as comparable to our own.

The expansion of the moral circle does not have to be powered by some mysterious drive toward goodness. It may come from the interaction between the selfish process of evolution and a law of complex systems. The biologists John Maynard Smith and Eörs Szathmáry and the journalist Robert Wright have explained how evolution can lead to greater and greater degrees of cooperation.[19] Repeatedly in the history of life, replicators have teamed up, specialized to divide the labor, and coordinated their behavior. It happens because replicators often find themselves in non-zero-sum games, in which particular strategies adopted by two players can leave them both better off (as opposed to a zero-sum game, where one player's profit is another player's loss). An exact analogy is found in the play by William Butler Yeats in which a blind man carries a lame man on his shoulders, allowing both of them to get around. During the evolution of life this dynamic has led replicating molecules to team up in chromosomes, organelles to team up in cells, cells to agglomerate into complex organisms, and organisms to hang out in societies. Independent agents repeatedly made their fate hostage to a larger system, not because they are inherently civic-minded but because they benefited from the division of labor and developed ways of damping conflicts among the agents making up the system.

Human societies, like living things, have become more complicated and cooperative over time. Again, it is because agents do better when they team up and specialize in pursuit of their shared interests, as long as they solve the problems of exchanging information and punishing cheaters. If I have more fruit than I can eat and you have more meat than you can eat, it pays each of us to trade our surplus with the other. If we face a common enemy, then, as Benjamin Franklin put it, "We must all hang together, or assuredly we shall all hang separately."

Wright argues that three features of human nature led to a steady expansion of the circle of human cooperators. One is the cognitive wherewithal to figure out how the world works. This yields know-how worth sharing and an ability to spread goods and information over larger territories, both of which expand opportunities for gains in trade. A second is language, which allows technology to be shared, bargains to be struck, and agreements to be enforced. A third is an emotional repertoire—sympathy, trust, guilt, anger, self-esteem— that impels us to seek new cooperators, maintain relationships with them, and safeguard the relationships against possible exploitation. Long ago these endowments put our species on a moral escalator. Our mental circle of respectworthy persons expanded in tandem with our physical circle of allies and trading partners. As technology accumulates and people in more parts of the planet become interdependent, the hatred between them tends to decrease, for the simple reason that you can't kill someone and trade with him too.

Non-zero-sum games arise not just from people's ability to help one another but from their ability to refrain from hurting one another. In many disputes, both sides come out ahead by dividing up the savings made available from not having to fight. That provides an incentive to develop technologies of conflict resolution, such as mediation, face-saving measures, measured restitution and retribution, and legal codes. The primatologist Frans de Waal has argued that the rudiments of conflict resolution may be found in many species of primates.[20] The human forms are ubiquitous across cultures, as universal as the conflicts of interest they are designed to defuse.[21]

Though the evolution of the expanding circle (its ultimate cause) may sound pragmatic or even cynical, the psychology of the expanding circle (its proximate cause) need not be. Once the sympathy knob is in place, having evolved to enjoy the benefits of cooperation and exchange, it can be cranked up by new kinds of information that other folks are similar to oneself. Words and images from erstwhile enemies can trigger the sympathy response. A historical record can warn against self-defeating cycles of vendetta. A cosmopolitan awareness may lead people to think, "There but for fortune go I." An expansion of sympathy may come from something as basic as the requirement to be logically consistent when imploring other people to behave in certain ways: people come to realize that they cannot force others to abide by rules that they themselves flout. Egoistic, sexist, racist, and xenophobic attitudes are logically inconsistent with the demand that everyone respect a single code of behavior.[22]

Peaceful coexistence, then, does not have to come from pounding selfish desires out of people. It can come from pitting some desires—the desire for safety, the benefits of cooperation, the ability to formulate and recognize universal codes of behavior—against the desire for immediate gain. These are just a few of the ways in which moral and social progress can ratchet upwards, not in spite of a fixed human nature but because of it.

~

WHEN YOU STOP to think about it, the idea of a pliant human nature does not deserve its reputation for optimism and uplift. If it did, B. F. Skinner would have been lauded as a great humanitarian when he argued that society should apply the technology of conditioning to humans, shaping people to use contraception, conserve energy, make peace, and avoid crowded cities.[23] Skinner was a staunch blank-slater and a passionate utopian. His uncommonly pure vision allows us to examine the implications of the "optimistic" denial of human nature. Given his premise that undesirable behavior is not in the genes but a product of the environment, it follows that we should control that environment—for all we would be doing is replacing haphazard schedules of reinforcement by planned ones.

Why are most people repelled by this vision? Critics of Skinner's *Beyond Freedom and Dignity* pointed out that no one doubts that behavior can be controlled; putting a gun to someone's head or threatening him with torture are time-honored techniques.[24] Even Skinner's preferred method of operant conditioning required starving the organism to 80 percent of its free-feeding weight and confining it to a box where schedules of reinforcement were carefully controlled. The issue is not whether we can change human behavior, but at what cost.

Since we are not just products of our environments, there will be costs. People have inherent desires such as comfort, love, family, esteem, autonomy, aesthetics, and self-expression, regardless of their history of reinforcement, and they suffer when the freedom to exercise the desires is thwarted. Indeed, it is difficult to *define* psychological pain without some notion of human nature. (Even the young Marx appealed to a "species character," with an impulse for creative activity, as the basis for his theory of alienation.) Sometimes we may choose to impose suffering to control behavior, as when we punish people who cause avoidable suffering in others. But we cannot pretend that we can reshape behavior without infringing in some way on other people's freedom and happiness. Human nature is the reason we do not surrender our freedom to behavioral engineers.

Inborn human desires are a nuisance to those with utopian and totalitarian visions, which often amount to the same thing. What stands in the way of most utopias is not pestilence and drought but human behavior. So utopians have to think of ways to control behavior, and when propaganda doesn't do the trick, more emphatic techniques are tried. The Marxist utopians of the twentieth century, as we saw, needed a tabula rasa free of selfishness and family ties and used totalitarian measures to scrape the tablets clean or start over with new ones. As Bertolt Brecht said of the East German government, "If the people did not do better the government would dismiss the people and elect a new one." Political philosophers and historians who have recently "reflected on our ravaged century," such as Isaiah Berlin, Kenneth Minogue, Robert Conquest, Jonathan Glover, James Scott, and Daniel Chirot, have pointed to utopian dreams as a major cause of twentieth-century nightmares.[25] For that matter, Wordsworth's revolutionary France, "thrilled with joy" while human nature was "born again," turned out to be no picnic either.

It's not just behaviorists and Stalinists who forgot that a denial of human nature may have costs in freedom and happiness. Twentieth-century Marxism was part of a larger intellectual current that has been called Authoritarian High Modernism: the conceit that planners could redesign society from the top down using "scientific" principles.[26] The architect Le Corbusier, for example, argued that urban planners should not be fettered by traditions and tastes, since they only perpetuated the overcrowded chaos of the cities of his day. "We must build places where mankind will be reborn," he wrote. "Each man will live in an ordered relation to the whole."[27] In Le Corbusier's utopia, planners would begin with a "clean tablecloth" (sound familiar?) and mastermind all buildings and public spaces to service "human needs." They had a minimalist conception of those needs: each person was thought to require a fixed amount of air, heat, light, and space for eating, sleeping, working, commuting, and a few other activities. It did not occur to Le Corbusier that intimate gatherings with family and friends might be a human need, so he proposed large communal dining halls to replace kitchens. Also missing from his list of needs was the desire to socialize in small groups in public places, so he planned his cities around freeways, large buildings, and vast open plazas, with no squares or crossroads in which people would feel comfortable hanging out to schmooze. Homes were "machines for living," free of archaic inefficiencies like gardens and ornamentation, and thus were efficiently packed together in large, rectangular housing projects.

Le Corbusier was frustrated in his aspiration to flatten Paris, Buenos Aires, and Rio de Janeiro and rebuild them according to his scientific principles. But in the 1950s he was given carte blanche to design Chandigarh, the capital of the Punjab, and one of his disciples was given a clean tablecloth for Brasília, the capital of Brazil. Today, both cities are notorious as uninviting wastelands detested by the civil servants who live in them. Authoritarian High Modernism also led to the "urban renewal" projects in many American cities during the 1960s that replaced vibrant neighborhoods with freeways, high-rises, and empty windswept plazas.

Social scientists, too, have sometimes gotten carried away with dreams of social engineering. The child psychiatrist Bruce Perry, concerned that ghetto mothers are not giving children the enriched environment needed by their plastic brains, believes we must "transform our culture": "We need to change our child rearing practices, we need to change the malignant and destructive view that children are the property of their biological parents. Human beings evolved not as individuals, but as communities. . . . Children *belong* to the community, they are *entrusted* to their parents."[28] Now, no one could object to rescuing children from neglect or cruelty, but if Perry's transformed culture came to pass, men with guns could break up any family that did not conform to the latest fad in parenting theory. As we will see in the chapter on children, most of these fads are based on flawed studies that treat every correlation between parents and children as proof of causation. Asian American and African American parents often flout the advice of the child-development gurus, using more traditional, authoritarian styles of childrearing that in all likelihood do their children no lasting harm.[29] The parenting police could strip them of their children.

Nothing in the concept of human nature is inconsistent with the ideals of feminism, or so I will argue in the chapter on gender. But some feminist theoreticians have embraced the Blank Slate and with it an authoritarian political philosophy that would give the government sweeping powers to implement their vision of gender-free minds. In a 1975 dialogue, Simone de Beauvoir said: "No woman should be authorized to stay at home to raise her children. Society should be totally different. Women should not have that choice, precisely because if there is such a choice, too many women will make that one."[30] Gloria Steinem was a bit more liberal; in a 1970 *Time* article she wrote: "The [feminist] revolution would not take away the option of being a housewife. A woman who prefers to be her husband's housekeeper and/or hostess would receive a percentage of his pay determined by the domestic-relations courts."[31] Betty Friedan has spoken out in favor of "compulsory preschool" for two-year-olds.[32] Catharine MacKinnon (who with Andrea Dworkin has pushed for laws against erotica) has said, "What you need is people who see through literature like Andrea Dworkin, who see through law like me, to see through art and create the uncompromised women's visual vocabulary"[33]—oblivious to the danger inherent in a few intellectuals' arrogating the role of deciding which art and literature the rest of society will enjoy.

In an interview in the *New York Times Magazine*, Carol Gilligan explained the implications of her (preposterous) theory that behavior problems in boys, such as stuttering and hyperactivity, are caused by cultural norms that pressure them to separate from their mothers:

Q: You would argue that men's biology is not so powerful that we can't change the culture of men?

A: Right. We have to build a culture that doesn't reward that separation from the person who raised them. . . .

Q: Everything you've said suggests that unless men change in fundamental ways, we're not going to have a sea change in the culture.

A: That seems right to me.[34]

An incredulous reader, hearing an echo of the attempt to engineer a "new socialist man," asked, "Does anyone, even in academia, still believe that this sort of thing turns out well?"[35] He was right to be concerned. In many schools, teachers have been told, falsely, that there is an "opportunity zone" in which a child's gender identification is malleable. They have used this zone to try to stamp out boyhood: banning same-sex play groups and birthday parties, forcing children to do gender-atypical activities, suspending boys who run during recess or play cops and robbers.[36] In her book *The War Against Boys,* the philosopher Christina Hoff Sommers rightly calls this agenda "meddlesome, abusive, and quite beyond what educators in a free society are mandated to do."[37]

Feminism, far from needing a blank slate, needs the opposite, a clear conception of human nature. One of the most pressing feminist causes today is the condition of women in the developing world. In many places female fetuses are selectively aborted, newborn girls are killed, daughters are malnourished and kept from school, adolescent girls have their genitals cut out, young women are cloaked from head to toe, adulteresses are stoned to death, and widows are expected to fall onto their husbands' funeral pyres. The relativist climate in many academic circles does not allow these horrors to be criticized because they are practices of other cultures, and cultures are superorganisms that, like people, have inalienable rights. To escape this trap, the feminist philosopher Martha Nussbaum has invoked "central functional capabilities" that all humans have a right to exercise, such as physical integrity, liberty of conscience, and political participation. She has been criticized in turn for taking on a colonial "civilizing mission" or "white woman's burden," in which arrogant Europeans would instruct the poor people of the world in what they want. But Nussbaum's moral argument is defensible if her "capabilities" are grounded, directly or indirectly, in a universal human nature. Human nature provides a yardstick to identify suffering in any member of our species.

The existence of a human nature is not a reactionary doctrine that dooms us to eternal oppression, violence, and greed. *Of course* we should try to reduce harmful behavior, just as we try to reduce afflictions like hunger, disease, and the elements. But we fight those afflictions not by denying the pesky facts of nature but by turning some of them against the others. For efforts at social change to be effective, they must identify the cognitive and moral resources that make some kinds of change possible. And for the efforts to be humane, they must acknowledge the universal pleasures and pains that make some kinds of change desirable.

Chapter 10

The Fear of Determinism

THIS CHAPTER IS not about the boo-word that is frequently (and inaccurately) hurled at any explanation of a behavioral tendency that mentions evolution or genetics. It is about determinism in its original sense, the concept that is opposed to "free will" in introductory philosophy courses. The fear of determinism in this sense is captured in a limerick:

> There was a young man who said: "Damn!
> It grieves me to think that I am
> Predestined to move
> In a circumscribed groove:
> In fact, not a bus, but a tram."

In the traditional conception of a ghost in the machine, our bodies are inhabited by a self or a soul that chooses the behavior to be executed by the body. These choices are not compelled by some prior physical event, like one billiard ball smacking into another and sending it into a corner pocket. The idea that our behavior is caused by the physiological activity of a genetically shaped brain would seem to refute the traditional view. It would make our behavior an automatic consequence of molecules in motion and leave no room for an uncaused behavior-chooser.

One fear of determinism is a gaping existential anxiety: that deep down we are not in control of our own choices. All our brooding and agonizing over the right thing to do is pointless, it would seem, because everything has already been preordained by the state of our brains. If you suffer from this anxiety, I suggest the following experiment. For the next few days, don't bother deliberating over your actions. It's a waste of time, after all; they have already been determined. Shoot from the hip, live for the moment, and if it feels good do it. No, I am not seriously suggesting that you try this! But a moment's reflection on what would happen if you *did* try to give up making decisions should serve as a Valium for the existential anxiety. The experience of choosing is not a fiction, regardless of how the brain works. It is a real neural process, with the obvious function of selecting behavior according to its foreseeable consequences. It responds to information from the senses, including the exhortations of other people. You cannot step outside it or let it go on without you because it *is* you. If the most ironclad form of determinism is real, you could not do anything about it anyway, because your anxiety about determinism, and how you would deal with it, would also be determined. It is the existential fear of determinism that is the real waste of time.

A more practical fear of determinism is captured in a saying by A. A. Milne: "No doubt Jack the Ripper excused himself on the grounds that it was human nature." The fear is that an understanding of human nature seems to eat away at the notion of personal responsibility. In the traditional view, the self or soul, having chosen what to do, takes responsibility when things turn out badly. As with the desk of Harry Truman, the buck stops here. But when we attribute an action to a person's brain, genes, or evolutionary history, it seems that we no longer hold the individual accountable. Biology becomes the perfect alibi, the get-out-of-jail-free card, the ultimate doctor's excuse note. As we have seen, this accusation has been made by the religious and cultural right, who want to preserve the soul, and the academic left, who want to preserve a "we" who can construct our own futures though in circumstances not of our own choosing.

Why is the notion of free will so closely tied to the notion of responsibility, and why is biology thought to threaten both? Here is the logic. We blame people for an evil act or bad decision only when they intended the consequences and could have chosen otherwise. We don't convict a hunter who shoots a friend he has mistaken for a deer, or the chauffeur who drove John F. Kennedy into the line of fire, because they could not foresee and did not intend the outcome of their actions. We show mercy to the victim of torture who betrays a comrade, to a delirious patient who lashes out at a nurse, or to a madman who strikes someone he believes to be a ferocious animal, because we feel they are not in command of their faculties. We don't put a small child on trial if he causes a death, nor do we try an animal or an inanimate object, because we believe them to be constitutionally incapable of making an informed choice.

A biology of human nature would seem to admit more and more people into the ranks of the blameless. A murderer may not literally be a raving lunatic, but our newfangled tools might pick up a shrunken amygdala or a hypometabolism in his frontal lobes or a defective gene for monoamine oxidase A, which renders him just as out of control. Or perhaps a test from the cognitive psychology lab will show that he has chronically limited foresight, rendering him oblivious to consequences, or that he has a defective theory of mind, making him incapable of appreciating the suffering of others. After all, if there is no ghost in the machine, *something* in the criminal's hardware must set him apart from the majority of people, those who would not hurt or kill in the same circumstances. Pretty soon we will find that something, and, it is feared, murderers will be excused from criminal punishment as surely as we now excuse madmen and small children.

Even worse, biology may show that we are *all* blameless. Evolutionary theory says that the ultimate rationale for our motives is that they perpetuated our ancestors' genes in the environment in which we evolved. Since none of us are aware of that rationale, none of us can be blamed for pursuing it, any more than we blame the mental patient who thinks he is subduing a mad dog but really is attacking a nurse. We scratch our heads when we learn of ancient customs that punished the soulless: the Hebrew rule of stoning an ox to death if it killed a man, the Athenian practice of putting an ax on trial if it injured a man (and hurling it over the city wall if found guilty), a medieval French case in which a sow was sentenced to be mangled for having mauled a child, and the whipping and burial of a church bell in 1685 for having assisted French heretics.[1] But evolutionary biologists insist we are not fundamentally different from animals, and molecular geneticists and neuroscientists insist we are not fundamentally different from inanimate matter. If people are soulless, why is it not just as silly to punish people? Shouldn't we heed the creationists, who say that if you teach children they are animals they will behave like animals? Should we go even farther than the National Rifle Association bumper sticker—GUNS DON'T KILL; PEOPLE KILL—and say that not even people kill, because people are just as mechanical as guns?

These concerns are by no means academic. Cognitive neuroscientists are sometimes approached by criminal defense lawyers hoping that a wayward pixel on a brain scan might exonerate their client (a scenario that is wittily played out in Richard Dooling's novel *Brain Storm*). When a team of geneticists found a rare gene that predisposed the men in one family to violent outbursts, a lawyer for an unrelated murder defendant argued that his client might have such a gene too. If so, the lawyer argued, "his actions may not have been a product of total free will."[2] When Randy Thornhill and Craig Palmer argued that rape is a consequence of male reproductive strategies, another lawyer contemplated using their theory to defend rape suspects.[3] (Insert your favorite lawyer joke here.) Biologically sophisticated legal scholars, such as Owen Jones, have argued that a "rape gene" defense would almost certainly fail, but the general threat remains that biological explanations will be used to exonerate wrongdoers.[4] Is this the bright future promised by the sciences of human nature—it wasn't me, it was my amygdala? Darwin made me do it? The genes ate my homework?

～

PEOPLE HOPING THAT an uncaused soul might rescue personal responsibility are in for a disappointment. In *Elbow Room: The Varieties of Free Will Worth*

Wanting, the philosopher Dan Dennett points out that the last thing we want in a soul is freedom to do anything it desires.[5] If behavior were chosen by an utterly free will, then we *really* couldn't hold people responsible for their actions. That entity would not be deterred by the threat of punishment, or be ashamed by the prospect of opprobrium, or even feel the twinge of guilt that might inhibit a sinful temptation in the future, because it could always choose to defy those causes of behavior. We could not hope to reduce evil acts by enacting moral and legal codes, because a free agent, floating in a different plane from the arrows of cause and effect, would be unaffected by the codes. Morality and law would be pointless. We could punish a wrongdoer, but it would be sheer spite, because it could have no predictable effect on the future behavior of the wrongdoer or of other people aware of the punishment.

On the other hand, if the soul *is* predictably affected by the prospect of esteem and shame or reward and punishment, it is no longer truly free, because it is compelled (at least probabilistically) to respect those contingencies. Whatever converts standards of responsibility into changes in the likelihood of behavior—such as the rule "If the community would think you're a boorish cad for doing X, don't do X"—can be programmed into an algorithm and implemented in neural hardware. The soul is superfluous.

Defensive scientists sometimes try to deflect the charge of determinism by pointing out that behavior is never perfectly predictable but always probabilistic, even in the dreams of the hardest-headed materialists. (In the heyday of Skinner's behaviorism, his students formulated the Harvard Law of Animal Behavior: "Under controlled experimental conditions of temperature, time, lighting, feeding, and training, the organism will behave as it damned well pleases.") Even identical twins reared together, who share all of their genes and most of their environment, are not identical in personality and behavior, just highly similar. Perhaps the brain amplifies random events at the molecular or quantum level. Perhaps brains are nonlinear dynamical systems subject to unpredictable chaos. Or perhaps the intertwined influences of genes and environment are so complicated that no mortal will ever trace them out with enough precision to predict behavior exactly.

The less-than-perfect predictability of behavior certainly gives the lie to the cliché that the sciences of human nature are "deterministic" in the mathematical sense. But it doesn't succeed in allaying the fear that science is eroding the concept of free will and personal responsibility. It is cold comfort to be told that a man's genes (or his brain or his evolutionary history) made him 99 percent likely to kill his landlady as opposed to 100 percent. Sure, the behavior was not strictly preordained, but why should the 1 percent chance of his having done otherwise suddenly make the guy "responsible"? In fact, there is *no* probability value that, by itself, ushers responsibility back in. One can always think that there is a 50 percent chance some molecules in Raskolnikov's brain went thisaway, compelling him to commit the murder, and a 50 percent chance they went thataway, compelling him not to. We still have nothing like free will, and no concept of responsibility that promises to reduce harmful acts. Hume noted the dilemma inherent in equating the problem of moral responsibility with the problem of whether behavior has a physical cause: either our actions are determined, in which case we are not responsible for them, or they are the result of random events, in which case we are not responsible for them.

PEOPLE WHO HOPE that a ban on biological explanations might restore personal responsibility are in for the biggest disappointment of all. The most risible pretexts for bad behavior in recent decades have come not from biological determinism but from *environmental* determinism: the abuse excuse, the Twinkie defense, black rage, pornography poisoning, societal sickness, media violence, rock lyrics, and different cultural mores (recently used by one lawyer to defend a Gypsy con artist and by another to defend a Canadian Indian woman who murdered her boyfriend).[6] Just in the week I wrote this paragraph, two new examples appeared in the newspapers. One is from a clinical psychologist who "seeks out a dialogue" with repeat murderers to help them win mitigation, clemency, or an appeal. It manages to pack the Blank Slate, the Noble Savage, the moralistic fallacy, and environmental determinism into a single passage:

> Most people don't commit horrendous crimes without profoundly damaging things happening to them. It isn't that monsters are being born right and left. It's that children are being born right and left and are being subjected to horrible things. As a consequence, they end up doing horrible things. And I would much rather live in that world than in a world where monsters are just born.[7]

The other is about a social work student in Manhattan:

> Tiffany F. Goldberg, a 25-year-old from Madison, Wis., was struck on the head with a chunk of concrete by a stranger this month. Afterward, she expressed concern for her attacker, speculating that he must have had a troubled childhood.
>
> Graduate students in social work at Columbia called Ms. Goldberg's attitude consistent with their outlook on violence. "Society is into blaming individuals," said Kristen Miller, 27, one of the students. "Violence is intergenerationally transmitted."[8]

Evolutionary psychologists are commonly chided for "excusing" men's promiscuity with the theory that a wandering eye in our ancestors was rewarded with a greater number of descendants. They can take heart from a recent biography that said Bruce Springsteen's "self-doubts made him frequently seek out the sympathy of groupies,"[9] a book review that said Woody Allen's sexual indiscretions "originated in trauma" and an "abusive" relationship with his mother,[10] and Hillary Clinton's explanation of her husband's libido in her infamous interview in *Talk:*

> He was so young, barely 4, when he was scarred by abuse that he can't even take it out and look at it. There was terrible conflict between his mother and grandmother. A psychologist once told me that for a boy being in the middle of a conflict between two women is the worst possible situation. There is always the desire to please each one.[11]

Mrs. Clinton was raked by the pundits for trying to excuse her husband's sexual escapades, though she said not a word about brains, genes, or evolution. The logic of the condemnation seems to be: If someone tries to explain an act as an effect of some cause, the explainer is saying that the act was not freely chosen and that the actor cannot be held responsible.

Environmental determinism is so common that a genre of satire has grown around it. In a *New Yorker* cartoon, a woman on a witness stand says, "True, my husband beat me because of his childhood; but I murdered him because of mine." In the comic strip *Non Sequitur,* the directory of a mental health clinic reads: "1st Floor: Mother's Fault. 2nd Floor: Father's Fault. 3rd Floor: Society's Fault." And who can forget the Jets in *West Side Story,* who imagined explaining to the local police sergeant, "We're depraved on accounta we're deprived"?

> Dear kindly Sergeant Krupke,
> You gotta understand,
> It's just our bringin' up-ke,
> That gets us out of hand.
> Our mothers all are junkies,
> Our fathers all are drunks.
> Golly Moses, natcherly we're punks!

~

SOMETHING HAS GONE terribly wrong. It is a confusion of *explanation* with *exculpation.* Contrary to what is implied by critics of biological *and* environmental theories of the causes of behavior, to explain behavior is not to exonerate the behaver. Hillary Clinton may have advanced the dumbest explanation in the history of psychobabble, but she does not deserve the charge of trying to excuse the president's behavior. (A *New York Times* story described Mr. Clinton's response to people's criticism of his wife: " 'I have not made any excuses for what was inexcusable, and neither has she, believe me,' he said, arching his eyebrows for emphasis.")[12] If behavior is not utterly random, it will have some explanation; if behavior *were* utterly random, we couldn't hold the person responsible in any case. So if we *ever* hold people responsible for their behavior, it will have to be in spite of any causal explanation we feel is warranted, whether it invokes genes, brains, evolution, media images, self-doubt, bringing up-ke, or being raised by bickering women. The difference between *explaining* behavior and *excusing* it is captured in the saying "To understand is not to forgive," and has been stressed in different ways by many philosophers, including Hume, Kant, and Sartre.[13] Most philosophers believe that unless a person was literally coerced (that is, someone held a gun to his head), we should consider his actions to have been freely chosen, even if they were caused by events inside his skull.

But *how* can we have both explanation, with its requirement of lawful causation, and responsibility, with its requirement of free choice? To have them both we don't need to resolve the ancient and perhaps unresolvable antinomy between free will and determinism. We only have to think clearly about what we want the notion of responsibility to achieve. Whatever may be its inherent abstract worth, responsibility has an eminently practical function: deterring harmful behavior. When we say that we hold someone responsible for a wrongful act, we expect him to punish himself—by compensating the victim, acquiescing to humiliation, incurring penalties, or expressing credible remorse—and we reserve the right to punish him ourselves. Unless a person is willing to suffer some unpleasant (and hence deterring) consequence, claims of responsibility are hollow. Richard Nixon was ridiculed when he bowed to pressure and finally "took responsibility" for the Watergate burglary but did not accept any costs such as apologizing, resigning, or firing his aides.

One reason to hold someone responsible is to deter the person from committing similar acts in the future. But that cannot be the whole story, because it is different only in degree from the contingencies of punishment used by behaviorists to modify the behavior of animals. In a social, language-using, reasoning organism, the policy can also deter similar acts by *other* organisms who learn of the contingencies and control their behavior so as not to incur the penalties. That is the ultimate reason we feel compelled to punish elderly Nazi war criminals, even though there is little danger that they would perpetrate another holocaust if we let them die in their beds in Bolivia. By holding them responsible—that is, by publicly enforcing a policy of rooting out and punishing evil wherever and whenever it occurs—we hope to deter others from committing comparable evils in the future.

This is not to say that the concept of responsibility is a recommendation by policy wonks for preventing the largest number of harmful acts at the least cost. Even if experts had determined that punishing a Nazi would prevent no future atrocities, or that we could save more lives by diverting the manpower to catching drunk drivers, we would still want to bring Nazis to justice. The demand for responsibility can come from a burning sense of just deserts, not only from literal calculations of how best to deter particular acts.

But punishment even in the pure sense of just deserts is *ultimately* a policy for deterrence. It follows from a paradox inherent to the logic of deterrence: though the *threat* of punishment can deter behavior, if the behavior does take place the punishment serves no purpose other than pure sadism or an illogical desire to make the threat credible retroactively. "It won't bring the victim back," say the opponents of capital punishment, but that can be said about *any* form of punishment. If we start the movie at the point at which a punishment is to be carried out, it looks like spite, because it is costly to the punisher and inflicts harm on the punishee without doing anyone any immediate good. In the middle decades of the twentieth century, the paradox of punishment and the rise of psychology and psychiatry led some intellectuals to argue that criminal punishment is a holdover from barbaric times and should be replaced by therapy and rehabilitation. The position was clear in the titles of books like George Bernard Shaw's *The Crime of Imprisonment* and the psychiatrist Karl Menninger's *The Crime of Punishment*. It was also articulated by leading jurists such as William O. Douglas, William Brennan, Earl Warren, and David Bazelon. These radical Krupkeists did not suffer from a fear of determinism; they welcomed it with open arms.

Few people today argue that criminal punishment is obsolete, even if they recognize that (other than incapacitating some habitual criminals) it is pointless in the short run. That is because if we ever did calculate the short-term effects in deciding whether to punish, potential wrongdoers could *anticipate* that calculation and factor it into their behavior. They could predict that we would not find it worthwhile to punish them once it was too late to prevent the crime, and could act with impunity, calling our bluff. The only solution is to adopt a resolute policy of punishing wrongdoers *regardless* of the immediate effects. If one is genuinely not bluffing about the threat of punishment, there is no bluff to call. As Oliver Wendell Holmes explained, "If I were having a philosophical talk with a man I was going to have hanged (or electrocuted) I should say, 'I don't doubt that your act was inevitable for you but to make it more avoidable by others we propose to sacrifice you to the common good. You may regard yourself as a soldier dying for your country if you like. But the law must keep its promises.' "[14] This promise-keeping underlies the policy of applying justice "as a matter of principle," regardless of the immediate costs or even of consistency with common sense. If a death-row inmate attempts suicide, we speed him to the emergency ward, struggle to resuscitate him, give him the best modern medicine to help him recuperate, and kill him. We do it as part of a policy that closes off all possibilities to "cheat justice."

Capital punishment is a vivid illustration of the paradoxical logic of deterrence, but the logic applies to lesser criminal punishments, to personal acts of revenge, and to intangible social penalties like ostracism and scorn. Evolutionary psychologists and game theorists have argued that the deterrence paradox led to the evolution of the emotions that undergird a desire for justice: the implacable need for retribution, the burning feeling that an evil act knocks the universe out of balance and can be canceled only by a commensurate punishment. People who are emotionally driven to retaliate against those who cross them, even at a cost to themselves, are more credible adversaries and less likely to be exploited.[15] Many judicial theorists argue that criminal law is simply a controlled implementation of the human desire for retribution, designed to keep it from escalating into cycles of vendetta. The Victorian jurist James Stephen said that "the criminal law bears the same relation to the urge for revenge as marriage does to the sexual urge."[16]

Religious conceptions of sin and responsibility simply extend this lever by implying that any wrongdoing that is undiscovered or unpunished by one's fellows will be discovered and punished by God. Martin Daly and Margo Wilson sum up the ultimate rationale of our intuitions about responsibility and godly retribution:

> From the perspective of evolutionary psychology, this almost mystical and seemingly irreducible sort of moral imperative is the output of a mental mechanism with a straightforward adaptive function: to reckon justice and administer punishment by a calculus which ensures that violators reap no advantage from their misdeeds. The enormous volume of mystico-religious bafflegab about atonement and penance and divine justice and the like is the attribution to higher, detached authority of what is actually a mundane, pragmatic matter: discouraging self-interested competitive acts by reducing their profitability to nil.[17]

～

THE DETERRENCE PARADOX also underlies the part of the logic of responsibility that makes us expand or contract it when we learn about a person's mental state. Modern societies do not just pick whatever policy is most effective at deterring wrongdoers. For example, if one's only value was to reduce crime, one could always make the punishments for it especially cruel, as most societies did until recently. One could convict people on the basis of an accusation, a guilty manner, or a forced confession. One could execute the entire family of a criminal, or his entire clan or village. One could say to one's adversaries, as Vito Corleone said to the heads of the other crime families in *The Godfather*, "I'm a superstitious man. And if some unlucky accident should befall my son, if my son is struck by a bolt of lightning, I will blame some of the people here."

The reason these practices strike us as barbaric is that they inflict more harm than is necessary to deter evil in the future. As the political writer Harold

Laski said, "Civilization means, above all, an unwillingness to inflict unnecessary pain." The problem with broad-spectrum deterrents is that they catch innocent people in their nets, people who could not have been deterred from committing an undesirable act to start with (such as the kin of the man who pulled the trigger, or a bystander during a lightning storm that kills the Godfather's son). Since punishment of these innocents could not possibly deter other people like them, the harm has no compensating benefit even in the long run, and we consider it unjustified. We seek to fine-tune our policy of punishment so that it applies only to people who *could have been* deterred by it. They are the ones we "hold responsible," the ones we feel "deserve" the punishment.

A fine-tuned deterrence policy explains why we exempt certain harm-causers from punishment. We don't punish those who were unaware that their acts would lead to harm, because such a policy would do nothing to prevent similar acts by them or by others in the future. (Chauffeurs cannot be deterred from driving a president into the line of fire if they have no way of knowing there will be a line of fire.) We don't apply criminal punishment to the delirious, the insane, small children, animals, or inanimate objects, because we judge that they—and entities similar to them—lack the cognitive apparatus that could be informed of the policy and could inhibit behavior accordingly. We exempt these entities from responsibility not because they follow predictable laws of biology while everyone else follows mysterious not-laws of free will. We exempt them because, unlike most adults, they lack a functioning brain system that can respond to public contingencies of punishment.

And this explains why the usual exemptions from responsibility should *not* be granted to all males or all abuse victims or all of humanity, even when we think we can explain what led them to act as they did. The explanations may help us understand the parts of the brain that made a behavior tempting, but they say nothing about the *other* parts of the brain (primarily in the prefrontal cortex) that could have inhibited the behavior by anticipating how the community would respond to it. We are that community, and our major lever of influence consists in appealing to that inhibitory brain system. Why should we discard our lever on the system for inhibition just because we are coming to understand the system for temptation? If you believe we shouldn't, that is enough to hold people responsible for their actions—without appealing to a will, a soul, a self, or any other ghost in the machine.

This argument parallels a long-running debate about the most blatant example of a psychological explanation that nullifies responsibility, the insanity defense.[18] Many legal systems in the English-speaking world follow the nineteenth-century M'Naughten rule:

> . . . the jurors ought to be told in all cases that every man is to be presumed to be sane, and to possess a sufficient degree of reason to be

responsible for his crimes, until the contrary be proved to their satisfaction; and that, to establish a defense on the ground of insanity, it must be clearly proved that, at the time of the committing of the act, the party accused was laboring under such a defect of reason, from disease of the mind, as not to know the nature and quality of the act he was doing, or, if he did know it, that he did not know he was doing what was wrong.

This is an excellent characterization of a person who cannot be deterred. If someone is too addled to know that an act would harm someone, he cannot be inhibited by the injunction "Don't harm people, or else!" The M'Naughten rule aims to forgo spiteful punishment—retribution that harms the perpetrator with no hope of deterring him or people similar to him.

The insanity defense achieved its present notoriety, with dueling rent-a-shrinks and ingenious abuse excuses, when it was expanded from a practical test of whether the cognitive system responding to deterrence is working to the more nebulous tests of what can be said to have produced the behavior. In the 1954 *Durham* decision, Bazelon invoked "the science of psychiatry" and "the science of psychology" to create a new basis for the insanity defense:

The rule we now hold is simply that an accused is not criminally responsible if his unlawful act was the product of mental disease or mental defect.

Unless one believes that ordinary acts are chosen by a ghost in the machine, *all* acts are products of cognitive and emotional systems in the brain. Criminal acts are relatively rare—if everyone in a defendant's shoes acted as he did, the law against what he did would be repealed—so heinous acts will often be products of a brain system that is in *some* way different from the norm, and the behavior can be construed as "a product of mental disease or mental defect." The *Durham* decision and similar insanity rules, by distinguishing behavior that is a product of a brain condition from behavior that is something else, threatens to turn every advance in our understanding of the mind into an erosion of responsibility.

Now, some discoveries about the mind and brain really could have an impact on our attitudes toward responsibility—but they may call for expanding the domain of responsibility, not contracting it. Suppose desires that sometimes culminate in the harassment and battering of women are present in many men. Does that really mean that men should be punished more leniently for such crimes, because they can't help it? Or does it mean they should be punished more surely and severely, because that is the best way to counteract a strong or widespread urge? Suppose a vicious psychopath is found to have a defective sense of sympathy, which makes it harder for him to appreciate the suffering of his victims. Should we mitigate the punishment because he has diminished capacity? Or should we make the punishment more sure and severe to teach him a lesson in the only language he understands?

Why do people's intuitions go in opposite directions—both "If he has trouble controlling himself, he should be punished more leniently" and "If he has trouble controlling himself, he should be punished more severely"? It goes back to the deterrence paradox. Suppose some people need a threat of one lash with a wet noodle to deter them from parking in front of a fire hydrant. Suppose people with a bad gene, a bad brain, or a bad childhood need the threat of ten lashes. A policy that punishes illegal parkers with nine lashes will cause unnecessary suffering and not solve the problem: nine lashes is more than necessary to deter ordinary people and less than necessary to deter defective people. Only a penalty of ten lashes can reduce both illegal parking and lashing: everyone will be deterred, no one will block hydrants, and no one will get whipped. So, paradoxically, the two extreme policies (harsh punishment and no punishment) are defensible and the intermediate ones are not. Of course, people's deterrence thresholds in real life aren't pinned at just two values but are broadly distributed (one lash for some people, two for others, and so on), so many intermediate levels of punishment will be defensible, depending on how one weights the benefits of deterring wrongdoing against the costs of inflicting harm.

Even for those who are *completely* undeterrable, because of frontal-lobe damage, genes for psychopathy, or any other putative cause, we do not have to allow lawyers to loose them on the rest of us. We already have a mechanism for those likely to harm themselves or others but who do not respond to the carrots and sticks of the criminal justice system: involuntary civil commitment, in which we trade off some guarantees of civil liberties against the security of being protected from likely predators. In all these decisions, the sciences of human nature can help estimate the distribution of deterrabilities, but they cannot weight the conflicting values of avoiding the greatest amount of unnecessary punishment and preventing the greatest amount of future wrongdoing.[19]

I do not claim to have solved the problem of free will, only to have shown that we don't need to solve it to preserve personal responsibility in the face of an increasing understanding of the causes of behavior. Nor do I argue that deterrence is the only way to encourage virtue, just that we should recognize it as the active ingredient that makes responsibility worth keeping. Most of all, I hope I have dispelled two fallacies that have allowed the sciences of human nature to sow unnecessary fear. The first fallacy is that biological explanations corrode responsibility in a way that environmental explanations do not. The second fallacy is that causal explanations (both biological and environmental) corrode responsibility in a way that a belief in an uncaused will or soul does not.

Chapter 11

The Fear of Nihilism

THE FINAL FEAR of biological explanations of the mind is that they may strip our lives of meaning and purpose. If we are just machines that let our genes make copies of themselves, if our joys and satisfactions are just biochemical events that will someday sputter out for good, if life was not created for a higher purpose and directed toward a noble goal, then why go on living? Life as we treasure it would be sham, a Potemkin village with only a façade of value and worth.

The fear comes in two versions, religious and secular. A sophisticated version of the religious concern was formulated by Pope John Paul II in a 1996 address to the Pontifical Academy of Sciences, "Truth Cannot Contradict Truth."[1] The Pope acknowledged that Darwin's theory of evolution is "more than just a hypothesis," because converging discoveries in many independent fields, "neither sought nor fabricated," argue in its favor. But he drew the line at "the spiritual soul," a transition in the evolution of humans that amounted to an "ontological leap" unobservable by science. The spirit could not have emerged "from the forces of living matter," because that cannot "ground the dignity of the person":

> Man is the only creature on earth that God has wanted for its own sake.... In other terms, the human individual cannot be subordinated as a pure means or a pure instrument, either to the species or to society; he has value per se. He is a person. With his intellect and his will, he is capable of forming a relationship of communion, solidarity and self-giving with his peers.... Man is called to enter into a relationship of knowledge and love with God himself, a relationship which will find its complete fulfillment beyond time, in eternity....
>
> It is by virtue of his spiritual soul that the whole person possesses such a dignity even in his body.... If the human body take its origin

from pre-existent living matter, the spiritual soul is immediately created by God. . . . Consequently, theories of evolution which, in accordance with the philosophies inspiring them, consider the spirit as emerging from the forces of living matter or as a mere epiphenomenon of this matter, are incompatible with the truth about man. Nor are they able to ground the dignity of the person.

In other words, if scientists are right that the mind emerged from living matter, we would have to give up the value and dignity of the individual, solidarity and selflessness with regard to our fellow humans, and the higher purpose of realizing these values through the love of God and knowledge of his plans. Nothing would keep us from a life of callous exploitation and cynical self-centeredness.

Needless to say, debating the Pope is the ultimate exercise in futility. The point of this section is not to refute his doctrines, nor is it to condemn religion or argue against the existence of God. Religions have provided comfort, community, and moral guidance to countless people, and some biologists argue that a sophisticated deism, toward which many religions are evolving, can be made compatible with an evolutionary understanding of the mind and human nature.[2] My goal is defensive: to refute the accusation that a materialistic view of the mind is inherently amoral and that religious conceptions are to be favored because they are inherently more humane.

Even the most atheistic scientists do not, of course, advocate a callous amorality. The brain may be a physical system made of ordinary matter, but that matter is organized in such a way as to give rise to a sentient organism with a capacity to feel pleasure and pain. And that in turn sets the stage for the emergence of morality. The reason is succinctly explained in the comic strip *Calvin and Hobbes* (see p. 188).

The feline Hobbes, like his human namesake, has shown why an amoral egoist is in an untenable position. He is better off if he never gets shoved into the mud, but he can hardly demand that others refrain from shoving him if he himself is not willing to forgo shoving others. And since one is better off not shoving and not getting shoved than shoving and getting shoved, it pays to insist on a moral code, even if the price is adhering to it oneself. As moral philosophers through the ages have pointed out, a philosophy of living based on "Not everyone, just me!" falls apart as soon as one sees oneself from an objective standpoint as a person just like others. It is like insisting that "here," the point in space one happens to be occupying at the moment, is a special place in the universe.[3]

The dynamic between Calvin and Hobbes (the cartoon characters) is inherent to social organisms, and there are reasons to believe that the solution

Calvin and Hobbes © Watterson. Reprinted with permission of Universal Press Syndicate.

to it—a moral sense—evolved in our species rather than having to be deduced from scratch by each of us after we've picked ourselves up out of the mud.[4] Children as young as a year and a half spontaneously give toys, proffer help, and try to comfort adults or other children who are visibly distressed.[5] People in all cultures distinguish right from wrong, have a sense of fairness, help one another, impose rights and obligations, believe that wrongs should be redressed, and proscribe rape, murder, and some kinds of violence.[6] These normal sentiments are conspicuous by their absence in the aberrant individuals we call psychopaths.[7] The alternative, then, to the religious theory of the source of values is that evolution endowed us with a moral sense, and we have expanded its circle of application over the course of history through reason (grasping the logical interchangeability of our interests and others'), knowledge (learning of the advantages of cooperation over the long term), and sympathy (having experiences that allow us to feel other people's pain).

How can we tell which theory is preferable? A thought experiment can pit them against each other. What would be the right thing to do if God had commanded people to be selfish and cruel rather than generous and kind? Those who root their values in religion would have to say that we ought to be selfish and cruel. Those who appeal to a moral sense would say that we ought to reject God's command. This shows—I hope—that it is our moral sense that deserves priority.[8]

This thought experiment is not just a logical brainteaser of the kind beloved by thirteen-year-old atheists, such as why God cares how we behave if he can see the future and already knows. The history of religion shows that God *has* commanded people to do all manner of selfish and cruel acts: massacre Midianites and abduct their women, stone prostitutes, execute homosexuals, burn witches, slay heretics and infidels, throw Protestants out of windows, withhold medicine from dying children, shoot up abortion clinics, hunt down Salman Rushdie, blow themselves up in marketplaces, and crash airplanes into skyscrapers. Recall that even Hitler thought he was carrying out the will of God.[9] The recurrence of evil acts committed in the name of God shows that they are not random perversions. An omnipotent authority that no one can see is a useful backer for malevolent leaders hoping to enlist holy warriors. And since unverifiable beliefs have to be passed along from parents and peers rather than discovered in the world, they differ from group to group and become divisive identity badges.

And who says the doctrine of the soul is more humane than the understanding of the mind as a physical organ? I see no dignity in letting people die of hepatitis or be ravaged by Parkinson's disease when a cure may lie in research on stem cells that religious movements seek to ban because it uses balls of cells that have made the "ontological leap" to "spiritual souls." Sources of immense misery such as Alzheimer's disease, major depression, and schizophrenia will be alleviated not by treating thought and emotion as manifestations of an immaterial soul but by treating them as manifestations of physiology and genetics.[10]

Finally, the doctrine of a soul that outlives the body is anything but righteous, because it necessarily devalues the lives we live on this earth. When Susan Smith sent her two young sons to the bottom of a lake, she eased her conscience with the rationalization that "my children deserve to have the best, and now they will." Allusions to a happy afterlife are typical in the final letters of parents who take their children's lives before taking their own,[11] and we have recently been reminded of how such beliefs embolden suicide bombers and kamikaze hijackers. This is why we should reject the argument that if people stopped believing in divine retribution they would do evil with impunity. Yes, if nonbelievers thought they could elude the legal system, the opprobrium of their communities, and their own consciences, they would not be deterred by the threat of spending eternity in hell. But they would also not be tempted to massacre thousands of people by the promise of spending eternity in heaven.

Even the emotional comfort of a belief in an afterlife can go both ways. Would life lose its purpose if we ceased to exist when our brains die? On the contrary, nothing invests life with more meaning than the realization that every moment of sentience is a precious gift. How many fights have been averted, how many friendships renewed, how many hours not squandered, how many gestures of affection offered, because we sometimes remind ourselves that "life is short"?

~

WHY DO SECULAR thinkers fear that biology drains life of meaning? It is because biology seems to deflate the values we most cherish. If the reason we love our children is that a squirt of oxytocin in the brain compels us to protect our genetic investment, wouldn't the nobility of parenthood be undermined and its sacrifices devalued? If sympathy, trust, and a yearning for justice evolved as a way to earn favors and deter cheaters, wouldn't that imply that there are really no such things as altruism and justice for their own sake? We sneer at the philanthropist who profits from his donation because of the tax savings, the televangelist who thunders against sin but visits prostitutes, the politician who defends the downtrodden only when the cameras are rolling, and the sensitive new-age guy who backs feminism because it's a good way to attract women. Evolutionary psychology seems to be saying that we are *all* such hypocrites, all the time.

The fear that scientific knowledge undermines human values reminds me of the opening scene in *Annie Hall,* in which the young Alvy Singer has been taken to the family doctor:

> MOTHER: He's been depressed. All of a sudden, he can't do anything.
> DOCTOR: Why are you depressed, Alvy?
> MOTHER: Tell Dr. Flicker. [Answers for him.] It's something he read.
> DOCTOR: Something he read, huh?
> ALVY: [Head down.] The universe is expanding.
> DOCTOR: The universe is expanding?
> ALVY: Well, the universe is everything, and if it's expanding, someday it
>   will break apart and that would be the end of everything!
> MOTHER: What is that your business? [To the doctor.] He stopped
>   doing his homework.
> ALVY: What's the point?

The scene is funny because Alvy has confused two levels of analysis: the scale of billions of years with which we measure the universe, and the scale of decades, years, and days with which we measure our lives. As Alvy's mother points out, "What has the universe got to do with it? You're here in Brooklyn! Brooklyn is not expanding!"

People who are depressed at the thought that all our motives are selfish are as confused as Alvy. They have mixed up ultimate causation (why something evolved by natural selection) with proximate causation (how the entity works here and now). The mix-up is natural because the two explanations can look so much alike.

Richard Dawkins showed that a good way to understand the logic of natural selection is to imagine that genes are agents with selfish motives. No one should begrudge him the metaphor, but it contains a trap for the unwary. The genes have metaphorical motives—making copies of themselves—and the organisms they design have real motives. But they are not the *same* motives. Sometimes the most selfish thing a gene can do is wire *unselfish* motives into a human brain—heartfelt, unstinting, deep-in-the-marrow unselfishness. The love of children (who carry one's genes into posterity), a faithful spouse (whose genetic fate is identical to one's own), and friends and allies (who trust you if you're trustworthy) can be bottomless and un-impeachable as far as we humans are concerned (proximate level), even if it is metaphorically self-serving as far as the genes are concerned (ultimate level).

I suspect there is another reason why the explanations are so easily confused. We all know that people sometimes have ulterior motives. They may be publicly generous but privately greedy, publicly pious but privately cynical, publicly platonic but privately lusting. Freud accustomed us to the idea that ulterior motives are pervasive in behavior, exerting their effects from an inaccessible stratum of the mind. Combine this with the common misconception that the genes are a kind of essence or core of the person, and you get a mongrel of Dawkins and Freud: the idea that the metaphorical motives of the genes are the deep, unconscious, ulterior motives of the person. That is an error. Brooklyn is not expanding.

Even people who can keep genes and people apart in their minds might find themselves depressed. Psychology has taught us that aspects of our experience may be figments, artifacts of how information is processed in the brain. The difference in kind between our experience of red and our experience of green does not mirror any difference in kind in lightwaves in the world—the wavelengths of light, which give rise to our perception of hue, form a smooth continuum. Red and green, perceived as qualitatively different properties, are constructs of the chemistry and circuitry of our nervous system. They could be absent in an organism with different photopigments or wiring; indeed, people with the most common form of colorblindness are just such organisms. And the emotional coloring of an object is as much a figment as its physical coloring. The sweetness of fruit, the scariness of heights, and the vileness of carrion are fancies of a nervous system that evolved to react to those objects in adaptive ways.

The sciences of human nature seem to imply that the same is true of right and wrong, merit and worthlessness, beauty and ugliness, holiness and baseness. They are neural constructs, movies we project onto the interior of our skulls, ways to tickle the pleasure centers of the brain, with no more reality than the difference between red and green. When Marley's ghost asked Scrooge why he doubted his senses, he said, "Because a little thing affects them. A slight disorder of the stomach makes them cheats. You may be an undigested bit of beef, a blot of mustard, a crumb of cheese, a fragment of an underdone potato. There's more of gravy than of grave about you, whatever you are!" Science seems to be saying that the same is true of everything we value.

But just because our brains are prepared to think in certain ways, it does not follow that the objects of those thoughts are fictitious. Many of our faculties evolved to mesh with real entities in the world. Our perception of depth is the product of complicated circuitry in the brain, circuitry that is absent from other species. But that does not mean that there aren't real trees and cliffs out there, or that the world is as flat as a pancake. And so it may be with more abstract entities. Humans, like many animals, appear to have an innate sense of number, which can be explained by the advantages of reasoning about numerosity during our evolutionary history. (For example, if three bears go into a cave and two come out, is it safe to enter?) But the mere fact that a number faculty evolved does not mean that numbers are hallucinations. According to the Platonist conception of number favored by many mathematicians and philosophers, entities such as numbers and shapes have an existence independent of minds. The number three is not invented out of whole cloth; it has real properties that can be discovered and explored. No rational creature equipped with circuitry to understand the concept "two" and the concept of addition could discover that two plus one equals anything other than three. That is why we expect similar bodies of mathematical results to emerge from different cultures or even different planets. If so, the number sense evolved to grasp abstract truths in the world that exist independently of the minds that grasp them.

Perhaps the same argument can be made for morality. According to the theory of moral realism, right and wrong exist, and have an inherent logic that licenses some moral arguments and not others.[12] The world presents us with non-zero-sum games in which it is better for both parties to act unselfishly than for both to act selfishly (better not to shove and not to be shoved than to shove and be shoved). Given the goal of being better off, certain conditions follow necessarily. No creature equipped with circuitry to understand that it is immoral for you to hurt me could discover anything but that it is immoral for me to hurt you. As with numbers and the number sense, we would expect moral systems to evolve toward similar conclusions in different cultures or even different planets. And in fact the Golden Rule has been rediscovered many times: by the authors of Leviticus and the Mahabharata; by Hillel, Jesus, and Confucius; by the Stoic philosophers of the Roman Empire; by social contract theorists such as Hobbes, Rousseau, and Locke; and by moral philosophers such as Kant in his categorical imperative.[13] Our moral sense may have evolved to mesh with an intrinsic logic of ethics rather than concocting it in our heads out of nothing.

But even if the Platonic existence of moral logic is too rich for your blood, you can still see morality as something more than a social convention or religious dogma. Whatever its ontological status may be, a moral sense is part of the standard equipment of the human mind. *It's the only mind we've got*, and we have no choice but to take its intuitions seriously. If we are so constituted that we cannot help but think in moral terms (at least some of the time and toward some people), then morality is as real *for us* as if it were decreed by the Almighty or written into the cosmos. And so it is with other human values like love, truth, and beauty. Could we ever know whether they are really "out there" or whether we just think they are out there because the human brain makes it impossible not to think they are out there? And how bad would it be if they *were* inherent to the human way of thinking? Perhaps we should reflect on our condition as Kant did in his *Critique of Practical Reason:* "Two things fill the mind with ever new and increasing admiration and awe, the oftener and more steadily we reflect on them: the starry heavens above and the moral law within."

~

IN THE PAST four chapters I have shown why new ideas from the sciences of human nature do not undermine humane values. On the contrary, they present opportunities to sharpen our ethical reasoning and put those values on a firmer foundation. In a nutshell:

- It is a bad idea to say that discrimination is wrong only because the traits of all people are indistinguishable.
- It is a bad idea to say that violence and exploitation are wrong only because people are not naturally inclined to them.
- It is a bad idea to say that people are responsible for their actions only because the causes of those actions are mysterious.
- And it is a bad idea to say that our motives are meaningful in a personal sense only because they are inexplicable in a biological sense.

These are bad ideas because they make our values hostages to fortune, implying that someday factual discoveries could make them obsolete. And they are bad ideas because they conceal the downsides of denying human nature: persecution of the successful, intrusive social engineering, the writing off of suffering in other cultures, an incomprehension of the logic of justice, and the devaluing of human life on earth.

PART IV

KNOW THYSELF

Now that I have attempted to make the very idea of human nature respectable, it is time to say something about what it is and what difference it makes for our public and private lives. The chapters in Part IV present some current ideas about the design specs of the basic human faculties. These are not just topics in a psychology curriculum but have implications for many arenas of public discourse. Ideas about the contents of cognition—concepts, words, and images—shed light on the roots of prejudice, on the media, and on the arts. Ideas about the capacity for reason can enter into our policies of education and applications of technology. Ideas about social relations are relevant to the family, to sexuality, to social organization, and to crime. Ideas about the moral sense inform the way we evaluate political movements and how we trade off one value against another.

In each of these arenas, people always appeal to some conception of human nature, whether they acknowledge it or not. The problem is that the conceptions are often based on gut feelings, folk theories, and archaic versions of biology. My goal is to make these conceptions explicit, to suggest what is right and wrong about them, and to spell out some of the implications. Ideas about human nature cannot, on their own, resolve perplexing controversies or determine public policy. But without such ideas we are not playing with a full deck and are vulnerable to unnecessary befuddlement. As the biologist Richard Alexander has noted, "Evolution is surely most deterministic for those still unaware of it."[1]

Chapter 12

In Touch with Reality

What a piece of work is a man!
How noble in reason!
How infinite in faculty!
In form, in moving, how express and admirable!
In action, how like an angel!
In apprehension, how like a god!
— William Shakespeare

THE STARTING POINT for acknowledging human nature is a sheer awe and humility in the face of the staggering complexity of its source, the brain. Organized by the three billion bases of our genome and shaped by hundreds of millions of years of evolution, the brain is a network of unimaginable intricacy: a hundred billion neurons linked by a hundred trillion connections, woven into a convoluted three-dimensional architecture. Humbling, too, is the complexity of what it does. Even the mundane talents we share with other primates—walking, grasping, recognizing—are solutions to engineering problems at or beyond the cutting edge of artificial intelligence. The talents that are human birthrights—speaking and understanding, using common sense, teaching children, inferring other people's motives—will probably not be duplicated by machines in our lifetime, if ever. All this should serve as a counterweight to the image of the mind as formless raw material and to people as insignificant atoms making up the complex being we call "society."

The human brain equips us to thrive in a world of objects, living things, and other people. Those entities have a large impact on our well-being, and one would expect the brain to be well suited to detecting them and their powers. Failing to recognize a steep precipice or a hungry panther or a jealous spouse can have significant negative consequences for biological fitness, to put it mildly. The fantastic complexity of the brain is there in part to register consequential facts about the world around us.

But this truism has been rejected by many sectors of modern intellectual life. According to the relativistic wisdom prevailing in much of academia today, reality is socially constructed by the use of language, stereotypes, and media images. The idea that people have access to facts about the world is naïve, say the proponents of social constructionism, science studies, cultural studies, critical theory, postmodernism, and deconstructionism. In their view, observations are always infected by theories, and theories are saturated with ideology and political doctrines, so anyone who claims to have the facts or know the truth is just trying to exert power over everyone else.

Relativism is entwined with the doctrine of the Blank Slate in two ways. One is that relativists have a penny-pinching theory of psychology in which the mind has no mechanisms designed to grasp reality; all it can do is passively download words, images, and stereotypes from the surrounding culture. The other is the relativists' attitude toward science. Most scientists regard their work as an extension of our everyday ability to figure out what is out there and how things work. Telescopes and microscopes amplify the visual system; theories formalize our hunches about cause and effect; experiments refine our drive to gather evidence about events we cannot witness directly. Relativist movements agree that science is perception and cognition writ large, but they draw the opposite conclusion: that scientists, like laypeople, are unequipped to grasp an objective reality. Instead, their advocates say, "Western science is only one way of describing reality, nature, and the way things work—a very effective way, certainly, for the production of goods and profits, but unsatisfactory in most other respects. It is an imperialist arrogance which ignores the sciences and insights of most other cultures and times."[1] Nowhere is this more significant than in the scientific study of politically charged topics such as race, gender, violence, and social organization. Appealing to "facts" or "the truth" in connection with these topics is just a ruse, the relativists say, because there *is* no "truth" in the sense of an objective yardstick independent of cultural and political presuppositions.

Skepticism about the soundness of people's mental faculties also determines whether one should respect ordinary people's tastes and opinions (even those we don't much like) or treat the people as dupes of an insidious commercial culture. According to relativist doctrines like "false consciousness," "inauthentic preferences," and "interiorized authority," people may be mistaken about their own desires. If so, it would undermine the assumptions behind democracy, which gives ultimate authority to the preferences of the majority of a population, and the assumptions behind market economies, which treat people as the best judges of how they should allocate their own resources. Perhaps not coincidentally, it elevates the scholars and artists who analyze the use of language and images in society, because only they can unmask the ways in which such media mislead and corrupt.

This chapter is about the assumptions about cognition—in particular, concepts, words, and images—that underlie recent relativistic movements in intellectual life. The best way to introduce the argument is with examples from the study of perception, our most immediate connection to the world. They immediately show that the question of whether reality is socially constructed or directly available has not been properly framed. Neither alternative is correct.

Relativists have a point when they say that we don't just open our eyes and apprehend reality, as if perception were a window through which the soul gazes at the world. The idea that we just see things as they are is called naïve realism, and it was refuted by skeptical philosophers thousands of years ago with the help of a simple phenomenon: visual illusions. Our visual systems can play tricks on us, and that is enough to prove they are gadgets, not pipelines to the truth. Here are two of my favorites. In Roger Shepard's "Turning the Tables"[2] (right), the two parallelograms are identical in size and shape. In Edward Adelson's "Checker Shadow Illusion"[3] (below) the light square in the middle of the shadow (B) is the same shade of gray as the dark squares outside the shadow (A):

But just because the world we know is a construct of our brain, that does not mean it is an *arbitrary* construct—a phantasm created by expectations or the social context. Our perceptual systems are designed to register aspects of the external world that were important to our survival, like the sizes, shapes, and materials of objects. They need a complex design to accomplish this feat because the retinal image is not a replica of the world. The projection of an object on the retina grows, shrinks, and warps as the object moves around; color and brightness fluctuate as the lighting changes from sun to clouds or from indoor to outdoor light. But somehow the brain solves these maddening problems. It works as if it were reasoning backwards from the retinal image to hypotheses about reality, using geometry, optics, probability theory, and assumptions about the world. Most of the time the system works: people don't usually bump into trees or bite into rocks.

But occasionally the brain is fooled. The ground stretching away from our feet projects an image from the bottom to the center of our visual field. As a result, the brain often interprets down-up in the visual field as near-far in the world, especially when reinforced by other perspective cues such as occluded parts (like the hidden table legs). Objects stretching away from the viewer get foreshortened by projection, and the brain compensates for this, so we tend to see a given distance running up-and-down in the visual field as coming from a longer object than the same distance running left-to-right. And that makes us see the lengths and widths differently in the turned tables. By similar logic, objects in shadow reflect less light onto our retinas than objects in full illumination. Our brains compensate, making us see a given shade of gray as lighter when it is in shadow than when it is in sunshine. In each case we may see the lines and patches on the page incorrectly, but that is only because our visual systems are working very hard to see them as coming from a real world. Like a policeman framing a suspect, Shepard and Adelson have planted evidence that would lead a rational but unsuspecting observer to an incorrect conclusion. If we *were* in a world of ordinary 3-D objects that had projected those images onto our retinas, our perceptual experience would be accurate. Adelson explains: "As with many so-called illusions, this effect really demonstrates the success rather than the failure of the visual system. The visual system is not very good at being a physical light meter, but that is not its purpose. The important task is to break the image information down into meaningful components, and thereby perceive the nature of the objects in view."[4]

It's not that expectations from past experience are irrelevant to perception. But their influence is to make our perceptual systems more accurate, not more arbitrary. In the two words below, we perceive the same shape as an "H" in the first word and as an "A" in the second:[5]

THE CAT

We see the shapes that way because experience tells us—correctly—that the odds are high that there really *is* an "H" in the middle of the first word and an "A" in the middle of the second, even if that is not true in an atypical case. The mechanisms of perception go to a lot of trouble to ensure that what we see corresponds to what is usually out there.

So the demonstrations that refute naïve realism most decisively also refute the idea that the mind is disconnected from reality. There is a third alternative:

that the brain evolved fallible yet intelligent mechanisms that work to keep us in touch with aspects of reality that were relevant to the survival and reproduction of our ancestors. And that is true not just of our perceptual faculties but of our cognitive faculties. The fact that our cognitive faculties (like our perceptual faculties) are attuned to the real world is most obvious from their *response* to illusions: they recognize the possibility of a breach with reality and find a way to get at the truth behind the false impression. When we see an oar that appears to be severed at the water's surface, we know how to tell whether it really is severed or just looks that way: we can palpate the oar, slide a straight object along it, or pull on it to see if the submerged part gets left behind. The concept of truth and reality behind such tests appears to be universal. People in all cultures distinguish truth from falsity and inner mental life from overt reality, and try to deduce the presence of unobservable objects from the perceptible clues they leave behind.[6]

~

VISUAL PERCEPTION IS the most piquant form of knowledge of the world, but relativists are less concerned with how we see objects than with how we *categorize* them: how we sort our experiences into conceptual categories like birds, tools, and people. The seemingly innocuous suggestion that the categories of the mind correspond to something in reality became a contentious idea in the twentieth century because some categories—stereotypes of race, gender, ethnicity, and sexual orientation—can be harmful when they are used to discriminate or oppress.

The word *stereotype* originally referred to a kind of printing plate. Its current sense as a pejorative and inaccurate image standing for a category of people was introduced in 1922 by the journalist Walter Lippmann. Lippmann was an important public intellectual who, among other things, helped to found *The New Republic,* influenced Woodrow Wilson's policies at the end of World War I, and wrote some of the first attacks on IQ testing. In his book *Public Opinion,* Lippmann fretted about the difficulty of achieving true democracy in an age in which ordinary people could no longer judge public issues rationally because they got their information in what we today call sound bites. As part of this argument, Lippmann proposed that ordinary people's concepts of social groups were stereotypes: mental pictures that are incomplete, biased, insensitive to variation, and resistant to disconfirming information.

Lippmann had an immediate influence on social science (though the subtleties and qualifications of his original argument were forgotten). Psychologists gave people lists of ethnic groups and lists of traits and asked them to pair them up. Sure enough, people linked Jews with "shrewd" and "mercenary," Germans with "efficient" and "nationalistic," Negroes with "superstitious" and "happy-go-lucky," and so on.[7] Such generalizations are pernicious when applied to individuals, and though they are still lamentably common in much of the world, they are now actively avoided by educated people and by mainstream public figures.

By the 1970s, many thinkers were not content to note that stereotypes about categories of people can be inaccurate. They began to insist that the categories themselves don't exist other than in our stereotypes. An effective way to fight racism, sexism, and other kinds of prejudice, in this view, is to deny that conceptual categories about people have any claim to objective reality. It would be impossible to believe that homosexuals are effeminate, blacks superstitious, and women passive if there were no such things as categories of homosexuals, blacks, or women to begin with. For example, the philosopher Richard Rorty has written, " 'The homosexual,' 'the Negro,' and 'the female' are best seen not as inevitable classifications of human beings but rather as inventions that have done more harm than good."[8]

For that matter, many writers think, why stop there? Better still to insist that *all* categories are social constructions and therefore figments, because that would *really* make invidious stereotypes figments. Rorty notes with approval that many thinkers today "go on to suggest that quarks and genes probably are [inventions] too." Postmodernists and other relativists attack truth and objectivity not so much because they are interested in philosophical problems of ontology and epistemology but because they feel it is the best way to pull the rug out from under racists, sexists, and homophobes. The philosopher Ian Hacking provides a list of almost forty categories that have recently been claimed to be "socially constructed." The prime examples are race, gender, masculinity, nature, facts, reality, and the past. But the list has been growing and now includes authorship, AIDS, brotherhood, choice, danger, dementia, illness, Indian forests, inequality, the Landsat satellite system, the medicalized immigrant, the nation-state, quarks, school success, serial homicide, technological systems, white-collar crime, women refugees, and Zulu nationalism. According to Hacking, the common thread is a conviction that the category is not determined by the nature of things and therefore is not inevitable. The further implication is that we would be much better off if it were done away with or radically transformed.[9]

This whole enterprise is based on an unstated theory of human concept formation: that conceptual categories bear no systematic relation to things in the world but are socially constructed (and can therefore be reconstructed). Is it a correct theory? In some cases it has a grain of truth. As we saw in Chapter 4, some categories really are social constructions: they exist only because people tacitly agree to act as if they exist. Examples include money, tenure, citizenship, decorations for bravery, and the presidency of the United States.[10] But that does not mean that *all* conceptual categories are socially constructed. Concept formation has been studied for decades by cognitive psychologists, and they conclude that most concepts pick out categories of objects in the world which had some kind of reality before we ever stopped to think about them.[11]

Yes, every snowflake is unique, and no category will do complete justice to every one of its members. But intelligence depends on lumping together things that share properties, so that we are not flabbergasted by every new thing we encounter. As William James wrote, "A polyp would be a conceptual thinker if a feeling of 'Hollo! thingumbob again!' ever flitted through its mind." We perceive some traits of a new object, place it in a mental category, and infer that it is likely to have the other traits typical of that category, ones we cannot perceive. If it walks like a duck and quacks like a duck, it probably is a duck. If it's a duck, it's likely to swim, fly, have a back off which water rolls, and contain meat that's tasty when wrapped in a pancake with scallions and hoisin sauce.

This kind of inference works because the world really does contain ducks, which really do share properties. If we lived in a world in which walking quacking objects were no more likely to contain meat than did any other object, the category "duck" would be useless and we probably would not have evolved the ability to form it. If you were to construct a giant spreadsheet in which the rows and columns were traits that people notice and the cells were filled in by objects that possess that combination of traits, the pattern of filled cells would be lumpy. You would find lots of entries at the intersection of the "quacks" row and the "waddles" column but none at the "quacks" row and the "gallops" column. Once you specify the rows and columns, the lumpiness comes from the world, not from society or language. It is no coincidence that the same living things tend to be classified together by the words in European cultures, the words for plant and animal kinds in other cultures (including preliterate cultures), and the Linnaean taxa of professional biologists equipped with calipers, dissecting tools, and DNA sequencers. Ducks, biologists say, are several dozen species in the subfamily Anatinae, each with a distinct anatomy, an ability to interbreed with other members of their species, and a common ancestor in evolutionary history.

Most cognitive psychologists believe that conceptual categories come from two mental processes.[12] One of them notices clumps of entries in the mental spreadsheet and treats them as categories with fuzzy boundaries, prototypical members, and overlapping similarities, like the members of a family. That's why our mental category "duck" can embrace odd ducks that don't match the prototypical duck, such as lame ducks, who cannot swim or fly, Muscovy ducks, which have claws and spurs on their feet, and Donald Duck, who talks and wears clothing. The other mental process looks for crisp rules and definitions and enters them into chains of reasoning. The second system can learn that true ducks molt twice a season and have overlapping scales on their legs and hence that certain birds that look like geese and are called geese really are ducks. Even when people don't know these facts from academic biology, they have a strong intuition that species are defined by an internal essence or hidden trait that lawfully gives rise to its visible features.[13]

Anyone who teaches the psychology of categorization has been hit with this question from a puzzled student: "You're telling us that putting things into categories is rational and makes us smart. But we've always been taught that putting *people* into categories is irrational and makes us sexist and racist. If categorization is so great when we think about ducks and chairs, why is it so terrible when we think about genders and ethnic groups?" As with many ingenuous questions from students, this one uncovers a shortcoming in the literature, not a flaw in their understanding.

The idea that stereotypes are inherently irrational owes more to a condescension toward ordinary people than it does to good psychological research. Many researchers, having shown that stereotypes existed in the minds of their subjects, assumed that the stereotypes had to be irrational, because they were uncomfortable with the possibility that some trait might be statistically true of some group. They never actually checked. That began to change in the 1980s, and now a fair amount is known about the accuracy of stereotypes.[14]

With some important exceptions, stereotypes are in fact *not* inaccurate when assessed against objective benchmarks such as census figures or the reports of the stereotyped people themselves. People who believe that African Americans are more likely to be on welfare than whites, that Jews have higher average incomes than WASPs, that business students are more conservative than students in the arts, that women are more likely than men to want to lose weight, and that men are more likely than women to swat a fly with their bare hands, are not being irrational or bigoted. Those beliefs are correct. People's stereotypes are generally consistent with the statistics, and in many cases their bias is to *underestimate* the real differences between sexes or ethnic groups.[15] This does not mean that the stereotyped traits are unchangeable, of course, or that people think they are unchangeable, only that people perceive the traits fairly accurately at the time.

Moreover, even when people believe that ethnic groups have characteristic traits, they are never mindless stereotypers who literally believe that each and every member of the group possesses those traits. People may think that Germans are, on average, more efficient than non-Germans, but no one believes that every last German is more efficient than every non-German.[16] And people have no trouble overriding a stereotype when they have good information about an individual. Contrary to a common accusation, teachers' impressions of their individual pupils are not contaminated by their stereotypes of race, gender, or socioeconomic status. The teachers' impressions accurately reflect the pupil's performance as measured by objective tests.[17]

Now for the important exceptions. Stereotypes can be downright inaccurate when a person has few or no firsthand encounters with the stereotyped group, or belongs to a group that is overtly hostile to the one being judged. During World War II, when the Russians were allies of the United States and the Germans were enemies, Americans judged Russians to have more positive traits than Germans. Soon afterward, when the alliances reversed, Americans judged Germans to have more positive traits than Russians.[18]

Also, people's ability to set aside stereotypes when judging an individual is accomplished by their conscious, deliberate reasoning. When people are distracted or put under pressure to respond quickly, they are more likely to judge that a member of an ethnic group has all the stereotyped traits of the group.[19] This comes from the two-part design of the human categorization system mentioned earlier. Our network of fuzzy associations naturally reverts to a stereotype when we first encounter an individual. But our rule-based categorizer can block out those associations and make deductions based on the relevant facts about that individual. It can do so either for practical reasons, when information about a group-wide average is less diagnostic than information about the individual, or for social and moral reasons, out of respect for the imperative that one *ought* to ignore certain group-wide averages when judging an individual.

The upshot of this research is not that stereotypes are always accurate but that they are not always false, or even usually false. This is just what we would expect if human categorization—like the rest of the mind—is an adaptation that keeps track of aspects of the world that are relevant to our long-term well-being. As the social psychologist Roger Brown pointed out, the main difference between categories of people and categories of other things is that when you use a prototypical exemplar to stand for a category of things, no one takes offense. When Webster's dictionary used a sparrow to stand for all birds, "emus and ostriches and penguins and eagles did not go on the attack." But just imagine what would have happened if Webster's had used a picture of a soccer mom to illustrate *woman* and a picture of a business executive to illustrate *man*. Brown remarks, "Of course, people would be right to take offense since a prototype can never represent the variation that exists in natural categories. It's just that birds don't care but people do."[20]

What are the implications of the fact that many stereotypes are statistically accurate? One is that contemporary scientific research on sex differences cannot be dismissed just because some of the findings are consistent with traditional stereotypes of men and women. Some parts of those stereotypes may be false, but the mere fact that they are stereotypes does not prove that they are false in every respect.

The partial accuracy of many stereotypes does not, of course, mean that racism, sexism, and ethnic prejudice are acceptable. Quite apart from the democratic principle that in the public sphere people should be treated as individuals, there are good reasons to be concerned about stereotypes.

Stereotypes based on hostile depictions rather than on firsthand experience are bound to be inaccurate. And some stereotypes are accurate only because of self-fulfilling prophecies. Forty years ago it may have been factually correct that few women and African Americans were qualified to be chief executives or presidential candidates. But that was only because of barriers that prevented them from attaining those qualifications, such as university policies that refused them admission out of a belief that they were not qualified. The institutional barriers had to be dismantled before the facts could change. The good news is that when the facts do change, people's stereotypes can change with them.

What about policies that go farther and actively compensate for prejudicial stereotypes, such as quotas and preferences that favor underrepresented groups? Some defenders of these policies assume that gatekeepers are incurably afflicted with baseless prejudices, and that quotas must be kept in place forever to neutralize their effects. The research on stereotype accuracy refutes that argument. Nonetheless, the research might support a different argument for preferences and other gender- and color-sensitive policies. Stereotypes, even when they are accurate, might be self-fulfilling, and not just in the obvious case of institutionalized barriers like those that kept women and African Americans out of universities and professions. Many people have heard of the Pygmalion effect, in which people perform as other people (such as teachers) expect them to perform. As it happens, the Pygmalion effect appears to be small or nonexistent, but there are more subtle forms of self-fulfilling prophecies.[21] If subjective decisions about people, such as admissions, hiring, credit, and salaries, are based in part on group-wide averages, they will conspire to make the rich richer and the poor poorer. Women are marginalized in academia, making them genuinely less influential, which increases their marginalization. African Americans are treated as poorer credit risks and denied credit, which makes them less likely to succeed, which makes them poorer credit risks. Race- and gender-sensitive policies, according to arguments by the psychologist Virginia Valian, the economist Glenn Loury, and the philosopher James Flynn, may be needed to break the vicious cycle.[22]

Pushing in the other direction is the finding that stereotypes are least accurate when they pertain to a coalition that is pitted against one's own in hostile competition. This should make us nervous about identity politics, in which public institutions identify their members in terms of their race, gender, and ethnic group and weigh every policy by how it favors one group over another. In many universities, for example, minority students are earmarked for special orientation sessions and encouraged to view their entire academic experience through the lens of their group and how it has been victimized. By implicitly pitting one group against another, such policies may cause each group to brew stereotypes about the other that are more pejorative than the ones they would develop in personal encounters. As with other policy issues I examine in this book, the data from the lab do not offer a thumbs-up or thumbs-down verdict on race- and gender-conscious policies. But by highlighting the features of our psychology that different policies engage, the findings can make the tradeoffs clearer and the debates better informed.

~

OF ALL THE faculties that go into the piece of work called man, language may be the most awe-inspiring. "Remember that you are a human being with a soul and the divine gift of articulate speech," Henry Higgins implored Eliza Doolittle. Galileo's alter ego, humbled by the arts and inventions of his day, commented on language in its written form:

> But surpassing all stupendous inventions, what sublimity of mind was his who dreamed of finding means to communicate his deepest thoughts to any other person, though distant by mighty intervals of place and time! Of talking with those who are in India; of speaking to those who are not yet born and will not be born for a thousand or ten thousand years; and with what facility, by the different arrangements of twenty characters upon a page![23]

But a funny thing happened to language in intellectual life. Rather than being appreciated for its ability to communicate thought, it was condemned for its power to *constrain* thought. Famous quotations from two philosophers capture the anxiety. "We have to cease to think if we refuse to do it in the prisonhouse of language," wrote Friedrich Nietzsche. "The limits of my language mean the limits of my world," wrote Ludwig Wittgenstein.

How could language exert this stranglehold? It would if words and phrases were the medium of thought itself, an idea that falls naturally out of the Blank Slate. If there is nothing in the intellect that was not first in the senses, then words picked up by the ears are the obvious source of any abstract thought that cannot be reduced to sights, smells, or other sounds. Watson tried to explain thinking as microscopic movements of the mouth and throat; Skinner hoped his 1957 book *Verbal Behavior,* which explained language as a repertoire of rewarded responses, would bridge the gap between pigeons and people.

The other social sciences also tended to equate language with thought. Boas's student Edward Sapir called attention to differences in how languages carve up the world into categories, and Sapir's student Benjamin Whorf stretched those observations into the famous Linguistic Determinism hypothesis: "We cut nature up, organize it into concepts, and ascribe significances as we do, largely because we are parties to an agreement to organize it in this way—an agreement that holds throughout our speech community and is codified in the patterns of our language. The agreement is, of course, an implicit and unstated one, *but its terms are absolutely obligatory*."[24] More recently, the anthropologist Clifford Geertz wrote that "thinking consists not of 'happenings in the head' (though happenings there and elsewhere are necessary for it to occur) but of a traffic in what have been called . . . significant symbols—words for the most part."[25]

As with so many ideas in social science, the centrality of language is taken to extremes in deconstructionism, postmodernism, and other relativist doctrines. The writings of oracles like Jacques Derrida are studded with such aphorisms as "No escape from language is possible," "Text is self-referential," "Language is power," and "There is nothing outside the text." Similarly, J. Hillis Miller wrote that "language is not an instrument or tool in man's hands, a submissive means of thinking. Language rather thinks man and his 'world' . . . if he will allow it to do so."[26] The prize for the most extreme statement must go to Roland Barthes, who declared, "Man does not exist prior to language, either as a species or as an individual."[27]

The ancestry of these ideas is said to be from linguistics, though most linguists believe that deconstructionists have gone off the deep end. The original observation was that many words are defined in part by their relationship to other words. For example, *he* is defined by its contrast with *I, you, they,* and *she,* and *big* makes sense only as the opposite of *little.* And if you look up words in a dictionary, they are defined by other words, which are defined by still other words, until the circle is completed when you get back to a definition containing the original word. Therefore, say the deconstructionists, language is a self-contained system in which words have no necessary connection to reality. And since language is an arbitrary instrument, not a medium for communicating thoughts or describing reality, the powerful can use it to manipulate and oppress others. This leads in turn to an agitation for linguistic reforms: neologisms like *co* or *na* that would serve as gender-neutral pronouns, a succession of new terms for racial minorities, and a rejection of standards of clarity in criticism and scholarship (for if language is no longer a window onto thought but the very stuff of thought, the metaphor of "clarity" no longer applies).

Like all conspiracy theories, the idea that language is a prisonhouse denigrates its subject by overestimating its power. Language is the magnificent faculty that we use to get thoughts from one head to another, and we can co-opt it in many ways to help our thoughts along. But it is not the same as thought, not the only thing that separates humans from other animals, not the basis of all culture, and not an inescapable prisonhouse, an obligatory agreement, the limits of our world, or the determiner of what is imaginable.[28]

We have seen that perception and categorization provide us with concepts that keep us in touch with the world. Language extends that lifeline by connecting the concepts to words. Children hear noises coming out of a family member's mouth, use their intuitive psychology and their grasp of the context to infer what the speaker is trying to say, and mentally link the words to the concepts and the grammatical rules to the relationships among them. Bowser upends a chair, Sister yells, "The dog knocked over the chair!" and Junior deduces that *dog* means dog, *chair* means chair, and the subject of the verb *knock over* is the agent doing the knocking over.[29] Now Junior can talk about other dogs, other chairs, and other knockings over. There is nothing self-referential or imprisoning about it. As the novelist Walker Percy quipped, a deconstructionist is an academic who claims that texts have no referents and then leaves a message on his wife's answering machine asking her to order a pepperoni pizza for dinner.

Language surely does affect our thoughts, rather than just labeling them for the sake of labeling them. Most obviously, language is the conduit through which people share their thoughts and intentions and thereby acquire the knowledge, customs, and values of those around them. In the song "Christmas" from their rock opera, The Who described the plight of a boy without language: "Tommy doesn't know what day it is; he doesn't know who Jesus was or what prayin' is."

Language can allow us to share thoughts not just directly, by its literal content, but also indirectly, via metaphors and metonyms that nudge listeners into grasping connections they may not have noticed before. For example, many expressions treat time as if it were a valuable resource, such as *waste time, spend time, valuable time,* and *time is money.*[30] Presumably on the first occasion a person used one of these expressions, her audience wondered why she was using a word for money to refer to time; after all, you can't literally spend time the way you spend pieces of gold. Then, by assuming that the speaker was not gibbering, they figured out the ways in which time indeed has something in common with money, and assumed that that was what the speaker intended to convey. Note that even in this clear example of language affecting thought, language is not *the same thing* as thought. The original coiner of the metaphor had to see the analogy without the benefit of the English expressions, and the first listeners had to make sense of it using a chain of ineffable thoughts about the typical intentions of speakers and the properties shared by time and money.

Aside from its use as a medium of communication, language can be pressed into service as one of the media used by the brain for storing and manipulating information.[31] The leading theory of human working memory, from the psychologist Alan Baddeley, captures the idea nicely.[32] The mind makes use of a "phonological loop": a silent articulation of words or numbers that persists for a few seconds and can be sensed by the mind's ear. The loop acts as a "slave system" at the service of a "central executive." By describing things to ourselves using snatches of language, we can temporarily store the result of a mental computation or retrieve chunks of data stored as verbal expressions. Mental arithmetic involving large numbers, for example, may be carried out by retrieving verbal formulas such as "Seven times eight is fifty-six."[33] But as the technical terms of the theory make clear, language is serving as a slave of an executive, not as the medium of all thought.

Why do virtually all cognitive scientists and linguists believe that language is not a prisonhouse of thought?[34] First, many experiments have plumbed the minds of creatures without language, such as infants and nonhuman primates, and have found the fundamental categories of thought working away: objects, space, cause and effect, number, probability, agency (the initiation of behavior by a person or animal), and the functions of tools.[35]

Second, our vast storehouse of knowledge is certainly not couched in the words and sentences in which we learned the individual facts. What did you read in the page before this one? I would like to think that you can give a reasonably accurate answer to the question. Now try to write down the exact words you read in those pages. Chances are you cannot recall a single sentence verbatim, probably not even a single phrase. What you remembered is the *gist* of those passages—their content, meaning, or sense—not the language itself. Many experiments on human memory have confirmed that what we remember over the long term is the content, not the wording, of stories and conversations. Cognitive scientists model this "semantic memory" as a web of logical propositions, images, motor programs, strings of sounds, and other data structures connected to one another in the brain.[36]

A third way to put language in its place is to think about how we use it. Writing and speaking do not consist of transcribing an interior monologue onto paper or playing it into a microphone. Rather, we engage in a constant give-and-take between the thoughts we try to convey and the means our language offers to convey them. We often grope for words, are dissatisfied with what we write because it does not express what we wanted to say, or discover when every combination of words seems wrong that we do not really *know* what we want to say. And when we get frustrated by a mismatch between our language and our thoughts, we don't give up, defeated and mum, but change the language. We concoct neologisms *(quark, meme, clone, deep structure),* invent slang *(to spam, to diss, to flame, to surf the web, a spin doctor),* borrow useful words from other languages *(joie de vivre, schlemiel, angst, machismo),* or coin new metaphors *(waste time, vote with your feet, push the outside of the envelope).* That is why every language, far from being an immutable penitentiary, is constantly under renovation. Despite the lamentations of language lovers and the coercion of tongue troopers, languages change unstoppably as people need to talk about new things or convey new attitudes.[37]

Finally, language itself could not function if it did not sit atop a vast infrastructure of tacit knowledge about the world and about the intentions of other people. When we understand language, we have to listen between the lines to winnow out the unintended readings of an ambiguous sentence, piece together fractured utterances, glide over slips of the tongue, and fill in the countless unsaid steps in a complete train of thought. When the shampoo bottle says "Lather, rinse, repeat," we don't spend the rest of our lives in the shower; we infer that it means "repeat once." And we know how to interpret ambiguous headlines such as "Kids Make Nutritious Snacks," "Prostitutes Appeal to Pope," and "British Left Waffles on Falkland Islands," because we effortlessly apply our background knowledge about the kinds of things that people are likely to convey in newspapers. Indeed, the very existence of ambiguous sentences, in which one string of words expresses two thoughts, proves that thoughts are not the same thing as strings of words.

~

LANGUAGE OFTEN MAKES the news precisely because it can part company with thoughts and attitudes. In 1998 Bill Clinton exploited the expectations behind ordinary comprehension to mislead prosecutors about his affair with Monica Lewinsky. He used words like *alone, sex,* and *is* in senses that were technically defensible but which deviated from charitable guesses about what people ordinarily mean by these terms. For example, he suggested he was not "alone" with Lewinsky, even though they were the only two people in the room, because other people were in the Oval Office complex at the time. He said that he did not have "sex" with her, because they did not engage in intercourse. His words, like all words, are certainly vague at their boundaries. Exactly how far away or hidden must the nearest person be before one is considered alone? At what point in the continuum of bodily contact—from an accidental brush in an elevator to tantric bliss—do we say that sex has occurred? Ordinarily we resolve the vagueness by guessing how our conversational partner would interpret words in the context, and we choose our words accordingly. Clinton's ingenuity in manipulating these guesses, and the outrage that erupted when he was forced to explain what he had done, show that people have an acute understanding of the difference between words and the thoughts they are designed to convey.

Language conveys not just literal meanings but also a speaker's attitude. Think of the difference between *fat* and *voluptuous, slender* and *scrawny, thrifty* and *stingy, articulate* and *slick.* Racial epithets, which are laced with contempt, are justifiably off-limits among responsible people, because using them conveys the tacit message that contempt for the people referred to by the epithet is acceptable. But the drive to adopt new terms for disadvantaged groups goes much further than this basic sign of respect; it often assumes that words and attitudes are so inseparable that one can reengineer people's attitudes by tinkering with the words. In 1994 the *Los Angeles Times* adopted a style sheet that banned some 150 words, including *birth defect, Canuck, Chinese fire drill, dark continent, divorcée, Dutch treat, handicapped, illegitimate, invalid, manmade, New World, stepchild,* and *to welsh.* The editors assumed that words register in the brain with their literal meanings, so that an *invalid* is understood as "someone who is not valid" and *Dutch treat* is understood as a slur on contemporary Netherlanders. (In fact, it is one of many idioms in which *Dutch* means "ersatz," such as *Dutch oven, Dutch door, Dutch uncle, Dutch courage,* and *Dutch auction,* the remnants of a long-forgotten rivalry between the English and the Dutch.)

But even the more reasonable attempts at linguistic reform are based on a dubious theory of linguistic determinism. Many people are puzzled by the replacement of formerly unexceptionable terms by new ones: *Negro* by *black* by *African American, Spanish-American* by *Hispanic* by *Latino, crippled* by *handicapped* by *disabled* by *challenged, slum* by *ghetto* by *inner city* by (according to the *Times*) *slum* once again. Occasionally the neologisms are defended with some rationale about their meaning. In the 1960s, the word *Negro* was replaced by the word *black,* because the parallel between the words *black* and *white* was meant to underscore the equality of the races. Similarly, *Native American* reminds us of who was here first and avoids the geographically inaccurate term *Indian*. But often the new terms replace ones that were perfectly congenial in their day, as we see in names for old institutions that are obviously sympathetic to the people being named: the United Negro College Fund, the National Association for the Advancement of Colored People, the Shriners Hospitals for Crippled Children. And sometimes a term can be tainted or unfashionable while a minor variant is fine: consider *colored people* versus *people of color, Afro-American* versus *African American, Negro*—Spanish for "black"—versus *black*. If anything, a respect for literal meaning should send us off looking for a new word for the descendants of Europeans, who are neither white nor Caucasian. Something else must be driving the replacement process.

Linguists are familiar with the phenomenon, which may be called the euphemism treadmill. People invent new words for emotionally charged referents, but soon the euphemism becomes tainted by association, and a new word must be found, which soon acquires its own connotations, and so on. *Water closet* becomes *toilet* (originally a term for any kind of body care, as in *toilet kit* and *toilet water*), which becomes *bathroom,* which becomes *restroom,* which becomes *lavatory. Undertaker* changes to *mortician,* which changes to *funeral director. Garbage collection* turns into *sanitation,* which turns into *environmental services. Gym* (from *gymnasium,* originally "high school") becomes *physical education,* which becomes (at Berkeley) *human biodynamics.* Even the word *minority*—the most neutral label conceivable, referring only to relative numbers—was banned in 2001 by the San Diego City Council (and nearly banned by the Boston City Council) because it was deemed disparaging to non-whites. "No matter how you slice it, *minority* means less than," said a semantically challenged official at Boston College, where the preferred term is AHANA (an acronym for African-American, Hispanic, Asian, and Native American).[38]

The euphemism treadmill shows that concepts, not words, are primary in people's minds. Give a concept a new name, and the name becomes colored by the concept; the concept does not become freshened by the name, at least not for long. Names for minorities will continue to change as long as people have negative attitudes toward them. We will know that we have achieved mutual respect when the names stay put.

~

"IMAGE IS NOTHING. Thirst is everything," screams a soft-drink ad that tries to create a new image for its product by making fun of soft-drink ads that try to create images for their products. Like words, images are salient tokens of our mental lives. And like words, images are said to have an insidious power over our consciousness, presumably because they are inscribed directly onto a blank slate. In postmodernist and relativist thinking, images are held to shape our view of reality, or to *be* our view of reality, or to be reality itself. This is especially true of images representing celebrities, politicians, women, and AHANAs. And as with language, the scientific study of imagery shows that the fear is misplaced.

A good description of the standard view of images within cultural studies and related disciplines may be found in the *Concise Glossary of Cultural Theory*. It defines *image* as a "mental or visual representation of an object or event as depicted in the mind, a painting, a photograph, or film." Having thus run together images in the world (such as paintings) with images in the mind, the entry lays out the centrality of images in postmodernism, cultural studies, and academic feminism.

First it notes, reasonably enough, that images can misrepresent reality and thereby serve the interests of an ideology. A racist caricature, presumably, is a prime example. But then it takes the concept further:

With what is called the "crisis of representation" brought about by . . . postmodernism, however, it is often questioned whether an image can be thought to simply represent, or misrepresent, a supposedly prior or external, image-free reality. Reality is seen rather as always subject to, or as the product of, modes of representation. In this view we inescapably inhabit a world of images or representations and not a "real world" and true or false images of it.

In other words, if a tree falls in a forest and there is no artist to paint it, not only did the tree make no sound, but it did not fall, and there was no tree there to begin with.

In a further move . . . we are thought to exist in a world of HYPERREAL-ITY, in which images are self-generating and entirely detached from any supposed reality. This accords with a common view of contemporary entertainment and politics as being all a matter of "image," or appearance, rather than of substantial content.

Actually, the doctrine of hyperreality *contradicts* the common view of contemporary politics and entertainment as being a matter of image and appearance. The whole point of the common view is that there *is* a reality separate from images, and that is what allows us to decry the images that are misleading. We can, for example, criticize an old movie that shows slaves leading happy lives, or an ad that shows a corrupt politician pretending to defend the environment. If there were no such thing as substantial content, we would have no basis for preferring an accurate documentary about slavery to an apologia for it, or preferring a good exposé of a politician to a slick campaign ad.

The entry notes that images are associated with the world of publicity, advertising, and fashion, and thereby with business and profits. An image may thus be tied to "an imposed stereotype or an alternative subjective or cultural identity." Media images become mental images: people cannot help but think that women or politicians or African Americans conform to the depictions in movies and advertisements. And this elevates cultural studies and postmodernist art into forces for personal and political liberation:

> The study of "images of women" or "women's images" sees this field as one in which stereotypes of women can be reinforced, parodied, or actively contested through critical analysis, alternative histories, or creative work in writing and the media committed to the production of positive counter-images.[39]

I have not hidden my view that this entire line of thinking is a conceptual mess. If we want to understand how politicians or advertisers manipulate us, the last thing we should do is blur distinctions among things in the world, our perception of those things when they are in front of our eyes, the mental images of those things that we construct from memory, and physical images such as photographs and drawings.

As we saw at the beginning of this chapter, the visual brain is an immensely complicated system that was designed by the forces of evolution to give us an accurate reading of the consequential things in front of us. The "intelligent eye," as perceptual psychologists call it, does not just compute the shapes and motions of people before us. It also guesses their thoughts and intentions by noticing how they gaze at, approach, avoid, help, or hinder other objects and people. And these guesses are then measured against everything else we know about people—what we infer from gossip, from a person's words and deeds, and from Sherlock Holmes–style deductions. The result is the knowledge base or semantic memory that also underlies our use of language.

Physical images such as photographs and paintings are devices that reflect light in patterns similar to those coming off real objects, thereby making the visual system respond as if it were really seeing those objects. Though people have long dreamed of illusions that *completely* fool the brain—Descartes's evil demon, the philosopher's thought experiment in which a person does not realize he is a brain in a vat, the science-fiction writer's prophecy of perfect virtual reality like in *The Matrix*—in actuality the illusions foisted upon us by physical images are never more than partially effective. Our perceptual systems pick up on the imperfections of an image—the brush strokes, pixels, or frame—and our conceptual systems pick up on the fact that we are entertaining a hypothetical world that is separate from the real world. It's not that people invariably distinguish fiction from reality: they can lose themselves in fiction, or misremember something they read in a novel as something they read in the newspapers or that happened to a friend, or mistakenly believe that a stylized portrayal of a time and place is an accurate portrayal. But all of us are *capable* of distinguishing fictitious worlds from real ones, as we see when a two-year-old pretends that a banana is a telephone for the fun of it but at the same time understands that a banana is not literally a telephone.[40] Cognitive scientists believe that the ability to entertain propositions without necessarily believing them—to distinguish "John believes there is a Santa Claus" from "There is a Santa Claus"—is a fundamental ability of human cognition.[41] Many believe that a breakdown of this ability underlies the thought disorder in the syndrome called schizophrenia.[42]

Finally, there are mental images, the visualizations of objects and scenes in the mind's eye. The psychologist Stephen Kosslyn has shown that the brain is equipped with a system capable of reactivating and manipulating memories of perceptual experience, a bit like Photoshop with its tools for assembling, rotating, and coloring images.[43] Like language, imagery may be used as a slave system—a "visuospatial sketchpad"—by the central executive of the brain, making it a valuable form of mental representation. We use mental imagery, for example, when we visualize how a chair might fit in a living room or whether a sweater would look good on a relative. Imagery is also an invaluable tool to novelists, who imagine scenes before describing them in words, and to scientists, who rotate molecules or play out forces and motions in their imagination.

Though mental images allow our experiences (including our experience of media images) to affect our thoughts and attitudes long after the original objects have gone, it is a mistake to think that raw images are downloaded into our minds and then constitute our mental lives. Images are not stored in the mind like snapshots in a shoebox; if they were, how could you ever find the one you want? Rather, they are labeled and linked to a vast database of knowledge, which allows them to be evaluated and interpreted in terms of what they stand for.[44] Chess masters, for example, are famous for their ability to remember games in progress, but their mental images of the board are not raw photographs. Rather, they are saturated with abstract information about the game, such as which piece is threatening which other one and which clusters of pieces form viable defenses. We know this because when a chessboard is sprinkled with pieces at random, chess masters are no better at remembering the arrangement than amateurs are.[45] When images represent real people, not just chessmen, there are even more possibilities for organizing and annotating them with information about people's goals and motives—for example, whether the person in an image is sincere or just acting.

The reason that images cannot constitute the contents of our thoughts is that images, like words, are inherently ambiguous. An image of Lassie could stand for Lassie, collies, dogs, animals, television stars, or family values. Some other, more abstract form of information must pick out the concept that an image is taken to exemplify. Or consider the sentence *Yesterday my uncle fired his lawyer* (an example suggested by Dan Dennett). When understanding the sentence, Brad might visualize his own ordeals of the day before and glimpse the "uncle" slot in a family tree, then picture courthouse steps and an angry man. Irene might have no image for "yesterday" but might visualize her uncle Bob's face, a slamming door, and a power-suited woman. Yet despite these very different image sequences, both people have understood the sentence in the same way, as we could see by questioning them or asking them to paraphrase the sentence. "Imagery *couldn't* be the key to comprehension," Dennett points out, "because you can't draw a picture of an uncle, or of yesterday, or firing, or a lawyer. Uncles, unlike clowns and firemen, don't look different in any characteristic way that can be visually represented, and yesterdays don't look like anything at all."[46]

Since images are interpreted in the context of a deeper understanding of people and their relationships, the "crisis of representation," with its paranoia about the manipulation of our mind by media images, is overblown. People are not helplessly programmed with images; they can evaluate and interpret what they see using everything else they know, such as the credibility and motives of the source.

The postmodernist equating of images with thoughts has not only made a hash of several scholarly disciplines but has laid waste to the world of contemporary art. If images are the disease, the reasoning goes, then art is the cure. Artists can neutralize the power of media images by distorting them or reproducing them in odd contexts (like the ad parodies in *Mad* magazine or on *Saturday Night Live,* only not funny). Anyone familiar with contemporary art has seen the countless works in which stereotypes of women, minorities, or gay people are "reinforced, parodied, or actively contested." A prototypical example is a 1994 exhibit at the Whitney Museum in New York called "Black Male: Representations of Masculinity in Contemporary Art." It aimed to take apart the way that African American men are culturally constructed in demonizing and marginalizing visual stereotypes such as the sex symbol, the athlete, the Sambo, and the photograph in a Wanted poster. According to the catalogue essay, "The real struggle is over the power to control images." The art critic Adam Gopnik (whose mother and sister are cognitive scientists) called attention to the simplistic theory of cognition behind this tedious formula:

> The show is intended to be socially therapeutic: its aim is to make you face the socially constructed images of black men, so that by confronting them—or, rather, seeing artists confront them on your behalf—you can make them go away. The trouble is that the entire enterprise of "disassembling social images" rests on an ambiguity in the way we use the word "image." Mental images are not really images at all, but instead consist of complicated opinions, positions, doubts, and passionately held convictions, rooted in experience and amendable by argument, by more experience, or by coercion. Our mental images of black men, white judges, the press, and so on do not take the form of pictures of the kind that you can hang up (or "deconstruct") on a museum wall. . . . Hitler did not hate Jews because there were pictures of swarthy Semites with big noses imprinted on his cerebellum; racism does not exist in America because the picture of O. J. Simpson on the cover of *Time* is too dark. The view that visual clichés shape beliefs is both too pessimistic, in that it supposes that people are helplessly imprisoned by received stereotypes, and too optimistic, in that it supposes that if you could change the images you could change the beliefs.[47]

Recognizing that we are equipped with sophisticated faculties that keep us in touch with reality does not entail ignoring the ways in which our faculties can be turned against us. People lie, sometimes baldly, sometimes through insinuation and presupposition (as in the question "When did you stop beating your wife?"). People disseminate disinformation about ethnic groups, not just pejorative stereotypes but tales of exploitation and perfidy that serve to stoke moralistic outrage against them. People try to manipulate social realities like status (which exist in the mind of the beholder) to make themselves look good or to sell products.

But we can best protect ourselves against such manipulation by pinpointing the vulnerabilities of our faculties of categorization, language, and imagery, not by denying their complexity. The view that humans are passive receptacles of stereotypes, words, and images is condescending to ordinary people and gives unearned importance to the pretensions of cultural and academic elites. And exotic pronouncements about the limitations of our faculties, such as that there is nothing outside the text or that we inhabit a world of images rather than a real world, make it impossible even to identify lies and misrepresentations, let alone to understand how they are promulgated.

Chapter 13

Out of Our Depths

A man has got to know his limitations.
—Clint Eastwood in *Magnum Force*

MOST PEOPLE ARE familiar with the idea that some of our ordeals come from a mismatch between the source of our passions in evolutionary history and the goals we set for ourselves today. People gorge themselves in anticipation of a famine that never comes, engage in dangerous liaisons that conceive babies they don't want, and rev up their bodies in response to stressors from which they cannot run away.

What is true for the emotions may also be true for the intellect. Some of our perplexities may come from a mismatch between the purposes for which our cognitive faculties evolved and the purposes to which we put them today. This is obvious enough when it comes to raw data processing. People do not try to multiply six-digit numbers in their heads or remember the phone number of everyone they meet, because they know their minds were not designed for the job. But it is not as obvious when it comes to the way we conceptualize the world. Our minds keep us in touch with aspects of reality—such as objects, animals, and people—that our ancestors dealt with for millions of years. But as science and technology open up new and hidden worlds, our untutored intuitions may find themselves at sea.

What are these intuitions? Many cognitive scientists believe that human reasoning is not accomplished by a single, general-purpose computer in the head. The world is a heterogeneous place, and we are equipped with different kinds of intuitions and logics, each appropriate to one department of reality. These ways of knowing have been called systems, modules, stances, faculties, mental organs, multiple intelligences, and reasoning engines.[1] They emerge early in life, are present in every normal person, and appear to be computed in partly distinct sets of networks in the brain. They may be installed by different combinations of genes, or they may emerge when brain tissue self-organizes in response to different problems to be solved and different patterns in the sensory input. Most likely they develop by some combination of these forces.

What makes our reasoning faculties different from the departments in a university is that they are not just broad areas of knowledge, analyzed with whatever tools work best. Each faculty is based on a core intuition that was suitable for analyzing the world in which we evolved. Though cognitive scientists have not agreed on a *Gray's Anatomy* of the mind, here is a tentative but defensible list of cognitive faculties and the core intuitions on which they are based:

- An intuitive physics, which we use to keep track of how objects fall, bounce, and bend. Its core intuition is the concept of the object, which occupies one place, exists for a continuous span of time, and follows laws of motion and force. These are not Newton's laws but something closer to the medieval conception of impetus, an "oomph" that keeps an object in motion and gradually dissipates.[2]
- An intuitive version of biology or natural history, which we use to understand the living world. Its core intuition is that living things house a hidden essence that gives them their form and powers and drives their growth and bodily functions.[3]
- An intuitive engineering, which we use to make and understand tools and other artifacts. Its core intuition is that a tool is an object with a purpose—an object designed by a person to achieve a goal.[4]
- An intuitive psychology, which we use to understand other people. Its core intuition is that other people are not objects or machines but are animated by the invisible entity we call the mind or the soul. Minds contain beliefs and desires and are the immediate cause of behavior.
- A spatial sense, which we use to navigate the world and keep track of where things are. It is based on a dead reckoner, which updates coordinates of the body's location as it moves and turns, and a network of mental maps. Each map is organized by a different reference frame: the eyes, the head, the body, or salient objects and places in the world.[5]
- A number sense, which we use to think about quantities and amounts. It is based on an ability to register exact quantities for small numbers of objects (one, two, and three) and to make rough relative estimates for larger numbers.[6]
- A sense of probability, which we use to reason about the likelihood of uncertain events. It is based on the ability to track the relative frequencies of events, that is, the proportion of events of some kind that turn out one way or the other.[7]

- An intuitive economics, which we use to exchange goods and favors. It is based on the concept of reciprocal exchange, in which one party confers a benefit on another and is entitled to an equivalent benefit in return.
- A mental database and logic, which we use to represent ideas and to infer new ideas from old ones. It is based on assertions about what's what, what's where, or who did what to whom, when, where, and why. The assertions are linked in a mind-wide web and can be recombined with logical and causal operators such as AND, OR, NOT, ALL, SOME, NECESSARY, POSSIBLE, and CAUSE.[8]
- Language, which we use to share the ideas from our mental logic. It is based on a mental dictionary of memorized words and a mental grammar of combinatorial rules. The rules organize vowels and consonants into words, words into bigger words and phrases, and phrases into sentences, in such a way that the meaning of the combination can be computed from the meanings of the parts and the way they are arranged.[9]

The mind also has components for which it is hard to tell where cognition leaves off and emotion begins. These include a system for assessing danger, coupled with the emotion called fear, a system for assessing contamination, coupled with the emotion called disgust, and a moral sense, which is complex enough to deserve a chapter of its own.

These ways of knowing and core intuitions are suitable for the lifestyle of small groups of illiterate, stateless people who live off the land, survive by their wits, and depend on what they can carry. Our ancestors left this lifestyle for a settled existence only a few millennia ago, too recently for evolution to have done much, if anything, to our brains. Conspicuous by their absence are faculties suited to the stunning new understanding of the world wrought by science and technology. For many domains of knowledge, the mind could not have evolved dedicated machinery, the brain and genome show no hints of specialization, and people show no spontaneous intuitive understanding either in the crib or afterward. They include modern physics, cosmology, genetics, evolution, neuroscience, embryology, economics, and mathematics.

It's not just that we have to go to school or read books to learn these subjects. It's that we have no mental tools to grasp them intuitively. We depend on analogies that press an old mental faculty into service, or on jerry-built mental contraptions that wire together bits and pieces of other faculties. Understanding in these domains is likely to be uneven, shallow, and contaminated by primitive intuitions. And that can shape debates in the border disputes in which science and technology make contact with everyday life. The point of this chapter is that together with all the moral, empirical, and political factors that go into these debates, we should add the cognitive factors: the way our minds naturally frame issues. Our own cognitive makeup is a missing piece of many puzzles, including education, bioethics, food safety, economics, and human understanding itself.

~

THE MOST OBVIOUS arena in which we confront native ways of thinking is the schoolhouse. Any theory of education must be based on a theory of human nature, and in the twentieth century that theory was often the Blank Slate or the Noble Savage.

Traditional education is based in large part on the Blank Slate: children come to school empty and have knowledge deposited in them, to be reproduced later on tests. (Critics of traditional education call this the "savings and loan" model.) The Blank Slate also underlies the common philosophy that the early school-age years are an opportunity zone in which social values are shaped for life. Many schools today use the early grades to instill desirable attitudes toward the environment, gender, sexuality, and ethnic diversity.

Progressive educational practice, for its part, is based on the Noble Savage. As A. S. Neill wrote in his influential book *Summerhill*, "A child is innately wise and realistic. If left to himself without adult suggestion of any kind, he will develop as far as he is capable of developing."[10] Neill and other progressive theorists of the 1960s and 1970s argued that schools should do away with examinations, grades, curricula, and even books. Though few schools went that far, the movement left a mark on educational practice. In the method of reading instruction known as Whole Language, children are not taught which letter goes with which sound but are immersed in a book-rich environment where reading skills are expected to blossom spontaneously.[11] In the philosophy of mathematics instruction known as constructivism, children are not drilled with arithmetic tables but are enjoined to rediscover mathematical truths themselves by solving problems in groups.[12] Both methods fare badly when students' learning is assessed objectively, but advocates of the methods tend to disdain standardized testing.

An understanding of the mind as a complex system shaped by evolution runs against these philosophies. The alternative has emerged from the work of cognitive scientists such as Susan Carey, Howard Gardner, and David Geary.[13] Education is neither writing on a blank slate nor allowing the child's nobility to come into flower. Rather, education is a technology that tries to make up for what the human mind is innately bad at. Children don't have to go to school to learn to walk, talk, recognize objects, or remember the personalities of their friends, even though these tasks are much harder than reading, adding, or remembering dates in history. They do have to go to school to learn written language, arithmetic, and science, because those bodies of knowledge and skill were invented too recently for any species-wide knack for them to have evolved.

Far from being empty receptacles or universal learners, then, children are equipped with a toolbox of implements for reasoning and learning in particular ways, and those implements must be cleverly recruited to master problems for which they were not designed. That requires not just inserting new facts and skills in children's minds but debugging and disabling old ones. Students cannot learn Newtonian physics until they unlearn their intuitive impetus-based physics.[14] They cannot learn modern biology until they unlearn their intuitive biology, which thinks in terms of vital essences. And they cannot learn evolution until they unlearn their intuitive engineering, which attributes design to the intentions of a designer.[15]

Schooling also requires pupils to expose and reinforce skills that are ordinarily buried in unconscious black boxes. When children learn to read, the vowels and consonants that are seamlessly woven together in speech must be forced into children's awareness before they can associate them with squiggles on a page.[16] Effective education may also require co-opting old faculties to deal with new demands. Snatches of language can be pressed into service to do calculation, as when we recall the stanza "Five times five is twenty-five."[17] The logic of grammar can be used to grasp large numbers: the expression *four thousand three hundred and fifty-seven* has the grammatical structure of an English noun phrase like *hat, coat, and mittens*. When a student parses the number phrase she can call to mind the mental operation of aggregation, which is related to the mathematical operation of addition.[18] Spatial cognition is drafted into understanding mathematical relationships through the use of graphs, which turn data or equations into shapes.[19] Intuitive engineering supports the learning of anatomy and physiology (organs are understood as gadgets with functions), and intuitive physics supports the learning of chemistry and biology (stuff, including living stuff, is made out of tiny, bouncy, sticky objects).[20]

Geary points out a final implication. Because much of the content of education is not cognitively natural, the process of mastering it may not always be easy and pleasant, notwithstanding the mantra that learning is fun. Children may be innately motivated to make friends, acquire status, hone motor skills, and explore the physical world, but they are not necessarily motivated to adapt their cognitive faculties to unnatural tasks like formal mathematics. A family, peer group, and culture that ascribe high status to school achievement may be needed to give a child the motive to persevere toward effortful feats of learning whose rewards are apparent only over the long term.[21]

THE LAYPERSON'S INTUITIVE psychology or "theory of mind" is one of the brain's most striking abilities. We do not treat other people as wind-up dolls but think of them as being animated by minds: nonphysical entities we cannot see or touch but that are as real to us as bodies and objects. Aside from allowing us to predict people's behavior from their beliefs and desires, our theory of mind is tied to our ability to empathize and to our conception of life and death. The difference between a dead body and a living one is that a dead body no longer contains the vital force we call a mind. Our theory of mind is the source of the concept of the soul. The ghost in the machine is deeply rooted in our way of thinking about people.

A belief in the soul, in turn, meshes with our moral convictions. The core of morality is the recognition that others have interests as we do—that they "feel want, taste grief, need friends," as Shakespeare put it—and therefore that they have a right to life, liberty, and the pursuit of their interests. But who are those "others"? We need a boundary that allows us to be callous to rocks and plants but forces us to treat other humans as "persons" that possess inalienable rights. Otherwise, it seems, we would place ourselves on a slippery slope that ends in the disposal of inconvenient people or in grotesque deliberations on the value of individual lives. As Pope John Paul II pointed out, the notion that every human carries infinite value by virtue of possessing a soul would seem to give us that boundary.

Until recently the intuitive concept of the soul served us pretty well. Living people had souls, which come into existence at the moment of conception and leave their bodies when they die. Animals, plants, and inanimate objects do not have souls at all. But science is showing that what we call the soul—the locus of sentience, reason, and will—consists of the information-processing activity of the brain, an organ governed by the laws of biology. In an individual person it comes into existence gradually through the differentiation of tissues growing from a single cell. In the species it came into existence gradually as the forces of evolution modified the brains of simpler animals. And though our concept of souls used to fit pretty well with natural phenomena—a woman was either pregnant or not, a person was either dead or alive—biomedical research is now presenting us with cases where the two are out of register. These cases are not just scientific curiosities but are intertwined with pressing issues such as contraception, abortion, infanticide, animal rights, cloning, euthanasia, and research involving human embryos, especially the harvesting of stem cells.

In the face of these difficult choices it is tempting to look to biology to find or ratify boundaries such as "when life begins." But that only highlights the clash between two incommensurable ways of conceiving life and mind. The intuitive and morally useful concept of an immaterial spirit simply cannot be reconciled with the scientific concept of brain activity emerging gradually in ontogeny and phylogeny. No matter where we try to draw the line between life and nonlife, or between mind and nonmind, ambiguous cases pop up to challenge our moral intuitions.

The closest event we can find to a thunderclap marking the entry of a soul into the world is the moment of conception. At that instant a new human genome is determined, and we have an entity destined to develop into a unique individual. The Catholic Church and certain other Christian denominations designate conception as the moment of ensoulment and the beginning of life (which, of course, makes abortion a form of murder). But just as a microscope reveals that a straight edge is really ragged, research on human reproduction shows that the "moment of conception" is not a moment at all. Sometimes several sperm penetrate the outer membrane of the egg, and it takes time for the egg to eject the extra chromosomes. What and where is the soul during this interval? Even when a single sperm enters, its genes remain separate from those of the egg for a day or more, and it takes yet another day or so for the newly merged genome to control the cell. So the "moment" of conception is in fact a span of twenty-four to forty-eight hours.[22] Nor is the conceptus destined to become a baby. Between two-thirds and three-quarters of them never implant in the uterus and are spontaneously aborted, some because they are genetically defective, others for no discernible reason.

Still, one might say that at whatever point during this interlude the new genome is formed, the specification of a unique new person has come into existence. The soul, by this reasoning, may be identified with the genome. But during the next few days, as the embryo's cells begin to divide, they can split into several embryos, which develop into identical twins, triplets, and so on. Do identical twins share a soul? Did the Dionne quintuplets make do with one-fifth of a soul each? If not, where did the four extra souls come from? Indeed, every cell in the growing embryo is capable, with the right manipulations, of becoming a new embryo that can grow into a child. Does a multicell embryo consist of one soul per cell, and if so, where do the other souls go when the cells lose that ability? And not only can one embryo become two people, but two embryos can become one person. Occasionally two fertilized eggs, which ordinarily would go on to become fraternal twins, merge into a single embryo that develops into a person who is a genetic chimera: some of her cells have one genome, others have another genome. Does her body house two souls?

For that matter, if human cloning ever became possible (and there appears to be no technical obstacle), every cell in a person's body would have the special ability that is supposedly unique to a conceptus, namely developing into a human being. True, the genes in a cheek cell can become a person only with unnatural intervention, but that is just as true for an egg that is fertilized in vitro. Yet no one would deny that children conceived by IVF have souls.

The idea that ensoulment takes place at conception is not only hard to reconcile with biology but does not have the moral superiority credited to it. It implies that we should prosecute users of intrauterine contraceptive devices and the "morning-after pill" for murder, because they prevent the conceptus from implanting. It implies that we should divert medical research from curing cancer and heart disease to preventing the spontaneous miscarriages of vast numbers of microscopic conceptuses. It impels us to find surrogate mothers for the large number of embryos left over from IVF that are currently sitting in fertility clinic freezers. It would outlaw research on conception and early embryonic development that promises to reduce infertility, birth defects, and pediatric cancer, and research on stem cells that could lead to treatments for Alzheimer's disease, Parkinson's disease, diabetes, and spinal-cord injuries. And it flouts the key moral intuition that other people are worthy of moral consideration because of their feelings—their ability to love, think, plan, enjoy, and suffer—all of which depend on a functioning nervous system.

The enormous moral costs of equating a person with a conceptus, and the cognitive gymnastics required to maintain that belief in the face of modern biology, can sometimes lead to an agonizing reconsideration of deeply held beliefs. In 2001, Senator Orrin Hatch of Utah broke with his longtime allies in the anti-abortion movement and came out in favor of stem-cell research after studying the science of reproduction and meditating on his Mormon faith. "I have searched my conscience," he said. "I just cannot equate a child living in the womb, with moving toes and fingers and a beating heart, with an embryo in a freezer."[23]

The belief that bodies are invested with souls is not just a product of religious doctrine but embedded in people's psychology and likely to emerge whenever they have not digested the findings of biology. The public reaction to cloning is a case in point. Some people fear that cloning would present us with the option of becoming immortal, others that it could produce an army of obedient zombies, or a source of organs for the original person to harvest when needed. In the recent Arnold Schwarzenegger movie *The Sixth Day*, clones are called "blanks," and their DNA gives them only a physical form, not a mind; they acquire a mind when a neural recording of the original person is downloaded into them. When Dolly the sheep was cloned in 1997, the cover of *Der Spiegel* showed a parade of Claudia Schiffers, Hitlers, and Einsteins, as if being a supermodel, fascist dictator, or scientific genius could be copied along with the DNA.

Clones, in fact, are just identical twins born at different times. If Einstein had a twin, he would not have been a zombie, would not have continued Einstein's stream of consciousness if Einstein had predeceased him, would not have given up his vital organs without a struggle, and probably would have been no Einstein (since intelligence is only partly heritable). The same would be true of a person cloned from a speck of Einstein. The bizarre misconceptions of cloning can be traced to the persistent belief that the body is suffused with a soul. One conception of cloning, which sets off a fear of an army of zombies, blanks, or organ farms, imagines the process to be the duplication of a body without a soul. The other, which sets off fears of a Faustian grab at immortality or of a resurrected Hitler, conceives of cloning as duplicating the body *together with* the soul. This conception may also underlie the longing of some bereaved parents for a dead child to be cloned, as if that would bring the child back to life. In fact, the clone would not only grow up in a different world from the one the dead sibling grew up in, but would have different brain tissue and would traverse a different line of sentient experience.

The discovery that what we call "the person" emerges piecemeal from a gradually developing brain forces us to reframe problems in bioethics. It would have been convenient if biologists had discovered a point at which the brain is fully assembled and is plugged in and turned on for the first time, but that is not how brains work. The nervous system emerges in the embryo as a simple tube and differentiates into a brain and spinal cord. The brain begins to function in the fetus, but it continues to wire itself well into childhood and even adolescence. The demand by both religious and secular ethicists that we identify the "criteria for personhood" assumes that a dividing line in brain development can be found. But any claim that such a line has been sighted leads to moral absurdities.

If we set the boundary for personhood at birth, we should be prepared to allow an abortion minutes before birth, despite the lack of any significant difference between a late-term fetus and a neonate. It seems more reasonable to draw the line at viability. But viability is a continuum that depends on the state of current biomedical technology and on the risks of impairment that parents are willing to tolerate in their child. And it invites the obvious rejoinder: if it is all right to abort a twenty-four-week fetus, then why not the barely distinguishable fetus of twenty-four weeks plus one day? And if that is permissible, why not a fetus of twenty-four weeks plus two days, or three days, and so on until birth? On the other hand, if it is impermissible to abort a fetus the day before its birth, then what about two days before, and three days, and so on, all the way back to conception?

We face the same problem in reverse when considering euthanasia and living wills at the end of life. Most people do not depart this world in a puff of smoke but suffer a gradual and uneven breakdown of the various parts of the brain and body. Many kinds and degrees of existence lie between the living and the dead, and that will become even more true as medical technology improves.

We face the problem again in grappling with demands for animal rights. Activists who grant the right to life to any sentient being must conclude that a hamburger eater is a party to murder and that a rodent exterminator is a perpetrator of mass murder. They must outlaw medical research that would sacrifice a few mice but save a million children from painful deaths (since no one would agree to drafting a few human beings for such experiments, and on this view mice have the rights we ordinarily grant to people). On the other hand, an opponent of animal rights who maintains that personhood comes from being a member of *Homo sapiens* is just a species bigot, no more thoughtful than the race bigots who value the lives of whites more than blacks. After all, other mammals fight to stay alive, appear to experience pleasure, and undergo pain, fear, and stress when their well-being is compromised. The great apes also share our higher pleasures of curiosity and love of kin, and our deeper aches of boredom, loneliness, and grief. Why should those interests be respected for our species but not for others?

Some moral philosophers try to thread a boundary across this treacherous landscape by equating personhood with cognitive traits that humans happen to possess. These include an ability to reflect upon oneself as a continuous locus of consciousness, to form and savor plans for the future, to dread death, and to express a choice not to die.[24] At first glance the boundary is appealing because it puts humans on one side and animals and conceptuses on the other. But it also implies that nothing is wrong with killing unwanted newborns, the senile, and the mentally handicapped, who lack the qualifying traits. Almost no one is willing to accept a criterion with those implications.

There is no solution to these dilemmas, because they arise out of a fundamental incommensurability: between our intuitive psychology, with its all-or-none concept of a person or soul, and the brute facts of biology, which tell us that the human brain evolved gradually, develops gradually, and can die gradually. And that means that moral conundrums such as abortion, euthanasia, and animal rights will never be resolved in a decisive and intuitively satisfying way. This does not mean that no policy is defensible and that the whole matter should be left to personal taste, political power, or religious dogma. As the bioethicist Ronald Green has pointed out, it just means we have to reconceptualize the problem: from *finding* a boundary in nature to *choosing* a boundary that best trades off the conflicting goods and evils for each policy dilemma.[25] We should make decisions in each case that can be practically implemented, that maximize happiness, and that minimize current and future suffering. Many of our current policies are already compromises of this sort: research on animals is permitted but regulated; a late-term fetus is not awarded full legal status as a person but may not be aborted unless it is necessary to protect the mother's life or health. Green notes that the shift from finding boundaries to choosing boundaries is a conceptual revolution of Copernican proportions. But the old conceptualization, which amounts to trying to pinpoint when the ghost enters the machine, is scientifically untenable and has no business guiding policy in the twenty-first century.

The traditional argument against pragmatic, case-by-case decisions is that they lead to slippery slopes. If we allow abortion, we will soon allow infanticide; if we permit research on stem cells, we will bring on a Brave New World of government-engineered humans. But here, I think, the nature of human cognition can get us out of the dilemma rather than pushing us into one. A slippery slope assumes that conceptual categories must have crisp boundaries that allow in-or-out decisions, or else anything goes. But that is not how human concepts work. As we have seen, many everyday concepts have fuzzy boundaries, and the mind distinguishes between a fuzzy boundary and no boundary at all. "Adult" and "child" are fuzzy categories, which is why we could raise the drinking age to twenty-one or lower the voting age to eighteen. But that did not put us on a slippery slope in which we eventually raised the drinking age to fifty or lowered the voting age to five. Those policies really would violate our concepts of "child" and "adult," fuzzy though their boundaries may be. In the same way, we can bring our concepts of life and mind into register with biological reality without necessarily slipping down a slope.

~

WHEN A 1999 CYCLONE in India left millions of people in danger of starvation, some activists denounced relief societies for distributing a nutritious grain meal because it contained genetically modified varieties of corn and soybeans (varieties that had been eaten without apparent harm in the United States). These activists are also opposed to "golden rice," a genetically modified variety that could prevent blindness in millions of children in the developing world and alleviate vitamin A deficiency in a quarter of a billion more.[26] Other activists have vandalized research facilities at which the safety of genetically modified foods is tested and new varieties are developed. For these people, even the *possibility* that such foods could be safe is unacceptable.

A 2001 report by the European Union reviewed eighty-one research projects conducted over fifteen years and failed to find any new risks to human health or to the environment posed by genetically modified crops.[27] This is no surprise to a biologist. Genetically modified foods are no more dangerous than "natural" foods because they are not fundamentally different from natural foods. Virtually every animal and vegetable sold in a health-food store has been "genetically modified" for millennia by selective breeding and hybridization. The wild ancestor of carrots was a thin, bitter white root; the ancestor of corn had an inch-long, easily shattered cob with a few small, rock-hard kernels. Plants are Darwinian creatures with no particular desire to be eaten, so they did not go out of their way to be tasty, healthy, or easy for us to grow and harvest. On the contrary: they *did* go out of their way to *deter* us from eating them, by evolving irritants, toxins, and bitter-tasting compounds.[28] So there is nothing especially safe about natural foods. The "natural" method of selective breeding for pest resistance simply increases the concentration of the plant's own poisons; one variety of natural potato had to be withdrawn from the market because it proved to be toxic to people.[29] Similarly, natural flavors—defined by one food scientist as "a flavor that's been derived with an out-of-date technology"—are often chemically indistinguishable from their artificial counterparts, and when they are distinguishable, sometimes the natural flavor is the more dangerous one. When "natural" almond flavor, benzaldehyde, is derived from peach pits, it is accompanied by traces of cyanide; when it is synthesized as an "artificial flavor," it is not.[30]

A blanket fear of all artificial and genetically modified foods is patently irrational on health grounds, and it could make food more expensive and hence less available to the poor. Where do these specious fears come from? Partly they arise from the carcinogen-du-jour school of journalism that uncritically reports any study showing elevated cancer rates in rats fed megadoses of chemicals. But partly they come from an intuition about living things that was first identified by the anthropologist James George Frazer in 1890 and has recently been studied in the lab by Paul Rozin, Susan Gelman, Frank Keil, Scott Atran, and other cognitive scientists.[31]

People's intuitive biology begins with the concept of an invisible essence residing in living things, which gives them their form and powers. These essentialist beliefs emerge early in childhood, and in traditional cultures they dominate reasoning about plants and animals. Often the intuitions serve people well. They allow preschoolers to deduce that a raccoon that looks like a skunk will have raccoon babies, that a seed taken from an apple and planted with flowers in a pot will produce an apple tree, and that an animal's behavior depends on its innards, not on its appearance. They allow traditional peoples to deduce that different-looking creatures (such as a caterpillar and a butterfly) can belong to the same kind, and they impel them to extract juices and powders from living things and try them as medicines, poisons, and food supplements. They can prevent people from sickening themselves by eating things that have been in contact with infectious substances such as feces, sick people, and rotting meat.[32]

But intuitive essentialism can also lead people into error.[33] Children falsely believe that a child of English-speaking parents will speak English even if brought up in a French-speaking family, and that boys will have short hair and girls will wear dresses even if they are brought up with no other member of their sex from which they can learn those habits. Traditional peoples believe in sympathetic magic, otherwise known as voodoo. They think similar-looking objects have similar powers, so that a ground-up rhinoceros horn is a cure for erectile dysfunction. And they think that animal parts can transmit their powers to anything they mingle with, so that eating or wearing a part of a fierce animal will make one fierce.

Educated Westerners should not feel too smug. Rozin has shown that we have voodoolike intuitions ourselves. Most Americans won't touch a sterilized cockroach, or even a plastic one, and won't drink juice that the roach has touched for even a fraction of a second.[34] And even Ivy League students believe that you are what you eat. They judge that a tribe that hunts turtles for their meat and wild boar for their bristles will be good swimmers, and that a tribe that hunts turtles for their shells and wild boar for their meat will be tough fighters.[35] In his history of biology, Ernst Mayr showed that many biologists originally rejected the theory of natural selection because of their belief that a species was a pure type defined by an essence. They could not wrap their minds around the concept that species are populations of variable individuals and that one can blend into another over evolutionary time.[36]

In this context, the fear of genetically modified foods no longer seems so strange: it is simply the standard human intuition that every living thing has an essence. Natural foods are thought to have the pure essence of the plant or animal and to carry with them the rejuvenating powers of the pastoral environment in which they grew. Genetically modified foods, or foods containing artificial additives, are thought of as being deliberately laced with a contaminant tainted by its origins in an acrid laboratory or factory. Arguments that invoke genetics, biochemistry, evolution, and risk analysis are likely to fall on deaf ears when pitted against this deep-rooted way of thinking.

Essentialist intuitions are not the only reason that perceptions of danger can be off the mark. Risk analysts have discovered to their bemusement that people's fears are often way out of line with objective hazards. Many people avoid flying, though car travel is eleven times more dangerous. They fear getting eaten by a shark, though they are four hundred times more likely to drown in their bathtub. They clamor for expensive measures to get chloroform and trichloroethylene out of drinking water, though they are hundreds of times more likely to get cancer from a daily peanut butter sandwich (since peanuts can carry a highly carcinogenic mold).[37] Some of these risks may be misestimated because they tap into our innate fears of heights, confinement, predation, and poisoning.[38] But even when people are presented with objective information about danger, they may not appreciate it because of the way the mind assesses probabilities.

A statement like "The chance of dying of botulism poisoning in a given year is .000001" is virtually incomprehensible. For one thing, magnitudes with lots of zeroes at the beginning or end are beyond the ken of our number sense. The psychologist Paul Slovic and his colleagues found that people are unmoved by a lecture on the hazards of not wearing a seat belt which mentions that a fatal collision occurs once in every 3.5 million person-trips. But they say they will buckle up when the odds are recalculated to show that their lifetime chance of dying in a collision is one percent.[39]

The other reason for the incomprehensibility of many statistics is that the probability of a single event, such as *my* dying in a plane crash (as opposed to the frequency of some events relative to others, such as the proportion of all airline passengers who die in crashes), is a genuinely puzzling concept, even to mathematicians. What sense can we make of the odds offered by expert bookmakers for particular events, such as that the Archbishop of Canterbury will confirm the second coming within a year (1000 to 1), that a Mr. Braham of Luton, England, will invent a perpetual motion machine (250 to 1), or that Elvis Presley is alive and well (1000 to 1)?[40] Either Elvis is alive or he isn't, so what does it mean to say that the probability that he is alive is .001? Similarly, what should we think when aviation safety analysts tell us that on average a single landing in a commercial airliner reduces one's life expectancy by fifteen minutes? When the plane comes down, either my life expectancy will be reduced by a lot more than fifteen minutes or it won't be reduced at all. Some mathematicians say that the probability of a single event is more like a gut feeling of confidence, expressed on a scale of 0 to 1, than a meaningful mathematical quantity.[41]

The mind is more comfortable in reckoning probabilities in terms of the relative frequency of remembered or imagined events.[42] That can make recent and memorable events—a plane crash, a shark attack, an anthrax infection—loom larger in one's worry list than more frequent and boring events, such as the car crashes and ladder falls that get printed beneath the fold on page B14. And it can lead risk experts to speak one language and ordinary people to hear another. In hearings for a proposed nuclear waste site, an expert might present a fault tree that lays out the conceivable sequences of events by which radioactivity might escape. For example, erosion, cracks in the bedrock, accidental drilling, or improper sealing might cause the release of radioactivity into groundwater. In turn, groundwater movement, volcanic activity, or an impact of a large meteorite might cause the release of radioactive wastes into the biosphere. Each train of events can be assigned a probability, and the aggregate probability of an accident from all the causes can be estimated. When people hear these analyses, however, they are not reassured but become more fearful than ever—they hadn't realized there are so many ways for something to go wrong! They mentally tabulate the *number* of disaster scenarios, rather than mentally aggregating the *probabilities* of the disaster scenarios.[43]

None of this implies that people are dunces or that "experts" should ram unwanted technologies down their throats. Even with a complete understanding of the risks, reasonable people might choose to forgo certain technological advances. If something is viscerally revolting, a democracy should allow people to reject it whether or not it is "rational" by some criterion that ignores our psychology. Many people would reject vegetables grown in sanitized human waste and would avoid an elevator with a glass floor, not because they believe these things are dangerous but because the thought gives them the willies. If they have the same reaction to eating genetically modified foods or living next to a nuclear power plant, they should have the option of rejecting them, too, as long as they do not try to force their preferences on others or saddle them with the costs.

Also, even if technocrats provide reasonable estimates of a risk (which is itself an iffy enterprise), they cannot dictate what level of risk people ought to accept. People might object to a nuclear power plant that has a minuscule risk of a meltdown not because they overestimate the risk but because they feel that the costs of the catastrophe, no matter how remote, are too dreadful. And of course any of these tradeoffs may be unacceptable if people perceive that the benefits would go to the wealthy and powerful while they themselves absorb the risks.

Nonetheless, understanding the difference between our best science and our ancient ways of thinking can only make our individual and collective decisions better informed. It can help scientists and journalists explain a new technology in the face of the most common misunderstandings. And it can help all of us understand the technology so that we can accept or reject it on grounds that we can justify to ourselves and to others.

~

IN *THE WEALTH OF NATIONS*, Adam Smith wrote that there is "a certain propensity in human nature . . . to truck, barter, and exchange one thing for another." The exchange of goods and favors is a human universal and may have an ancient history. In archaeological sites tens of millennia old, pretty seashells and sharp flints are found hundreds of miles from their sources, which suggests that they got there by networks of trade.[44]

The anthropologist Alan Fiske has surveyed the ethnographic literature and found that virtually all human transactions fall into four patterns, each with a distinctive psychology.[45] The first is Communal Sharing: groups of people, such as the members of a family, share things without keeping track of who gets what. The second is Authority Ranking: dominant people confiscate what they want from lower-ranking ones. But the other two types of transactions are defined by exchanges.

The most common kind of exchange is what Fiske calls Equality Matching. Two people exchange goods or favors at different times, and the traded items are identical or at least highly similar or easily comparable. The trading partners assess their debts by simple addition or subtraction and are satisfied when the favors even out. The partners feel that the exchange binds them in a relationship, and often people will consummate exchanges just to maintain it. For example, in the trading rings of the Pacific Islands, gifts circulate from chief to chief, and the original giver may eventually get his gift back. (Many Americans suspect that this is what happens to Christmas fruitcakes.) When someone violates an Equality Matching relationship by taking a benefit without returning it in kind, the other party feels cheated and may retaliate aggressively. Equality Matching is the only mechanism of trade in most hunter-gatherer societies. Fiske notes that it is supported by a mental model of tit-for-tat reciprocity, and Leda Cosmides and John Tooby have shown that this way of thinking comes easily to Americans as well.[46] It appears to be the core of our intuitive economics.

Fiske contrasts Equality Matching with a very different system called Market Pricing, the system of rents, prices, wages, and interest rates that underlies modern economies. Market Pricing relies on the mathematics of multiplication, division, fractions, and large numbers, together with the social institutions of money, credit, written contracts, and complex divisions of labor. Market Pricing is absent in hunter-gatherer societies, and we know it played no role in our evolutionary history because it relies on technologies like writing, money, and formal mathematics, which appeared only recently. Even today the exchanges carried out by Market Pricing may involve causal chains that are impossible for any individual to grasp in full. I press some keys to enter characters into this manuscript today and entitle myself to receive some groceries years from now, not because I will barter a copy of *The Blank Slate* to a banana grower but because of a tangled web of third and fourth and fifth parties (publishers, booksellers, truckers, commodity brokers) that I depend on without fully understanding what they do.

When people have different ideas about which of these four modes of interacting applies to a current relationship, the result can range from blank incomprehension to acute discomfort or outright hostility. Think about a dinner guest offering to pay the host for her meal, a person barking an order to a friend, or an employee helping himself to a shrimp off the boss's plate. Misunderstandings in which one person thinks of a transaction in terms of Equality Matching and another thinks in terms of Market Pricing are even more pervasive and can be even more dangerous. They tap into very different psychologies, one of them intuitive and universal, the other rarefied and learned, and clashes between them have been common in economic history.

Economists refer to "the physical fallacy": the belief that an object has a true and constant value, as opposed to being worth only what someone is willing to pay for it at a given place and time.[47] This is simply the difference between the Equality Matching and Market Pricing mentalities. The physical fallacy may not arise when three chickens are exchanged for one knife, but when the exchanges are mediated by money, credit, and third parties, the fallacy can have ugly consequences. The belief that goods have a "just price" implies that it is avaricious to charge anything higher, and the result has been mandatory pricing schemes in medieval times, communist regimes, and many Third World countries. Such attempts to work around the law of supply and demand have usually led to waste, shortages, and black markets. Another consequence of the physical fallacy is the widespread practice of outlawing interest, which comes from the intuition that it is rapacious to demand additional money from someone who has paid back exactly what he borrowed. Of course, the only reason people borrow at one time and repay it later is that the money is worth more to them at the time they borrow it than it will be at the time they repay it. So when regimes enact sweeping usury laws, people who could put money to productive use cannot get it, and everyone's standards of living go down.[48]

Just as the value of something may change with time, which creates a niche for lenders who move valuable things around in time, so it may change with space, which creates a niche for middlemen who move valuable things around in space. A banana is worth more to me in a store down the street than it is in a warehouse a hundred miles away, so I am willing to pay more to the grocer than I would to the importer—even though by "eliminating the middleman" I could pay less per banana. For similar reasons, the importer is willing to charge the grocer less than he would charge me.

But because lenders and middlemen do not cause tangible objects to come into being, their contributions are difficult to grasp, and they are often thought of as skimmers and parasites. A recurring event in human history is the outbreak of ghettoization, confiscation, expulsion, and mob violence against middlemen, often ethnic minorities who learned to specialize in the middleman niche.[49] The Jews in Europe are the most familiar example, but the expatriate Chinese, the Lebanese, the Armenians, and the Gujeratis and Chettyars of India have suffered similar histories of persecution.

One economist in an unusual situation showed how the physical fallacy does not depend on any unique historical circumstance but easily arises from human psychology. He watched the entire syndrome emerge before his eyes when he spent time in a World War II prisoner-of-war camp. Every month the prisoners received identical packages from the Red Cross. A few prisoners circulated through the camp, trading and lending chocolates, cigarettes, and other commodities among prisoners who valued some items more than others or who had used up their own rations before the end of the month. The middlemen made a small profit from each transaction, and as a result they were deeply resented—a microcosm of the tragedy of the middleman minority. The economist wrote: "[The middleman's] function, and his hard work in bringing buyer and seller together, were ignored; profits were not regarded as a reward for labour, but as the result of sharp practises. Despite the fact that his very existence was proof to the contrary, the middleman was held to be redundant."[50]

The obvious cure for the tragic shortcomings of human intuition in a high-tech world is education. And this offers priorities for educational policy: to provide students with the cognitive tools that are most important for grasping the modern world and that are most unlike the cognitive tools they are born with. The perilous fallacies we have seen in this chapter, for example, would give high priority to economics, evolutionary biology, and probability and statistics in any high school or college curriculum. Unfortunately, most curricula have barely changed since medieval times, and are barely changeable, because no one wants to be the philistine who seems to be saying that it is unimportant to learn a foreign language, or English literature, or trigonometry, or the classics. But no matter how valuable a subject may be, there are only twenty-four hours in a day, and a decision to teach one subject is also a decision not to teach another one. The question is not whether trigonometry is important, but whether it is more important than statistics; not whether an educated person should know the classics, but whether it is more important for an educated person to know the classics than to know elementary economics. In a world whose complexities are constantly challenging our intuitions, these tradeoffs cannot responsibly be avoided.

$\sim$

"OUR NATURE IS an illimitable space through which the intelligence moves without coming to an end," wrote the poet Wallace Stevens in 1951.[51] The limitlessness of intelligence comes from the power of a combinatorial system. Just as a few notes can combine into any melody and a few characters can combine into any printed text, a few ideas—PERSON, PLACE, THING, CAUSE, CHANGE, MOVE, AND, OR, NOT—can combine into an illimitable space of thoughts.[52] The ability to conceive an unlimited number of new combinations of ideas is the powerhouse of human intelligence and a key to our success as a species. Tens of thousands of years ago our ancestors conceived new sequences of actions that could drive game, extract a poison, treat an illness, or secure an alliance. The modern mind can conceive of a substance as a combination of atoms, the plan for a living thing as the combination of DNA nucleotides, and a relationship among quantities as a combination of mathematical symbols. Language, itself a combinatorial system, allows us to share these intellectual fruits.

The combinatorial powers of the human mind can help explain a paradox about the place of our species on the planet. Two hundred years ago the economist Thomas Malthus (1766–1834) called attention to two enduring features of human nature. One is that "food is necessary for the existence of man." The other is that "the passion between the sexes is necessary and will remain nearly in its present state." He famously deduced:

> The power of population is indefinitely greater than the power in the earth to produce subsistence for man. Population, when unchecked, increases in a geometrical ratio. Subsistence increases only in an arithmetic ratio. A slight acquaintance with numbers will show the immensity of the first power in comparison with the second.

Malthus depressingly concluded that an increasing proportion of humanity would starve, and that efforts to aid them would only lead to more misery because the poor would breed children doomed to hunger in their turn. Many recent prophets of gloom reiterated his argument. In 1967 William and Paul

Paddock wrote a book called *Famine 1975!* and in 1970 the biologist Paul Ehrlich, author of *The Population Bomb,* predicted that sixty-five million Americans and four billion other people would starve to death in the 1980s. In 1972 a group of big thinkers known as the Club of Rome predicted that either natural resources would suffer from catastrophic declines in the ensuing decades or that the world would choke in pollutants.

The Malthusian predictions of the 1970s have been disconfirmed. Ehrlich was wrong both about the four billion victims of starvation and about declining resources. In 1980 he bet the economist Julian Simon that five strategic metals would become increasingly scarce by the end of the decade and would thus rise in price. He lost five out of five bets. The famines and shortages never happened, despite increases both in the number of people on Earth (now six billion and counting) and in the amount of energy and resources consumed by each one.[53] Horrific famines still occur, of course, but not because of a worldwide discrepancy between the number of mouths and the amount of food. The economist Amartya Sen has shown that they can almost always be traced to short-lived conditions or to political and military upheavals that prevent food from reaching the people who need it.[54]

The state of our planet is a vital concern, and we need the clearest possible understanding of where the problems lie so as not to misdirect our efforts. The repeated failure of simple Malthusian thinking shows that it cannot be the best way to analyze environmental challenges. Still, Malthus's logic seems impeccable. Where did it go wrong?

The immediate problem with Malthusian prophecies is that they underestimate the effects of technological change in increasing the resources that support a comfortable life.[55] In the twentieth century food supplies increased exponentially, not linearly. Farmers grew more crops on a given plot of land. Processors transformed more of the crops into edible food. Trucks, ships, and planes got the food to more people before it spoiled or was eaten by pests. Reserves of oil and minerals increased, rather than decreased, because engineers could find more of them and figure out new ways to get at them.

Many people are reluctant to grant technology this seemingly miraculous role. A technology booster sounds too much like the earnest voiceover in a campy futuristic exhibit at the world's fair. Technology may have bought us a temporary reprieve, one might think, but it is not a source of inexhaustible magic. It cannot refute the laws of mathematics, which pit exponential population growth against finite, or at best arithmetically increasing, resources. Optimism would seem to require a faith that the circle can be squared.

But recently the economist Paul Romer has invoked the combinatorial nature of cognitive information processing to show how the circle might be squared after all.[56] He begins by pointing out that human material existence is limited by *ideas,* not by stuff. People don't need coal or copper wire or paper

per se; they need ways to heat their homes, communicate with other people, and store information. Those needs don't have to be satisfied by increasing the availability of physical resources. They can be satisfied by using new ideas—recipes, designs, or techniques—to rearrange existing resources to yield more of what we want. For example, petroleum used to be just a contaminant of water wells; then it became a source of fuel, replacing the declining supply of whale oil. Sand was once used to make glass; now it is used to make microchips and optical fiber.

Romer's second point is that ideas are what economists call "nonrival goods." Rival goods, such as food, fuel, and tools, are made of matter and energy. If one person uses them, others cannot, as we recognize in the saying "You can't eat your cake and have it." But ideas are made of *information*, which can be duplicated at negligible cost. A recipe for bread, a blueprint for a building, a technique for growing rice, a formula for a drug, a useful scientific law, or a computer program can be given away without anything being subtracted from the giver. The seemingly magical proliferation of nonrival goods has recently confronted us with new problems concerning intellectual property, as we try to adapt a legal system that was based on owning stuff to the problem of owning information—such as musical recordings—that can easily be shared over the Internet.

The power of nonrival goods may have been a presence throughout human evolutionary history. The anthropologists John Tooby and Irven DeVore have argued that millions of years ago our ancestors occupied the "cognitive niche" in the world's ecosystem. By evolving mental computations that can model the causal texture of the environment, hominids could play out scenarios in their mind's eye and figure out new ways of exploiting the rocks, plants, and animals around them. Human practical intelligence may have coevolved with language (which allows know-how to be shared at low cost) and with social cognition (which allows people to cooperate without being cheated), yielding a species that literally lives by the power of ideas.

Romer points out that the combinatorial process of creating new ideas can circumvent the logic of Malthus:

> Every generation has perceived the limits to growth that finite resources and undesirable side effects would pose if no new recipes or ideas were discovered. And every generation has underestimated the potential for finding new recipes and ideas. We consistently fail to grasp how many ideas remain to be discovered. The difficulty is the same one we have with compounding. Possibilities do not add up. They multiply.[57]

For example, a hundred chemical elements, combined serially four at a time and in ten different proportions, can yield 330 billion compounds. If scientists evaluated them at a rate of a thousand a day, it would take them a million years to work through the possibilities. The number of ways of assembling instructions into computer programs or parts into machines is equally mind-boggling. At least in principle, the exponential power of human cognition works on the same scale as the growth of the human population, and we can resolve the paradox of the Malthusian disaster that never happened. None of this licenses complacency about our use of natural resources, of course. The fact that the space of possible ideas is staggeringly large does not mean that the solution to a given problem lies in that space or that we will find it by the time we need it. It only means that our understanding of humans' relation to the material world has to acknowledge not just our bodies and our resources but also our minds.

~

THE TRUISM THAT all good things come with costs as well as benefits applies in full to the combinatorial powers of the human mind. If the mind is a biological organ rather than a window onto reality, there should be truths that are literally inconceivable, and limits to how well we can ever grasp the discoveries of science.

The possibility that we might come to the end of our cognitive rope has been brought home by modern physics. We have every reason to believe that the best theories in physics are true, but they present us with a picture of reality that makes no sense to the intuitions about space, time, and matter that evolved in the brains of middle-sized primates. The strange ideas of physics—for instance, that time came into existence with the Big Bang, that the universe is curved in the fourth dimension and possibly finite, and that a particle may act like a wave—just make our heads hurt the more we ponder them. It's impossible to stop thinking thoughts that are literally incoherent, such as "What was it like before the Big Bang?" or "What lies beyond the edge of the universe?" or "How does the damn particle manage to pass through two slits at the same time?" Even the physicists who discovered the nature of reality claim not to understand their theories. Murray Gell-Mann described quantum mechanics as "that mysterious, confusing discipline which none of us really understands but which we know how to use."[58] Richard Feynman wrote, "I think I can safely say that no one understands quantum mechanics. . . . Do not keep asking yourself, if you can possibly avoid it, 'But how can it be like that?' . . . Nobody knows how it can be like that."[59] In another interview, he added, "If you think you understand quantum theory, you don't understand quantum theory!"[60]

Our intuitions about life and mind, like our intuitions about matter and space, may have run up against a strange world forged by our best science. We have seen how the concept of life as a magical spirit united with our bodies doesn't get along with our understanding of the mind as the activity of a gradually developing brain. Other intuitions about the mind find themselves just as flat-footed in pursuit of the advancing frontier of cognitive neuroscience. We have every reason to believe that consciousness and decision making arise from the electrochemical activity of neural networks in the brain. But how moving molecules should throw off subjective feelings (as opposed to mere intelligent computations) and how they bring about choices that we freely make (as opposed to behavior that is caused) remain deep enigmas to our Pleistocene psyches.

These puzzles have an infuriatingly holistic quality to them. Consciousness and free will seem to suffuse the neurobiological phenomena at every level, and cannot be pinpointed to any combination or interaction among parts. The best analyses from our combinatorial intellects provide no hooks on which we can hang these strange entities, and thinkers seem condemned either to denying their existence or to wallowing in mysticism. For better or worse, our world might always contain a wisp of mystery, and our descendants might endlessly ponder the age-old conundrums of religion and philosophy, which ultimately hinge on concepts of matter and mind.[61] Ambrose Bierce's *The Devil's Dictionary* contains the following entry:

> **Mind**, *n.* A mysterious form of matter secreted by the brain. Its chief activity consists in the endeavor to ascertain its own nature, the futility of the attempt being due to the fact that it has nothing but itself to know itself with.

Chapter 14

The Many Roots of Our Suffering

THE FIRST EDITION of Richard Dawkins's *The Selfish Gene* contained a foreword by the biologist who originated some of its key ideas, Robert Trivers. He closed with a flourish:

> Darwinian social theory gives us a glimpse of an underlying symmetry and logic in social relationships which, when more fully comprehended by ourselves, should revitalize our political understanding and provide the intellectual support for a science and medicine of psychology. In the process it should also give us a deeper understanding of the many roots of our suffering.[1]

These were arresting claims for a book on biology, but Trivers knew he was onto something. Social psychology, the science of how people behave toward one another, is often a mishmash of interesting phenomena that are "explained" by giving them fancy names. Missing is the rich deductive structure of other sciences, in which a few deep principles can generate a wealth of subtle predictions—the kind of theory that scientists praise as "beautiful" or "elegant." Trivers derived the first theory in social psychology that deserves to be called elegant. He showed that a deceptively simple principle—follow the genes—can explain the logic of each of the major kinds of human relationships: how we feel toward our parents, our children, our siblings, our lovers, our friends, and ourselves.[2] But Trivers knew that the theory did something else as well. It offered a scientific explanation for the tragedy of the human condition.

"Nature is a hanging judge," goes an old saying. Many tragedies come from our physical and cognitive makeup. Our bodies are extraordinarily improbable arrangements of matter, with many ways for things to go wrong and only a few ways for things to go right. We are certain to die, and smart enough to know it. Our minds are adapted to a world that no longer exists, prone to misunderstandings correctable only by arduous education, and condemned to perplexity about the deepest questions we can entertain.

But some of the most painful shocks come from the social world—from the manipulations and betrayals of other people. According to the fable, a scorpion asked a frog to carry him across a river, reassuring the frog that he wouldn't sting him because if he did, he would drown too. Halfway across, the scorpion did sting him, and when the sinking frog asked why, the scorpion replied, "It's in my nature." Technically speaking, a scorpion with this nature could not have evolved, but Trivers has explained why it sometimes *seems* as if human nature is like the fabled scorpion nature, condemned to apparently pointless conflict.

It's no mystery why organisms sometimes harm one another. Evolution has no conscience, and if one creature hurts another to benefit itself, such as by eating, parasitizing, intimidating, or cuckolding it, its descendants will come to predominate, complete with those nasty habits. All this is familiar from the vernacular sense of "Darwinian" as a synonym for "ruthless" and from Tennyson's depiction of nature as red in tooth and claw. If that were all there was to the evolution of the human condition, we would have to agree with the rock song: Life sucks, then you die.

But of course life doesn't always suck. Many creatures cooperate, nurture, and make peace, and humans in particular find comfort and joy in their families, friends, and communities. This, too, should be familiar to readers of *The Selfish Gene* and the other books on the evolution of altruism that have appeared in the years since.[3] There are several reasons why organisms may evolve a willingness to do good deeds. They may help other creatures while pursuing their own interests, say, when they form a herd that confuses predators or live off each other's by-products. This is called mutualism, symbiosis, or cooperation. Among humans, friends who have common tastes, hobbies, or enemies are a kind of symbiont pair. The two parents of a brood of children are an even better example. Their genes are tied up in the same package, their children, so what is good for one is good for the other, and each has an interest in keeping the other alive and healthy. These shared interests set the stage for companionate love and marital love to evolve.

And in some cases organisms may benefit other organisms at a *cost* to themselves, which biologists call altruism. Altruism in this technical sense can evolve in two main ways. First, since relatives share genes, any gene that inclines an organism toward helping a relative will increase the chance of survival of a copy of itself that sits inside that relative, even if the helper sacrifices its own fitness in the generous act. Such genes will, on average, come to predominate, as long as the cost to the helper is less than the benefit to the recipient discounted by their degree of relatedness. Family love—the cherishing of children, siblings, parents, grandparents, uncles and aunts, nieces and nephews, and cousins—can evolve. This is called nepotistic altruism.

Altruism can also evolve when organisms trade favors. One helps another by grooming, feeding, protecting, or backing him, and is helped in turn when the needs reverse. This is called reciprocal altruism, and it can evolve when the parties recognize each other, interact repeatedly, can confer a large benefit on others at small cost to themselves, keep a memory for favors offered or denied, and are impelled to reciprocate accordingly. Reciprocal altruism can evolve because cooperators do better than hermits or misanthropes. They enjoy the gains of trading their surpluses, pulling ticks out of one another's hair, saving each other from drowning or starvation, and baby-sitting each other's children. Reciprocators can also do better over the long run than the cheaters who take favors without returning them, because the reciprocators will come to recognize the cheaters and shun or punish them.

The demands of reciprocal altruism can explain why the social and moralistic emotions evolved. Sympathy and trust prompt people to extend the first favor. Gratitude and loyalty prompt them to repay favors. Guilt and shame deter them from hurting or failing to repay others. Anger and contempt prompt them to avoid or punish cheaters. And among humans, any tendency of an individual to reciprocate or cheat does not have to be witnessed firsthand but can be recounted by language. This leads to an interest in the reputation of others, transmitted by gossip and public approval or condemnation, and a concern with one's own reputation. Partnerships, friendships, alliances, and communities can emerge, cemented by these emotions and concerns.

Many people start to get nervous at this point, but the discomfort is not from the tragedies that Trivers explained. It comes instead from two misconceptions, each of which we have encountered before. First, all this talk about genes that influence behavior does not mean that we are cuckoo clocks or player pianos, mindlessly executing the dictates of DNA. The genes in question are those that endow us with the neural systems for conscience, deliberation, and will, and when we talk about the selection of such genes, we are talking about the various ways those faculties could have evolved. The error comes from the Blank Slate and the Ghost in the Machine: if one starts off thinking that our higher mental faculties are stamped in by society or inhere in a soul, then when biologists mention genetic influence the first alternatives that come to mind are puppet strings or trolley tracks. But if higher faculties, including learning, reason, and choice, are products of a nonrandom organization of the brain, there have to be genes that help do the organizing, and that raises the question of how those genes would have been selected in the course of human evolution.

The second misconception is to imagine that talk about costs and benefits implies that people are Machiavellian cynics, coldly calculating the genetic advantages of befriending and marrying. To fret over this picture, or denounce it because it is ugly, is to confuse proximate and ultimate causation. People don't care about their genes; they care about happiness, love, power, respect, and other passions. The cost-benefit calculations are a metaphorical way of describing the selection of alternative genes over millennia, not a literal description of what takes place in a human brain in real time. Nothing prevents the amoral process of natural selection from evolving a brain with genuine big-hearted emotions. It is said that those who appreciate legislation and sausages should not see them being made. The same is true for human emotions.

So if love and conscience can evolve, where's the tragedy? Trivers noticed that the confluence of genetic interests that gave rise to the social emotions is only *partial*. Because we are not clones, or even social insects (who can share up to three-quarters of their genes), what ultimately is best for one person is not identical to what ultimately is best for another. Thus every human relationship, even the most devoted and intimate, carries the seeds of conflict. In the movie *AntZ*, an ant with the voice of Woody Allen complains to his psychoanalyst:

> It's this whole gung-ho superorganism thing that I *just can't get.* I try, but I just don't get it. What is it, I'm supposed to do everything for the colony and . . . what about *my* needs?

The humor comes from the clash between ant psychology, which originates in a genetic system that makes workers more closely related to one another than they would be to their offspring, and human psychology, in which our genetic distinctness leads us to ask, "What about *my* needs?" Trivers, following on the work of William Hamilton and George Williams, did some algebra that predicts the extent to which people should ask themselves that question.[4]

The rest of this chapter is about that deceptively simple algebra and how its implications overturn many conceptions of human nature. It discredits the Blank Slate, which predicts that people's regard for their fellows is determined by their "role," as if it were a part assigned arbitrarily to an actor. But it also discredits some naïve views of evolution that are common among people who *don't* believe in the Blank Slate. Most people have intuitions about the natural state of affairs. They may believe that if we acted as nature "wants" us to, families would function as harmonious units, or individuals would act for the good of the species, or people would show the true selves beneath their social masks, or, as Newt Gingrich said in 1995, the male of our species would hunt giraffes and wallow in ditches like little piglets.[5] Understanding the patterns of genetic overlap that bind and divide us can replace simplistic views of all kinds with a more subtle understanding of the human condition. Indeed, it can illuminate the human condition in ways that complement the insights of artists and philosophers through the millennia.

THE MOST OBVIOUS human tragedy comes from the difference between our feelings toward kin and our feelings toward non-kin, one of the deepest divides in the living world. When it comes to love and solidarity among people, the relative viscosity of blood and water is evident in everything from the clans and dynasties of traditional societies to the clogging of airports during holidays with people traveling across the world to be with their families.[6] It has also been borne out by quantitative studies. In traditional foraging societies, genetic relatives are more likely to live together, work in each other's gardens, protect each other, and adopt each other's needy or orphaned children, and are less likely to attack, feud with, and kill each other.[7] Even in modern societies, which tend to sunder ties of kinship, the more closely two people are genetically related, the more inclined they are to come to one another's aid, especially in life-or-death situations.[8]

But love and solidarity are relative. To say that people are more caring toward their relatives is to say that they are more callous toward their nonrelatives. The epigraph to Robert Wright's book on evolutionary psychology is an excerpt from Graham Greene's *The Power and the Glory* in which the protagonist broods about his daughter: "He said, 'Oh god, help her. Damn me, I deserve it, but let her live forever.' This was the love he should have felt for every soul in the world: all the fear and the wish to save concentrated unjustly on the one child. He began to weep. . . . He thought: This is what I should feel all the time for everyone."

Family love indeed subverts the ideal of what we should feel for every soul in the world. Moral philosophers play with a hypothetical dilemma in which people can run through the left door of a burning building to save some number of children or through the right door to save their own child.[9] If you are a parent, ponder this question: Is there *any* number of children that would lead you to pick the left door? Indeed, all of us reveal our preference with our pocketbooks when we spend money on trifles for our own children (a bicycle, orthodontics, an education at a private school or university) instead of saving the lives of unrelated children in the developing world by donating the money to charity. Similarly, the practice of parents bequeathing their wealth to their children is one of the steepest impediments to an economically egalitarian society. Yet few people would allow the government to confiscate 100 percent of their estate, because most people see their children as an extension of themselves and thus as the proper beneficiaries of their lifelong striving.

Nepotism is a universal human bent and a universal scourge of large organizations. It is notorious for sapping countries led by hereditary dynasties and for bogging down governments and businesses in the Third World. A recurring historic solution was to give positions of local power to people who had no family ties, such as eunuchs, celibates, slaves, or people a long way from home.[10] A more recent solution is to outlaw or regulate nepotism, though the regulations always come with tradeoffs and exceptions. Small businesses—or, as they are often called, "family businesses" or "Mom-and-Pop businesses"— are highly nepotistic, and thereby can conflict with principles of equal opportunity and earn the resentment of the surrounding community.

B. F. Skinner, ever the Maoist, wrote in the 1970s that people should be rewarded for eating in large communal dining halls rather than at home with their families, because large pots have a lower ratio of surface area to volume than small pots and hence are more energy efficient. The logic is impeccable, but this mindset collided with human nature many times in the twentieth century—horrifically in the forced collectivizations in the Soviet Union and China, and benignly in the Israeli kibbutzim, which quickly abandoned their policy of rearing children separately from their parents. A character in a novel by the Israeli writer Batya Gur captures the kind of sentiment that led to this change: "I want to tuck in my children at night myself . . . and when they have a nightmare I want them to come to my bed, not to some intercom, and not to make them go out at night in the dark looking for our room, stumbling over stones, thinking that every shadow is a monster, and in the end standing in front of a closed door or being dragged back to the children's house."[11]

It is not just recent dreams of collectivism that are subverted by kin solidarity. The journalist Ferdinand Mount has documented that the family has been a subversive institution throughout history. Family ties cut across the bonds connecting comrades and brethren and thus are a nuisance to governments, cults, gangs, revolutionary movements, and established religions. But even a thinker as sympathetic to human nature as Noam Chomsky does not acknowledge that people feel differently about their children from how they feel about acquaintances and strangers. Here is an excerpt of an interview with the lead guitarist of the rap metal group Rage Against the Machine:

> RAGE: Another unquestionable idea is that people are naturally competitive, and that therefore, capitalism is the only proper way to organize society. Do you agree?
>
> CHOMSKY: Look around you. In a family for example, if the parents are hungry do they steal food from the children? They would if they were competitive. In most social groupings that are even semi-sane people support each other and are sympathetic and helpful and care about other people and so on. Those are normal human emotions. It takes plenty of training to drive those feelings out of people's heads, and they show up all over the place.[12]

Unless people treat other members of society the way they treat their own children, the answer is a non sequitur: people could care deeply about their children but feel differently about the millions of other people who make up society. The very framing of the question and answer assumes that humans are competitive or sympathetic across the board, rather than having different emotions toward people with whom they have different genetic relationships.

Chomsky implies that people are born with fraternal feelings toward their social groups and that the feelings are driven out of their heads by training. But it seems to be the other way around. Throughout history, when leaders have tried to unite a social group they have trained their members to think of it as a family and to redirect their familial emotions inside it.[13] The names used by groups that strive for solidarity—brethren, brotherhoods, fraternal organizations, sisterhood, sororities, crime families, the family of man—concede in their metaphors that kinship is the paradigm to which they aspire. (No society tries to strengthen the family by likening it to a trade union, political party, or church group.) The tactic is provably effective. Several experiments have shown that people are more convinced by a political speech if the speaker appeals to their hearts and minds with kinship metaphors.[14]

Verbal metaphors are one way to nudge people to treat acquaintances like family, but usually stronger tactics are needed. In his ethnographic survey, Alan Fiske showed that the ethos of Communal Sharing (one of his four universal social relations) arises spontaneously among the members of a family but is extended to other groups only with the help of elaborate customs and ideologies.[15] Unrelated people who want to share like a family create mythologies about a common flesh and blood, a shared ancestry, and a mystical bond to a territory (tellingly called a natal land, fatherland, motherland, or mother country). They reinforce the myths with sacramental meals, blood sacrifices, and repetitive rituals, which submerge the self into the group and create an impression of a single organism rather than a federation of individuals. Their religions speak of possession by spirits and other kinds of mind melds, which, according to Fiske, "suggest that people may often want to have more intense or pure Communal Sharing relationships than they are able to realize with ordinary human beings."[16] The dark side of this cohesion is groupthink, a cult mentality, and myths of racial purity—the sense that outsiders are contaminants who pollute the sanctity of the group.

None of this means that nonrelatives are ruthlessly competitive toward one another, only that they are not as spontaneously cooperative as kin. And ironically, for all this talk of solidarity and sympathy and common blood, we shall soon see that families are not such harmonious units either.

∼

TOLSTOY'S FAMOUS REMARK that happy families are all alike but every unhappy family is unhappy in its own way is not true at the level of ultimate (evolutionary) causation. Trivers showed how the seeds of unhappiness in every family have the same underlying source.[17] Though relatives have common interests because of their common genes, the degree of overlap is not identical within all the permutations and combinations of family members. Parents are related to all of their offspring by an equal factor, 50 percent, but each child is related to himself or herself by a factor of 100 percent. And that has a subtle but profound implication for the currency of family life, parents' investment in their children.

Parental investment is a limited resource. A day has only twenty-four hours, short-term memory can hold only four chunks of information, and, as many a frazzled mother has pointed out, "I only have two hands!" At one end of the lifespan, children learn that a mother cannot pump out an unlimited stream of milk; at the other, they learn that parents do not leave behind infinite inheritances.

To the extent that emotions among people reflect their typical genetic relatedness, Trivers argued, the members of a family should disagree on how parental investment should be divvied up. Parents should want to split their investment equitably among the children—if not in absolutely equal parts, then according to each child's ability to prosper from the investment. But each child should to want the parent to dole out twice as much of the investment to himself or herself as to a sibling, because children share half their genes with each full sibling but share all their genes with themselves. Given a family with two children and one pie, each child should want to split it in a ratio of two-thirds to one-third, while parents should want it to be split fifty-fifty. The result is that no distribution will make everyone happy. Of course, it's not that parents and children literally fight over pie or milk or inheritances (though they may), and they certainly don't fight over genes. In our evolutionary history, parental investment affected a child's survival, which affected the probability that the genes for various familial emotions in parents and in children would have been passed on to us today. The prediction is that family members' expectations of one another are not perfectly in sync.

Parent-offspring conflict and its obverse, sibling-sibling conflict, can be seen throughout the animal kingdom.[18] Littermates or nestmates fight among themselves, sometimes lethally, and fight with their mothers over access to milk, food, and care. (As Woody Allen's character in *AntZ* pointed out, "When you're the middle child in a family of five million, you don't get much attention.") The conflict also plays out in the physiology of prenatal human development. Fetuses tap their mothers' bloodstreams to mine the most nutrients possible from her body, while the mother's body resists to keep it in good shape for future children.[19] And it continues to play itself out after birth. Until recently, in most cultures, mothers who had poor prospects for sustaining a newborn to maturity cut their losses and abandoned it to die.[20] The fat cheeks and precocious responsiveness in a baby's face may be an advertisement of health designed to tilt the decision in its favor.[21]

But the most interesting conflicts are the psychological ones, played out in family dramas. Trivers touted the liberatory nature of sociobiology by invoking an "underlying symmetry in our social relationships" and "submerged actors in the social world."[22] He was referring to women, as we will see in the chapter on gender, and to children. The theory of parent-offspring conflict says that families do not contain all-powerful, all-knowing parents and their passive, grateful children. Natural selection should have equipped children with psychological tactics allowing them to hold their own in a struggle with their parents, with neither party having a permanent upper hand. Parents have a short-lived advantage in sheer brawn, but children can fight back by being cute, whining, throwing tantrums, pulling guilt trips, tormenting their siblings, getting between their parents, and holding themselves hostage with the threat of self-destructive behavior.[23] As they say, insanity is hereditary: you get it from your children.

Most profoundly, children do not allow their personalities to be shaped by their parents' nagging, blandishments, or attempts to serve as role models.[24] As we shall see in the chapter on children, the effect of being raised by a given pair of parents within a culture is surprisingly small: children who grow up in the same home end up no more alike in personality than children who were separated at birth; adopted siblings grow up to be no more similar than strangers. The findings flatly contradict the predictions of every theory in the history of psychology but one. Trivers alone had predicted:

> The offspring cannot rely on its parents for disinterested guidance. One expects the offspring to be preprogrammed to resist some parental manipulation while being open to other forms. When the parent imposes an arbitrary system of reinforcement (punishment and reward) in order to manipulate the offspring to act against its own best interests, selection will favor offspring that resist such schedules of reinforcement.[25]

That children don't turn out the way their parents want is, for many people, one of the bittersweet lessons of parenthood. "Your children are not your children," wrote the poet Kahlil Gibran. "You may give them your love but not your thoughts, for they have their own thoughts."[26]

The most obvious prediction of the theory of parent-offspring conflict is that parents and siblings should all have different perceptions of how the parents treated the siblings. Indeed, studies of the grown members of families show that most parents claim they treated their children equitably, while a majority of siblings claim they did not get their fair share.[27] Researchers call it the Smothers Brothers effect, after the comedy pair whose duller member had the signature line "Mom always liked you best."

But the logic of parent-offspring conflict does not apply only to contemporaneous siblings. Offspring of any age tacitly compete against the unborn descendants that parents *might* have if they were ceded the time and energy. Since men can always father children (especially in the polygynous systems that until recently characterized most societies), and since both sexes can lavish investment on grandchildren, potential conflicts of interest between parents and offspring hang over them for life. When parents arrange a marriage, they may cut a deal that sacrifices a child's interest for future considerations benefiting a sibling or the father. Children and adults may hold different opinions on whether a child should stick around to help the family or strike out on his or her own reproductive career. Married children have to decide how to allocate time and energy between the nuclear family they have created and the extended family they were born into. Parents have to decide whether to distribute their resources in equal parts or to the child who can make the best use of them.

The logic of parent-offspring and sibling-sibling conflict casts a new light on the doctrine of "family values" that is prominent in the contemporary religious and cultural right. According to this doctrine, the family is a haven of nurturance and benevolence, allowing parents to convey values to children that best serve their interests. Modern cultural forces, by allowing women to spend less time with young children and by expanding the world of older children beyond the family circle, have supposedly thrown a grenade into this nest, harming children and society alike. Part of this theory is surely accurate; parents and other relatives have a stronger interest in the well-being of a child than any third party does. But parent-offspring conflict implies that there is more to the picture.

If one could ask young children what they want, it would undoubtedly be the undivided attention of their mothers twenty-four hours a day. But that does not mean that nonstop mothering is the biological norm. The need to find a balance between investing in an offspring and staying healthy (ultimately to invest in other offspring) is inherent to all living things. Human mothers are no exception, and often have to resist the demands of their pint-sized tyrants so as not to compromise their own survival and the survival of their other born and unborn children. The anthropologist Sarah Blaffer Hrdy has shown that the tradeoff between working and mothering was not invented by power-suited Yuppies of the 1980s. Women in foraging societies use a variety of arrangements to raise their children without starving in the process, including seeking status within the group (which improves the children's well-being) and sharing childcare duties with other women in the band. Fathers, of course, are usually the main providers other than the mother herself, but they have bad habits like dying, deserting, and not making a living, and mothers have never depended on them alone.[28]

The weakening of parents' hold over their older children is also not just a recent casualty of destructive forces. It is part of a long-running expansion of freedom in the West that has granted children their always-present desire for more autonomy than parents are willing to cede. In traditional societies, children were shackled to the family's land, betrothed in arranged marriages, and under the thumb of the family patriarch.[29] That began to change in medieval Europe, and some historians argue it was the first steppingstone in the extension of rights that we associate with the Enlightenment and that culminated in the abolition of feudalism and slavery.[30] Today it is no doubt true that some children are led astray by a bad crowd or popular culture. But some children are rescued from abusive or manipulative families by peers, neighbors, and teachers. Many children have profited from laws, such as compulsory schooling and the ban on forced marriages, that may override the preferences of their parents. Some may profit from information, such as about contraception or careers, that their parents try to withhold. And some must escape a stifling cultural ghetto to discover the cosmopolitan delights of the modern world. Isaac Bashevis Singer's novel *Shosha* begins with a reminiscence of the protagonist's childhood in the Jewish section of Warsaw at the beginning of the twentieth century:

> I was brought up on three dead languages—Hebrew, Aramaic, and Yiddish . . . —and in a culture that developed in Babylon: the Talmud. The cheder [schoolroom] where I studied was a room in which the teacher ate and slept, and his wife cooked. There I studied not arithmetic, geography, physics, chemistry, or history, but the laws governing an egg laid on a holiday and sacrifices in a temple destroyed two thousand years ago. Although my ancestors had settled in Poland some six or seven hundred years before I was born, I knew only a few words of the Polish language. . . . I was an anachronism in every way, but I didn't know it.

Singer's reminiscence is more nostalgic than bitter, and of course most families offer far more nurturance than repression or strife. At the proximate level, Tolstoy was surely right that there are happy and unhappy families and that unhappy families are unhappy in different ways, depending on the chemistry of the people thrown together by genetics and fate. The conflict inherent to families does not make family ties any less central to human existence. It only implies that the balancing of competing interests that governs all human interactions does not end at the door of the family home.

～

AMONG THE COMBINATIONS of people that Trivers considered is the pair consisting of a man and a woman. The logic of their relationship is rooted in the most fundamental difference between the sexes: not their chromosomes, not their plumbing, but their parental investment.[31] In mammals, the minimal parental investments of a male and a female differ dramatically. A male can get away with a few minutes of copulation and a tablespoon of semen, but a female carries an offspring for months inside her body and nourishes it before and after it is born. As they say of the respective contributions of the chicken and the pig to eggs and bacon, the first is involved, but the second is committed. Since it takes one member of each sex to make a baby, access to females is the limiting resource for males in reproduction. For a male to maximize the number of his descendants, he should mate with as many females as possible; for a female to maximize the number of her descendants, she should mate with the best-quality male available. This explains the two widespread sex differences in many species in the animal kingdom: males compete, females choose; males seek quantity, females quality.

Humans are mammals, and our sexual behavior is consistent with our Linnaean class. Donald Symons sums up the ethnographic record on sex differences in sexuality: "Among all peoples it is primarily men who court, woo, proposition, seduce, employ love charms and love magic, give gifts in exchange for sex, and use the services of prostitutes."[32] Among Western peoples, studies have shown that men seek a greater number of sexual partners than women, are less picky in their choice of a short-term partner, and are far more likely to be customers for visual pornography.[33] But the male of *Homo sapiens* differs from the male of most other mammals in a crucial way: men invest in their offspring rather than leaving all the investing to the female. Though deprived of organs that can siphon nutrients directly into his children, a man can help them indirectly by feeding, protecting, teaching, and nurturing them. The minimum investments of a man and a woman are still unequal, because a child can be born to a single mother whose husband has fled but not to a single father whose wife has fled. But the investment of the man is greater than zero, which means that women are also predicted to compete in the mate market, though they should compete over the males most likely to invest (and the males with the highest genetic quality) rather than the males most willing to mate.

The genetic economics of sex also predicts that both sexes have a genetic incentive to commit adultery, though for partly different reasons. A philandering man can have additional offspring by impregnating women other than his wife. A philandering woman can have better offspring by conceiving a child by a man with better genes than her husband while having her husband around to help nurture the child. But when a wife gets the best of both worlds from her affair, the husband gets the worst of both worlds, because he is investing in another man's genes that have usurped the place of his own. We thus get the flip side of the evolution of fatherly feelings: the evolution of male sexual jealousy, designed to prevent his wife from having another man's child.

Women's jealousy is tilted more toward preventing the alienation of a man's affections, a sign of his willingness to invest in another woman's children at the expense of her own.[34]

The biological tragedy of the sexes is that the genetic interests of a man and a woman can be so close that they almost count as a single organism, but the possibilities for their interests to diverge are never far away. The biologist Richard Alexander points out that if a couple marry for life, are perfectly monogamous, and favor their nuclear family above each spouse's extended family, their genetic interests are identical, tied up in the single basket containing their children.[35] Under that idealization, the love between a man and a woman should be the strongest emotional bond in the living world—"two hearts beating as one"—and of course for some lucky couples it is. Unfortunately, the ifs in the deduction are big ifs. The power of nepotism means that spouses are always being tugged apart by in-laws and, if there are any, by stepchildren. And the incentives of adultery mean that spouses can always be tugged apart by cuckolds and home-wreckers. It is no surprise to an evolutionary biologist that infidelity, stepchildren, and in-laws are among the main causes of marital strife.

Nor is it a surprise that the act of love itself should be fraught with conflict. Sex is the most concentrated source of physical pleasure granted by our nervous system, so why is it such an emotional bramble bush? In all societies, sex is at least somewhat "dirty." It is conducted in private, pondered obsessively, regulated by custom and taboo, the subject of gossip and teasing, and a trigger for jealous rage.[36] For a brief period in the 1960s and 1970s people dreamed of an erotopia in which men and women could engage in sex without hang-ups and inhibitions. The protagonist of Erica Jong's *Fear of Flying* fantasized about "the zipless fuck": anonymous, casual, and free of guilt and jealousy. "If you can't be with the one you love, love the one you're with," sang Stephen Stills. "If you love somebody, set them free," sang Sting.

But Sting also sang, "Every move you make, I'll be watching you." And Isadora Wing concluded that fastener-free copulation is "rarer than the unicorn." Even in a time when seemingly anything goes, most people do not partake in sex as casually as they partake in food or conversation. That includes today's college campuses, which are reportedly hotbeds of the brief sexual encounters known as "hooking up." The psychologist Elizabeth Paul sums up her research on the phenomenon: "Casual sex is not casual. Very few people are coming out unscathed."[37] The reasons are as deep as anything in biology. One of the hazards of sex is a baby, and a baby is not just any seven-pound object but, from an evolutionary point of view, our reason for being. Every time a woman has sex with a man she is taking a chance at sentencing herself to years of motherhood, with the additional gamble that the whims of her partner could make it *single* motherhood. She is committing a chunk of her finite reproductive output to the genes and intentions of that man, forgoing the opportunity to use it with some other man who may have better endowments of either or both. The man, for his part, may be either implicitly committing his sweat and toil to the incipient child or deceiving his partner about such intentions.

And that covers only the immediate participants. As Jong lamented elsewhere, there are never just two people in bed. They are always accompanied in their minds by parents, former lovers, and real and imagined rivals. In other words, third parties have an interest in the possible outcome of a sexual liaison. The romantic rivals of the man or woman, who are being cuckolded or rendered celibate or bereft by their act of love, have reasons to want to be in their places. The interests of third parties help us understand why sex is almost universally conducted in private. Symons points out that because a man's reproductive success is strictly limited by his access to women, in the minds of men sex is always a rare commodity. People may have sex in private for the same reason that people during a famine eat in private: to avoid inciting dangerous envy.[38]

As if the bed weren't crowded enough, every child of a man and a woman is also the grandchild of two other men and two other women. Parents take an interest in their children's reproduction because in the long run it is their reproduction too. Worse, the preciousness of female reproductive capacity makes it a valuable resource for the men who control her in traditional patriarchal societies, namely her father and brothers. They can trade a daughter or sister for additional wives or resources for themselves, and thus they have an interest in protecting their investment by keeping her from becoming pregnant by men other than the ones they want to sell her to. It is not just the husband or boyfriend who takes a proprietary interest in a woman's sexual activity, then, but also her father and brothers.[39] Westerners were horrified by the treatment of women under the regime of the Taliban in Afghanistan from 1995 to 2001, when women were cloaked in burqas and forbidden to work, attend school, and leave their homes unaccompanied. Wilson and Daly have shown that laws and customs with the same intent—giving men control over their wives' and daughters' sexuality—have been common throughout history and in many societies, including our own.[40] Many a father of a teenage girl has had the fleeting thought that the burqa is not such a bad idea after all.

On strictly rational grounds, the volatility of sex is a paradox, because in an era with contraception and women's rights these archaic entanglements should have no claim on our feelings. We should be ziplessly loving the one we're with, and sex should inspire no more gossip, music, fiction, raunchy humor, or strong emotions than eating or talking does. The fact that people are tormented by the Darwinian economics of babies they are no longer having is testimony to the long reach of human nature.

WHAT ABOUT PEOPLE who are not tied by blood or children? No one doubts that human beings make sacrifices for people who are unrelated to them. But they could do so in two different ways.

Humans, like ants, could have a gung-ho superorganism thing that prompts them to do everything for the colony. The idea that people are instinctively communal is an important precept of the romantic doctrine of the Noble Savage. It figured in the theory of Engels and Marx that "primitive communism" was the first social system, in the anarchism of Peter Kropotkin (who wrote, "The ants and termites have renounced the 'Hobbesian war,' and they are the better for it"), in the family-of-man utopianism of the 1960s, and in the writings of contemporary radical scientists such as Lewontin and Chomsky.[41] Some radical scientists imagine that the only alternative is an Ayn Randian individualism in which every man is an island. Steven Rose and the sociologist Hilary Rose, for instance, call evolutionary psychology a "right-wing libertarian attack on collectivity."[42] But the accusation is factually incorrect—as we shall see in the chapter on politics, many evolutionary psychologists are on the political left—and it is conceptually incorrect. The real alternative to romantic collectivism is not "right-wing libertarianism" but a recognition that social generosity comes from a complex suite of thoughts and emotions rooted in the logic of *reciprocity*. That gives it a very different psychology from the communal sharing practiced by social insects, human families, and cults that try to pretend they are families.[43]

Trivers built on arguments by Williams and Hamilton that pure, public-minded altruism—a desire to benefit the group or species at the expense of the self—is unlikely to evolve among nonrelatives, because it is vulnerable to invasion by cheaters who prosper by enjoying the good deeds of others without contributing in turn. But as I mentioned, Trivers also showed that a measured reciprocal altruism *can* evolve. Reciprocators who help others who have helped them, and who shun or punish others who have failed to help them, will enjoy the benefits of gains in trade and outcompete individualists, cheaters, and pure altruists.[44] Humans are well equipped for the demands of reciprocal altruism. They remember each other as individuals (perhaps with the help of dedicated regions of the brain), and have an eagle eye and a flypaper memory for cheaters.[45] They feel moralistic emotions—liking, sympathy, gratitude, guilt, shame, and anger—that are uncanny implementations of the strategies for reciprocal altruism in computer simulations and mathematical models. Experiments have confirmed the prediction that people are most inclined to help a stranger when they can do so at low cost, when the stranger is in need, and when the stranger is in a position to reciprocate.[46] They like people who grant them favors, grant favors to those they like, feel guilty when they have withheld a possible favor, and punish those who withhold favors from them.[47]

An ethos of reciprocity can pilot not just one-on-one exchanges but contributions to the public good, such as hunting animals that are too large for the hunter to eat himself, building a lighthouse that keeps everyone's ships off the rocks, or banding together to invade neighbors or to repel their invasions. The inherent problem with public goods is captured in Aesop's fable "Who Will Bell the Cat?" The mice in a household agree they would be better off if the cat had a bell around its neck to warn them of its approach, but no mouse will risk life and limb to attach the bell. A willingness to bell the cat—that is, to contribute to the public good—can nonetheless evolve, if it is accompanied by a willingness to reward those who shoulder the burden or to punish the cheaters who shirk it.[48]

The tragedy of reciprocal altruism is that sacrifices on behalf of nonrelatives cannot survive without a web of disagreeable emotions like anxiety, mistrust, guilt, shame, and anger. As the journalist Matt Ridley puts it in his survey of the evolution of cooperation:

> Reciprocity hangs, like a sword of Damocles, over every human head. He's only asking me to his party so I'll give his book a good review. They've been to dinner twice and never asked us back once. After all I did for him, how could he do that to me? If you do this for me, I promise I'll make it up later. What did I do to deserve that? You owe it to me. Obligation; debt; favour; bargain; contract; exchange; deal. . . . Our language and our lives are permeated with ideas of reciprocity.[49]

Studies of altruism by behavioral economists have thrown a spotlight on this sword of Damocles by showing that people are neither the amoral egoists of classical economic theory nor the all-for-one-and-one-for-all communalists of utopian fantasies. In the Ultimatum Game, for example, one participant gets a large sum of money to divide between himself and another participant, and the second one can take it or leave it. If he leaves it, neither side gets anything. A selfish proposer would keep the lion's share; a selfish respondent would accept the remaining crumbs, no matter how small, because part of a loaf is better than none. In reality the proposer tends to offer almost half of the total sum, and the respondent doesn't settle for much less than half, even though turning down a smaller share is an act of spite that deprives both participants. The respondent seems to be driven by a sense of righteous anger and punishes a selfish proposer accordingly; the proposer anticipates this and makes an offer that is just generous enough to be accepted. We know that the proposer's generosity is driven by the fear of a spiteful response because of the outcome of two variants of the experiment. In the Dictator game, the proposer simply divides the sum between the two players and there is nothing the respondent can do about it. With no fear of reprisal, the proposer makes a far stingier offer. The offer still tends to be more generous than it has to be, because the proposer worries about getting a reputation for stinginess that could come back to bite him in the long run. We know *this* because of the outcome of the Double-Blind Dictator game, where proposals from many players are sealed and neither the respondent nor the experimenter knows who offered how much. In this variant, generosity plummets; a majority of the proposers keep everything for themselves.[50]

And then there is the Public Good game, in which everyone makes a voluntary contribution to a common pot of money, the experimenter doubles it, and the pot is divided evenly among the participants regardless of what they contributed. The optimal strategy for each player acting individually is to be a free rider and contribute nothing, hoping that others will contribute something and he can get a share of their contribution. Of course, if every player thinks that way, the pot stays empty and no one earns a dime. The optimum for the group is for all the players to contribute everything they have so they can all double their money. When the game is played repeatedly, however, everyone tries to become a free rider, and the pot dwindles to a self-defeating zero. On the other hand, if people are allowed both to contribute to the pot *and* to levy fines on those who don't contribute, conscience doth make cowards of them all, and almost everyone contributes to the common good, allowing everyone to make a profit.[51] The same phenomenon has been independently documented by social psychologists, who call it "social loafing." When people are part of a group, they pull less hard on a rope, clap less enthusiastically, and think up fewer ideas in a brainstorming session—unless they think their contributions to the group effort are being monitored.[52]

These experiments may be artificial, but the motives they expose played themselves out in the real-life experiments known as utopian communities. In the nineteenth century and early decades of the twentieth, self-contained communes based on a philosophy of communal sharing sprang up throughout the United States. All of them collapsed from internal tensions, the ones guided by socialist ideology after a median of two years, the ones guided by religious ideology after a median of twenty years.[53] The Israeli kibbutzim, originally galvanized by socialism and Zionism, steadily dismantled their collectivist philosophy over the decades. It was undermined by their members' desire to live with their families, to own their own clothing, and to keep small luxuries or sums of money acquired outside the kibbutz. And the kibbutzim were dragged down by inefficiencies because of the free-rider problem—they were, in the words of one kibbutznik, a "paradise for parasites."[54]

In other cultures, too, generosity is doled out according to a complex mental calculus. Remember Fiske's ethnographic survey, which shows that the ethic of Communal Sharing arises spontaneously mainly within families (and on circumscribed occasions such as feasts). Equality Matching—that is, reciprocal altruism—is the norm for everyday interactions among more distant relatives and nonrelatives.[55] A possible exception is the distribution of meat by bands of foragers, who pool the risks of hunting large game (with its big but unpredictable windfalls) by sharing their catch.[56] Even here, the ethic is far from unstinting generosity, and the sharing is described as having "an edge of hostility."[57] Hunters generally have no easy way of keeping their catch from others, so they don't so much *share* their catch as stand by while others confiscate it. Their hunting effort is treated as a public good, and they are punished by gossip and ostracism if they resist the confiscation, are rewarded by prestige (which earns them sexual partners) if they tolerate it, and may be entitled to payback when the tables turn. A similar psychology may be found among the last hunter-gatherers in our own culture, commercial fishermen. In *The Perfect Storm,* Sebastian Junger writes:

> Sword[fish] boat captains help each other out on the high seas whenever they can; they lend engine parts, offer technical advice, donate food or fuel. The competition between a dozen boats rushing a perishable commodity to market fortunately doesn't kill an inherent sense of concern for each other. This may seem terrifically noble, but it's not—or at least not entirely. It's also self-interested. Each captain knows that he may be the next one with the frozen injector or the leaking hydraulics.[58]

Beginning with Ashley Montagu in 1952, thinkers with collectivist sympathies have tried to eke out a place for unmeasured generosity by invoking group selection, a Darwinian competition among groups of organisms rather than among individual organisms.[59] The hope is that groups whose members sacrifice their interests for the common good will outcompete those in which every man is for himself, and as a result generous impulses will come to prevail in the species. Williams dashed the dream in 1966 when he pointed out that unless a group is genetically fixed and hermetically sealed, mutants or immigrants constantly infiltrate it.[60] A selfish infiltrator would soon take over the group with its descendants, who are more numerous because they have reaped the advantages of others' sacrifices without making their own. This would happen long before the group could parlay its internal cohesion into victory over neighboring groups and bud off new offspring groups to repeat the process.

The term "group selection" survives in evolutionary biology, but usually with different meanings from the one Montagu had in mind. Groups were certainly part of our evolutionary environment, and our ancestors evolved traits, such as a concern with one's reputation, that led them to prosper in groups. Sometimes the interests of an individual and the interests of a group can coincide; for example, both do better when the group is not exterminated by enemies. Some theorists invoke group selection to explain a willingness to punish free riders who do not contribute to the public good.[61] The biologist David Sloan Wilson and the philosopher Elliot Sober recently redefined "group" as a set of mutual reciprocators, providing an alternative language in which to describe Trivers's theory but not an alternative to the theory itself.[62] But no one believes the original idea that selection among groups led to the evolution of unstinting self-sacrifice. Even putting aside the theoretical difficulties explained by Williams, we know empirically that people in all cultures do things that lead them to prosper at the *expense* of their group, such as lying, competing for mates, having affairs, getting jealous, and fighting for dominance.

Group selection, in any case, does not deserve its feel-good reputation. Whether or not it endowed us with generosity toward the members of our group, it would certainly have endowed us with a hatred of the members of *other* groups, because it favors whatever traits lead one group to prevail over its rivals. (Recall that group selection was the version of Darwinism that got twisted into Nazism.) This does not mean that group selection is incorrect, only that subscribing to a scientific theory for its apparent political palatability can backfire. As Williams put it, "To claim that [natural selection at the level of competing groups] is morally superior to natural selection at the level of competing individuals would imply, in its human application, that systematic genocide is morally superior to random murder."[63]

～

PEOPLE DO MORE for their fellows than return favors and punish cheaters. They often perform generous acts without the slightest hope for payback, ranging from leaving a tip in a restaurant they will never visit again to throwing themselves on a live grenade to save their brothers in arms. Trivers, together with the economists Robert Frank and Jack Hirshleifer, has pointed out that pure magnanimity can evolve in an environment of people seeking to discriminate fair-weather friends from loyal allies.[64] Signs of heartfelt loyalty and generosity serve as guarantors of one's promises, reducing a partner's worry that you will default on them. The best way to convince a skeptic that you are trustworthy and generous is to *be* trustworthy and generous.

Of course, such virtue cannot be the dominant mode of human interaction or else we could dispense with the gargantuan apparatus designed to keep exchanges fair—money, cash registers, banks, accounting firms, billing departments, courts—and base our economy on the honor system. At the other extreme, people also commit acts of outright treachery, including larceny, fraud, extortion, murder, and other ways of taking a benefit at someone else's expense. Psychopaths, who lack all traces of a conscience, are the most extreme example, but social psychologists have documented what they call Machiavellian traits in many individuals who fall short of outright psychopathy.[65] Most people, of course, are in the middle of the range, displaying mixtures of reciprocity, pure generosity, and greed.

Why do people range across such a wide spectrum? Perhaps all of us are capable of being saints or sinners, depending on the temptations and threats at hand. Perhaps we are set on one of these paths early in life by our upbringing or by the mores of our peer group. Perhaps we *choose* these paths early in life because we are endowed with a deck of conditional strategies on how to develop a personality: if you discover that you are attractive and charming, try being a manipulator; if you are large and commanding, try being a bully; if you are surrounded by generous people, be generous in kind; and so on. Perhaps we are predisposed to being nastier or nicer by our genes. Perhaps human development is a lottery, and fate assigns us a personality at random. Most likely, our differences come from several of these forces or from hybrids among them. For example, we may all develop a sense of generosity if enough of our friends and neighbors are generous, but the threshold or the multiplier of that function may differ among us genetically or at random: some people need only a few nice neighbors to grow up nice, others need a majority.

Genes are certainly a factor. Conscientiousness, agreeableness, neuroticism, psychopathy, and criminal behavior are substantially (though by no means completely) heritable, and altruism may be as well.[66] But this only replaces the original question—Why do people vary in their selfishness?—with another one. Natural selection tends to make the members of a species alike in their adaptive traits, because whichever version of a trait is better than the others will be selected and the alternative versions will die out. That is why most evolutionary psychologists attribute systematic differences among people to their *environments* and attribute only random differences to the genes. This genetic noise can come from at least two sources. Inside the genome, rust never sleeps: random mutations constantly creep in and are only slowly and unevenly eliminated by selection.[67] And selection can favor molecular variability for its own sake to keep us one step ahead of the parasites that constantly evolve to infiltrate our cells and tissues. Differences in the functioning of whole bodies and brains could be a by-product of this churning of protein sequences.[68]

But the theory of reciprocal altruism raises another possibility: that some of the genetic differences among people in their social emotions are systematic. One exception to the rule that selection reduces variability arises when the best strategy depends on what *other* organisms are doing. The child's game of scissors-paper-rock is one analogy, and another may be found in the decision of which route to take to work. As commuters begin to avoid a congested highway and opt for a less traveled route, the new one will no longer be less traveled, so many will choose the first one, until congestion builds up there, which will induce still other commuters to choose the second route, and so on. The commuters will eventually distribute themselves in some ratio between the two roads. The same thing can happen in evolution, where it is called frequency-dependent selection.

One corollary of reciprocal altruism, shown in a number of simulations, is that frequency-dependent selection can produce temporary or permanent *mixtures* of strategies. For example, even if reciprocators predominate in a population, a minority of cheaters can sometimes survive, taking advantage of the generosity of the reciprocators as long as they don't grow so numerous as to meet other cheaters too often or to be recognized and punished by the reciprocators. Whether the population ends up homogeneous or with a mixture of strategies depends on which strategies are competing, which start off more numerous, how easily they enter and leave the population, and the payoffs for cooperation and defection.[69]

We have an intriguing parallel. In the real world, people differ genetically in their selfish tendencies. And in models of the evolution of altruism, actors may evolve differences in their selfish tendencies. It could be a coincidence, but it probably is not. Several biologists have adduced evidence that psychopathy is a cheating strategy that evolved by frequency-dependent selection.[70] Statistical analyses show that a psychopath, rather than merely falling at the end of a continuum for one or two traits, has a distinct cluster of traits (superficial charm, impulsivity, irresponsibility, callousness, guiltlessness, mendacity, and exploitiveness) that sets him off from the rest of the population.[71] And many psychopaths show none of the subtle physical abnormalities produced by biological noise, suggesting that psychopathy is not always a biological mistake.[72] The psychologist Linda Mealey has argued that frequency-dependent selection has produced at least two kinds of psychopaths. One kind consists of people who are genetically predisposed to psychopathy regardless of how they grow up. The other kind is made up of people who are predisposed to psychopathy only in certain circumstances, namely when they perceive themselves to be competitively disadvantaged in society and find themselves at home in a group of other antisocial peers.

The possibility that some individuals are born with a weak conscience runs squarely against the doctrine of the Noble Savage. It calls to mind the old-fashioned notions of born criminals and bad seeds, and it was blotted out by twentieth-century intellectuals and replaced with the belief that all wrongdoers are victims of poverty or bad parenting. In the late 1970s Norman Mailer received a letter from a prisoner named Jack Henry Abbott, who had spent most of his life behind bars for crimes ranging from passing bad checks to killing a fellow prisoner. Mailer was writing a book about the murderer Gary Gilmore, and Abbott offered to help him get into the mindset of a killer by sharing his prison diaries and his radical critique of the criminal justice system. Mailer was dazzled by Abbott's prose and proclaimed him to be a brilliant new writer and thinker—"an intellectual, a radical, a potential leader, a man obsessed with a vision of more elevated human relations in a better world that revolution could forge." He arranged for Abbott's letters to be published in the *New York Review of Books* and then as a 1980 book, *In the Belly of the Beast*. Here is an excerpt, in which Abbott describes what it is like to stab someone to death:

> You can feel his life trembling through the knife in your hand. It almost overcomes you, the gentleness of the feeling at the center of a coarse act of murder. . . . You go to the floor with him to finish him. It is like cutting hot butter, no resistance at all. They always whisper one thing at the end: "Please." You get the odd impression he is not imploring you not to harm him, but to do it right.

Over the objections of prison psychiatrists who saw that Abbott had PSYCHOPATH written all over his face, Mailer and other New York literati helped him win an early parole. Abbott was soon feted at literary dinners, likened to Solzhenitsyn and Jacobo Timerman, and interviewed on *Good Morning America* and in *People* magazine. Two weeks later he got into an argument with an aspiring young playwright who was working as a waiter in a restaurant and had asked Abbott not to use the employees' restroom. Abbott asked him to step outside, stabbed him in the chest, and left him to bleed to death on the sidewalk.[73]

Psychopaths can be clever and charming, and Mailer was only the latest in a series of intellectuals from all over the political spectrum who were conned in the 1960s and 1970s. In 1973 William F. Buckley helped win the early release of Edgar Smith, a man who had been convicted of molesting a fifteen-year-old cheerleader and crushing her head with a rock. Smith won his freedom in exchange for confessing to the crime, and then, as Buckley was interviewing him on his national television program, he recanted the confession. Three years later he was arrested for beating another young woman with a rock, and he is now serving a life sentence for attempted murder.[74]

Not everyone was conned. The comedian Richard Pryor described his experience at the Arizona State Penitentiary during the filming of *Stir Crazy*:

> It made my heart ache, you know, to see all these beautiful black men in the joint. Goddam; the warriors should be out there helping the *masses*. I *felt* that way, I was real naïve. Six weeks I was up there and I *talked* to the brothers. I talked to 'em, and . . . [Looks around, frightened] . . . *Thank God we got penitentiaries!* I asked one, "Why did you kill *everybody in the house?*" He says, "They was home." . . . I met one dude, kidnap-murdered *four times.* And I thought, three times, that was your last, right? I says, "What happened?" [Answers in falsetto] "I can't get this shit right! But I'm getting paroled in two years."

Pryor was not, of course, denying the inequities that continue to put dispro-portionate numbers of African Americans in prison. He was only contrasting the common sense of ordinary people with the romanticism of intellectuals—and perhaps exposing their condescending attitude that poor people can't be expected to refrain from murder, and that they should not be alarmed by the murderers in their midst.

The romantic notion that all malefactors are depraved on accounta they're deprived has worn thin among experts and laypeople alike. Many psy-chopaths had difficult lives, of course, but that does not mean that having a difficult life turns one into a psychopath. There is an old joke about two social workers discussing a problematic child: "Johnny came from a broken home." "Yes, Johnny could break any home." Machiavellian personalities can be found in all social classes—there are kleptocrats, robber barons, military dictators, and rogue financiers—and some psychopaths, such as the cannibal Jeffrey Dahmer, have come from decent, upper-middle-class homes. And none of this means that all people who resort to violence or crime are psychopaths, only that some of the worst ones are.

Psychopaths, as far as we know, cannot be "cured." Indeed, the psycholo-gist Marnie Rice has shown that certain harebrained ideas for therapy, such as boosting their self-esteem and teaching them social skills, can make them even more dangerous.[75] But that does not mean there is nothing we can do about them. For example, Mealey shows that of the two kinds of psychopaths she dis-tinguished, inveterate psychopaths are unmoved by programs that try to get them to appreciate the harm they do, but they may be responsive to surer pun-ishments that induce them to behave more responsibly out of sheer self-interest. Conditional psychopaths, on the other hand, may respond better to social changes that prevent them from slipping through society's cracks. Whether or not these are the best prescriptions, they are examples of how sci-ence and policy might come to grips with a problem that many intellectuals tried to wish away in the twentieth century but that has long been a concern of religion, philosophy, and fiction: the existence of evil.

~

ACCORDING TO TRIVERS, every human relationship—our ties to our par-ents, siblings, romantic partners, and friends and neighbors—has a distinct psychology forged by a pattern of converging and diverging interests. What about the relationship that is, according to the pop song, "the greatest love of all"—the relationship with the self? In a pithy and now-famous passage, Trivers wrote:

If . . . deceit is fundamental to animal communication, then there must be strong selection to spot deception and this ought, in turn, to select for

a degree of self-deception, rendering some facts and motives unconscious so as not to betray—by the subtle signs of self-knowledge—the
deception being practiced. Thus, the conventional view that natural selection favors nervous systems which produce ever more accurate images of the world must be a very naïve view of mental evolution.[76]

The conventional view may be largely correct when it comes to the physical
world, which allows for reality checks by multiple observers and where misconceptions are likely to harm the perceiver. But as Trivers notes, it may not be
correct when it comes to the self, which one can access in a way that others
cannot and where misconceptions may be helpful. Sometimes parents may
want to convince a child that what they are doing is for the child's own good,
children may want to convince parents that they are needy rather than greedy,
lovers may want to convince each other that they will always be true, and unrelated folks may want to convince one another that they are worthy cooperators. These opinions are often embellishments, if not tall tales, and to slip them
beneath a partner's radar a speaker should believe in them so as not to stam-
mer, sweat, or trip himself up in contradictions. Ice-veined liars might, of
course, get away with telling bald fibs to strangers, but they would also have
trouble keeping friends, who could never take their promises seriously. The
price of looking credible is being unable to lie with a straight face, and that
means a part of the mind must be designed to believe its own propaganda—
while another part registers just enough truth to keep the self-concept in touch
with reality.

The theory of self-deception was foreshadowed by the sociologist Erving
Goffman in his 1959 book *The Presentation of Self in Everyday Life,* which disputed the romantic notion that behind the masks we show other people is the
one true self. No, said Goffman; it's masks all the way down. Many discoveries
in the ensuing decades have borne him out.[77]

Though modern psychologists and psychiatrists tend to reject orthodox
Freudian theory, many acknowledge that Freud was right about the defense
mechanisms of the ego. Any therapist will tell you that people protest too
much, deny or repress unpleasant facts, project their flaws onto others, turn
their discomfort into abstract intellectual problems, distract themselves with
time-consuming activities, and rationalize away their motives. The psychiatrists Randolph Nesse and Alan Lloyd have argued that these habits do not
safeguard the self against bizarre sexual wishes and fears (like having sex with
one's mother) but are tactics of self-deception: they suppress evidence that we
are not as beneficent or competent as we would like to think.[78] As Jeff Gold-
blum said in *The Big Chill,* "Rationalizations are more important than sex."
When his friends demurred, he asked, "Have you ever gone a week without a
rationalization?"

As we saw in Chapter 3, when a person suffers neurological damage, the healthy parts of the brain engage in extraordinary confabulations to explain away the foibles caused by the damaged parts (which are invisible to the self because they are *part of* the self) and to present the whole person as a capable, rational actor. A patient who fails to experience a visceral click of recognition when he sees his wife, but who acknowledges that she looks and acts just like his wife, may deduce that an amazing impostor is living in his house. A patient who believes she is at home and is shown the hospital elevator may say without missing a beat, "You wouldn't believe what it cost us to have that installed."[79] After the Supreme Court justice William O. Douglas suffered a stroke that left him paralyzed on one side and confined to a wheelchair, he invited reporters on a hike and told them he wanted to try out for the Washington Redskins. He was soon forced to step down when he refused to acknowledge that anything was wrong with his judgment.[80]

In social psychology experiments, people consistently overrate their own skill, honesty, generosity, and autonomy. They overestimate their contribution to a joint effort, chalk up their successes to skill and their failures to luck, and always feel that the other side has gotten the better deal in a compromise.[81] People keep up these self-serving illusions even when they are wired to what they think is an accurate lie-detector. This shows that they are not lying to the experimenter but lying to themselves. For decades every psychology student has learned about "cognitive dissonance reduction," in which people change whatever opinion it takes to maintain a positive self-image.[82] The cartoonist Scott Adams illustrates it well:

Dilbert reprinted by permission of United Feature Syndicate, Inc.

If the cartoon were completely accurate, though, life would be a cacophony of spoinks.

Self-deception is among the deepest roots of human strife and folly. It implies that the faculties that ought to allow us to settle our differences—seeking the truth and discussing it rationally—are miscalibrated so that all parties assess themselves to be wiser, abler, and nobler than they really are. Each party to a dispute can *sincerely* believe that the logic and evidence are on his side and that his opponent is deluded or dishonest or both.[83] Self-deception is one of the reasons that the moral sense can, paradoxically, often do more harm than good, a human misfortune we will explore in the next chapter.

~

THE MANY ROOTS of our suffering illuminated by Trivers are not a cause for lamentations and wailings. The genetic overlaps that unite and divide us are tragic not in the everyday sense of a catastrophe but in the dramatic sense of a stimulus that encourages us to ponder our condition. According to a definition in the *Cambridge Encyclopedia,* "The fundamental purpose of tragedy . . . was claimed by Aristotle to be the awakening of pity and fear, of a sense of wonder and awe at the human potential, including the potential for suffering; it makes an assertion of human value in the face of a hostile universe." Trivers's accounts of the inherent conflicts within families, couples, societies, and the self can reinforce that purpose.

Nature may have played a cruel trick by slightly mistuning the emotions of people who share their flesh and blood, but in doing so she provided steady work for generations of authors and playwrights. Endless are the dramatic possibilities inherent in the fact that two people can be bound by the strongest emotional bonds in the living world and at the same time not always want the best for each other. Aristotle was perhaps the first to note that tragic narratives focus on family relations. A story about two strangers who fight to the death, he pointed out, is nowhere near as interesting as a story about two *brothers* who fight to the death. Cain and Abel, Jacob and Esau, Oedipus and Laius, Michael and Fredo, JR and Bobby, Frasier and Niles, Joseph and his brothers, Lear and his daughters, Hannah and her sisters . . . As cataloguers of dramatic plots have noted for centuries, "enmity of kinsmen" and "rivalry of kinsmen" are enduring formulas.[84]

In his book *Antigones,* the literary critic George Steiner showed that the Antigone legend has a singular place in Western literature. Antigone was the daughter of Oedipus and Jocasta, but the fact that her father was her brother and her sister was her mother was only the beginning of her family troubles. In defiance of King Creon, she buried her slain brother Polynices, and when the king found out, he ordered her buried alive. She cheated him by killing herself first, whereupon the king's son, who was madly in love with her and unable to get her a pardon, killed himself on her grave. Steiner observes that *Antigone* is widely considered "not only the finest of the Greek tragedies, but a work of art nearer to perfection than any other produced by the human spirit."[85] It has been performed for more than two millennia and has inspired countless variations and spinoffs. Steiner explains its enduring resonance:

> It has, I believe, been given to only one literary text to express all the principal constants of conflict in the condition of man. These constants are

fivefold: the confrontation of men and of women; of age and of youth; of society and of the individual; of the living and the dead; of men and of god(s). The conflicts which come of these five orders of confrontation are not negotiable. Men and women, old and young, the individual and the community or state, the quick and the dead, mortals and immortals, define themselves in the conflictual process of defining each other.[86] . . . Because Greek myths encode certain primary biological and social confrontations and self-perceptions in the history of man, they endure as an animate legacy in collective memory and recognition.[87]

The bittersweet process of defining ourselves by our conflicts with others is not just a subject for literature but can illuminate the nature of our emotions and the content of our consciousness. If a genie offered us the choice between belonging to a species that could achieve perfect egalitarianism and solidarity and belonging to a species like ours in which relationships with parents, siblings, and children are uniquely precious, it is not so clear that we would choose the former. Our close relatives have a special place in our hearts only because the place for every other human being, by definition, is less special, and we have seen that many social injustices fall out of that bargain. So, too, is social friction a product of our individuality and of our pursuit of happiness. We may envy the harmony of an ant colony, but when Woody Allen's alter ego Z complained to his psychiatrist that he felt insignificant, the psychiatrist replied, "You've made a real breakthrough, Z. You *are* insignificant."

Donald Symons has argued that we have genetic conflict to thank for the fact that we have feelings toward other people at all.[88] Consciousness is a manifestation of the neural computations necessary to figure out how to get the rare and unpredictable things we need. We feel hunger, savor food, and have a palate for countless fascinating tastes because food was hard to get during most of our evolutionary history. We don't normally feel longing, delight, or fascination regarding oxygen, even though it is crucial for survival, because it was never hard to obtain. We just breathe.

The same may be true of conflicts over kin, mates, and friends. I mentioned that if a couple were guaranteed to be faithful, to favor each other over their kin, and to die at the same time, their genetic interests would be identical, wrapped up in their common children. One can even imagine a species in which every couple was marooned on an island for life and their offspring dispersed at maturity, never to return. Since the genetic interests of the two mates are identical, one might at first think that evolution would endow them with a blissful perfection of sexual, romantic, and companionate love.

But, Symons argues, nothing of the sort would happen. The relation between the mates would evolve to be like the relation among the cells of a single body, whose genetic interests are also identical. Heart cells and lung cells don't have to fall in love to get along in perfect harmony. Likewise, the couples in this species would have sex only for the purpose of procreation (why waste energy?), and sex would bring no more pleasure than the rest of reproductive physiology such as the release of hormones or the formation of the gametes:

> There would be no falling in love, because there would be no alternative mates to choose among, and falling in love would be a huge waste. You would literally love your mate as yourself, but that's the point: you *don't* really love yourself, except metaphorically; you *are* yourself. The two of you would be, as far as evolution is concerned, one flesh, and your relationship would be governed by mindless physiology. . . . You might feel pain if you observed your mate cut herself, but all the feelings we have about our mates that make a relationship so wonderful when it is working well (and so painful when it is not) would never evolve. Even if a species had them when they took up this way of life, they would be selected out as surely as the eyes of a cave-dwelling fish are selected out, because they would be all cost and no benefit.[89]

The same is true for our emotions toward family and friends: the richness and intensity of the feelings in our minds are proof of the preciousness and fragility of those bonds in life. In short, without the possibility of suffering, what we would have is not harmonious bliss, but rather, no consciousness at all.

Chapter 15

The Sanctimonious Animal

ONE OF THE deepest fears people have of a biological understanding of the mind is that it would lead to moral nihilism. If we are not created by God for a higher purpose, say the critics on the right, or if we are products of selfish genes, say the critics on the left, then what would prevent us from becoming amoral egoists who look out only for number one? Wouldn't we have to see ourselves as venal mercenaries who cannot be expected to care for the less fortunate? Both sides point to Nazism as the outcome of accepting biological theories of human nature.

The preceding chapter showed that this fear is misplaced. Nothing prevents the godless and amoral process of natural selection from evolving a big-brained social species equipped with an elaborate moral sense.[1] Indeed, the problem with *Homo sapiens* may not be that we have too little morality. The problem may be that we have too much.

What leads people to deem an action immoral ("Killing is wrong") as opposed to disliked ("I hate broccoli"), unfashionable ("Don't wear stripes with plaids"), or imprudent ("Avoid wine on long flights")? People feel that moral rules are universal. Injunctions against murder and rape, for example, are not matters of taste or fashion but have a transcendent and universal warrant. People feel that others who commit immoral acts ought to be punished: not only is it right to inflict harm on people who have committed a moral infraction, it is wrong *not* to, that is, to "let them get away with it." One can easily say, "I don't like broccoli, but I don't care if you eat it," but no one would say, "I don't like killing, but I don't care if you murder someone." That is why pro-choice advocates are missing the point when they say, in the words of the bumper sticker, "If you're against abortion, don't have one." If someone believes abortion is immoral, then allowing other people to engage in it is not an option, any more than allowing people to rape or murder is an option. People therefore feel justified in invoking divine retribution or the coercive power of the state to enact the punishments. Bertrand Russell wrote, "The infliction of cruelty with a good conscience is a delight to moralists—that is why they invented hell."

Our moral sense licenses aggression against others as a way to prevent or punish immoral acts. That is fine when the act deemed immoral truly *is* immoral by any standard, such as rape and murder, and when the aggression is meted out fairly and serves as a deterrent. The point of this chapter is that the human moral sense is not guaranteed to pick out those acts as the targets of its righteous indignation. The moral sense is a gadget, like stereo vision or intuitions about number. It is an assembly of neural circuits cobbled together from older parts of the primate brain and shaped by natural selection to do a job. That does not mean that morality is a figment of our imagination, any more than the evolution of depth perception means that 3-D space is a figment of our imagination. (As we saw in Chapters 9 and 11, morality has an internal logic, and possibly even an external reality, that a community of reflective thinkers may elucidate, just as a community of mathematicians can elucidate truths about number and shape.) But it does mean that the moral sense is laden with quirks and prone to systematic error—moral illusions, as it were— just like our other faculties.

Consider this story:

> Julie and Mark are brother and sister. They are traveling together in France on summer vacation from college. One night they are staying alone in a cabin near the beach. They decide that it would be interesting and fun if they tried making love. At the very least it would be a new experience for each of them. Julie was already taking birth control pills, but Mark uses a condom too, just to be safe. They both enjoy making love, but they decide not to do it again. They keep the night as a special secret, which makes them feel even closer to each other. What do you think about that; was it OK for them to make love?

The psychologist Jonathan Haidt and his colleagues have presented the story to many people.[2] Most immediately declare that what Julie and Mark did was wrong, and then they grope for reasons *why* it was wrong. They mention the dangers of inbreeding, but they are reminded that the siblings used two forms of contraception. They suggest that Julie and Mark will be emotionally hurt, but the story makes it clear that they were not. They venture that the act would offend the community, but then they recall that it was kept secret. They submit that it might interfere with future relationships, but they acknowledge that Julie and Mark agreed never to do it again. Eventually many of the respondents admit, "I don't know, I can't explain it, I just know it's wrong." Haidt calls this "moral dumbfounding" and has evoked it by other disagreeable but victimless scenarios:

A woman is cleaning out her closet, and she finds her old American flag. She doesn't want the flag anymore, so she cuts it up into pieces and uses the rags to clean her bathroom.

A family's dog was killed by a car in front of their house. They had heard that dog meat was delicious, so they cut up the dog's body and cooked it and ate it for dinner.

A man goes to the supermarket once a week and buys a dead chicken. But before cooking the chicken, he has sexual intercourse with it. Then he cooks it and eats it.

Many moral philosophers would say that there is nothing wrong with these acts, because private acts among consenting adults that do not harm other sentient beings are not immoral. Some might criticize the acts using a more subtle argument having to do with commitments to policies, but the infractions would still be deemed minor compared with the truly heinous acts of which people are capable. But for everyone else, such argumentation is beside the point. People have gut feelings that give them emphatic moral convictions, and they struggle to rationalize the convictions after the fact.[3] These convictions may have little to do with moral judgments that one could justify to others in terms of their effects on happiness or suffering. They arise instead from the neurobiological and evolutionary design of the organs we call moral emotions.

~

HAIDT HAS RECENTLY compiled a natural history of the emotions making up the moral sense.[4] The four major families are just what we would expect from Trivers's theory of reciprocal altruism and the computer models of the evolution of cooperation that followed. The other-condemning emotions—contempt, anger, and disgust—prompt one to punish cheaters. The other-praising emotions—gratitude and an emotion that may be called elevation, moral awe, or being moved—prompt one to reward altruists. The other-suffering emotions—sympathy, compassion, and empathy—prompt one to help a needy beneficiary. And the self-conscious emotions—guilt, shame, and embarrassment—prompt one to avoid cheating or to repair its effects.

Cutting across these sets of emotions we find a distinction among three spheres of morality, each of which frames moral judgments in a different way. The ethic of *autonomy* pertains to an individual's interests and rights. It emphasizes fairness as the cardinal virtue, and is the core of morality as it is understood by secular educated people in Western cultures. The ethic of *community* pertains to the mores of the social group; it includes values like duty, respect, adherence to convention, and deference to a hierarchy. The ethic of *divinity* pertains to a sense of exalted purity and holiness, which is opposed to a sense of contamination and defilement.

The autonomy-community-divinity trichotomy was first developed by the anthropologist Richard Shweder, who noted that non-Western traditions have rich systems of beliefs and values with all the hallmarks of moralizing but without the Western concept of individual rights.[5] The elaborate Hindu beliefs surrounding purification are a prime example. Haidt and the psychologist Paul Rozin have built on Shweder's work, but they have interpreted the moral spheres not as arbitrary cultural variants but as universal mental faculties with different evolutionary origins and functions.[6] They show that the moral spheres differ in their cognitive content, their homologues in other animals, their physiological correlates, and their neural underpinnings.

Anger, for example, which is the other-condemning emotion in the sphere of autonomy, evolved from systems for aggression and was recruited to implement the cheater-punishment strategy demanded by reciprocal altruism. Disgust, the other-condemning emotion in the sphere of divinity, evolved from a system for avoiding biological contaminants like disease and spoilage. It may have been recruited to demarcate the moral circle that divides entities that we engage morally (such as peers) from those we treat instrumentally (such as animals) and those we actively avoid (such as people with a contagious disease). Embarrassment, the self-conscious emotion in the sphere of community, is a dead ringer for the gestures of appeasement and submission found in other primates. The reason that dominance got melded with morality in the first place is that reciprocity depends not only on a person's willingness to grant and return favors but on that person's ability to do so, and dominant people have that ability.

Relativists might interpret the three spheres of morality as showing that individual rights are a parochial Western custom and that we should respect other cultures' ethics of community and divinity as equally valid alternatives. I conclude instead that the design of the moral sense leaves people in all cultures vulnerable to confusing defensible moral judgments with irrelevant passions and prejudices. The ethic of autonomy or fairness is in fact not uniquely Western; Amartya Sen and the legal scholar Mary Ann Glendon have shown that it also has deep roots in Asian thought.[7] Conversely, the ethic of community and the ethic of divinity are pervasive in the West. The ethic of community, which equates morality with a conformity to local norms, underlies the cultural relativism that has become boilerplate on college campuses. Several scholars have noticed that their students are unequipped to explain why Nazism was wrong, because the students feel it is impermissible to criticize the values of another culture.[8] (I can confirm that students today reflexively hedge their moral judgments, saying things like, "Our society puts a high value on being good to other people.") Donald Symons comments on the way that people's judgments can do a backflip when they switch from autonomy- to community-based morality:

> If only one person in the world held down a terrified, struggling, screaming little girl, cut off her genitals with a septic blade, and sewed her back up, leaving only a tiny hole for urine and menstrual flow, the only question would be how severely that person should be punished, and whether the death penalty would be a sufficiently severe sanction. But when millions of people do this, instead of the enormity being magnified millions-fold, suddenly it becomes "culture," and thereby magically becomes less, rather than more, horrible, and is even defended by some Western "moral thinkers," including feminists.[9]

The ethic of community also includes a deference to an established hierarchy, and the mind (including the Western mind) all too easily conflates prestige with morality. We see it in words that implicitly equate status with virtue—*chivalrous, classy, gentlemanly, honorable, noble*—and low rank with sin—*low-class, low-rent, mean, nasty, shabby, shoddy, villain* (originally meaning "peasant"), *vulgar*. The Myth of the Noble Noble is obvious in contemporary celebrity worship. Members of the royalty like Princess Diana and her American equivalent, John F. Kennedy Jr., are awarded the trappings of sainthood even though they were morally unexceptional people (yes, Diana supported charities, but that's pretty much the job description of a princess in this day and age). Their good looks brighten their halos even more, because people judge attractive men and women to be more virtuous.[10] Prince Charles, who also supports charities, will never be awarded the trappings of sainthood, even if he dies a tragic death.

People also confuse morality with purity, even in the secular West. Remember from Chapter 1 that many words for cleanliness and dirt are also words for virtue and sin (*pure, unblemished, tainted,* and so on). Haidt's subjects seem to have conflated contamination with sin when they condemned eating a dog, having sex with a dead chicken, and enjoying consensual incest (which reflects our instinctive repulsion toward sex with siblings, an emotion that evolved to deter inbreeding).

The mental mix-up of the good and the clean can have ugly consequences. Racism and sexism are often expressed as a desire to avoid pollutants, as in the ostracism of the "untouchable" caste in India, the sequestering of menstruating women in Orthodox Judaism, the fear of contracting AIDS from casual contact with gay men, the segregated facilities for eating, drinking, bathing, and sleeping under the Jim Crow and apartheid policies, and the "racial hygiene" laws in Nazi Germany. One of the haunting questions of twentieth-century history is how so many ordinary people committed wartime atrocities. The philosopher Jonathan Glover has documented that a common denominator is degradation: a diminution of the victim's status or cleanliness or both. When someone strips a person of dignity by making jokes about his suffering, giving him a humiliating appearance (a dunce cap, awkward prison garb, a crudely shaved head), or forcing him to live in filthy conditions, ordinary people's compassion can evaporate and they find it easy to treat him like an animal or object.[11]

The peculiar mixture of fairness, status, and purity constituting the moral sense should make us suspicious of appeals to raw sentiment in resolving difficult moral issues. In an influential essay called "The Wisdom of Repugnance," Leon Kass (now the chair of George W. Bush's Council on Bioethics) argued that we should abandon moral reasoning when it comes to cloning and go with our gut feelings:

> We are repelled by the prospect of cloning human beings not because of the strangeness or novelty of the undertaking, but because we intuit and feel, immediately and without argument, the violation of things that we rightfully hold dear. Repugnance, here as elsewhere, revolts against the excesses of human willfulness, warning us not to transgress what is unspeakably profound. Indeed, in this age in which everything is held to be permissible so long as it is freely done, in which our given human nature no longer commands respect, in which our bodies are regarded as mere instruments of our autonomous rational wills, repugnance may be the only voice left that speaks up to defend the central core of our humanity. Shallow are the souls that have forgotten how to shudder.[12]

There may be good arguments against human cloning, but the shudder test is not one of them. People have shuddered at all kinds of morally irrelevant violations of standards of purity in their culture: touching an untouchable, drinking from the same water fountain as a person of color, allowing Jewish blood to mix with Aryan blood, tolerating sodomy between consenting men. As recently as 1978, many people (including Kass) shuddered at the new technology of in vitro fertilization, or, as it was then called, "test-tube babies." But now it is morally unexceptionable and, for hundreds of thousands of people, a source of immeasurable happiness or of life itself.

The difference between a defensible moral position and an atavistic gut feeling is that with the former we can give *reasons* why our conviction is valid. We can explain why torture and murder and rape are wrong, or why we should oppose discrimination and injustice. On the other hand, no good reasons can be produced to show why homosexuality should be suppressed or why the races should be segregated. And the good reasons for a moral position are not pulled out of thin air: they always have to do with what makes people better off or worse off, and are grounded in the logic that we have to treat other people in the way that we demand they treat us.

~

ANOTHER STRANGE FEATURE of the moral emotions is that they can be turned on and off like a switch. These mental spoinks are called moralization and amoralization, and have recently been studied in the lab by Rozin.[13] They consist in flipping between a mindset that judges behavior in terms of *preference* with a mindset that judges behavior in terms of *value*.

There are two kinds of vegetarians: those who avoid meat for health reasons, namely reducing dietary fat and toxins, and those who avoid meat for moral reasons, namely respecting the rights of animals. Rozin has shown that compared with health vegetarians, moral vegetarians offer more reasons for their meat avoidance, have a greater emotional reaction to meat, and are more likely to treat it as a contaminant—they refuse, for example, to eat a bowl of soup into which a drop of meat broth has fallen. Moral vegetarians are more likely to think that other people should be vegetarians, and they are more likely to invest their dietary habit with bizarre virtues, like believing that meat eating makes people more aggressive and animalistic. But it is not just vegetarians who associate eating habits with moral value. When college students are given descriptions of people and asked to rate their character, they judge that a person who eats cheeseburgers and milkshakes is less nice and considerate than a person who eats chicken and salad!

Rozin notes that smoking has recently been moralized. For many years the decision of whether to smoke was treated as a matter of preference or prudence: some people simply didn't enjoy smoking or avoided it because it was hazardous to their health. But with the discovery of the harmful effects of secondhand smoke, smoking is now treated as an immoral act. Smokers are banished and demonized, and the psychology of disgust and contamination is brought into play. Nonsmokers avoid not just smoke but anything that has ever been in contact with smoke: in hotels, they demand smoke-free rooms or even smoke-free floors. Similarly, the desire for retribution has been awakened: juries have slapped tobacco companies with staggering financial penalties, appropriately called "punitive damages." This is not to say that these decisions are unjustified, only that we should be aware of the emotions that may be driving them.

At the same time, many behaviors have been amoralized, switching (in the eyes of many people) from moral flaws to lifestyle choices. The amoralized acts include divorce, illegitimacy, working motherhood, marijuana use, homosexuality, masturbation, sodomy, oral sex, atheism, and any practice of a non-Western culture. Similarly, many afflictions have been reassigned from the wages of sin to the vagaries of bad luck and have been redubbed accordingly.

The homeless used to be called bums and tramps; sexually transmitted diseases were formerly known as venereal diseases. Most of the professionals who work with drug addiction insist that it is not a bad choice but a kind of illness.

To the cultural right, all this shows that morality has been under assault from the cultural elite, as we see in the sect that calls itself the Moral Majority. To the left, it shows that the desire to stigmatize private behavior is archaic and repressive, as in H. L. Mencken's definition of Puritanism as "the haunting fear that someone, somewhere, may be happy." Both sides are wrong. As if to compensate for all the behaviors that have been amoralized in recent decades, we are in the midst of a campaign to moralize new ones. The Babbitts and the bluenoses have been replaced by the activists for a nanny state and the college towns with a foreign policy, but the psychology of moralization is the same. Here are some examples of things that have acquired a moral coloring only recently:

advertising to children • automobile safety • Barbie dolls • "big box" chain stores • cheesecake photos • clothing from Third World factories • consumer product safety • corporate-owned farms • defense-funded research • disposable diapers • disposable packaging • ethnic jokes • executive salaries • fast food • flirtation in the workplace • food additives • fur • hydroelectric dams • IQ tests • logging • mining • nuclear power • oil drilling • owning certain stocks • poultry farms • public holidays (Columbus Day, Martin Luther King Day) • research on AIDS • research on breast cancer • spanking • suburbia ("sprawl") • sugar • tax cuts • toy guns • violence on television • weight of fashion models

Many of these things can have harmful consequences, of course, and no one would want them trivialized. The question is whether they are best handled by the psychology of moralization (with its search for villains, elevation of accusers, and mobilization of authority to mete out punishment) or in terms of costs and benefits, prudence and risk, or good and bad taste. Pollution, for example, is often treated as a crime of defiling the sacred, as in the song by the rock group Traffic: "Why don't we . . . try to save this land, and make a promise not to hurt again this holy ground." This can be contrasted with the attitude of economists like Robert Frank, who (alluding to the costs of cleanups) said, "There is an optimal amount of pollution in the environment, just as there is an optimal amount of dirt in your house."

Moreover, all human activities have consequences, often with various degrees of benefit and harm to different parties, but not all of them are conceived as immoral. We don't show contempt to the man who fails to change the batteries in his smoke alarms, takes his family on a driving vacation (multiplying their risk of accidental death), or moves to a rural area (increasing pollution and fuel use in commuting and shopping). Driving a gas-guzzling SUV is seen as morally dubious, but driving a gas-guzzling Volvo is not; eating a Big Mac is suspect, but eating imported cheese or tiramisù is not. Becoming aware of the psychology of moralization need not make us morally obtuse. On the contrary, it can alert us to the possibility that a decision to treat an act in terms of virtue and sin as opposed to cost and benefit has been made on morally irrelevant grounds—in particular, whether the saints and sinners would be in one's own coalition or someone else's. Much of what is today called "social criticism" consists of members of the upper classes denouncing the tastes of the lower classes (bawdy entertainment, fast food, plentiful consumer goods) while considering themselves egalitarians.

~

THERE IS ANOTHER bit of moral psychology that is commonly associated with primitive thinking but is alive and well in modern minds: concepts of the sacred and the taboo. Some values are considered not just worthy but sacrosanct. They have infinite or transcendental worth, trumping all other considerations. One is not permitted even to *think* of trading them off against other values, because the very thought is self-evidently sinful and deserves only condemnation and outrage.

The psychologist Philip Tetlock elicited the psychology of the sacred and the taboo in the students of American universities.[14] He asked them whether people should be allowed to buy and sell organs for transplantation, auction licenses to adopt orphans, pay for the right to become a citizen, sell their vote in an election, or pay someone to serve in their stead in prison or the military. Not surprisingly, most of the students thought that the practices were unethical and should be outlawed. But their responses went well beyond disagreement: they were outraged that anyone would consider legalizing these practices, were insulted to have been asked, and wanted to punish anyone who tolerated them. When they were asked to justify their opinion, all they could say was that the practices were "degrading, dehumanizing, and unacceptable." The students even sought to cleanse themselves by volunteering to campaign against a (fictitious) movement to legalize the auctioning of adoption rights. Their outrage was reduced a bit, but was still potent, after hearing arguments in favor of the taboo policies, such as that a market in orphans would put more children in loving homes and that lower-income people would be given vouchers to participate.

Another study asked about a hospital administrator who had to decide whether to spend a million dollars on a liver transplant for a child or use it on other hospital needs. (Administrators implicitly face this kind of choice all the time, because there are lifesaving procedures that are astronomically expensive and cannot be carried out on everyone who needs them.) Not only did respondents want to punish an administrator who chose to spend the money on the hospital, they wanted to punish an administrator who chose to save the child but *thought for a long time* before making the decision (like the frugal comedian Jack Benny when a mugger said, "Your money or your life").

The taboo on thinking about core values is not totally irrational. We judge people not just on what they *do* but on what they *are*—not just on whether someone has given more than he has taken, but on whether he is the kind of person who would sell you down the river or knife you in the back if it were ever in his interests to do so. To determine whether someone is emotionally committed to a relationship, guaranteeing the veracity of his promises, one should ascertain how he thinks: whether he holds your interests sacred or constantly weighs them against the profits to be made by selling you out. The notion of *character* joins the moral picture, and with it the notion of moral identity: the concept of one's own character that is maintained internally and projected to others.

Tetlock points out that it is in the very nature of our commitments to other people to deny that we can put a price on them: "To transgress these normative boundaries, to attach a monetary value to one's friendships or one's children or one's loyalty to one's country, is to disqualify oneself from certain societal roles, to demonstrate that one just 'doesn't get it'—one does not understand what it means to be a true friend or parent or citizen."[15] Taboo tradeoffs, which pit a sacred value against a secular one (such as money), are "morally corrosive: the longer one contemplates indecent proposals, the more irreparably one compromises one's moral identity."[16]

Unfortunately, a psychology that treats some desiderata as having infinite value can lead to absurdities. Tetlock reviews some examples. The Delaney Clause of the Food and Drug Act of 1958 sought to improve public health by banning all new food additives for which there was any risk of carcinogenicity. That sounded good but wasn't. The policy left people exposed to more-dangerous food additives that were already on the market, it created an incentive for manufacturers to introduce new dangerous additives as long as they were not carcinogenic, and it outlawed products that could have saved more lives than they put at risk, such as the saccharin used by diabetics. Similarly after the discovery of hazardous waste at the Love Canal in 1978, Congress passed the Superfund Act, which required the *complete* cleanup of *all* hazardous waste sites. It turned out to cost millions of dollars to clean up the last 10 percent of the waste at a given site—money that could have been spent on cleaning up other sites or reducing other health risks. So the lavish fund went bankrupt before even a fraction of its sites could be decontaminated, and its effect on Americans' health was debatable. After the *Exxon Valdez* oil spill, four-fifths of the respondents in one poll said that the country should pursue greater environmental protection "regardless of cost." Taken literally, that meant they were prepared to shut down all schools, hospitals, and police and fire stations, stop funding social programs, medical research, foreign aid, and national defense, or raise the income tax rate to 99 percent, if that is what it would have cost to protect the environment.

Tetlock observes that these fiascoes came about because any politician who honestly presented the inexorable tradeoffs would be crucified for violating a taboo. He would be guilty of "tolerating poisons in our food and water," or worse, "putting a dollar value on human life." Policy analysts note that we are stuck with wasteful and inegalitarian entitlement programs because any politician who tried to reform them would be committing political suicide. Savvy opponents would frame the reform in the language of taboo: "breaking our faith with the elderly," "betraying the sacred trust of veterans who risked their lives for their country," "scrimping on the care and education of the young."

In the Preface, I called the Blank Slate a sacred doctrine and human nature a modern taboo. This can now be stated as a technical hypothesis. The thrust of the radical science movement was to moralize the scientific study of the mind and to engage the mentality of taboo. Recall, from Part II, the indignant outrage, the punishment of heretics, the refusal to consider claims as they were actually stated, the moral cleansing through demonstrations and manifestos and public denunciations. Weizenbaum condemned ideas "whose very contemplation ought to give rise to feelings of disgust" and denounced the less-than-human scientists who "can even think of such a thing." But of course it is the job of scholars to think about things, even if only to make it clear why they are wrong. Moralization and scholarship thus often find themselves on a collision course.

~

THIS RUTHLESS DISSECTION of the human moral sense does not mean that morality is a sham or that every moralist is a self-righteous prig. Moral psychology may be steeped in emotion, but then many philosophers have argued that morality cannot be grounded in reason alone anyway. As Hume wrote, " 'Tis not contrary to reason to prefer the destruction of the whole world to the scratching of my finger."[17] The emotions of sympathy, gratitude, and guilt are the source of innumerable acts of kindness great and small, and a measured righteous anger and ethical certitude must have sustained great moral leaders throughout history.

Glover notes that many twentieth-century atrocities were set in motion when the moral emotions were disabled. Decent people were lulled into committing appalling acts by a variety of amoralizing causes, such as utopian ideologies, phased decisions (in which the targets of bombing might shift from isolated factories to factories near neighborhoods to the neighborhoods themselves), and the diffusion of responsibility within a bureaucracy. It was often raw moral sentiment—feeling empathy for victims, or asking oneself the moral-identity question "Am I the kind of person who could do this?"—that stopped people in mid-atrocity. The moral sense, amplified and extended by reasoning and a knowledge of history, is what stands between us and a Mad Max nightmare of ruthless psychopaths.

But there is still much to be wary of in human moralizing: the confusion of morality with status and purity, the temptation to overmoralize matters of judgment and thereby license aggression against those with whom we disagree, the taboos on thinking about unavoidable tradeoffs, and the ubiquitous vice of self-deception, which always manages to put the self on the side of the angels. Hitler was a moralist (indeed, a moral vegetarian) who, by most accounts, was convinced of the rectitude of his cause. As the historian Ian Buruma wrote, "This shows once again that true believers can be more dangerous than cynical operators. The latter might cut a deal; the former have to go to the end—and drag the world down with them."[18]

PART V

HOT BUTTONS

S ome debates are so entwined with people's moral identity that one might despair that they can ever be resolved by reason and evidence. Social psychologists have found that with divisive moral issues, especially those on which liberals and conservatives disagree, all combatants are intuitively certain they are correct and that their opponents have ugly ulterior motives. They argue out of respect for the social convention that one should always provide reasons for one's opinions, but when an argument is refuted, they don't change their minds but work harder to find a replacement argument. Moral debates, far from resolving hostilities, can escalate them, because when people on the other side don't immediately capitulate, it only proves they are impervious to reason.[1]

Nowhere is this more obvious than in the topics I will explore in this part of the book. People's opinions on politics, violence, gender, children, and the arts help define the kind of person they think they are and the kind of person they want to be. They prove that the person is opposed to oppression, violence, sexism, philistinism, and the abuse or neglect of children. Unfortunately, folded into these opinions are assumptions about the psychological makeup of *Homo sapiens*. Conscientious people may thus find themselves unwittingly staked to positions on empirical questions in biology or psychology. When scientific facts come in they rarely conform exactly to our expectations; if they did, we would not have to do science in the first place. So when facts tip over a sacred cow, people are tempted to suppress the facts and to clamp down on debate because the facts threaten everything they hold sacred. And this can leave us unequipped to deal with just those problems for which new facts and analyses are most needed.

The landscape of the sciences of human nature is strewn with these third rails, hot zones, black holes, and Chernobyls. I have picked five of them to explore in the next few chapters, while necessarily leaving out many others (for instance, race, sexual orientation, education, drug abuse, and mental illness).

Social psychologists have discovered that even in heated ideological battles, common ground can sometimes be found.[2] Each side must acknowledge that the other is arguing out of principle, too, and that they both share certain values and disagree only over which to emphasize in cases where they conflict. Finding such common ground is my goal in the discussions to follow.

Chapter 16

Politics

I often think it's comical
How nature always does contrive
That every boy and every gal,
. That's born into the world alive,
Is either a little Liberal,
Or else a little Conservative![1]

GILBERT AND SULLIVAN got it mostly right in 1882: liberal and conservative political attitudes are largely, though far from completely, heritable. When identical twins who were separated at birth are tested in adulthood, their political attitudes turn out to be similar, with a correlation coefficient of .62 (on a scale from −1 to +1).[2] Liberal and conservative attitudes are heritable not, of course, because attitudes are synthesized directly from DNA but because they come naturally to people with different temperaments. Conservatives, for example, tend to be more authoritarian, conscientious, traditional, and rule-bound. But whatever its immediate source, the heritability of political attitudes can explain some of the sparks that fly when liberals and conservatives meet. When it comes to attitudes that are heritable, people react more quickly and emotionally, are less likely to change their minds, and are more attracted to like-minded people.[3]

Liberalism and conservatism have not just genetic roots, of course, but historical and intellectual ones. The two political philosophies were articulated in the eighteenth century in terms that would be familiar to readers of the editorial pages today, and their foundations can be traced back millennia to the political controversies of ancient Greece. During the past three centuries, many revolutions and uprisings were fought over these philosophies, as are the major elections in modern democracies.

This chapter is about the intellectual connections between the sciences of human nature and the political rift between right-wing and left-wing political philosophies. The connection is not a secret. As philosophers have long noted, the two sides are not just political belief systems but empirical ones, rooted in different conceptions of human nature. Small wonder that the sciences of human nature have been so explosive. Evolutionary psychology, behavioral genetics, and some parts of cognitive neuroscience are widely seen as falling on the political right, which in a modern university is about the worst thing you can say about something. No one can make sense of the controversies surrounding mind, brain, genes, and evolution without understanding their alignment with ancient political fault lines. E. O. Wilson learned this too late:

> I had been blindsided by the attack [on *Sociobiology*]. Having expected some frontal fire from social scientists on primarily evidential grounds, I had received instead a political enfilade from the flank. A few observers were surprised that I was surprised. John Maynard Smith, a senior British evolutionary biologist and former Marxist, said that he disliked the last chapter of *Sociobiology* himself and "it was also absolutely obvious to me—I cannot believe Wilson didn't know—that this was going to provoke great hostility from American Marxists, and Marxists everywhere." But it was true. . . . In 1975 I was a political naïf: I knew almost nothing about Marxism as either a political belief or a mode of analysis, I had paid little attention to the dynamism of the activist left, and I had never heard of Science for the People. I was not even an intellectual in the European or New York–Cambridge sense.[4]

As we shall see, the new sciences of human nature really do resonate with assumptions that historically were closer to the right than to the left. But today the alignments are not as predictable. The accusation that these sciences are irredeemably conservative comes from the Left Pole, the mythical place from which all directions are right. The political associations of a belief in human nature now crosscut the liberal-conservative dimension, and many political theorists invoke evolution and genetics to argue for policies on the left.

～

THE SCIENCES OF human nature are pressing on two political hot buttons, not just one. The first is how we conceptualize the entity known as "society." The political philosopher Roger Masters has shown how sociobiology (and related theories invoking evolution, genetics, and brain science) inadvertently took sides in an ancient dispute between two traditions of understanding the social order.[5]

In the *sociological* tradition, a society is a cohesive organic entity and its individual citizens are mere parts. People are thought to be social by their very nature and to function as constituents of a larger superorganism. This is the tradition of Plato, Hegel, Marx, Durkheim, Weber, Kroeber, the sociologist

Talcott Parsons, the anthropologist Claude Lévi-Strauss, and postmodernism in the humanities and social sciences.

In the *economic* or *social contract* tradition, society is an arrangement negotiated by rational, self-interested individuals. Society emerges when people agree to sacrifice some of their autonomy in exchange for security from the depredations of others wielding *their* autonomy. It is the tradition of Thrasymachus in Plato's *Republic,* and of Machiavelli, Hobbes, Locke, Rousseau, Smith, and Bentham. In the twentieth century it became the basis for the rational actor or "economic man" models in economics and political science, and for cost-benefit analyses of public choices.

The modern theory of evolution falls smack into the social contract tradition. It maintains that complex adaptations, including behavioral strategies, evolved to benefit the individual (indeed, the genes for those traits within an individual), not the community, species, or ecosystem.[6] Social organization evolves when the long-term benefits to the individual outweigh the immediate costs. Darwin was influenced by Adam Smith, and many of his successors analyze the evolution of sociality using tools that come right out of economics, such as game theory and other optimization techniques.

Reciprocal altruism, in particular, is just the traditional concept of the social contract restated in biological terms. Of course, humans were never solitary (as Rousseau and Hobbes incorrectly surmised), and they did not inaugurate group living by haggling over a contract at a particular time and place. Bands, clans, tribes, and other social groups are central to human existence and have been so for as long as we have been a species. But the *logic* of social contracts may have propelled the evolution of the mental faculties that keep us in these groups. Social arrangements are evolutionarily contingent, arising when the benefits of group living exceed the costs.[7] With a slightly different ecosystem and evolutionary history, we could have ended up like our cousins the orangutans, who are almost entirely solitary. And according to evolutionary biology, all societies—animal and human—seethe with conflicts of interest and are held together by shifting mixtures of dominance and cooperation.

Throughout the book we have seen how the sciences of human nature have clashed with the sociological tradition. The social sciences were taken over by the doctrine that social facts live in their own universe, separate from the universe of individual minds. In Chapter 4 we saw an alternative conception in which cultures and societies arise from individual people pooling their discoveries and negotiating the tacit agreements that underlie social reality. We saw how a departure from the sociological paradigm was a major heresy of Wilson's *Sociobiology,* and that the primacy of society was a foundation of Marxism and played a role in its disdain for the interests of individual people.

The division between the sociological and economic traditions is aligned with the division between the political left and the political right, but only roughly. Marxism is obviously in the sociological tradition, and free-market conservatism is obviously in the economic tradition. In the liberal 1960s, Lyndon Johnson wanted to forge a Great Society, Pierre Trudeau a Just Society. In the conservative 1980s, Margaret Thatcher said, "There is no such thing as society. There are individual men and women, and there are families."

But as Masters points out, Durkheim and Parsons were in the sociological tradition, yet they were conservatives. One can easily see how conservative beliefs can favor the preservation of society as an entity and thereby downplay the desires of individuals. Conversely, Locke was in the social contract tradition, but he is a patron saint of liberalism, and Rousseau, who *coined* the expression "social contract," was an inspiration for liberal and revolutionary thinkers. Social contracts, like any contract, can become unfair to some of the signatories, and may have to be renegotiated progressively or redrawn from scratch in a revolution.

So the clash between the sociological and economic traditions can explain some of the heat ignited by the sciences of human nature, but it is not identical to the firefight between the political left and the political right. The rest of the chapter will scrutinize that second and hotter button.

~

THE RIGHT-LEFT AXIS aligns an astonishing collection of beliefs that at first glance seem to have nothing in common. If you learn that someone is in favor of a strong military, for example, it is a good bet that the person is also in favor of judicial restraint rather than judicial activism. If someone believes in the importance of religion, chances are she will be tough on crime and in favor of lower taxes. Proponents of a laissez-faire economic policy tend to value patriotism and the family, and they are more likely to be old than young, pragmatic than idealistic, censorious than permissive, meritocratic than egalitarian, gradualist than revolutionary, and in a business rather than a university or government agency. The opposing positions cluster just as reliably: if someone is sympathetic to rehabilitating offenders, or to affirmative action, or to generous welfare programs, or to a tolerance of homosexuality, chances are good that he will also be a pacifist, an environmentalist, an activist, an egalitarian, a secularist, and a professor or student.

Why on earth should people's beliefs about sex predict their beliefs about the size of the military? What does religion have to do with taxes? Whence the linkage between strict construction of the Constitution and disdain for shocking art? Before we can understand why beliefs about an innate human nature might cluster with liberal beliefs or with conservative beliefs, we have to understand why liberal beliefs cluster with other liberal beliefs and conservative beliefs cluster with other conservative beliefs.

The meanings of the words are of no help. Marxists in the Soviet Union and its aftermath were called conservatives; Reagan and Thatcher were called revolutionaries. Liberals are liberal about sexual behavior but not about business practices; conservatives want to conserve communities and traditions, but they also favor the free market economy that subverts them. People who call themselves "classical liberals" are likely to be called "conservatives" by adherents of the version of leftism known as political correctness.

Nor can most contemporary liberals and conservatives articulate the cores of their belief systems. Liberals think that conservatives are just amoral plutocrats, and conservatives think that if you are not a liberal before you are twenty you have no heart but if you are a liberal after you are twenty you have no brain (attributed variously to Georges Clemenceau, Dean Inge, Benjamin Disraeli, and Maurice Maeterlinck). Strategic alliances—such as the religious fundamentalists and free-market technocrats on the right, or the identity politicians and civil libertarians on the left—may frustrate the search for any intellectual common denominator. Everyday political debates, such as whether tax rates should be exactly what they are or a few points higher or lower, are just as uninformative.

The most sweeping attempt to survey the underlying dimension is Thomas Sowell's *A Conflict of Visions*.[8] Not every ideological struggle fits his scheme, but as we say in social science, he has identified a factor that can account for a large proportion of the variance. Sowell explains two "visions" of the nature of human beings that were expressed in their purest forms by Edmund Burke (1729–1797), the patron of secular conservatism, and William Godwin (1756–1836), the British counterpart to Rousseau. In earlier times they might have been referred to as different visions of the perfectibility of man. Sowell calls them the Constrained Vision and the Unconstrained Vision; I will refer to them as the Tragic Vision (a term he uses in a later book) and the Utopian Vision.[9]

In the Tragic Vision, humans are inherently limited in knowledge, wisdom, and virtue, and all social arrangements must acknowledge those limits. "Mortal things suit mortals best," wrote Pindar; "from the crooked timber of humanity no truly straight thing can be made," wrote Kant. The Tragic Vision is associated with Hobbes, Burke, Smith, Alexander Hamilton, James Madison, the jurist Oliver Wendell Holmes Jr., the economists Friedrich Hayek and Milton Friedman, the philosophers Isaiah Berlin and Karl Popper, and the legal scholar Richard Posner.

In the Utopian Vision, psychological limitations are artifacts that *come from* our social arrangements, and we should not allow them to restrict our gaze from what is possible in a better world. Its creed might be "Some people see things as they are and ask 'why?'; I dream things that never were and ask 'why not?'" The quotation is often attributed to the icon of 1960s liberalism, Robert F. Kennedy, but it was originally penned by the Fabian socialist George

Bernard Shaw (who also wrote, "There is nothing that can be changed more completely than human nature when the job is taken in hand early enough").[10] The Utopian Vision is also associated with Rousseau, Godwin, Condorcet, Thomas Paine, the jurist Earl Warren, the economist John Kenneth Galbraith, and to a lesser extent the political philosopher Ronald Dworkin.

In the Tragic Vision, our moral sentiments, no matter how beneficent, overlie a deeper bedrock of selfishness. That selfishness is not the cruelty or aggression of the psychopath, but a concern for our well-being that is so much a part of our makeup that we seldom reflect on it and would waste our time lamenting it or trying to erase it. In his book *The Theory of Moral Sentiments*, Adam Smith remarked:

> Let us suppose that the great empire of China, with all its myriads of in-habitants, was suddenly swallowed up by an earthquake, and let us con-sider how a man of humanity in Europe, who had no sort of connection with that part of the world, would react upon receiving intelligence of this dreadful calamity. He would, I imagine, first of all express very strongly his sorrow for the misfortune of that unhappy people, he would make many melancholy reflections upon the precariousness of human life, and the vanity of all the labours of man, which could thus be anni-hilated in a moment. He would, too, perhaps, if he was a man of specu-lation, enter into many reasonings concerning the effects which this disaster might produce upon the commerce of Europe, and the trade and business of the world in general. And when all this fine philosophy was over, when all these humane sentiments had been once fairly ex-pressed, he would pursue his business or his pleasure, take his repose or his diversion, with the same ease and tranquillity as if no such accident had happened. The most frivolous disaster which could befall himself would occasion a more real disturbance. If he was to lose his little finger tomorrow, he would not sleep to-night; but provided he never saw them, he would snore with the most profound security over the ruin of a hundred million of his brethren.[11]

In the Tragic Vision, moreover, human nature has not changed. Traditions such as religion, the family, social customs, sexual mores, and political institu-tions are a distillation of time-tested techniques that let us work around the shortcomings of human nature. They are as applicable to humans today as they were when they developed, even if no one today can explain their ratio-nale. However imperfect society may be, we should measure it against the cru-elty and deprivation of the actual past, not the harmony and affluence of an imagined future. We are fortunate enough to live in a society that more or less works, and our first priority should be not to screw it up, because human nature always leaves us teetering on the brink of barbarism. And since no one is smart enough to predict the behavior of a single human being, let alone millions of them interacting in a society, we should distrust any formula for changing society from the top down, because it is likely to have unintended consequences that are worse than the problems it was designed to fix. The best we can hope for are incremental changes that are continuously adjusted according to feedback about the sum of their good and bad consequences. It also follows that we should not aim to *solve* social problems like crime or poverty, because in a world of competing individuals one person's gain may be another person's loss. The best we can do is trade off one cost against another. In Burke's famous words, written in the aftermath of the French Revolution:

> [One] should approach to the faults of the state as to the wounds of a father, with pious awe and trembling solicitude. By this wise prejudice we are taught to look with horror on those children of their country who are prompt rashly to hack that aged parent in pieces, and put him into the kettle of magicians, in hopes that by their poisonous weeds, and wild incantations, they may regenerate the paternal constitution, and renovate their father's life.[12]

In the Utopian Vision, human nature changes with social circumstances, so traditional institutions have no inherent value. That was then, this is now. Traditions are the dead hand of the past, the attempt to rule from the grave. They must be stated explicitly so their rationale can be scrutinized and their moral status evaluated. And by that test, many traditions fail: the confinement of women to the home, the stigma against homosexuality and premarital sex, the superstitions of religion, the injustice of apartheid and segregation, the dangers of patriotism as exemplified in the mindless slogan "My country, right or wrong." Practices such as absolute monarchy, slavery, war, and patriarchy once seemed inevitable but have disappeared or faded from many parts of the world through changes in institutions that were once thought to be rooted in human nature. Moreover, the existence of suffering and injustice presents us with an undeniable moral imperative. We don't know what we can achieve until we try, and the alternative, resigning ourselves to these evils as the way of the world, is unconscionable. At Robert Kennedy's funeral, his brother Edward quoted from one of his recent speeches:

> All of us will ultimately be judged and as the years pass we will surely judge ourselves, on the effort we have contributed to building a new world society and the extent to which our ideals and goals have shaped that effort.
> The future does not belong to those who are content with today,

apathetic toward common problems and their fellow man alike, timid and fearful in the face of new ideas and bold projects. Rather it will belong to those who can blend vision, reason and courage in a personal commitment to the ideals and great enterprises of American Society.

Our future may lie beyond our vision, but it is not completely beyond our control. It is the shaping impulse of America that neither fate nor nature nor the irresistible tides of history, but the work of our own hands, matched to reason and principle, will determine our destiny. There is pride in that, even arrogance, but there is also experience and truth. In any event, it is the only way we can live.[13]

Those with the Tragic Vision are unmoved by ringing declarations attributed to the first-person plural *we, our,* and *us*. They are more likely to use the pronouns as the cartoon possum Pogo did: We have met the enemy, and he is us. We are all members of the same flawed species. Putting our moral vision into practice means imposing our will on others. The human lust for power and esteem, coupled with its vulnerability to self-deception and self-righteousness, makes that an invitation to a calamity, all the worse when that power is directed at a goal as quixotic as eradicating human self-interest. As the conservative philosopher Michael Oakshott wrote, "To try to do something which is inherently impossible is always a corrupting enterprise."

The two kinds of visionaries thereby line up on opposite sides of many issues that would seem to have little in common. The Utopian Vision seeks to articulate social goals and devise policies that target them directly: economic inequality is attacked in a war on poverty, pollution by environmental regulations, racial imbalances by preferences, carcinogens by bans on food additives. The Tragic Vision points to the self-interested motives of the people who would implement these policies—namely, the expansion of their bureaucratic fiefdoms—and to their ineptitude at anticipating the myriad consequences, especially when the social goals are pitted against millions of people pursuing their own interests. Thus, say the Tragic Visionaries, the Utopians fail to anticipate that welfare might encourage dependency, or that a restriction on one pollutant might force people to use another.

Instead, the Tragic Vision looks to systems that produce desirable outcomes even when no member of the system is particularly wise or virtuous. Market economies, in this vision, accomplish that goal: remember Smith's butcher, brewer, and baker providing us with dinner out of self-interest rather than benevolence. No mastermind has to understand the intricate flow of goods and services that make up an economy in order to anticipate who needs what, and when and where. Property rights give people an incentive to work and produce; contracts allow them to enjoy gains in trade. Prices convey information about scarcity and demand to producers and consumers, so they can react by following a few simple rules—make more of what is profitable, buy less of what is expensive—and the "invisible hand" will do the rest. The intelligence of the system is distributed across millions of not-necessarily-intelligent producers and consumers, and cannot be articulated by anyone in particular.

People with the Utopian Vision point to market failures that can result from having a blind faith in free markets. They also call attention to the unjust distribution of wealth that tends to be produced by free markets. Opponents with the Tragic Vision argue that the notion of justice makes sense only when applied to human decisions within a framework of laws, not when applied to an abstraction called "society." Friedrich Hayek wrote, "The manner in which the benefits and burdens are apportioned by the market mechanism would in many instances have to be regarded as very unjust *if* it were the result of a deliberate allocation to particular people." But that concern with social justice rests on a confusion, he claimed, because "the particulars of [a spontaneous order] cannot be just or unjust."[14]

Some of today's battles between left and right fall directly out of these different philosophies: big versus small government, high versus low taxes, protectionism versus free trade, measures that aim to reduce undesirable outcomes (poverty, inequality, racial imbalance) versus measures that merely level the playing field and enforce the rules. Other battles follow in a less obvious way from the opposing visions of human potential. The Tragic Vision stresses fiduciary duties, even when the person executing them cannot see their immediate value, because they allow imperfect beings who cannot be sure of their virtue or foresight to participate in a tested system. The Utopian Vision stresses social responsibility, where people hold their actions to a higher ethical standard. In Lawrence Kohlberg's famous theory of moral development, a willingness to ignore rules in favor of abstract principles was literally identified as a "higher stage" (which, perhaps tellingly, most people never reach).

The most obvious example is the debate on strict constructionism and judicial restraint on one side and judicial activism in pursuit of social justice on the other. Earl Warren, the chief justice of the U.S. Supreme Court from 1954 to 1969, was the prototypical judicial activist, who led the court to implement desegregation and expand the rights of the accused. He was known for interrupting lawyers in mid-argument by asking, "Is it *right?* Is it *good?*" The opposing view was stated by Oliver Wendell Holmes, who said his job was "to see that the game is played according to the rules whether I like them or not." He conceded that "to improve conditions of life and the race is the main thing," and added, "But how the devil can I tell whether I am not pulling it down more in some other place?"[15] Those with the Tragic Vision see judicial activism as an invitation to egotism and caprice and as unfair to those who have played by the rules as they were publicly stated. Those with the Utopian Vision see judicial restraint as the mindless preservation of arbitrary injustices—as Dickens's Mr.

Bumble put it, "The law is an ass."[16] An infamous example is the Dred Scott decision of 1856, in which the Supreme Court ruled on narrow legalistic grounds that a freed slave could not sue to make his freedom official and that Congress could not prohibit slavery in federal territories.

Radical political reform, like radical judicial reform, will be more or less appealing depending on one's confidence in human intelligence and wisdom. In the Utopian Vision, solutions to social problems are readily available. Speaking in 1967 about the conditions that breed violence, Lyndon Johnson said, "All of us know what those conditions are: ignorance, discrimination, slums, poverty, disease, not enough jobs."[17] If we already know the solutions, all we have to do is choose to implement them, and that requires only sincerity and dedication. By the same logic, anyone opposing the solutions must be motivated by blindness, dishonesty, and callousness. Those with the Tragic Vision say instead that solutions to social problems are elusive. The inherent conflicts of interest among people leave us with few options, all of them imperfect. Opponents of radical reform are showing a wise distrust of human hubris.

The political orientation of the universities is another manifestation of conflicting visions of human potential. Adherents of the Tragic Vision distrust knowledge stated in explicitly articulated and verbally justified propositions, which is the stock-in-trade of academics, pundits, and policy analysts. Instead they trust knowledge that is distributed diffusely throughout a system (such as a market economy or set of social mores) and which is tuned by adjustments by many simple agents using feedback from the world. (Cognitive scientists will be reminded of the distinction between symbolic representations and distributed neural networks, and that is no coincidence: Hayek, the foremost advocate of distributed intelligence in societies, was an early neural network modeler.)[18] For much of the twentieth century, political conservatism had an anti-intellectual streak, until conservatives decided to play catch-up in the battle for hearts and minds and funded policy think tanks as a counterweight to universities.

Finally, the disagreements on crime and war fall right out of the conflicting theories of human nature. Given the obvious waste and cruelty of war, those with the Utopian Vision see it as a kind of pathology that arises from misunderstandings, shortsightedness, and irrational passions. War is to be prevented by public expressions of pacifist sentiments, better communication between potential enemies, less saber-rattling rhetoric, fewer weapons and military alliances, a de-emphasis on patriotism, and negotiating to avert war at any cost. Adherents of the Tragic Vision, with their cynical view of human nature, see war as a rational and tempting strategy for people who think they can gain something for themselves or their nation. The calculations might be mistaken in any instance, and they may be morally deplorable because they give no weight to the suffering of the losers, but they are not literally pathological or irrational. On this view the only way to ensure peace is to raise the cost of war to potential aggressors by developing weaponry, arousing patriotism, rewarding bravery, flaunting one's might and resolve, and negotiating from strength to deter blackmail.

The same arguments divide the visions on crime. Those with the Utopian Vision see crime as inherently irrational and seek to prevent it by identifying the root causes. Those with the Tragic Vision see crime as inherently rational and believe that the root cause is all too obvious: people rob banks because that's where the money is. The most effective crime-prevention programs, they say, strike directly at the rational incentives. A high probability of unpleasant punishment raises the anticipated cost of crime. A public emphasis on personal responsibility helps enforce the incentives by closing any loopholes left open by the law. And strict parenting practices allow children to internalize these contingencies early in life.[19]

~

AND ONTO THIS battlefield strode an innocent E. O. Wilson. The ideas from evolutionary biology and behavioral genetics that became public in the 1970s could not have been more of an insult to those with the Utopian Vision. That vision was, after all, based on the Blank Slate (no permanent human nature), the Noble Savage (no selfish or evil instincts), and the Ghost in the Machine (an unfettered "we" that can choose better social arrangements). And here were scientists talking about *selfish genes!* And saying that adaptations are not for the good of the species but for the good of individuals and their kin (as if to vindicate Thatcher's claim that "there is no such thing as society"). That people scrimp on altruism because it is vulnerable to cheaters. That in pre-state societies men go to war even when they are well fed, because status and women are permanent Darwinian incentives. That the moral sense is riddled with biases, including a tendency to self-deception. And that conflicts of genetic interest are built in to social animals and leave us in a state of permanent tragedy. It looked as if the scientists were saying to proponents of the Tragic Vision: You're right, they're wrong.

The Utopians, particularly those in the radical science movement, replied that current findings on human intelligence and motivation are irrelevant. They can tell us only about what we have achieved in today's society, not what we might achieve in tomorrow's. Since we know that social arrangements can change if we decide to change them, any scientist who speaks of constraints on human nature must *want* oppression and injustice to continue.

My own view is that the new sciences of human nature really do vindicate some version of the Tragic Vision and undermine the Utopian outlook that until recently dominated large segments of intellectual life. The sciences say nothing, of course, about differences in values that are associated with particular right-wing and left-wing positions (such as in the tradeoffs between unemployment and environmental protection, diversity and economic efficiency, or individual freedom and community cohesion). Nor do they speak directly to policies that are based on a complex mixture of assumptions about the world. But they do speak to the parts of the visions that are general claims about how the mind works. Those claims may be evaluated against the facts, just like any empirical hypothesis. The Utopian vision that human nature might radically change in some imagined society of the remote future is, of course, literally unfalsifiable, but I think that many of the discoveries recounted in preceding chapters make it unlikely. Among them I would include the following:

- The primacy of family ties in all human societies and the consequent appeal of nepotism and inheritance.[20]
- The limited scope of communal sharing in human groups, the more common ethos of reciprocity, and the resulting phenomena of social loafing and the collapse of contributions to public goods when reciprocity cannot be implemented.[21]
- The universality of dominance and violence across human societies (including supposedly peaceable hunter-gatherers) and the existence of genetic and neurological mechanisms that underlie it.[22]
- The universality of ethnocentrism and other forms of group-against-group hostility across societies, and the ease with which such hostility can be aroused in people within our own society.[23]
- The partial heritability of intelligence, conscientiousness, and antisocial tendencies, implying that some degree of inequality will arise even in perfectly fair economic systems, and that we therefore face an inherent trade-off between equality and freedom.[24]
- The prevalence of defense mechanisms, self-serving biases, and cognitive dissonance reduction, by which people deceive themselves about their autonomy, wisdom, and integrity.[25]
- The biases of the human moral sense, including a preference for kin and friends, a susceptibility to a taboo mentality, and a tendency to confuse morality with conformity, rank, cleanliness, and beauty.[26]

It is not just conventional scientific data that tell us the mind is not infinitely malleable. I think it is no coincidence that beliefs that were common among intellectuals in the 1960s—that democracies are obsolete, revolution is desirable, the police and armed forces dispensable, and society designable from the top down—are now rarer. The Tragic Vision and the Utopian Vision inspired historical events whose interpretations are much clearer than they were just a few decades ago. Those events can serve as additional data to test the visions' claims about human psychology.

The visions contrast most sharply in the political revolutions they spawned. The first revolution with a Utopian Vision was the French Revolution—recall Wordsworth's description of the times, with "human nature seeming born again." The revolution overthrew the ancien régime and sought to begin from scratch with the ideals of liberty, equality, and fraternity and a belief that salvation would come from vesting authority in a morally superior breed of leaders. The revolution, of course, sent one leader after another to the guillotine as each failed to measure up to usurpers who felt they had a stronger claim to wisdom and virtue. No political structure survived the turnover of personnel, leaving a vacuum that would be filled by Napoleon. The Russian Revolution was also animated by the Utopian Vision, and it also burned through a succession of leaders before settling into the personality cult of Stalin. The Chinese Revolution, too, put its faith in the benevolence and wisdom of a man who displayed, if anything, a particularly strong dose of human foibles like dominance, lust, and self-deception. The perennial limitations of human nature prove the futility of political revolutions based only on the moral aspirations of the revolutionaries. In the words of the song about revolution by The Who: Meet the new boss; same as the old boss.

Sowell points out that Marxism is a hybrid of the two visions.[27] It invokes the Tragic Vision to interpret the past, when earlier modes of production left no choice but the forms of social organization known as feudalism and capitalism. But it invokes a Utopian Vision for the future, in which we can shape our nature in dialectical interaction with the material and social environment. In that new world, people will be motivated by self-actualization rather than self-interest, allowing us to realize the ideal, "From each according to his abilities, to each according to his needs." Marx wrote that a communist society would be

> the genuine resolution of the antagonism between man and nature and between man and man; it is the true resolution of the conflict between existence and essence, objectification and self-affirmation, freedom and necessity, individual and species. It is the riddle of history solved.[28]

It doesn't get any less tragic or more utopian than that. Marx dismissed the worry that selfishness and dominance would corrupt those carrying out the general will. For example, he waved off the anarchist Mikhail Bakunin's fear that the workers in charge would become despotic: "If Mr. Bakunin were familiar just with the position of a manager in a workers' cooperative, he could send all his nightmares about authority to the devil."[29]

In the heyday of radical science, any proposal about human nature that conflicted with the Marxist vision was dismissed as self-evidently wrong. But history is a kind of experiment, albeit an imperfectly controlled one, and its data suggest that it was the radical assessment that got it wrong. Marxism is now almost universally recognized as an experiment that failed, at least in its worldly implementations.[30] The nations that adopted it either collapsed, gave it up, or languish in backward dictatorships. As we saw in earlier chapters, the ambition to remake human nature turned its leaders into totalitarian despots and mass murderers. And the assumption that central planners were morally disinterested and cognitively competent enough to direct an entire economy led to comical inefficiencies with serious consequences. Even the more humane forms of European socialism have been watered down to the point where so-called Communist Parties have platforms that not long ago would have been called reactionary. Wilson, the world's expert on ants, may have had the last laugh in his verdict on Marxism: "Wonderful theory. Wrong species."[31]

∼

"TWO CHEERS FOR democracy," proclaimed E. M. Forster. "Democracy is the worst form of government except all those other forms that have been tried," said Winston Churchill. These are encomiums worthy of the Tragic Vision. For all their flaws, liberal democracies appear to be the best form of large-scale social organization our sorry species has come up with so far. They provide more comfort and freedom, more artistic and scientific vitality, longer and safer lives, and less disease and pollution than any of the alternatives. Modern democracies never have famines, almost never wage war on one another, and are the top choice of people all over the world who vote with their feet or with their boats. The moderate success of democracies, like the failures of radical revolutions and of Marxist governments, is now widely enough agreed upon that it may serve as another empirical test for rival theories of human nature.

The modern concept of democracy emerged in seventeenth- and eighteenth-century England and was refined in the frenzy of theorizing that surrounded the American independence movement. It is no coincidence that the major theoreticians of the social contract, such as Hobbes, Locke, and Hume, were also major armchair psychologists. As Madison wrote, "What is government itself but the greatest of all reflections on human nature?"[32]

The brains behind the American Revolution (which is sometimes labeled with the oxymoron "conservative revolution") inherited the tragic vision of thinkers like Hobbes and Hume.[33] (Significantly, the founders appear not to have been influenced by Rousseau at all, and the popular belief that they got the idea of democracy from the Iroquois Federation is just 1960s granola.)[34] The legal scholar John McGinnis has argued that their theory of human nature could have come right out of modern evolutionary psychology.[35] It acknowledges the desire of individuals to further their interests in the form of an inalienable right to "life, liberty, and the pursuit of happiness." The state emerges from an agreement instituted to protect those rights, rather than being the embodiment of an autonomous superorganism. Rights need to be protected because when people live together their different talents and circumstances will lead some of them to possess things that others want. ("Men have different and unequal faculties for acquiring property," noted Madison.)[36] There are two ways to get something you want from other people: steal it or trade for it. The first involves the psychology of dominance; the second, the psychology of reciprocal altruism. The goal of a peaceful and prosperous society is to minimize the use of dominance, which leads to violence and waste, and to maximize the use of reciprocity, which leads to gains in trade that make everyone better off.

The Constitution, McGinnis shows, was consciously designed to implement these goals. It encouraged reciprocal exchanges through the Commerce Clause, which authorized Congress to remove barriers to trade imposed by the states. It protected them from the danger of cheaters through the Contracts Clause, which prevented states from impairing the enforcement of contracts. And it precluded rulers from confiscating the fruits of the more productive citizens via the Takings Clause, which forbids the government to expropriate private property without compensation.

The feature of human nature that most impressed the framers was the drive for dominance and esteem, which, they feared, imperils all forms of government. Someone must be empowered to make decisions and enforce laws, and that someone is inherently vulnerable to corruption. How to anticipate and limit that corruption became an obsession of the framers. John Adams wrote, "The desire for the esteem of others is as real a want of *nature* as hunger. It is the principal end of government to regulate this passion."[37] Alexander Hamilton wrote, "The love of fame [is] the ruling passion of the noblest minds."[38] James Madison wrote, "If men were angels, no government would be necessary. If angels were to govern men, neither external nor internal controls on government would be necessary."[39]

So external and internal controls there would be. "Parchment barriers," said Madison, were not enough; rather, "ambition must be made to counteract ambition."[40] Checks and balances were instituted to stalemate any faction that grew too powerful. They included the division of authority between federal and state governments, the separation of powers among the executive, legislative, and judiciary branches, and the splitting of the legislative branch into two houses.

Madison was especially adamant that the Constitution rein in the part of human nature that encourages war, which is not a primitive lust for blood, he claimed, but an advanced lust for esteem:

> War is in fact the true nurse of executive aggrandizement. In war a physical force is created, and it is the executive will to direct it. In war the public treasures are to be unlocked, and it is the executive hand which is to dispense them. In war the honors and emoluments of office are to be

multiplied; and it is the executive patronage under which they are to be enjoyed. It is in war finally that laurels are to be gathered, and it is the executive brow they are to encircle. The strongest passions and the most dangerous weakness of the human breast—ambition, avarice, vanity, the honorable or venial love of fame—are all in conspiracy against the desire and duty of peace.[41]

This inspired the War Powers Clause, which gave Congress, not the president, the power to declare war. (It was infamously circumvented in the years of the Vietnam conflict, during which Johnson and Nixon never formally declared a state of war.)

McGinnis notes that even the freedoms of speech, assembly, and the press were motivated by features of human nature. The framers justified them as means of preventing tyranny: a network of freely communicating citizens can counteract the might of the individuals in government. As we now say, they can "speak truth to power." The dynamic of power sharing protected by these rights might go way back in evolutionary history. The primatologists Frans de Waal, Robin Dunbar, and Christopher Boehm have shown how a coalition of lower-ranking primates can depose a single alpha male.[42] Like McGinnis, they suggest this may be a crude analogue of political democracy.

None of this means that the American Constitution was a guarantee of a happy and moral society, of course. By working within the glaringly undersized moral circle of the day, the Constitution failed to stand in the way of the genocide of native peoples, the slavery and segregation of African Americans, and the disenfranchisement of women. It said little about the conduct of foreign affairs, which (except with regard to strategic allies) has generally been guided by a cynical realpolitik. The first failing has been addressed by explicit measures to expand the legal circle, such as the Equal Protection clause of the Fourteenth Amendment; the second is unsolved and perhaps unsolvable, because other countries are necessarily outside any circle delineated by a national document. The Constitution also lacked any principled compassion for those at the bottom of the meritocracy, assuming that equality of opportunity was the only mechanism needed to address the distribution of wealth. And it is incapable of stipulating the suite of values and customs that appear to be necessary for a democracy to function in practice.

Acknowledging the relative success of constitutional democracy does not require one to be a flag-waving patriot. But it does suggest that something may have been right about the theory of human nature that guided its architects.

～

The left needs a new paradigm.

—Peter Singer, *A Darwinian Left* (1999)[43]

Conservatives need Charles Darwin.
  —Larry Arnhart, "Conservatives, Design, and Darwin" (2000)[44]

What's going on? That voices of the contemporary left and the contemporary right are both embracing evolutionary psychology after decades of reviling it shows two things. One is that biological facts are beginning to box in plausible political philosophies. The belief on the left that human nature can be changed at will, and the belief on the right that morality rests on God's endowing us with an immaterial soul, are becoming rearguard struggles against the juggernaut of science. A popular bumper sticker in the 1990s urged, QUESTION AUTHORITY. Another bumper sticker replied, QUESTION GRAVITY. All political philosophies have to decide when their arguments are turning into the questioning of gravity.

The second development is that an acknowledgment of human nature can no longer be associated with the political right. Once the Utopian Vision is laid to rest, the field of political positions is wide open. The Tragic Vision, after all, has not been vindicated in anything like its most lugubrious form. For all its selfishness, the human mind is equipped with a moral sense, whose circle of application has expanded steadily and might continue to expand as more of the world becomes interdependent. And for all its limitations, human cognition is an open-ended combinatorial system, which in principle can increase its mastery over human affairs, just as it has increased its mastery of the physical and living worlds.

Traditions, for their part, are adapted not to human nature alone but to human nature in the context of an infrastructure of technology and economic exchange (one does not have to be a Marxist to accept this insight from Marx). Some traditional institutions, like families and the rule of law, may be adapted to eternal features of human psychology. Others, such as primogeniture, were obviously adapted to the demands of a feudal system that required keeping the family lands intact, and became obsolete when the economic system changed in the wake of industrialization. More recently, feminism was in part a response to improved reproductive technologies and the shift to a service economy. Because social conventions are not adapted to human nature alone, a respect for human nature does not require preserving all of them.

For these reasons I think political beliefs will increasingly cut across the centuries-old divide between the Tragic and Utopian Visions. They will diverge by invoking different aspects of human nature, by giving different weightings to conflicting goals, or by offering different assessments of the likely outcomes of particular courses of action.

I end the chapter with a tour of some thinkers on the left who are scrambling the traditional alignment between human nature and right-wing politics. As its title suggests, *A Darwinian Left* is the most systematic attempt to map out the new alignment.[45] Singer writes, "It is time for the left to take seriously the fact that we are evolved animals, and that we bear the evidence of our inheritance, not only in our anatomy and our DNA, but in our behavior too."[46] For Singer this means acknowledging the limits of human nature, which makes the perfectibility of humankind an impossible goal. And it means acknowledging specific components of human nature. They include self-interest, which implies that competitive economic systems will work better than state monopolies; the drive for dominance, which makes powerful governments vulnerable to overweening autocrats; ethnocentrism, which puts nationalist movements at risk of committing discrimination and genocide; and differences between the sexes, which should temper measures for rigid gender parity in all walks of life.

So what's left of the left? an observer might ask. Singer replies, "If we shrug our shoulders at the avoidable suffering of the weak and the poor, of those who are getting exploited and ripped off, or who simply do not have enough to sustain life at a decent level, we are not of the left. If we say that that is just the way the world is, and always will be, and there is nothing we can do about it, we are not part of the left. The left wants to do something about this situation."[47] Singer's leftism, like traditional leftism, is defined by a contrast with a defeatist Tragic Vision. But its goal—"doing something"—has been downsized considerably from Robert Kennedy's goal in the 1960s of "building a new world society."

The Darwinian left has ranged from vague expressions of values to wonkish policy initiatives. We have already met two theoreticians at the vaguer end. Chomsky has been the most vocal defender of an innate cognitive endowment since he nailed his thesis of an inborn language faculty to the behaviorists' door in the late 1950s. He has also been a fierce left-wing critic of American society and has recently inspired a whole new generation of campus radicals (as we saw in his interview with Rage Against the Machine). Chomsky insists that the connections between his science and his politics are slender but real:

> A vision of a future social order is . . . based on a concept of human nature. If, in fact, man is an indefinitely malleable, completely plastic being, with no innate structures of mind and no intrinsic needs of a cultural or social character, then he is a fit subject for the "shaping of behavior" by the State authority, the corporate manager, the technocrat, or the central committee. Those with some confidence in the human species will hope this is not so and will try to determine the intrinsic characteristics that provide the framework for intellectual development, the growth of moral consciousness, cultural achievement, and participation in a free community.[48]

He describes his political vision as "libertarian socialist" and "anarcho-syndi-calist," the kind of anarchism that values spontaneous cooperation (as opposed to anarcho-capitalism, the kind that values individualism).[49] This vision, he suggests, lies in a Cartesian tradition that includes "Rousseau's opposition to tyranny, oppression, and established authority, . . . Kant's defense of freedom, Humboldt's precapitalist liberalism with its emphasis on the basic human need for free creation under conditions of voluntary association, and Marx's critique of alienated fragmented labor that turns men into machines, depriving them of their 'species character' of 'free conscious activity' and 'productive life' in asso-ciation with their fellows."[50] Chomsky's political beliefs, then, resonate with his scientific belief that humans are innately endowed with a desire for community and a drive for creative free expression, language being the paradigm example. That holds out the hope for a society organized by cooperation and natural productivity rather than by hierarchical control and the profit motive.

Chomsky's theory of human nature, though strongly innatist, is innocent of modern evolutionary biology, with its demonstration of ubiquitous con-flicts of genetic interest. These conflicts lead to a darker view of human nature, one that has always been a headache for those with anarchist dreams. But the thinker who first elucidated these conflicts, Robert Trivers, was a left-wing radical as well, and one of the rare *white* Black Panthers. As we saw in Chapter 6, Trivers viewed sociobiology as a subversive discipline. A sensitivity to con-flicts of interest can illuminate the interests of repressed agents, such as women and younger generations, and it can expose the deception and self-deception that elites use to justify their dominance.[51] In that way sociobiology follows in the liberal tradition of Locke by using science and reason to debunk the rationalizations of rulers. Reason was used in Locke's time to question the divine right of kings, and may be used in our time to question the pretension that current political arrangements serve everyone's interests.

Though it may come as a shock to many people, the use of IQ tests and a recognition of innate differences in intelligence can support—and in the past did support—left-wing political goals. In his article "Bell Curve Liberals," the journalist Adrian Wooldridge points out that IQ testing was welcomed by the British left as the ultimate subverter of a caste society ruled by inbred upper-class twits.[52] Together with other liberals and socialists, Sidney and Beatrice Webb hoped to turn the educational system into a "capacity-catching ma-chine" that could "rescue talented poverty from the shop or the plough" and direct them into the ruling elite. They were opposed by conservatives such as T. S. Eliot, who worried that a system that sorts people by ability would disor-ganize civil society by breaking the bonds of class and tradition at both ends of the ladder. At one end it would fragment working-class communities, di-viding them by talent. At the other it would remove the ethic of noblesse oblige from the upper classes, who now would have "earned" their success and be responsible to no one, rather than inheriting it and being obligated to help the less fortunate. Wooldridge argues that "the left can hardly afford to ignore I.Q. tests, which, for all their inadequacies, are still the best means yet devised for spotting talent wherever it occurs, in the inner cities as well as the plush housing estates, and ensuring that talent is matched to the appropriate educational streams and job opportunities."

For their part, Richard Herrnstein and Charles Murray (the authors of *The Bell Curve*) argued that the heritability of intelligence ought to galvanize the left into a greater commitment to Rawlsian social justice.[53] If intelligence were entirely acquired, then policies for equal opportunity would suffice to guarantee an equitable distribution of wealth and power. But if some souls have the misfortune of being born into brains with lower ability, they could fall into poverty through no fault of their own, even in a perfectly fair system of economic competition. If social justice consists of seeing to the well-being of the worst off, then recognizing genetic differences calls for an active redistribution of wealth. Indeed, though Herrnstein was a conservative and Murray a right-leaning libertarian and communitarian, they were not opposed to simple redistributive measures such as a negative income tax for the lowest wage earners, which would give a break to those who play by the rules but still can't scrape by. Murray's libertarianism leads him to oppose government programs that are more activist than that, but he and Herrnstein noted that a hereditarian left is a niche waiting to be filled.

An important challenge to conservative political theory has come from behavioral economists such as Richard Thaler and George Akerlof, who were influenced by the evolutionary cognitive psychology of Herbert Simon, Amos Tversky, Daniel Kahneman, Gerd Gigerenzer, and Paul Slovic.[54] These psychologists have argued that human thinking and decision making are biological adaptations rather than engines of pure rationality. These mental systems work with limited amounts of information, have to reach decisions in a finite amount of time, and ultimately serve evolutionary goals such as status and security. Conservatives have always invoked limitations on human reason to rein in the pretense that we can understand social behavior well enough to redesign society. But those limitations also undermine the assumption of rational self-interest that underlies classical economics and secular conservatism. Ever since Adam Smith, classical economists have argued that in the absence of outside interference, individuals making decisions in their own interests will do what is best for themselves and for society. But if people do not always calculate what is best for themselves, they might be better off with the taxes and regulations that classical economists find so perverse.

For example, rational agents informed by interest rates and their life expectancies should save the optimal proportion of their wages for comfort in their old age. Social security and mandatory savings plans should be unnecessary—indeed, harmful—because they take away choice and hence the opportunity to find the best balance between consuming now and saving for the future. But economists repeatedly find that people spend their money like drunken sailors. They act as if they think they will die in a few years, or as if the future is completely unpredictable, which may be closer to the reality of our evolutionary ancestors than it is to life today.[55] If so, then allowing people to manage their own savings (for example, letting them keep their entire paycheck and investing it as they please) may work against their interests. Like Odysseus approaching the island of the Sirens, people might rationally agree to let their employer or the government tie them to the mast of forced savings.

The economist Robert Frank has appealed to the evolutionary psychology of status to point out other shortcomings of the rational-actor theory and, by extension, laissez-faire economics.[56] Rational actors should eschew not only forced retirement savings but other policies that ostensibly protect them, such as mandatory health benefits, workplace safety regulations, unemployment insurance, and union dues. All of these cost money that would otherwise go into their paychecks, and workers could decide for themselves whether to take a pay cut to work for a company with the most paternalistic policies or go for the biggest salary and take higher risks on the job. Companies, in their competition for the best employees, should find the balance demanded by the employees they want.

The rub, Frank points out, is that people are endowed with a craving for status. Their first impulse is to spend money in ways that put themselves ahead of the Joneses (houses, cars, clothing, prestigious educations), rather than in ways that only they know about (health care, job safety, retirement savings). Unfortunately, status is a zero-sum game, so when everyone has more money to spend on cars and houses, the houses and cars get bigger but people are no happier than they were before. Like hockey players who agree to wear helmets only if a rule forces their opponents to wear them too, people might agree to regulations that force everyone to pay for hidden benefits like health care that make them happier in the long run, even if the regulations come at the expense of disposable income. For the same reason, Frank argues, we would be better off if we implemented a steeply graduated tax on consumption, replacing the current graduated tax on income. A consumption tax would damp down the futile arms race for ever more lavish cars, houses, and watches and compensate people with resources that provably increase happiness, such as leisure time, safer streets, and more pleasant commuting and working conditions.

Finally, Darwinian leftists have been examining the evolutionary psychology of economic inequality. The economists Samuel Bowles and Herbert Gintis, formerly Marxists and now Darwinians, have reviewed the literature from ethnography and behavioral economics which suggests that people are neither antlike altruists nor self-centered misers.[57] As we saw in Chapter 14, people share with others who they think are willing to share, and punish those who are not. (Gintis calls this "strong reciprocity," which is like reciprocal altruism or "weak reciprocity" but is aimed at other people's willingness to contribute to public goods rather than at tit-for-tat exchanges.)[58] This psychology makes people oppose indiscriminate welfare and expansive social programs not because they are callous or greedy but because they think such programs reward the indolent and punish the industrious. Bowles and Gintis note that even in today's supposedly antiwelfare climate, polls show that most people are willing to pay higher taxes for some kinds of universal social insurance. They are willing to pay to guarantee basic needs such as food, shelter, and health care, to aid the victims of bad luck, and to help people who are down and out become self-sufficient. In other words, people are opposed to a blanket welfare state not out of *greed* but out of *fairness*. A welfare system that did not try to rewrite the public consciousness, and which distinguished between the deserving and the undeserving poor, would, they argue, be perfectly consonant with human nature.

The politics of economic inequality ultimately hinge on a tradeoff between economic freedom and economic equality. Though scientists cannot dictate how these desiderata should be weighted, they can help assess the morally relevant costs and thereby enable us to make a more informed decision. Once again the psychology of status and dominance has a role to play in this assessment. In absolute terms, today's poor are materially better off than the aristocracy of just a century ago. They live longer, are better fed, and enjoy formerly unimaginable luxuries such as central heating, refrigerators, telephones, and round-the-clock entertainment from television and radio. Conservatives say this makes it hard to argue that the station of lower-income people is an ethical outrage that ought to be redressed at any cost.

But if people's sense of well-being comes from an assessment of their social status, and social status is relative, then extreme inequality can make people on the lower rungs feel defeated even if they are better off than most of humanity. It is not just a matter of hurt feelings: people with lower status are less healthy and die younger, and communities with greater inequality have poorer health and shorter life expectancies.[59] The medical researcher Richard Wilkinson, who documented these patterns, argues that low status triggers an ancient stress reaction that sacrifices tissue repair and immune function for an immediate fight-or-flight response. Wilkinson, together with Martin Daly and Margo Wilson, have pointed to another measurable cost of economic inequality. Crime rates are much higher in regions with greater disparities of wealth (even after controlling for absolute levels of wealth), partly because chronic low status leads men to become obsessed with rank and to kill one another over trivial insults.[60] Wilkinson argues that reducing economic inequality would make millions of lives happier, safer, and longer.

This well-populated gallery of left-wing innatists should not come as a surprise, even after centuries in which human nature was a preserve of the right. Mindful both of science and of history, the Darwinian left has abandoned the Utopian Vision that brought so many unintended disasters. Whether this non-Utopian left is really all that different from the contemporary secular right, and whether its particular policies are worth their costs, is not for me to argue here. The point is that traditional political alignments ought to change as we learn more about human beings. The ideologies of the left and the right took shape before Darwin, before Mendel, before anyone knew what a gene or a neuron or a hormone was. Every student of political science is taught that political ideologies are based on theories of human nature. Why must they be based on theories that are three hundred years out of date?

Chapter 17

Violence

> The story of the human race is war. Except for brief and precarious interludes there has never been peace in the world; and long before history began murderous strife was universal and unending.[1]

WINSTON CHURCHILL'S SUMMARY of our species could be dismissed as the pessimism of a man who fought history's most awful war and was present at the birth of a cold war that could have destroyed humanity altogether. In fact it has sadly stood the test of time. Though the cold war is a memory, and hot wars between major nations are rare, we still do not have peace in the world. Even before the infamous year of 2001, with its horrific terrorist attacks on the United States and subsequent war in Afghanistan, the World Conflict List catalogued sixty-eight areas of systematic violence, from Albania and Algeria through Zambia and Zimbabwe.[2]

Churchill's speculation about prehistory has also been borne out. Modern foragers, who offer a glimpse of life in prehistoric societies, were once thought to engage only in ceremonial battles that were called to a halt as soon as the first man fell. Now they are known to kill one another at rates that dwarf the casualties from our world wars.[3] The archaeological record is no happier. Buried in the ground and hidden in caves lie silent witnesses to a bloody prehistory stretching back hundreds of thousands of years. They include skeletons with scalping marks, ax-shaped dents, and arrowheads embedded in them; weapons like tomahawks and maces that are useless for hunting but specialized for homicide; fortification defenses such as palisades of sharpened sticks; and paintings from several continents showing men firing arrows, spears, or boomerangs at one another and being felled by these weapons.[4] For decades, "anthropologists of peace" denied that any human group had ever practiced cannibalism, but evidence to the contrary has been piling up and now includes a smoking gun. In an 850-year-old site in the American Southwest, archaeologists have found human bones that were hacked up like the bones of animals used for food. They also found traces of human myoglobin (a muscle protein) on pot shards, and—damningly—in a lump of fossilized human excrement.[5] Members of *Homo antecessor,* relatives of the common ancestor of Neanderthals and modern humans, bashed and butchered one another too, suggesting that violence and cannibalism go back at least 800,000 years.[6]

War is only one of the ways in which people kill other people. In much of the world, war shades into smaller-scale violence such as ethnic strife, turf battles, blood feuds, and individual homicides. Here too, despite undeniable improvements, we do not have anything like peace. Though Western societies have seen murder rates fall between tenfold and a hundredfold in the past millennium, the United States lost a million people to homicide in the twentieth century, and an American man has about a one-half percent lifetime chance of being murdered.[7]

History indicts our species not just with the number of killings but with the manner. Hundreds of millions of Christians decorate their homes and adorn their bodies with a facsimile of a device that inflicted an unimaginably agonizing death on people who were a nuisance to Roman politicians. It is just one example of the endless variations of torture that the human mind has devised over the millennia, many of them common enough to have become words in our lexicon: *to crucify, to draw and quarter, to flay, to press, to stone; the garrote, the rack, the stake, the thumbscrew.* Dostoevsky's Ivan Karamazov, learning of the atrocities committed by the Turks in Bulgaria, said, "No animal could ever be so cruel as a man, so artfully, so artistically cruel." The annual reports of Amnesty International show that artistic cruelty is by no means a thing of the past.

~

THE REDUCTION OF violence on scales large and small is one of our greatest moral concerns. We ought to use every intellectual tool available to understand what it is about the human mind and human social arrangements that leads people to hurt and kill so much. But as with the other moral concerns examined in this part of the book, the effort to figure out what is going on has been hijacked by an effort to legislate the correct answer. In the case of violence, the correct answer is that violence has nothing to do with human nature but is a pathology inflicted by malign elements outside us. Violence is a behavior taught by the culture, or an infectious disease endemic to certain environments.

This hypothesis has become the central dogma of a secular faith, repeatedly avowed in public proclamations like a daily prayer or pledge of allegiance. Recall Ashley Montagu's UNESCO resolution that biology supports an ethic of "universal brotherhood" and the anthropologists who believed that "nonviolence and peace were likely the norm throughout most of human prehistory." In the 1980s, many social science organizations endorsed the Seville

Statement, which declared that it is "scientifically incorrect" to say that humans have a "violent brain" or have undergone selection for violence.[8] "War is not an instinct but an invention," wrote Ortega y Gasset, paralleling his claim that man has no nature but only history.[9] A recent United Nations Declaration on the Elimination of Violence Against Women announced that "violence is part of an historical process, and is not natural or born of biological determinism." A 1999 ad by the National Funding Collaborative on Violence Prevention declared that "violence is learned behavior."[10]

Another sign of this faith-based approach to violence is the averred *certainty* that particular environmental explanations are correct. We *know* the causes of violence, it is repeatedly said, and we also know how to eliminate it. Only a failure of commitment has prevented us from doing so. Remember Lyndon Johnson saying that "all of us know" that the conditions that breed violence are ignorance, discrimination, poverty, and disease. A 1997 article on violence in a popular science magazine quoted a clinical geneticist who echoed LBJ:

> We know what causes violence in our society: poverty, discrimination, the failure of our educational system. It's not the genes that cause violence in our society. It's our social system.[11]

The authors of the article, the historians Betty and Daniel Kevles, agreed:

> We need better education, nutrition, and intervention in dysfunctional homes and in the lives of abused children, perhaps to the point of removing them from the control of their incompetent parents. But such responses would be expensive and socially controversial.[12]

The creed that violence is learned behavior often points to particular elements of American culture as the cause. A member of a toy-monitoring group recently told a reporter, "Violence is a learned behavior. Every toy is educational. The question is, what do you want your children to learn?"[13] Media violence is another usual suspect. As two public health experts recently wrote:

> The reality is that children learn to value and use violence to solve their problems and deal with strong feelings. They learn it from role models in their families and communities. They learn it from the heroes we put in front of them on television, the movies, and video games.[14]

Childhood abuse, recently implicated in Richard Rhodes's *Why They Kill*, is a third putative cause. "The tragedy is that people who have been victimized often become victimizers themselves," said the president of the Criminal Justice Policy Foundation. "It's a cycle we could break, but it involves some expense. As a society, we haven't put our resources there."[15] Note in these statements the mouthing of the creed ("Violence is a learned behavior"), the certainty that it is true ("The reality is"), and the accusation that we suffer from a lack of commitment ("We haven't put our resources there") rather than an ignorance of how to solve the problem.

Many explanations blame "culture," conceived as a superorganism that teaches, issues commands, and doles out rewards and punishments. A *Boston Globe* columnist must have been oblivious to the circularity of his reasoning when he wrote:

> So why is America more violent than other industrialized Western democracies? It's our cultural predisposition to violence. We pummel each other, maul each other, stab each other and shoot each other because it's our cultural imperative to do so.[16]

When culture is seen as an entity with beliefs and desires, the beliefs and desires of actual people are unimportant. After Timothy McVeigh blew up a federal office building in Oklahoma City in 1995, killing 168 people, the journalist Alfie Kohn ridiculed Americans who "yammer about individual responsibility" and attributed the bombing to American individualism: "We have a cultural addiction to competition in this country. We're taught in classrooms and playing fields that other people are obstacles to our own success."[17] A related explanation for the bombing put the blame on American symbols, such as the arrow-clutching eagle on the national seal, and state mottoes, including "Live Free or Die" (New Hampshire) and "With the sword, we seek peace, but under liberty" (Massachusetts).[18]

A popular recent theory attributes American violence to a toxic and peculiarly American conception of maleness inculcated in childhood. The social psychologist Alice Eagly explained sprees of random shootings by saying, "This sort of behavior has been part of the male role as it has been construed in US culture, from the frontier tradition on."[19] According to the theory, popularized in bestsellers like Dan Kindlon's *Raising Cain* and William Pollack's *Real Boys*, we are going through a "national crisis of boyhood in America," caused by the fact that boys are forced to separate from their mothers and stifle their emotions. "What's the matter with men?" asked an article in the *Boston Globe Magazine*. "Violent behavior, emotional distance, and higher rates of drug addiction can't be explained by hormones," it answers. "The problem, experts say, is cultural beliefs about masculinity—everything packed into the phrase 'a real man.'"[20]

THE STATEMENT THAT "violence is learned behavior" is a mantra repeated by right-thinking people to show that they believe that violence should be reduced. It is not based on any sound research. The sad fact is that despite the repeated assurances that "we know the conditions that breed violence," we barely have a clue. Wild swings in crime rates—up in the 1960s and late 1980s, down in the late 1990s—continue to defy any simple explanation. And the usual suspects for understanding violence are completely unproven and sometimes patently false. This is most blatant in the case of factors like "nutrition" and "disease" that are glibly thrown into lists of the social ills that allegedly bring on violence. There is no evidence, to put it mildly, that violence is caused by a vitamin deficiency or a bacterial infection. But the other putative causes suffer from a lack of evidence as well.

Aggressive parents often have aggressive children, but people who conclude that aggression is learned from parents in a "cycle of violence" never consider the possibility that violent tendencies could be inherited as well as learned. Unless one looks at *adopted* children and shows that they act more like their adoptive parents than like their biological parents, cycles of violence prove nothing. Similarly, the psychologists who note that men commit more acts of violence than women and then blame it on a culture of masculinity are wearing intellectual blinkers that keep them from noticing that men and women differ in their biology as well as in their social roles. American children are exposed to violent role models, of course, but they are also exposed to clowns, preachers, folk singers, and drag queens; the question is why children find some people more worthy of imitation than others.

To show that violence is caused by special themes of American culture, a bare minimum of evidence would be a correlation in which the cultures that have those themes also tend to be more violent. Even that correlation, if it existed, would not prove that the cultural themes cause the violence rather than the other way around. But there may be no such correlation in the first place.

To begin with, American culture is not uniquely violent. All societies have violence, and America is not the most violent one in history or even in today's world. Most countries in the Third World, and many of the former republics of the Soviet Union, are considerably more violent, and they have nothing like the American tradition of individualism.[21] As for cultural norms of masculinity and sexism, Spain has its machismo, Italy its braggadocio, and Japan its rigid gender roles, yet their homicide rates are a fraction of that of the more feminist-influenced United States. The archetype of a masculine hero prepared to use violence in a just cause is one of the most common motifs in mythology, and it can be found in many cultures with relatively low rates of violent crime. James Bond, for example—who actually has a *license to kill*—is British, and martial arts films are popular in many industrialized Asian countries. In any case, only a bookworm who has never actually seen an American movie or television program could believe that they glorify murderous fanatics like Timothy McVeigh or teenagers who randomly shoot classmates in high school cafeterias. Masculine heroes in the mass media are highly moralistic: they fight bad guys.

Among conservative politicians and liberal health professionals alike it is an article of faith that violence in the media is a major cause of American violent crime. The American Medical Association, the American Psychological Association, and the American Academy of Pediatrics testified before Congress that over 3,500 studies had investigated the connection and only 18 failed to find one. Any social scientist can smell fishy numbers here, and the psychologist Jonathan Freedman decided to look for himself. In fact, only *two hundred* studies have looked for a connection between media violence and violent behavior, and *more than half* failed to find one.[22] The others found correlations that are small and readily explainable in other ways—for example, that violent children seek out violent entertainment, and that children are temporarily aroused (but not permanently affected) by action-packed footage. Freedman and several other psychologists who have reviewed the literature have concluded that exposure to media violence has little or no effect on violent behavior in the world.[23] Reality checks from recent history suggest the same thing. People were more violent in the centuries *before* television and movies were invented. Canadians watch the same television shows as Americans but have a fourth their homicide rate. When the British colony of St. Helena installed television for the first time in 1995, its people did not become more violent.[24] Violent computer games took off in the 1990s, a time when crime rates plummeted.

What about the other usual suspects? Guns, discrimination, and poverty play a role in violence, but in no case is it a simple or decisive one. Guns surely make it easier for people to kill, and harder for them to de-escalate a fight before a death occurs, and thus multiply the lethality of conflicts large and small. Nonetheless, many societies had sickening rates of violence before guns were invented, and people do not automatically kill one another just because they have access to guns. The Israelis and Swiss are armed to the teeth but have low rates of violent personal crime, and among American states, Maine and North Dakota have the lowest homicide rates but almost every home has a gun.[25] The idea that guns increase lethal crime, though certainly plausible, has been so difficult to prove that in 1998 the legal scholar John Lott published a book of statistical analyses with a title that flaunts the opposite conclusion: *More Guns, Less Crime.* Even if he is wrong, as I suspect he is, it is not so easy to show that more guns mean *more* crime.

As for discrimination and poverty, again it is hard to show a direct cause-and-effect relationship. Chinese immigrants to California in the nineteenth century and Japanese-Americans in World War II faced severe discrimination, but they did not react with high rates of violence. Women are poorer than men and are more likely to need money to feed children, but they are less likely to steal things by force. Different subcultures that are equally impoverished can vary radically in their rates of violence, and as we shall see, in many cultures relatively affluent men can be quick to use lethal force.[26] Though no one could object to a well-designed program that was shown to reduce crime, one cannot simply blame crime rates on a lack of commitment to social programs. These programs first flourished in the 1960s, the decade in which rates of violent crime skyrocketed.

Scientifically oriented researchers on violence chant a different mantra: "Violence is a public health problem." According to the National Institute of Mental Health, "Violent behavior can best be understood—and prevented—if it is attacked as if it were a contagious disease that flourishes in vulnerable individuals and resource-poor neighborhoods." The public health theory has been echoed by many professional organizations, such as the American Psychological Society and the Centers for Disease Control, and by political figures as diverse as the surgeon general in the Clinton administration and the Republican senator Arlen Specter.[27] The public health approach tries to identify "risk factors" that are more common in poor neighborhoods than affluent ones. They include neglect and abuse in childhood, harsh and inconsistent discipline, divorce, malnutrition, lead poisoning, head injuries, untreated attention deficit hyperactivity disorder, and the use of alcohol and crack cocaine during pregnancy.

Researchers in this tradition are proud that their approach is both "biological"—they measure bodily fluids and take pictures of the brain—and "cultural"—they look for environmental causes of the brain conditions that might be ameliorated by the equivalent of public health measures. Unfortunately, there is a rather glaring flaw in the whole analogy. A good definition of a disease or disorder is that it consists of suffering experienced by an individual because of a malfunction of a mechanism in the individual's body.[28] But as a writer for *Science* recently pointed out, "Unlike most diseases, it's usually not the perpetrator who defines aggression as a problem; it's the environment. Violent people may feel they are functioning normally, and some may even enjoy their occasional outbursts and resist treatment."[29] Other than the truism that violence is more common in some people and places than others, the public health theory has little to recommend it. As we shall see, violence is not a disease in anything like the medical sense.

~

PURE ENVIRONMENTAL THEORIES of violence remain an article of faith because they embody the Blank Slate and the Noble Savage. Violence, according to these theories, isn't a natural strategy in the human repertoire; it's learned behavior, or poisoning by a toxic substance, or the symptom of an infectious illness. In earlier chapters we saw the moral appeal of such doctrines: to differentiate the doctrine-holders from jingoists of earlier periods and ruffians of different classes; to reassure audiences that they do not think violence is "natural" in the sense of "good"; to express an optimism that violence can be eliminated, particularly by benign social programs rather than punitive deterrence; to stay miles away from the radioactive position that some individuals, classes, or races are innately more violent than others.

Most of all, the learned-behavior and public health theories are moral declarations, public avowals that the declarer is opposed to violence. Condemning violence is all to the good, of course, but not if it is disguised as an empirical claim about our psychological makeup. Perhaps the purest example of this wishful confusion comes from Ramsey Clark, attorney general in the Johnson administration and the author of the 1970 bestseller *Crime in America*. In arguing that the criminal justice system should replace punishment with rehabilitation, Clark explained:

> The theory of rehabilitation is based on the belief that healthy, rational people will not injure others, that they will understand that the individual and his society are best served by conduct that does not inflict injury, and that a just society has the ability to provide health and purpose and opportunity for all its citizens. Rehabilitated, an individual will not have the capacity—cannot bring himself—to injure another or take or destroy property.[30]

Would that it were so! This theory is a fine example of the moralistic fallacy: it would be so nice *if* the idea were true that we should all believe *that* it is true. The problem is that it is not true. History has shown that plenty of healthy, rational people can bring themselves to injure others and destroy property because, tragically, an individual's interests sometimes *are* served by hurting others (especially if criminal penalties for hurting others are eliminated, an irony that Clark seems to have missed). Conflicts of interest are inherent to the human condition, and as Martin Daly and Margo Wilson point out, "Killing one's adversary is the ultimate conflict resolution technique."[31]

Admittedly, it is easy to equate health and rationality with morality. The metaphors pervade the English language, as when we call an evildoer *crazy, degenerate, depraved, deranged, mad, malignant, psycho, sick,* or *twisted.* But the metaphors are bound to mislead us when we contemplate the causes of violence and ways to reduce it. Termites are not malfunctioning when they eat the wooden beams in houses, nor are mosquitoes when they bite a victim and spread the malaria parasite. They are doing exactly what evolution designed them to do, even if the outcome makes people suffer. For scientists to moralize about these creatures or call their behavior pathological would only send us all down blind alleys, such as a search for the "toxic" influences on these creatures or a "cure" that would restore them to health. For the same reason, human violence does not have to be a disease for it to be worth combating. If anything, it is the belief that violence is an aberration that is dangerous, because it lulls us into forgetting how easily violence may erupt in quiescent places.

The Blank Slate and the Noble Savage owe their support not just to their moral appeal but to enforcement by ideology police. The blood libel against Napoleon Chagnon for documenting warfare among the Yanomamö is the most lurid example of the punishment of heretics, but it is not the only one. In 1992 a Violence Initiative in the Alcohol, Drug Abuse, and Mental Health Administration was canceled because of false accusations that the research aimed to sedate inner-city youth and to stigmatize them as genetically prone to violence. (In fact, it advocated the public health approach.) A conference and book on the legal and moral issues surrounding the biology of violence, which was to include advocates of all viewpoints, was canceled by Bernadine Healey, director of the National Institutes of Health, who overruled a unanimous peer-review decision because of concerns "associated with the sensitivity and validity of the proposed conference."[32] The university sponsoring the conference appealed and won, but when the conference was held three years later, protesters invaded the hall and, as if to provide material for comedians, began a shoving match with the participants.[33]

What was everyone so sensitive about? The stated fear was that the government would define political unrest in response to inequitable social conditions as a psychiatric disease and silence the protesters by drugging them or worse. The radical psychiatrist Peter Breggin called the Violence Initiative "the most terrifying, most racist, most hideous thing imaginable" and "the kind of plan one would associate with Nazi Germany."[34] The reasons included "the medicalization of social issues, the declaration that the victim of oppression, in this case the Jew, is in fact a genetically and biologically defective person, the mobilization of the state for eugenic purposes and biological purposes, the heavy use of psychiatry in the development of social-control programs."[35] This is a fanciful, indeed paranoid, reading, but Breggin has tirelessly repeated it, especially to African American politicians and media outlets. Anyone using the words "violence" and "biology" in the same paragraph may be put under a cloud of suspicion for racism, and this has affected the intellectual climate regarding violence. No one has ever gotten into trouble for saying that violence is completely learned.

∼

THERE ARE MANY reasons to believe that violence in humans is not literally a sickness or poisoning but part of our design. Before presenting them, let me allay two fears.

The first fear is that examining the roots of violence in human nature consists of reducing violence to the bad genes of violent individuals, with the unsavory implication that ethnic groups with higher rates of violence must have more of these genes.

There can be little doubt that some individuals are constitutionally more prone to violence than others. Take men, for starters: across cultures, men kill men twenty to forty times more often than women kill women.[36] And the lion's share of the killers are *young* men, between the ages of fifteen and thirty.[37] Some young men, moreover, are more violent than others. According to one estimate, 7 percent of young men commit 79 percent of repeated violent offenses.[38] Psychologists find that individuals prone to violence have a distinctive personality profile. They tend to be impulsive, low in intelligence, hyperactive, and attention-deficit. They are described as having an "oppositional temperament": they are vindictive, easily angered, resistant to control, deliberately annoying, and likely to blame everything on other people.[39] The most callous among them are psychopaths, people who lack a conscience, and they make up a substantial percentage of murderers.[40] These traits emerge in early childhood, persist through the lifespan, and are largely heritable, though nowhere near completely so.

Sadists, hotheads, and other natural-born killers are part of the problem of violence, not just because of the harm they wreak but because of the aggressive posture they force *others* into for deterrence and self-defense. But my point here is that they are not the major part of the problem. Wars start and stop, crime rates yo-yo, societies go from militant to pacifist or vice versa within a generation, all without any change in the frequencies of the local genes. Though ethnic groups differ today in their average rates of violence, the differences do not call for a genetic explanation, because the rate for a group at one historical period may be matched to that of any other group at another period. Today's docile Scandinavians descended from bloodthirsty Vikings, and Africa, wracked by war after the fall of colonialism, is much like Europe after the fall of the Roman Empire. Any ethnic group that has made it into the present probably had pugnacious ancestors in the not-too-distant past.

The second fear is that if people are endowed with violent motives, they can't help being violent, or must be violent all the time, like the Tasmanian Devil in *Looney Tunes* who tears through an area leaving a swath of destruction in his wake. This fear is a reaction to archaic ideas of killer apes, a thirst for blood, a death wish, a territorial imperative, and a violent brain. In fact, if the brain is equipped with strategies for violence, they are *contingent* strategies, connected to complicated circuitry that computes when and where they should be deployed. Animals deploy aggression in highly selective ways, and humans, whose limbic systems are enmeshed with outsize frontal lobes, are of course even more calculating. Most people today live their adult lives without ever pressing their violence buttons.

So what is the evidence that our species may have evolved mechanisms for discretionary violence? The first thing to keep in mind is that aggression is an organized, goal-directed activity, not the kind of event that could come from a random malfunction. If your lawnmower continued to run after you released the handle and it injured your foot, you might suspect a sticky switch or other breakdown. But if the lawnmower lay in wait until you emerged from the garage and then chased you around the yard, you would have to conclude that someone had installed a chip that programmed it to do so.

The presence of deliberate chimpicide in our chimpanzee cousins raises the possibility that the forces of evolution, not just the idiosyncrasies of a particular human culture, prepared us for violence. And the ubiquity of violence in human societies throughout history and prehistory is a stronger hint that we are so prepared.

When we look at human bodies and brains, we find more direct signs of design for aggression. The larger size, strength, and upper-body mass of men is a zoological giveaway of an evolutionary history of violent male-male competition.[41] Other signs include the effects of testosterone on dominance and violence (which we will encounter in the chapter on gender), the emotion of anger (complete with reflexive baring of the canine teeth and clenching of the fists), the revealingly named fight-or-flight response of the autonomic nervous system, and the fact that disruptions of inhibitory systems of the brain (by alcohol, damage to the frontal lobe or amygdala, or defective genes involved in serotonin metabolism) can lead to aggressive attacks, initiated by circuits in the limbic system.[42]

Boys in all cultures spontaneously engage in rough-and-tumble play, which is obviously practice for fighting. They also divide themselves into coalitions that compete aggressively (calling to mind the remark attributed to the Duke of Wellington that "the Battle of Waterloo was won upon the playing fields of Eton").[43] And children are violent well before they have been infected by war toys or cultural stereotypes. The most violent age is not adolescence but toddlerhood: in a recent large study, almost half the boys just past the age of two, and a slightly smaller percentage of the girls, engaged in hitting, biting, and kicking. As the author pointed out, "Babies do not kill each other, because we do not give them access to knives and guns. The question . . . we've been trying to answer for the past 30 years is how do children learn to aggress. [But] that's the wrong question. The right question is how do they learn not to aggress."[44]

Violence continues to preoccupy the mind throughout life. According to independent surveys in several countries by the psychologists Douglas Kenrick and David Buss, more than 80 percent of women and 90 percent of men fantasize about killing people they don't like, especially romantic rivals, stepparents, and people who have humiliated them in public.[45] People in all cultures take pleasure in thinking about killings, if we are to judge by the popularity of murder mysteries, crime dramas, spy thrillers, Shakespearean tragedies, biblical stories, hero myths, and epic poems. (A character in Tom Stoppard's *Rosencrantz and Guildenstern Are Dead* asks, "You're familiar with the great tragedies of antiquity, are you? The great homicidal classics?") People also enjoy watching the stylized combat we call "sports," which are contests of aiming, chasing, or fighting, complete with victors and the vanquished. If language is a guide, many other efforts are conceptualized as forms of aggression: intellectual argument (*to shoot down, defeat*, or *destroy* an idea or its proponent), social reform (*to fight crime, to combat prejudice, the War on Poverty, the War on Drugs*), and medical treatment (*to fight cancer, painkillers, to defeat AIDS, the War on Cancer*).

In fact, the entire question of what went wrong (socially or biologically) when a person engages in violence is badly posed. Almost everyone recognizes the need for violence in defense of self, family, and innocent victims. Moral philosophers point out that there are even circumstances in which torture is justified—say, when a captured terrorist has planted a time bomb in a crowded place and refuses to say where it is. More generally, whether a violent mindset is called heroic or pathological often depends on whose ox has been gored. Freedom fighter or terrorist, Robin Hood or thief, Guardian Angel or vigilante, nobleman or warlord, martyr or kamikaze, general or gang leader—these are value judgments, not scientific classifications. I doubt that the brains or genes of most of the lauded protagonists would differ from those of their vilified counterparts.

In this way I find myself in agreement with the radical scientists who insist that we will never understand violence by looking only at the genes or brains of violent people. Violence is a social and political problem, not just a biological and psychological one. Nonetheless, the phenomena we call "social" and "political" are not external happenings that mysteriously affect human affairs like sunspots; they are shared understandings among individuals at a given time and place. So one cannot understand violence without a thorough understanding of the human mind.

In the rest of this chapter I explore the logic of violence, and why emotions and thoughts devoted to it may have evolved. This is necessary to disentangle the knot of biological and cultural causes that make violence so puzzling. It can help explain why people are prepared for violence but act on those inclinations only in particular circumstances; when violence is, at least in some sense, rational and when it is blatantly self-defeating; why violence is more prevalent in some times and places than in others, despite a lack of any genetic difference among the actors; and, ultimately, how we might reduce and prevent violence.

THE FIRST STEP in understanding violence is to set aside our abhorrence of it long enough to examine why it can sometimes pay off in personal or evolutionary terms. This requires one to invert the statement of the problem—not why violence occurs, but why it is avoided. Morality, after all, did not enter the universe with the Big Bang and then pervade it like background radiation. It was discovered by our ancestors after billions of years of the morally indifferent process known as natural selection.

In my view, the consequences of this background amorality were best worked out by Hobbes in *Leviathan*. Unfortunately, Hobbes's pithy phrase "nasty, brutish, and short" and his image of an all-powerful leviathan keeping us from each other's throats have led people to misunderstand his argument. Hobbes is commonly interpreted as proposing that man in a state of nature was saddled with an irrational impulse for hatred and destruction. In fact his analysis is more subtle, and perhaps even more tragic, for he showed how the dynamics of violence fall out of interactions among rational and self-interested agents. Hobbes's analysis has been rediscovered by evolutionary biology, game theory, and social psychology, and I will use it to organize my discussion of the logic of violence before turning to the ways in which humans deploy peaceable instincts to counteract their violent ones.

Here is the analysis that preceded the famous "life of man" passage:

> So that in the nature of man, we find three principal causes of quarrel.
> First, competition; secondly, diffidence; thirdly, glory. The first maketh
> men invade for gain; the second, for safety; and the third, for reputation.
> The first use violence, to make themselves masters of other men's per-
> sons, wives, children, and cattle; the second, to defend them; the third,
> for trifles, as a word, a smile, a different opinion, and any other sign of
> undervalue, either direct in their persons or by reflection in their kin-
> dred, their friends, their nation, their profession, or their name.[46]

First, competition. Natural selection is powered by competition, which means that the products of natural selection—survival machines, in Richard Dawkins's metaphor—should, by default, do whatever helps them survive and reproduce. He explains:

> To a survival machine, another survival machine (which is not its own
> child or another close relative) is part of its environment, like a rock or
> a river or a lump of food. It is something that gets in the way, or some-
> thing that can be exploited. It differs from a rock or a river in one im-
> portant respect: it is inclined to hit back. This is because it too is a
> machine that holds its immortal genes in trust for the future, and it too

will stop at nothing to preserve them. Natural selection favors genes that control their survival machines in such a way that they make the best use of their environment. This includes making the best use of other survival machines, both of the same and of different species.[47]

If an obstacle stands in the way of something an organism needs, it should neutralize the obstacle by disabling or eliminating it. This includes obstacles that happen to be other human beings—say, ones that are monopolizing desirable land or sources of food. Even among modern nation-states, raw self-interest is a major motive for war. The political scientist Bruce Bueno de Mesquita analyzed the instigators of 251 real-world conflicts of the past two centuries and concluded that in most cases the aggressor correctly calculated that a successful invasion would be in its national interest.[48]

Another human obstacle consists of men who are monopolizing women who could otherwise be taken as wives. Hobbes called attention to the phenomenon without knowing the evolutionary reason, which was provided centuries later by Robert Trivers: the difference in the minimal parental investments of males and females makes the reproductive capacity of females a scarce commodity over which males compete.[49] This explains why men are the violent gender, and also why they always have something to fight over, even when their survival needs have been met. Studies of warfare in pre-state societies have confirmed that men do not have to be short of food or land to wage war.[50] They often raid other villages to abduct women, to retaliate for past abductions, or to defend their interests in disputes over exchanges of women for marriage. In societies in which women have more say in the matter, men still compete for women by competing for the status and wealth that tend to attract them. The competition can be violent because, as Daly and Wilson point out, "Any creature that is recognizably on track toward complete reproductive failure must somehow expend effort, often at risk of death, to try to improve its present life trajectory."[51] Impoverished young men on this track are therefore likely to risk life and limb to improve their chances in the sweepstakes for status, wealth, and mates.[52] In all societies they are the demographic sector in which the firebrands, delinquents, and cannon fodder are concentrated. One of the reasons the crime rate shot up in the 1960s is that boys from the baby boom began to enter their crime-prone years.[53] Though there are many reasons why countries differ in their willingness to wage war, one factor is simply the proportion of the population that consists of men between the ages of fifteen and twenty-nine.[54]

This whole cynical analysis may not ring true to modern readers, because we cannot think of other people as mere parts of our environment that may have to be neutralized like weeds in a garden. Unless we are psychopaths, we *sympathize* with other people and cannot blithely treat them as obstacles or prey. Such sympathy, however, has not prevented people from committing all manner of atrocities throughout history and prehistory. The contradiction may be resolved by recalling that people discern a moral circle that may not embrace all human beings but only the members of their clan, village, or tribe.[55] Inside the circle, fellow humans are targets of sympathy; outside, they are treated like a rock or a river or a lump of food. In a previous book I mentioned that the language of the Wari people of the Amazon has a set of noun classifiers that distinguish edible from inedible objects, and that the edible class includes anyone who is not a member of the tribe. This prompted the psychologist Judith Rich Harris to observe:

> In the Wari dictionary
> Food's defined as "Not a Wari."
> Their dinners are a lot of fun
> For all but the un-Wari one.

Cannibalism is so repugnant to us that for years even anthropologists failed to admit that it was common in prehistory. It is easy to think: could other human beings really be capable of such a depraved act? But of course animal rights activists have a similarly low opinion of meat eaters, who not only cause millions of preventable deaths but do so with utter callousness: castrating and branding cattle without an anesthetic, impaling fish by the mouth and letting them suffocate in the hold of a boat, boiling lobsters alive. My point is not to make a moral case for vegetarianism but to shed light on the mindset of human violence and cruelty. History and ethnography suggest that people can treat strangers the way we now treat lobsters, and our incomprehension of such deeds may be compared with animal rights activists' incomprehension of ours. It is no coincidence that Peter Singer, the author of *The Expanding Circle,* is also the author of *Animal Liberation.*

The observation that people may be morally indifferent to other people who are outside a mental circle immediately suggests an opening for the effort to reduce violence: understand the psychology of the circle well enough to encourage people to put all of humanity inside it. In earlier chapters we saw how the moral circle has been growing for millennia, pushed outward by the expanding networks of reciprocity that make other human beings more valuable alive than dead.[56] As Robert Wright has put it, "Among the many reasons I don't think we should bomb the Japanese is that they built my minivan." Other technologies have contributed to a cosmopolitan view that makes it easy to imagine trading places with other people. These include literacy, travel, a knowledge of history, and realistic art that helps people project themselves into the daily lives of people who in other times might have been their mortal enemies.

We have also seen how the circle can shrink. Recall that Jonathan Glover showed that atrocities are often accompanied by tactics of dehumanization such as the use of pejorative names, degrading conditions, humiliating dress, and "cold jokes" that make light of suffering.[57] These tactics can flip a mental switch and reclassify an individual from "person" to "nonperson," making it as easy for someone to torture or kill him as it is for us to boil a lobster alive. (Those who poke fun at politically correct names for ethnic minorities, including me, should keep in mind that they originally had a humane rationale.) The social psychologist Philip Zimbardo has shown that even among the students of an elite university, tactics of dehumanization can easily push one person outside another's moral circle. Zimbardo created a mock prison in the basement of the Stanford University psychology department and randomly assigned students to the role of prisoner or guard. The "prisoners" had to wear smocks, leg irons, and nylon-stocking caps and were referred to by serial numbers. Before long the "guards" began to brutalize them—standing on their backs while they did push-ups, spraying them with fire extinguishers, forcing them to clean toilets with their bare hands—and Zimbardo called off the experiment for the subjects' safety.[58]

In the other direction, signs of a victim's humanity can occasionally break through and flip the switch back to the sympathy setting. When George Orwell fought in the Spanish Civil War, he once saw a man running for his life half-dressed, holding up his pants with one hand. "I refrained from shooting at him," Orwell wrote. "I did not shoot partly because of that detail about the trousers. I had come here to shoot at 'Fascists'; but a man who is holding up his trousers isn't a 'Fascist,' he is visibly a fellow creature, similar to your self."[59] Glover recounts another example, reported by a South African journalist:

> In 1985, in the old apartheid South Africa, there was a demonstration in Durban. The police attacked the demonstrators with customary violence. One policeman chased a black woman, obviously intending to beat her with his club. As she ran, her shoe slipped off. The brutal policeman was also a well-brought-up young Afrikaner, who knew that when a woman loses her shoe you pick it up for her. Their eyes met as he handed her the shoe. He then left her, since clubbing her was no longer an option.[60]

We should not, however, delude ourselves into thinking that the reaction of Orwell (one of the twentieth century's greatest moral voices) and of the "well-brought-up" Afrikaner is typical. Many intellectuals believe that the majority of soldiers cannot bring themselves to fire their weapons in battle. The claim is incredible on the face of it, given the tens of millions of soldiers who were shot in the wars of the last century. (I am reminded of the professor in Stoppard's *Jumpers* who noted that Zeno's Paradox prevents an arrow from ever reaching its target, so Saint Sebastian must have died of fright.) The belief turns out to be traceable to a single, dubious study of infantrymen in World War II. In follow-up interviews, the men denied having even been *asked* whether they had fired their weapons, let alone having claimed they hadn't.[61] Recent surveys of soldiers in battle and of rioters in ethnic massacres find that they often kill with gusto, sometimes in a state they describe as "joy" or "ecstasy."[62]

Glover's anecdotes reinforce the hope that people are capable of putting strangers inside a violence-proof moral circle. But they also remind us that the default setting may be to keep them out.

~

SECONDLY, DIFFIDENCE, IN its original sense of "distrust." Hobbes had translated Thucydides' *History of the Peloponnesian War* and was struck by his observation that "what made war inevitable was the growth of Athenian power and the fear which this caused in Sparta." If you have neighbors, they may covet what you have, in which case you have become an obstacle to their desires. Therefore you must be prepared to defend yourself. Defense is an iffy matter even with technologies such as castle walls, the Maginot Line, or antiballistic missile defenses, and it is even iffier without them. The only option for self-protection may be to wipe out potentially hostile neighbors first in a preemptive strike. As Yogi Berra advised, "The best defense is a good offense and vice versa."

Tragically, you might arrive at this conclusion even if you didn't have an aggressive bone in your body. All it would take is the realization that others might covet what you have and a strong desire not to be massacred. Even more tragically, your neighbors have every reason to be cranking through the same deduction, and if they are, it makes your fears all the more compelling, which makes a preemptive strike all the more tempting, which makes a preemptive strike by *them* all the more tempting, and so on.

This "Hobbesian trap," as it is now called, is a ubiquitous cause of violent conflict.[63] The political scientist Thomas Schelling offered the analogy of an armed homeowner who surprises an armed burglar. Each might be tempted to shoot first to avoid being shot, even if neither wanted to kill the other. A Hobbesian trap pitting one man against another is a recurring theme in fiction, such as the desperado in Hollywood westerns, spy-versus-spy plots in cold-war thrillers, and the lyrics to Bob Marley's "I Shot the Sheriff."

But because we are a social species, Hobbesian traps more commonly pit groups against groups. There is safety in numbers, so humans, bound by shared genes or reciprocal promises, form coalitions for protection. Unfortunately, the logic of the Hobbesian trap means there is also *danger* in numbers, because neighbors may fear they are becoming outnumbered and form alliances in their turn to contain the growing menace. Since one man's containment is another man's encirclement, this can send the spiral of danger upward. Human sociality is the original "entangling alliance," in which two parties with no prior animus can find themselves at war when the ally of one attacks the ally of the other. It is the reason I discuss homicide and war in a single chapter. In a species whose members form bonds of loyalty, the first can easily turn into the second.

The danger is particularly acute for humans because, unlike most mammals, we tend to be patrilocal, with related males living together instead of dispersing from the group when they become sexually mature.[64] (Among chimpanzees and dolphins, related males also live together, and they too form aggressive coalitions.) What we call "ethnic groups" are very large extended families, and though in a modern ethnic group the family ties are too distant for kin-based altruism to be significant, this was not true of the smaller coalitions in which we evolved. Even today ethnic groups often *perceive* themselves as large families, and the role of ethnic loyalties in group-against-group violence is all too obvious.[65]

The other distinctive feature of *Homo sapiens* as a species is, of course, toolmaking. Competitiveness can channel toolmaking into weaponry, and diffidence can channel weaponry into an arms race. An arms race, like an alliance, can make war more likely by accelerating the spiral of fear and distrust. Our species' vaunted ability to make tools is one of the reasons we are so good at killing one another.

The vicious circle of a Hobbesian trap can help us understand why the escalation from friction to war (and occasionally, the de-escalation to detente) can happen so suddenly. Mathematicians and computer simulators have devised models in which several players acquire arms or form alliances in response to what the other players are doing. The models often display chaotic behavior, in which small differences in the values of the parameters can have large and unpredictable consequences.[66]

As we can infer from Hobbes's allusion to the Peloponnesian War, Hobbesian traps among groups are far from hypothetical. Chagnon describes how Yanomamö villages obsess over the danger of being massacred by other villages (with good reason) and occasionally engage in preemptive assaults, giving other villages good reason to engage in their own preemptive assaults, and prompting groups of villages to form alliances that make their neighbors ever more nervous.[67] Street gangs and Mafia families engage in similar machinations. In the past century, World War I, the Six-Day Arab-Israeli War, and the Yugoslavian wars in the 1990s arose in part from Hobbesian traps.[68]

The political scientist John Vasquez has made the point quantitatively. Using a database of hundreds of conflicts from the past two centuries, he concludes that the ingredients of a Hobbesian trap—concern with security, entangling alliances, and arms races—can statistically predict the escalation of friction into war.[69] The most conscious playing-out of the logic of Hobbesian traps took place among nuclear strategists during the cold war, when the fate of the world literally hinged on it. The logic produced some of the maddening paradoxes of nuclear strategy: why it is extraordinarily dangerous to have enough missiles to destroy an enemy but not enough to destroy him after he has attacked those missiles (because the enemy would have a strong incentive to strike preemptively), and why erecting an impregnable defense against enemy missiles could make the world a *more* dangerous place (because the enemy has an incentive to launch a preemptive strike before the completed defense turns him into a sitting duck).

When a stronger group overpowers a weaker one in a surprise raid, it should come as no surprise to a Hobbesian cynic. But when one side defeats another in a battle that both have joined, the logic is not so clear. Given that both the victor and the vanquished have much to lose in a battle, one would expect each side to assess the strength of the other and the weaker to cede the contested resource without useless bloodshed that would only lead to the same outcome. Most behavioral ecologists believe that rituals of appeasement and surrender among animals evolved for this reason (and not for the good of the species, as Lorenz had supposed). Sometimes the two sides are so well matched, and the stakes of a battle are so high, that they engage in a battle because it is the only way to find out who is stronger.[70]

But at other times a leader will march—or march his men—into the valley of death without any reasonable hope of prevailing. Military incompetence has long puzzled historians, and the primatologist Richard Wrangham suggests that it might grow out of the logic of bluff and self-deception.[71] Convincing an adversary to avoid a battle does not depend on *being* stronger but on *appearing* stronger, and that creates an incentive to bluff and to be good at detecting bluffs. Since the most effective bluffer is the one who believes his own bluff, a limited degree of self-deception in hostile escalations can evolve. It has to be limited, because having one's bluff called can be worse than folding on the first round, but when the limits are miscalibrated and both sides go to the brink, the result can be a human disaster. The historian Barbara Tuchman has highlighted the role of self-deception in calamitous wars throughout history in her books *The Guns of August* (about World War I) and *The March of Folly: From Troy to Vietnam*.

~

A READINESS TO inflict a preemptive strike is a double-edged sword, because it makes one an inviting *target* for a preemptive strike. So people have invented, and perhaps evolved, an alternative defense: the advertised deterrence policy known as *lex talionis*, the law of retaliation, familiar from the biblical injunction "An eye for an eye, a tooth for a tooth."[72] If you can credibly say to potential adversaries, "We won't attack first, but if we are attacked, we will survive and strike back," you remove Hobbes's first two incentives for quarrel, gain and mistrust. The policy that you will inflict as much harm on others as they inflicted on you cancels their incentive to raid for gain, and the policy that you will not strike first cancels their incentive to raid for mistrust. This is reinforced by the policy to retaliate with *no more harm* than they inflicted on you, because it allays the fear that you will use a flimsy pretext to justify a massive opportunistic raid.

The nuclear strategy of "Mutual Assured Destruction" is the most obvious contemporary example of the law of retaliation. But it is an explicit version of an ancient impulse, the emotion of vengeance, that may have been installed in our brains by natural selection. Daly and Wilson observe, "In societies from every corner of the world, we can read of vows to avenge a slain father or brother, and of rituals that sanctify those vows—of a mother raising her son to avenge a father who died in the avenger's infancy, of graveside vows, of drinking the deceased kinsman's blood as a covenant, or keeping his bloody garment as a relic."[73] Modern states often find themselves at odds with their citizens' craving for revenge. They prosecute vigilantes—people who "take the law into their own hands"—and, with a few recent exceptions, ignore the clamoring of crime victims and their relatives for a say in decisions to prosecute, plea-bargain, or punish.

As we saw in Chapter 10, for revenge to work as a deterrent it has to be implacable. Exacting revenge is a risky business, because if an adversary was dangerous enough to have hurt you in the first place, he is not likely to take punishment lying down. Since the damage has already been done, a coolly rational victim may not see it in his interests to retaliate. And since the aggressor can anticipate this, he could call the victim's bluff and abuse him with impunity. If, on the other hand, potential victims and their kin would be so consumed with the lust for retribution as to raise a son to avenge a slain father, drink the kinsman's blood as a covenant, and so on, an aggressor might think twice before aggressing.[74]

The law of retaliation requires that the vengeance have a moralistic pretext to distinguish it from a raw assault. The avenger must have been provoked by a prior act of aggression or other injustice. Studies of feuds, wars, and ethnic violence show that the perpetrators are almost always inflamed by some grievance against their targets.[75] The danger inherent in this psychology is obvious: two sides may disagree over whether an initial act of violence was justified (perhaps as an act of self-defense, the recovery of ill-gotten gains, or retribution for an earlier offense) or was an act of unprovoked aggression. One side may count an even number of reprisals and feel that the scales of justice have been balanced, while the other side counts an odd number and feels that they still have a score to settle.[76] Self-deception may embolden each side's belief in the rectitude of its cause and make reconciliation almost impossible.

Also necessary for vengeance to work as a deterrent is that the willingness to pursue it be made public, because the whole point of deterrence is to give would-be attackers second thoughts *beforehand*. And this brings us to Hobbes's final reason for quarrel.

~

THIRDLY, GLORY—THOUGH a more accurate word would be "honor." Hobbes's observation that men fight over "a word, a smile, a different opinion, and any other sign of undervalue" is as true now as it was in the seventeenth century. For as long as urban crime statistics have been recorded, the most frequent cause of homicide has been "argument"—what police blotters classify as "altercation of relatively trivial origin; insult, curse, jostling, etc."[77] A Dallas homicide detective recalls, "Murders result from little ol' arguments over nothing at all. Tempers flare. A fight starts, and somebody gets stabbed or shot. I've worked on cases where the principals had been arguing over a 10 cent record on a juke box, or over a one dollar gambling debt from a dice game."[78]

Wars between nation-states are often fought over national honor, even when the material stakes are small. In the late 1960s and early 1970s, most Americans had become disenchanted over their country's involvement in the war in Vietnam, which they thought was immoral or unwinnable or both. But rather than agreeing to withdraw American forces unconditionally, as the peace movement had advocated, a majority supported Richard Nixon and his slogan "Peace with Honor." In practice this turned into a slow withdrawal of American troops that prolonged the military presence until 1973 at a cost of twenty thousand American lives and the lives of many more Vietnamese—and with the same outcome, defeat of the South Vietnamese government. A defense of national honor was behind other recent wars, such as the British retaking of the Falkland Islands in 1982 and the American invasion of Grenada in 1983. A ruinous 1969 war between El Salvador and Honduras began with a disputed game between their national soccer teams.

Because of the logic of deterrence, fights over personal or national honor are not as idiotic as they seem. In a hostile milieu, people and countries must advertise their willingness to retaliate against anyone who would profit at their expense, and that means maintaining a reputation for avenging any slight or trespass, no matter how small. They must make it known that, in the words of the Jim Croce song, "You don't tug on Superman's cape; you don't spit into the wind; you don't pull the mask off the old Lone Ranger; and you don't mess around with Jim."

The mentality is foreign to those of us who can get Leviathan to show up by dialing 911, but that option is not always available. It was not available to people in pre-state societies, or on the frontier in the Appalachians or the Wild West, or in the remote highlands of Scotland, the Balkans, or Indochina. It is not available to people who are unwilling to bring in the police because of the nature of their work, such as Prohibition rum-runners, inner-city drug dealers, and Mafia wise guys. And it is not available to nation-states in their dealings with one another. Daly and Wilson comment on the mentality that applies in all these arenas:

> In chronically feuding and warring societies, an essential manly virtue is the capacity for violence; head-hunting and coup counting may then become prestigious, and the commission of a homicide may even be an obligatory rite of passage. To turn the other cheek is not saintly but stupid. Or contemptibly weak.[79]

So the social constructionists I cited earlier are not wrong in pointing to a culture of combative masculinity as a major cause of violence. But they are wrong in thinking that it is peculiarly American, that it is caused by separation from one's mother or an unwillingness to express one's emotions, and that it is an arbitrary social construction that can be "deconstructed" by verbal commentary. And fans of the public health approach are correct that rates of violence vary with social conditions, but they are wrong in thinking that violence is a pathology in anything like the medical sense. Cultures of honor spring up all over the world because they amplify universal human emotions like pride, anger, revenge, and the love of kith and kin, and because they appear at the time to be sensible responses to local conditions.[80] Indeed, the emotions themselves are thoroughly familiar even when they don't erupt in violence, such as in road rage, office politics, political mudslinging, academic backstabbing, and email flame wars.

In *Culture of Honor*, the social psychologists Richard Nisbett and Dov Cohen show that violent cultures arise in societies that are beyond the reach of the law and in which precious assets are easily stolen.[81] Societies that herd animals meet both conditions. Herders tend to live in territories that are unsuitable for growing crops and thus far from the centers of government. And their major asset, livestock, is easier to steal than the major asset of farmers, land. In herding societies a man can be stripped of his wealth (and of his ability to acquire wealth) in an eyeblink. Men in that milieu cultivate a hair trigger for violent retaliation, not just against rustlers, but against anyone who would test their resolve by signs of disrespect that could reveal them to be easy pickings for rustlers. Scottish highlanders, Appalachian mountain men, Western cowboys, Masai warriors, Sioux Indians, Druze and Bedouin tribesmen, Balkan clansmen, and Indochinese Montagnards are familiar examples.

A man's honor is a kind of "social reality" in John Searle's sense: it exists because everyone agrees it exists, but it is no less real for that, since it resides in a shared granting of power. When the lifestyle of a people changes, their culture of honor can stay with them for a long time, because it is difficult for anyone to be the first to renounce the culture. The very act of renouncing it can be a concession of weakness and low status even when the sheep and mountains are a distant memory.

The American South has long had higher rates of violence than the North, including a tradition of dueling among "men of honor" such as Andrew Jackson. Nisbett and Cohen note that much of the South was originally settled by Scottish and Irish herdsman, whereas the North was settled by English farmers. Also, for much of its history the mountainous frontier of the South was beyond the reach of the law. The resulting Southern culture of honor is, remarkably, alive at the turn of the twenty-first century in laws and social attitudes. Southern states place fewer restrictions on gun ownership, allow people to shoot an assailant or burglar without having to retreat first, are tolerant of spanking by parents and corporal punishment by schools, are more hawkish on issues of national defense, and execute more of their criminals.[82]

These attitudes do not float in a cloud called "culture" but are visible in the psychology of individual Southerners. Nisbett and Cohen advertised a fake psychology experiment at the liberal University of Michigan. To get to the lab, respondents had to squeeze by a stooge who was filing papers in a hallway. As a respondent brushed past him, the stooge slammed the drawer shut and muttered, "Asshole." Students from Northern states laughed him off, but students from Southern states were visibly upset. The Southerners had elevated levels of testosterone and cortisol (a stress hormone) and reported lower levels of self-esteem. They compensated by giving a firmer handshake and acting more dominant toward the experimenter, and on the way out of the lab they refused to back down when another stooge approached in a narrow hallway and one of the two had to step aside. It's not that Southerners walk around chronically fuming: a control group who had not been insulted were as cool and collected as the Northerners. And Southerners do not approve of violence in the abstract, only of violence provoked by an insult or trespass.

African American inner-city neighborhoods are among the more conspicuously violent environments in Western democracies, and they too have an entrenched culture of honor. In his insightful essay "The Code of the Streets," the sociologist Elijah Anderson describes the young men's obsession with respect, their cultivation of a reputation for toughness, their willingness to engage in violent retaliation for any slight, and their universal acknowledgment of the rules of this code.[83] Were it not for giveaways in their dialect, such as "If someone disses you, you got to straighten them out," Anderson's description of the code would be indistinguishable from accounts of the culture of honor among white Southerners.

Inner-city African Americans were never goatherds, so why did they develop a culture of honor? One possibility is that they brought it with them from the South when they migrated to large cities after the two world wars—a nice irony for Southern racists who would blame inner-city violence on something distinctively African American. Another factor is that the young men's wealth is easily stealable, since it is often in the form of cash or drugs. A third is that the ghettos are a kind of frontier in which police protection is unreliable—the gangsta rap group Public Enemy has a recording called "911 Is a Joke." A fourth is that poor people, especially young men, cannot take pride in a prestigious job, a nice house, or professional accomplishments, and this may be doubly true for African Americans after centuries of slavery and discrimination. Their reputation on the streets is their only claim to status. Finally, Anderson points out that the code of the streets is self-perpetuating. A majority of African American families in the inner city subscribe to peaceable middle-class values they refer to as "decent."[84] But that is not enough to end the culture of honor:

> Everybody knows that if the rules are violated, there are penalties. Knowledge of the code is thus largely defensive; it is literally necessary for operating in public. Therefore, even though families with a decency orientation are usually opposed to the values of the code, they often reluctantly encourage their children's familiarity with it to enable them to negotiate the inner-city environment.[85]

Studies of the dynamics of ghetto violence are consistent with Anderson's analysis. The jump in American urban crime rates between 1985 and 1993 can be tied in part to the appearance of crack cocaine and the underground economy it spawned. As the economist Jeff Grogger points out, "Violence is a way to enforce property rights in the absence of legal recourse."[86] The emergence of violence within the new drug economy then set off the expected Hobbesian trap. As the criminologist Jeffrey Fagan noted, gun use spread contagiously as "young people who otherwise wouldn't carry guns felt that they had to in order to avoid being victimized by their armed peers."[87] And as we saw in the chapter on politics, conspicuous economic inequality is a good predictor of violence (better than poverty itself), presumably because men deprived of legitimate means of acquiring status compete for status on the streets instead.[88] It is not surprising, then, that when African American teenagers are taken out of underclass neighborhoods they are no more violent or delinquent than white teenagers.[89]

∼

HOBBES'S ANALYSIS OF the causes of violence, borne out by modern data on crime and war, shows that violence is not a primitive, irrational urge, nor is it a "pathology" except in the metaphorical sense of a condition that everyone would like to eliminate. Instead, it is a near-inevitable outcome of the dynamics of self-interested, rational social organisms.

But Hobbes is famous for presenting not just the causes of violence but a means of preventing it: "a common power to keep them all in awe." His commonwealth was a means of implementing the principle "that a man be willing, when others are so too . . . to lay down this right to all things; and be contented with so much liberty against other men, as he would allow other men against himself."[90] People vest authority in a sovereign person or assembly who can use the collective force of the contractors to hold each one to the agreement, because "covenants, without the sword, are but words, and of no strength to secure a man at all."[91]

A governing body that has been granted a monopoly on the legitimate use of violence can neutralize each of Hobbes's reasons for quarrel. By inflicting penalties on aggressors, the governing body eliminates the profitability of invading for gain. That in turn defuses the Hobbesian trap in which mutually distrustful peoples are each tempted to inflict a preemptive strike to avoid being invaded for gain. And a system of laws that defines infractions and penalties and metes them out disinterestedly can obviate the need for a hair trigger for retaliation and the accompanying culture of honor. People can rest assured that someone *else* will impose disincentives on their enemies, making it unnecessary for them to maintain a belligerent stance to prove they are not punching bags. And having a third party measure the infractions and the punishments circumvents the hazard of self-deception, which ordinarily convinces those on each side that they have suffered the greater number of offenses. These advantages of third-party intercession can also come from nongovernmental methods of conflict resolution, in which mediators try to help the hostile parties negotiate an agreement or arbitrators render a verdict but cannot enforce it.[92] The problem with these toothless measures is that the parties can always walk away when the outcome doesn't come out the way they want.

Adjudication by an armed authority appears to be the most effective general violence-reduction technique ever invented. Though we debate whether tweaks in criminal policy, such as executing murderers versus locking them up for life, can reduce violence by a few percentage points, there can be no debate on the massive effects of having a criminal justice system as opposed to living in anarchy. The shockingly high homicide rates of pre-state societies, with 10 to 60 percent of the men dying at the hands of other men, provide one kind of evidence.[93] Another is the emergence of a violent culture of honor in just about any corner of the world that is beyond the reach of the law.[94] Many historians argue that people acquiesced to centralized authorities during the Middle Ages and other periods to relieve themselves of the burden of having to retaliate against those who would harm them and their kin.[95] And the growth of those authorities may explain the *hundredfold* decline in homicide rates in European societies since the Middle Ages.[96] The United States saw a dramatic reduction in urban crime rates from the first half of the nineteenth century to the second half, which coincided with the formation of professional police forces in the cities.[97] The causes of the decline in American crime in the 1990s are controversial and probably multifarious, but many criminologists trace it in part to more intensive community policing and higher incarceration rates of violent criminals.[98]

The inverse is true as well. When law enforcement vanishes, all manner of violence breaks out: looting, settling old scores, ethnic cleansing, and petty warfare among gangs, warlords, and mafias. This was obvious in the remnants of Yugoslavia, the Soviet Union, and parts of Africa in the 1990s, but can also happen in countries with a long tradition of civility. As a young teenager in proudly peaceable Canada during the romantic 1960s, I was a true believer in Bakunin's anarchism. I laughed off my parents' argument that if the government ever laid down its arms all hell would break loose. Our competing predictions were put to the test at 8:00 A.M. on October 17, 1969, when the Montreal police went on strike. By 11:20 A.M. the first bank was robbed. By noon most downtown stores had closed because of looting. Within a few more hours, taxi drivers burned down the garage of a limousine service that had competed with them for airport customers, a rooftop sniper killed a provincial police officer, rioters broke into several hotels and restaurants, and a doctor slew a burglar in his suburban home. By the end of the day, six banks had been robbed, a hundred shops had been looted, twelve fires had been set, forty carloads of storefront glass had been broken, and three million dollars in property damage had been inflicted, before city authorities had to call in the army and, of course, the Mounties to restore order.[99] This decisive empirical test left my politics in tatters (and offered a foretaste of life as a scientist).

The generalization that anarchy in the sense of a lack of government leads to anarchy in the sense of violent chaos may seem banal, but it is often overlooked in today's still-romantic climate. Government in general is anathema to many conservatives, and the police and prison system are anathema to many liberals. Many people on the left, citing uncertainty about the deterrent value of capital punishment compared to life imprisonment, maintain that deterrence is not effective in general. And many oppose more effective policing of inner-city neighborhoods, even though it may be the most effective way for their decent inhabitants to abjure the code of the streets. Certainly we must combat the racial inequities that put too many African American men in prison, but as the legal scholar Randall Kennedy has argued, we must also combat the racial inequities that leave too many African Americans exposed to criminals.[100] Many on the right oppose decriminalizing drugs, prostitution, and gambling without factoring in the costs of the zones of anarchy that, by their own free-market logic, are inevitably spawned by prohibition policies. When demand for a commodity is high, suppliers will materialize, and if they cannot protect their property rights by calling the police, they will do so with a violent culture of honor. (This is distinct from the moral argument that our current drug policies incarcerate multitudes of nonviolent people.) School-children are currently fed the disinformation that Native Americans and other peoples in pre-state societies were inherently peaceable, leaving them uncomprehending, indeed contemptuous, of one of our species' greatest inventions, democratic government and the rule of law.

Where Hobbes fell short was in dealing with the problem of policing the police. In his view, civil war was such a calamity that any government—monarchy, aristocracy, or democracy—was preferable to it. He did not seem to appreciate that in practice a leviathan would not be an otherworldly sea monster but a human being or group of them, complete with the deadly sins of greed, mistrust, and honor. (As we saw in the preceding chapter, this became the obsession of the heirs of Hobbes who framed the American Constitution.) Armed men are always a menace, so police who are not under tight democratic control can be a far worse calamity than the crime and feuding that go on without them. In the twentieth century, according to the political scientist R. J. Rummel in *Death by Government,* 170 million people were killed by their own governments. Nor is murder-by-government a relic of the tyrannies of the middle of the century. The World Conflict List for the year 2000 reported:

> *The stupidest conflict in this year's count is Cameroon.* Early in the year, Cameroon was experiencing widespread problems with violent crime. The government responded to this crisis by creating and arming militias and paramilitary groups to stamp out the crime extrajudicially. Now, while violent crime has fallen, the militias and paramilitaries have created far more chaos and death than crime ever would have. Indeed, as the year wore on mass graves were discovered that were tied to the paramilitary groups.[101]

The pattern is familiar from other regions of the world (including our own) and shows that civil libertarians' concern about abusive police practices is an indispensable counterweight to the monopoly on violence we grant the state.

~

DEMOCRATIC LEVIATHANS HAVE proven to be an effective antiviolence measure, but they leave much to be desired. Because they fight violence with violence or the threat of violence, they can be a danger themselves. And it would be far better if we could find a way to get people to abjure violence to begin with rather than punishing them after the fact. Worst of all, no one has yet figured out how to set up a worldwide democratic leviathan that would penalize the aggressive competition, defuse the Hobbesian traps, and eliminate the cultures of honor that hold between the most dangerous perpetrators of violence of all, nation-states. As Kant noted, "The depravity of human nature is displayed without disguise in the unrestricted relations which obtain between the various nations."[102] The great question is how to get people and nations to repudiate violence from the start, preempting escalations of hostility before they can take off.

In the 1960s it all seemed so simple. War is unhealthy for children and other living things. What if they gave a war and nobody came? War: What is it good for? Absolutely nothing! The problem with these sentiments is that the other side has to feel the same way at the same time. In 1939 Neville Chamberlain offered his own antiwar slogan, "Peace in our time." It was followed by a world war and a holocaust, because his adversary did not agree that war is good for absolutely nothing. Chamberlain's successor, Churchill, explained why peace is not a simple matter of unilateral pacifism: "Nothing is worse than war? Dishonor is worse than war. Slavery is worse than war." A popular bumper sticker captures a related sentiment: IF YOU WANT PEACE, WORK FOR JUSTICE. The problem is that what one side sees as honor and justice the other side may see as dishonor and injustice. Also, "honor" can be a laudable willingness to defend life and liberty, but it can also be a reckless refusal to deescalate.

Sometimes all sides really do see that they would be better off beating their swords into plowshares. Scholars such as John Keegan and Donald Horowitz have noted a general decline in the taste for violence as a means of settling disputes within most Western democracies in the last half-century.[103] Civil wars, corporal and capital punishment, deadly ethnic riots, and foreign wars requiring face-to-face killing have declined or vanished. And as I have mentioned, though some decades in recent centuries have been more violent than others, the overall trend in crime has been downward.

One possible reason is the cosmopolitan forces that work to expand people's moral circle. Another may be the long-term effects of living with a leviathan. Today's civility in Europe, after all, followed centuries of beheadings and public hangings and exiles to penal colonies. And Canada may be more peaceable than its neighbor in part because its government outraced its people to the land. Unlike the United States, where settlers fanned out over a vast two-dimensional landscape with innumerable nooks and crannies, the habitable portion of Canada is a one-dimensional ribbon along the American border without remote frontiers and enclaves in which cultures of honor could fester. According to the Canadian studies scholar Desmond Morton, "Our west expanded in an orderly, peaceful fashion, with the police arriving before the settlers."[104]

But people can become less truculent without the external incentives of dollars and cents or governmental brute force. People all over the world have reflected on the futility of violence (at least when they are evenly enough matched with their adversaries that no one can prevail). A New Guinean native laments, "War is bad and nobody likes it. Sweet potatoes disappear, pigs disappear, fields deteriorate, and many relatives and friends get killed. But one cannot help it."[105] Chagnon reports that some Yanomamö men reflect on the futility of their feuds and a few make it known that they will have nothing to do with raiding.[106] In such cases it can become clear that both sides would come out ahead by splitting the differences between them rather than continuing to fight over them. During the trench warfare of World War I, weary British and German soldiers would probe each other's hostile intent with momentary respites in shelling. If the other side responded with a respite in kind, long periods of unofficial peace broke out beneath the notice of their bellicose commanders.[107] As a British soldier said, "We don't want to kill you, and you don't want to kill us, so why shoot?"[108]

The most consequential episode in which belligerents sought a way to release their deadly embrace was the Cuban Missile Crisis of 1962, when the United States discovered Soviet nuclear missiles in Cuba and demanded that they be removed. Khrushchev and Kennedy were both reminded of the human costs of the nuclear brink they were approaching, Khrushchev by memories of two world wars fought on his soil, Kennedy by a graphic briefing of the aftermath of an atomic bomb. And each understood they were in a Hobbesian trap. Kennedy had just read *The Guns of August* and saw how the leaders of great nations could blunder into a pointless war. Khrushchev wrote to Kennedy:

> You and I should not now pull on the ends of the rope in which you have tied a knot of war, because the harder you and I pull, the tighter this knot will become. And a time may come when this knot is tied so tight that the person who tied it is no longer capable of untying it, and then the knot will have to be cut.[109]

By identifying the trap, they could formulate a shared goal of escaping it. In the teeth of opposition from many of their advisers and large sectors of their publics, both made concessions that averted a catastrophe.

The problem with violence, then, is that the advantages of deploying it or renouncing it depend on what the other side does. Such scenarios are the province of game theory, and game theorists have shown that the best decision for each player individually is sometimes the worst decision for both collectively. The most famous example is the Prisoner's Dilemma, in which partners in crime are held in separate cells. Each is promised freedom if he is the first to implicate his partner (who then will get a harsh sentence), a light sentence if neither implicates the other, and a moderate sentence if each implicates the other. The optimal strategy for each prisoner is to defect from their partnership, but when both do so they end up with a worse outcome than if each stayed loyal. Yet neither can stay loyal out of fear that his partner might defect and leave him with the worst outcome of all. The Prisoner's Dilemma is similar to the pacifist's dilemma: what is good for one (belligerence) is bad for both, but what is good for both (pacifism) is unattainable when neither can be sure the other is opting for it.

The only way to win a Prisoner's Dilemma is to change the rules or find a way out of the game. The World War I soldiers changed the rules in a way that has been much discussed in evolutionary psychology: play it repeatedly and apply a strategy of reciprocity, remembering the other player's last action and repaying him in kind.[110] But in many antagonistic encounters that is not an option, because when the other player defects he can destroy you—or, in the case of the Cuban Missile Crisis, destroy the world. In that case the players had to recognize they were in a futile game and mutually decide to get out of it.

Glover draws an important conclusion about how the cognitive component of human nature might allow us to reduce violence even when it appears to be a rational strategy at the time:

> Sometimes, apparently rational self-interested strategies turn out (as in the prisoners' dilemma . . .) to be self-defeating. This may look like a defeat for rationality, but it is not. Rationality is saved by its own open-endedness. If a strategy of following accepted rules of rationality is sometimes self-defeating, this is not the end. We revise the rules to take account of this, so producing a higher-order rational strategy. This in turn may fail, but again we go up a level. At whatever level we fail, there is always the process of standing back and going up a further level.[111]

The process of "standing back and going up a further level" might be necessary to overcome the emotional impediments to peace as well as the intellectual ones. Diplomatic peacemakers try to hurry along the epiphanies that prompt adversaries to extricate themselves from a deadly game. They try to blunt competition by carefully fashioning compromises over the disputed resources. They try to defuse Hobbesian traps via "confidence-building measures" such as making military activities transparent and bringing in third parties as guarantors. And they try to bring the two sides into each other's moral circles by facilitating trade, cultural exchanges, and people-to-people activities.

This is fine as far as it goes, but the diplomats are sometimes frustrated that at the end of the day the two sides seem to hate each other as much as they did at the beginning. They continue to demonize their opponents, warp the facts, and denounce the conciliators on their own side as traitors. Milton J. Wilkinson, a diplomat who failed to get the Greeks and Turks to bury the hatchet over Cyprus, suggests that peacemakers must understand the emotional faculties of adversaries and not just neutralize the current rational incentives. The best-laid plans of peacemakers are often derailed by the adversaries' ethnocentrism, sense of honor, moralization, and self-deception.[112] These mindsets evolved to deal with hostilities in the ancestral past, and we must bring them into the open if we are to work around them in the present.

An emphasis on the open-endedness of human rationality resonates with the finding from cognitive science that the mind is a combinatorial, recursive system.[113] Not only do we have thoughts, but we have thoughts about our thoughts, and thoughts about our thoughts about our thoughts. The advances in human conflict resolution we have encountered in this chapter—submitting to the rule of law, figuring out a way for both sides to back down without losing face, acknowledging the possibility of one's own self-deception, accepting the equivalence of one's own interests and other people's—depend on this ability.

Many intellectuals have averted their gaze from the evolutionary logic of violence, fearing that acknowledging it is tantamount to accepting it or even to approving it. Instead they have pursued the comforting delusion of the Noble Savage, in which violence is an arbitrary product of learning or a pathogen that bores into us from the outside. But denying the logic of violence makes it easy to forget how readily violence can flare up, and ignoring the parts of the mind that ignite violence makes it easy to overlook the parts that can extinguish it. With violence, as with so many other concerns, human nature is the problem, but human nature is also the solution.

Chapter 18

Gender

Now that its namesake year has come and gone, the movie *2001: A Space Odyssey* provides an opportunity to measure imagination against reality. Arthur C. Clarke's 1968 sci-fi classic traced out the destiny of our species from ape-men on the savanna to a transcendence of time, space, and bodies that we can only dimly comprehend. Clarke and the director, Stanley Kubrick, contrived a radical vision of life in the third millennium, and in some ways it has come to pass. A permanent space station is being built, and voice mail and the Internet are a routine part of our lives. In other regards Clarke and Kubrick were overoptimistic about the march of progress. We still don't have suspended animation, missions to Jupiter, or computers that read lips and plot mutinies. And in still other regards they missed the boat completely. In their vision of the year 2001, people recorded their words on typewriters; Clarke and Kubrick did not anticipate word processors or laptop computers. And in their depiction of the new millennium, the American women were "girl assistants": secretaries, receptionists, and flight attendants.

That these visionaries did not anticipate the revolution in women's status of the 1970s is a pointed reminder of how quickly social arrangements can change. It was not so long ago that women were seen as fit only to be housewives, mothers, and sexual partners, were discouraged from entering the professions because they would be taking the place of a man, and were routinely subjected to discrimination, condescension, and sexual extortion. The ongoing liberation of women after millennia of oppression is one of the great moral achievements of our species, and I consider myself fortunate to have lived through some of its major victories.

The change in the status of women has several causes. One is the inexorable logic of the expanding moral circle, which led also to the abolition of despotism, slavery, feudalism, and racial segregation.[1] In the midst of the Enlightenment, the early feminist Mary Astell (1688–1731) wrote:

If absolute Sovereignty be not necessary in a State how comes it to be so in a Family? or if in a Family why not in a State? since no reason can be alleg'd for the one that will not hold more strongly for the other.

If *all Men are born free*, how is it that all Women are born slaves? As they must be if the being subjected to the *inconstant, uncertain, unknown, arbitrary Will* of Men, be the perfect Condition of Slavery?[2]

Another cause is the technological and economic progress that made it possible for couples to have sex and raise children without a pitiless division of labor in which a mother had to devote every waking moment to keeping the children alive. Clean water, sanitation, and modern medicine lowered infant mortality and reduced the desire for large broods of children. Baby bottles and pasteurized cow's milk, and then breast pumps and freezers, made it possible to feed babies without their mothers being chained to them around the clock. Mass production made it cheaper to buy things than to make them by hand, and plumbing, electricity, and appliances reduced the domestic workload even more. The increased value of brains over brawn in the economy, the extension of the human lifespan (with the prospect of decades of life after childrearing), and the affordability of extended education changed the values of women's options in life. Contraception, amniocentesis, ultrasound, and reproductive technologies made it possible for women to defer childbearing to the optimal points in their lives.

And of course the other major cause of women's progress is feminism: the political, literary, and academic movements that channeled these advances into tangible changes in policies and attitudes. The first wave of feminism, bookended in the United States by the Seneca Falls convention of 1848 and the ratification of the Nineteenth Amendment to the Constitution in 1920, gave women the right to vote, to serve as jurors, to hold property in marriage, to divorce, and to receive an education. The second wave, flowering in the 1970s, brought women into the professions, changed the division of labor in the home, exposed sexist biases in business, government, and other institutions, and threw a spotlight on women's interests in all walks of life. The recent progress in women's rights has not drained feminism of its raison d'être. In much of the Third World, women's position has not improved since the Middle Ages, and in our own society women are still subjected to discrimination, harassment, and violence.

Feminism is widely seen as being opposed to the sciences of human nature. Many of those scientists believe that the minds of the two sexes differ at birth, and feminists have pointed out that such beliefs have long been used to justify the unequal treatment of women. Women were thought to be designed for childrearing and home life and to be incapable of the reason necessary for politics and the professions. Men were believed to harbor irresistible urges that made them harass and rape women, and that belief served to excuse the perpetrators and to license fathers and husbands to control women in the guise of protecting them. Therefore, it might seem, the theories that are most friendly to women are the Blank Slate—if nothing is innate, differences between the sexes cannot be innate—and the Noble Savage—if we harbor no ignoble urges, sexual exploitation can be eliminated by changing our institutions.

The belief that feminism requires a blank slate and a noble savage has become a powerful impetus for spreading disinformation. A 1994 headline in the *New York Times* science section, for example, proclaimed, "Sexes Equal on South Sea Isle."[3] It was based on the work of the anthropologist Maria Lepowsky, who (perhaps channeling the ghost of Margaret Mead) said that gender relations on the island of Vanatinai prove that "the subjugation of women by men is not a human universal, and it is not inevitable." Only late in the story do we learn what this supposed "equality" amounts to: that men must do bride service to pay for wives, that warfare had been waged exclusively by men (who raided neighboring islands for brides), that women spend more time caring for children and sweeping up pig excrement, and that men spend more time building their reputations and hunting wild boar (which is accorded more prestige by both sexes). A similar disconnect between headline and fact appeared in a 1998 *Boston Globe* story entitled "Girls Appear to Be Closing Aggression Gap with Boys." How much have they "closed this gap"? According to the story, they now commit murder at *one-tenth* the rate of boys.[4] And in a 1998 op-ed, the co-producer of *Ms.* magazine's "Take Our Daughters to Work Day" explained recent high school shootings with the remarkable assertion that boys in America "are being trained by their parents, other adults, and our culture and media to harass, assault, rape, and murder girls."[5]

On the other side, some conservatives are confirming feminists' worst fears by invoking dubious sex differences to condemn the choices of women. In a *Wall Street Journal* editorial, the political scientist Harvey Mansfield wrote that "the protective element of manliness is endangered by women having equal access to jobs outside the home."[6] A book by F. Carolyn Graglia called *Domestic Tranquility: A Brief Against Feminism* theorized that women's maternal and sexual instincts are being distorted by the assertiveness and analytical mind demanded by a career. The journalists Wendy Shalit and Danielle Crittenden recently advised women to marry young, postpone their careers, and care for children in traditional marriages, even though they could not have written their books if they had followed their own advice.[7] Leon Kass has taken it upon himself to inform young women what they want: "For the first time in human history, mature women by the tens of thousands live the entire decade of their twenties—their most fertile years—neither in the homes of their fathers nor in the homes of their husbands; unprotected, lonely, and out of sync with their inborn nature. Some women positively welcome this state of affairs, but most do not."[8]

There is, in fact, no incompatibility between the principles of feminism and the possibility that men and women are not psychologically identical. To repeat: equality is not the empirical claim that all groups of humans are interchangeable; it is the moral principle that individuals should not be judged or constrained by the average properties of their group. In the case of gender, the barely defeated Equal Rights Amendment put it succinctly: "Equality of Rights under the law shall not be denied or abridged by the United States or any state on account of sex." If we recognize this principle, no one has to spin myths about the indistinguishability of the sexes to justify equality. Nor should anyone invoke sex differences to justify discriminatory policies or to hector women into doing what they don't want to do.

In any case, what we do know about the sexes does not call for any action that would penalize or constrain one sex or the other. Many psychological traits relevant to the public sphere, such as general intelligence, are the same on average for men and women, and virtually all psychological traits may be found in varying degrees among the members of each sex. No sex difference yet discovered applies to every last man compared with every last woman, so generalizations about a sex will always be untrue of many individuals. And notions like "proper role" and "natural place" are scientifically meaningless and give no grounds for restricting freedom.

Despite these principles, many feminists vehemently attack research on sexuality and sex differences. The politics of gender is a major reason that the application of evolution, genetics, and neuroscience to the human mind is bitterly resisted in modern intellectual life. But unlike other human divisions such as race and ethnicity, where any biological differences are minor at most and scientifically uninteresting, gender cannot possibly be ignored in the science of human beings. The sexes are as old as complex life and are a fundamental topic in evolutionary biology, genetics, and behavioral ecology. To disregard them in the case of our own species would be to make a hash of our understanding of our place in the cosmos. And of course differences between men and women affect every aspect of our lives. We all have a mother and a father, are attracted to members of the opposite sex (or notice our contrast with the people who are), and are never unaware of the sex of our siblings, children, and friends. To ignore gender would be to ignore a major part of the human condition.

The goal of this chapter is to clarify the relation between the biology of human nature and current controversies on the sexes, including the two most incendiary, the gender gap and sexual assault. With both of these hot buttons, I will argue against the conventional wisdom associated with certain people who claim to speak on behalf of feminism. That may create an illusion that the arguments go against feminism in general, or even against the interests of women. They don't in the least, and I must begin by showing why.

～

FEMINISM IS OFTEN derided because of the arguments of its lunatic fringe— for example, that all intercourse is rape, that all women should be lesbians, or that only 10 percent of the population should be allowed to be male.[9] Feminists reply that proponents of women's rights do not speak with one voice, and that feminist thought comprises many positions, which have to be evaluated independently.[10] That is completely legitimate, but it cuts both ways. To criticize a *particular* feminist proposal is not to attack feminism in general.

Anyone familiar with academia knows that it breeds ideological cults that are prone to dogma and resistant to criticism. Many women believe that this has now happened to feminism. In her book *Who Stole Feminism?* the philosopher Christina Hoff Sommers draws a useful distinction between two schools of thought.[11] *Equity feminism* opposes sex discrimination and other forms of unfairness to women. It is part of the classical liberal and humanistic tradition that grew out of the Enlightenment, and it guided the first wave of feminism and launched the second wave. *Gender feminism* holds that women continue to be enslaved by a pervasive system of male dominance, the gender system, in which "bi-sexual infants are transformed into male and female gender personalities, the one destined to command, the other to obey."[12] It is opposed to the classical liberal tradition and allied instead with Marxism, postmodernism, social constructionism, and radical science. It has became the credo of some women's studies programs, feminist organizations, and spokespeople for the women's movement.

Equity feminism is a moral doctrine about equal treatment that makes no commitments regarding open empirical issues in psychology or biology. Gender feminism is an empirical doctrine committed to three claims about human nature. The first is that the differences between men and women have nothing to do with biology but are socially constructed in their entirety. The second is that humans possess a single social motive—power—and that social life can be understood only in terms of how it is exercised. The third is that human interactions arise not from the motives of people dealing with each other as individuals but from the motives of *groups* dealing with other groups—in this case, the male gender dominating the female gender.

In embracing these doctrines, the genderists are handcuffing feminism to railroad tracks on which a train is bearing down. As we shall see, neuroscience, genetics, psychology, and ethnography are documenting sex differences that almost certainly originate in human biology. And evolutionary psychology is documenting a web of motives other than group-against-group dominance (such as love, sex, family, and beauty) that entangle us in many conflicts and confluences of interest with members of the same sex and of the opposite sex.

Gender feminists want either to derail the train or to have other women join them in martyrdom, but the other women are not cooperating. Despite their visibility, gender feminists do not speak for all feminists, let alone for all women.

To begin with, research on the biological basis of sex differences has been led by women. Because it is so often said that this research is a plot to keep women down, I will have to name names. Researchers on the biology of sex differences include the neuroscientists Raquel Gur, Melissa Hines, Doreen Kimura, Jerre Levy, Martha McClintock, Sally Shaywitz, and Sandra Witelson and the psychologists Camilla Benbow, Linda Gottfredson, Diane Halpern, Judith Kleinfeld, and Diane McGuinness. Sociobiology and evolutionary psychology, sometimes stereotyped as a "sexist discipline," is perhaps the most bi-gendered academic field I am familiar with. Its major figures include Laura Betzig, Elizabeth Cashdan, Leda Cosmides, Helena Cronin, Mildred Dickeman, Helen Fisher, Patricia Gowaty, Kristen Hawkes, Sarah Blaffer Hrdy, Magdalena Hurtado, Bobbie Low, Linda Mealey, Felicia Pratto, Marnie Rice, Catherine Salmon, Joan Silk, Meredith Small, Barbara Smuts, Nancy Wilmsen Thornhill, and Margo Wilson.

It is not just gender feminism's collision with science that repels many feminists. Like other inbred ideologies, it has produced strange excrescences, like the offshoot known as difference feminism. Carol Gilligan has become a gender-feminist icon because of her claim that men and women guide their moral reasoning by different principles: men think about rights and justice; women have feelings of compassion, nurturing, and peaceful accommodation.[13] If true, it would disqualify women from becoming constitutional lawyers, Supreme Court justices, and moral philosophers, who make their living by reasoning about rights and justice. But it is not true. Many studies have tested Gilligan's hypothesis and found that men and women differ little or not at all in their moral reasoning.[14] So difference feminism offers women the worst of both worlds: invidious claims without scientific support. Similarly, the gender-feminist classic called *Women's Ways of Knowing* claims that the sexes differ in their styles of reasoning. Men value excellence and mastery in intellectual matters and skeptically evaluate arguments in terms of logic and evidence; women are spiritual, relational, inclusive, and credulous.[15] With sisters like these, who needs male chauvinists?

Gender feminism's disdain for analytical rigor and classical liberal principles has recently been excoriated by equity feminists, among them Jean Bethke Elshtain, Elizabeth Fox-Genovese, Wendy Kaminer, Noretta Koertge, Donna Laframboise, Mary Lefkowitz, Wendy McElroy, Camille Paglia, Daphne Patai, Virginia Postrel, Alice Rossi, Sally Satel, Christina Hoff Sommers, Nadine Strossen, Joan Kennedy Taylor, and Cathy Young.[16] Well before them, prominent women writers demurred from gender-feminist ideology, including Joan

Didion, Doris Lessing, Iris Murdoch, Cynthia Ozick, and Susan Sontag.[17] And ominously for the movement, a younger generation has rejected the gender feminists' claims that love, beauty, flirtation, erotica, art, and heterosexuality are pernicious social constructs. The title of the book *The New Victorians: A Young Woman's Challenge to the Old Feminist Order* captures the revolt of such writers as Rene Denfeld, Karen Lehrman, Katie Roiphe, and Rebecca Walker, and of the movements called Third Wave, Riot Grrrl Movement, Pro-Sex Feminism, Lipstick Lesbians, Girl Power, and Feminists for Free Expression.[18]

The difference between gender feminism and equity feminism accounts for the oft-reported paradox that most women do not consider themselves feminists (about 70 percent in 1997, up from about 60 percent a decade before), yet they agree with every major feminist position.[19] The explanation is simple: the word "feminist" is often associated with gender feminism, but the positions in the polls are those of equity feminism. Faced with these signs of slipping support, gender feminists have tried to stipulate that only they can be considered the true advocates of women's rights. For example, in 1992 Gloria Steinem said of Paglia, "Her calling herself a feminist is sort of like a Nazi saying they're not anti-Semitic."[20] And they have invented a lexicon of epithets for what in any other area would be called disagreement: "backlash," "not getting it," "silencing women," "intellectual harassment."[21]

All this is an essential background to the discussions to come. To say that women and men do not have interchangeable minds, that people have desires other than power, and that motives belong to individual people and not just to entire genders is not to attack feminism or to compromise the interests of women, despite the misconception that gender feminism speaks in their name. All the arguments in the remainder of this chapter have been advanced most forcefully by women.

～

WHY ARE PEOPLE so afraid of the idea that the minds of men and women are not identical in every respect? Would we really be better off if everyone were like Pat, the androgynous nerd from *Saturday Night Live*? The fear, of course, is that different implies unequal—that if the sexes differed in any way, then men would have to be better, or more dominant, or have all the fun.

Nothing could be farther from biological thinking. Trivers alluded to a "symmetry in human relationships," which embraced a "genetic equality of the sexes."[22] From a gene's point of view, being in the body of a male and being in the body of a female are equally good strategies, at least on average (circumstances can nudge the advantage somewhat in either direction).[23] Natural selection thus tends toward an equal investment in the two sexes: equal numbers, an equal complexity of bodies and brains, and equally effective designs for survival. Is it better to be the size of a male baboon and have six-inch canine teeth or to be the size of a female baboon and not have them? Merely to ask the question is to reveal its pointlessness. A biologist would say that it's better to have the male adaptations to deal with male problems and the female adaptations to deal with female problems.

So men are not from Mars, nor are women from Venus. Men and women are from Africa, the cradle of our evolution, where they evolved together as a single species. Men and women have all the same genes except for a handful on the Y chromosome, and their brains are so similar that it takes an eagle-eyed neuroanatomist to find the small differences between them. Their average levels of general intelligence are the same, according to the best psychometric estimates,[24] and they use language and think about the physical and living world in the same general way. They feel the same basic emotions, and both enjoy sex, seek intelligent and kind marriage partners, get jealous, make sacrifices for their children, compete for status and mates, and sometimes commit aggression in pursuit of their interests.

But of course the minds of men and women are not identical, and recent reviews of sex differences have converged on some reliable differences.[25] Sometimes the differences are large, with only slight overlap in the bell curves. Men have a much stronger taste for no-strings sex with multiple or anonymous partners, as we see in the almost all-male consumer base for prostitution and visual pornography.[26] Men are far more likely to compete violently, sometimes lethally, with one another over stakes great and small (as in the recent case of a surgeon and an anesthesiologist who came to blows in the operating room while a patient lay on the table waiting to have her gall bladder removed).[27] Among children, boys spend far more time practicing for violent conflict in the form of what psychologists genteelly call "rough-and-tumble play."[28] The ability to manipulate three-dimensional objects and space in the mind also shows a large difference in favor of men.[29]

With some other traits the differences are small on average but can be large at the extremes. That happens for two reasons. When two bell curves partly overlap, the farther out along the tail you go, the larger the discrepancies between the groups. For example, men on average are taller than women, and the discrepancy is greater for more extreme values. At a height of five foot ten, men outnumber women by a ratio of thirty to one; at a height of six feet, men outnumber women by a ratio of two thousand to one. Also, confirming an expectation from evolutionary psychology, for many traits the bell curve for males is flatter and wider than the curve for females. That is, there are proportionally more males at the extremes. Along the left tail of the curve, one finds that boys are far more likely to be dyslexic, learning disabled, attention deficient, emotionally disturbed, and mentally retarded (at least for some types of retardation).[30] At the right tail, one finds that in a sample of talented students who score above 700 (out of 800) on the mathematics section of the Scholastic Assessment Test, boys outnumber girls by thirteen to one, even though the scores of boys and girls are similar within the bulk of the curve.[31]

With still other traits, the average values for the two sexes differ by smaller amounts and in different directions for different traits.[32] Though men, on average, are better at mentally rotating objects and maps, women are better at remembering landmarks and the positions of objects. Men are better throwers; women are more dexterous. Men are better at solving mathematical word problems, women at mathematical calculation. Women are more sensitive to sounds and smells, have better depth perception, match shapes faster, and are much better at reading facial expressions and body language. Women are better spellers, retrieve words more fluently, and have a better memory for verbal material.

Women experience basic emotions more intensely, except perhaps anger.[33] Women have more intimate social relationships, are more concerned about them, and feel more empathy toward their friends, though not toward strangers. (The common view that women are more empathic toward everyone is both evolutionarily unlikely and untrue.) They maintain more eye contact, and smile and laugh far more often.[34] Men are more likely to compete with one another for status using violence or occupational achievement, women more likely to use derogation and other forms of verbal aggression.

Men have a higher tolerance for pain and a greater willingness to risk life and limb for status, attention, and other dubious rewards. The Darwin Awards, given annually to "the individuals who ensure the long-term survival of our species by removing themselves from the gene pool in a sublimely idiotic fashion," almost always go to men. Recent honorees include the man who squashed himself under a Coke machine after tipping it forward to get a free can, three men who competed over who could stomp the hardest on an anti-tank mine, and the would-be pilot who tied weather balloons to his lawn chair, shot two miles into the air, and drifted out to sea (earning just an Honorable Mention because he was rescued by helicopter).

Women are more attentive to their infants' everyday cries (though both sexes respond equally to cries of extreme distress) and are more solicitous toward their children in general.[35] Girls play more at parenting and trying on social roles, boys more at fighting, chasing, and manipulating objects. And men and women differ in their patterns of sexual jealousy, their mate preferences, and their incentives to philander.

Many sex differences, of course, have nothing to do with biology. Hair styles and dress vary capriciously across centuries and cultures, and in recent decades participation in universities, professions, and sports has switched from mostly male to fifty-fifty or mostly female. For all we know, some of the current sex differences may be just as ephemeral. But gender feminists argue that *all* sex differences, other than the anatomical ones, come from the expectations of parents, playmates, and society. The radical scientist Anne Fausto-Sterling wrote:

> The key biological fact is that boys and girls have different genitalia, and it is this biological difference that leads adults to interact differently with different babies whom we conveniently color-code in pink or blue to make it unnecessary to go peering into their diapers for information about gender.[36]

But the pink-and-blue theory is becoming less and less credible. Here are a dozen kinds of evidence that suggest that the difference between men and women is more than genitalia-deep.

- Sex differences are not an arbitrary feature of Western culture, like the decision to drive on the left or on the right. In all human cultures, men and women are seen as having different natures. All cultures divide their labor by sex, with more responsibility for childrearing by women and more control of the public and political realms by men. (The division of labor emerged even in a culture where everyone had been committed to stamping it out, the Israeli kibbutz.) In all cultures men are more aggressive, more prone to stealing, more prone to lethal violence (including war), and more likely to woo, seduce, and trade favors for sex. And in all cultures one finds rape, as well as proscriptions against rape.[37]
- Many of the psychological differences between the sexes are exactly what an evolutionary biologist who knew only their physical differences would predict.[38] Throughout the animal kingdom, when the female has to invest more calories and risk in each offspring (in the case of mammals, through pregnancy and nursing), she also invests more in nurturing the offspring after birth, since it is more costly for a female to replace a child than for a male to replace one. The difference in investment is accompanied by a greater competition among males over opportunities to mate, since mating with many partners is more likely to multiply the number of offspring of a male than the number of offspring of a female. When the average male is larger than the average female (as is true of men and women), it bespeaks an evolutionary history of greater violent competition by males over mating opportunities. Other physical traits of men, such as later puberty, greater adult strength, and shorter lives, also indicate a history of selection for high-stakes competition.
- Many of the sex differences are found widely in other primates, indeed, throughout the mammalian class.[39] The males tend to compete more aggressively and to be more polygamous; the females tend to invest more in

parenting. In many mammals a greater territorial range is accompanied by an enhanced ability to navigate using the geometry of the spatial layout (as opposed to remembering individual landmarks). More often it is the male who has the greater range, and that is true of human hunter-gatherers. Men's advantage in using mental maps and performing 3-D mental rotation may not be a coincidence.[40]

- Geneticists have found that the diversity of the DNA in the mitochondria of different people (which men and women inherit from their mothers) is far greater than the diversity of the DNA in Y chromosomes (which men inherit from their fathers). This suggests that for tens of millennia men had greater variation in their reproductive success than women. Some men had many descendants and others had none (leaving us with a small number of distinct Y chromosomes), whereas a larger number of women had a more evenly distributed number of descendants (leaving us with a larger number of distinct mitochondrial genomes). These are precisely the conditions that cause sexual selection, in which males compete for opportunities to mate and females choose the best-quality males.[41]

- The human body contains a mechanism that causes the brains of boys and the brains of girls to diverge during development.[42] The Y chromosome triggers the growth of testes in a male fetus, which secrete androgens, the characteristically male hormones (including testosterone). Androgens have lasting effects on the brain during fetal development, in the months after birth, and during puberty, and they have transient effects at other times. Estrogens, the characteristically female sex hormones, also affect the brain throughout life. Receptors for the sex hormones are found in the hypothalamus, the hippocampus, and the amygdala in the limbic system of the brain, as well as in the cerebral cortex.

- The brains of men differ visibly from the brains of women in several ways.[43] Men have larger brains with more neurons (even correcting for body size), though women have a higher percentage of gray matter. (Since men and women are equally intelligent overall, the significance of these differences is unknown.) The interstitial nuclei in the anterior hypothalamus, and a nucleus of the stria terminalis, also in the hypothalamus, are larger in men; they have been implicated in sexual behavior and aggression. Portions of the cerebral commissures, which link the left and right hemispheres, appear to be larger in women, and their brains may function in a less lopsided manner than men's. Learning and socialization can affect the microstructure and functioning of the human brain, of course, but probably not the size of its visible anatomical structures.

- Variation in the level of testosterone among different men, and in the same man in different seasons or at different times of day, correlates with libido, self-confidence, and the drive for dominance.[44] Violent criminals have higher levels than nonviolent criminals; trial lawyers have higher levels than those who push paper. The relations are complicated for a number of reasons. Over a broad range of values, the concentration of testosterone in the bloodstream doesn't matter. Some traits, such as spatial abilities, peak at moderate rather than high levels. The effects of testosterone depend on the number and distribution of receptors for the molecule, not just on its concentration. And one's psychological state can affect testosterone levels as well as the other way around. But there is a causal relation, albeit a complicated one. When women preparing for a sex-change operation are given androgens, they improve on tests of mental rotation and get worse on tests of verbal fluency. The journalist Andrew Sullivan, whose medical condition had lowered his testosterone levels, describes the effects of injecting it: "The rush of a T shot is not unlike the rush of going on a first date or speaking before an audience. I feel braced. After one injection, I almost got in a public brawl for the first time in my life. There is always a lust peak—every time it takes me unaware."[45] Though testosterone levels in men and women do not overlap, variations in level have similar kinds of effects in the two sexes. High-testosterone women smile less often and have more extramarital affairs, a stronger social presence, and even a stronger handshake.

- Women's cognitive strengths and weaknesses vary with the phase of their menstrual cycle.[46] When estrogen levels are high, women get even better at tasks on which they typically do better than men, such as verbal fluency. When the levels are low, women get better at tasks on which men typically do better, such as mental rotation. A variety of sexual motives, including their taste in men, vary with the menstrual cycle as well.[47]

- Androgens have permanent effects on the developing brain, not just transient effects on the adult brain.[48] Girls with congenital adrenal hyperplasia overproduce androstenedione, the androgen hormone made famous by the baseball slugger Mark McGwire. Though their hormone levels are brought to normal soon after birth, the girls grow into tomboys, with more rough-and-tumble play, a greater interest in trucks than dolls, better spatial abilities, and, when they get older, more sexual fantasies and attractions involving other girls. Those who are treated with hormones only later in childhood show male patterns of sexuality when they become young adults, including quick arousal by pornographic images, an autonomous sex drive centered on genital stimulation, and the equivalent of wet dreams.[49]

- The ultimate fantasy experiment to separate biology from socialization would be to take a baby boy, give him a sex-change operation, and have his parents raise him as a girl and other people treat him as one. If gender is socially constructed, the child should have the mind of a normal girl; if it depends on prenatal hormones, the child should feel like a boy trapped in a girl's body. Remarkably, the experiment has been done in real life—not out of scientific curiosity, of course, but as a result of disease and accidents. One study looked at twenty-five boys who were born without a penis (a birth defect known as cloacal exstrophy) and who were then castrated and raised as girls. *All* of them showed male patterns of rough-and-tumble play and had typically male attitudes and interests. More than half of them spontaneously declared they were boys, one when he was just five years old.[50]

In a famous case study, an eight-month-old boy lost his penis in a botched circumcision (not by a mohel, I was relieved to learn, but by a bungling doctor). His parents consulted the famous sex researcher John Money, who had maintained that "Nature is a political strategy of those committed to maintaining the status quo of sex differences." He advised them to let the doctors castrate the baby and build him an artificial vagina, and they raised him as a girl without telling him what had happened.[51] I learned about the case as an undergraduate in the 1970s, when it was offered as proof that babies are born neuter and acquire a gender from the way they are raised. A *New York Times* article from the era reported that Brenda (née Bruce) "has been sailing contentedly through childhood as a genuine girl."[52] The facts were suppressed until 1997, when it was revealed that from a young age Brenda felt she was a boy trapped in a girl's body and gender role.[53] She ripped off frilly dresses, rejected dolls in favor of guns, preferred to play with boys, and even insisted on urinating standing up. At fourteen she was so miserable that she decided either to live her life as a male or to end it, and her father finally told her the truth. She underwent a new set of operations, assumed a male identity, and today is happily married to a woman.

• Children with Turner's syndrome are genetically neuter. They have a single X chromosome, inherited from either their mother or their father, instead of the usual two X chromosomes of a girl (one from her mother, the other from her father) or the X and Y of a boy (the X from his mother, the Y from his father). Since a female body plan is the default among mammals, they look and act like girls. Geneticists have discovered that parents' bodies can molecularly imprint genes on the X chromosome so they become more or less active in the developing bodies and brains of their children. A Turner's syndrome girl who gets her X chromosome from her father may have genes that are evolutionarily optimized for girls (since a paternal X always ends up in a daughter). A Turner's girl who gets her X from her mother may have genes that are evolutionarily optimized for boys (since a maternal X, though it can end up in either sex, will act unopposed only in a son, who has no counterpart to the X genes on his puny

Y chromosome). And in fact Turner's girls do differ psychologically depending on which parent gave them their X. The ones with an X from their father (which is destined for a girl) were better at interpreting body language, reading emotions, recognizing faces, handling words, and getting along with other people compared to the ones with an X from their mother (which is fully active only in a boy).[54]

- Contrary to popular belief, parents in contemporary America do not treat their sons and daughters very differently.[55] A recent assessment of 172 studies involving 28,000 children found that boys and girls are given similar amounts of encouragement, warmth, nurturance, restrictiveness, discipline, and clarity of communication. The only substantial difference was that about two-thirds of the boys were discouraged from playing with dolls, especially by their fathers, out of a fear that they would become gay. (Boys who prefer girls' toys often do turn out gay, but forbidding them the toys does not change the outcome.) Nor do differences between boys and girls depend on their observing masculine behavior in their fathers and feminine behavior in their mothers. When Hunter has two mommies, he acts just as much like a boy as if he had a mommy and a daddy.

Things are not looking good for the theory that boys and girls are born identical except for their genitalia, with all other differences coming from the way society treats them. If that were true, it would be an amazing coincidence that in every society the coin flip that assigns each sex to one set of roles would land the same way (or that one fateful flip at the dawn of the species should have been maintained without interruption across all the upheavals of the past hundred thousand years). It would be just as amazing that, time and again, society's arbitrary assignments matched the predictions that a Martian biologist would make for our species based on our anatomy and the distribution of our genes. It would seem odd that the hormones that make us male and female in the first place also modulate the characteristically male and female mental traits, both decisively in early brain development and in smaller degrees throughout our lives. It would be all the more odd that a second genetic mechanism differentiating the sexes (genomic imprinting) also installs characteristic male and female talents. Finally, two key predictions of the social construction theory—that boys treated as girls will grow up with girls' minds, and that differences between boys and girls can be traced to differences in how their parents treat them—have gone down in flames.

Of course, just because many sex differences are rooted in biology does not mean that one sex is superior, that the differences will emerge for all people in all circumstances, that discrimination against a person based on sex is justified, or that people should be coerced into doing things typical of their sex. But neither are the differences without consequences.

$\sim$

BY NOW MANY people are happy to say what was unsayable in polite company a few years ago: that males and females do not have interchangeable minds. Even the comic pages have commented on the shift in the debate, as we see in this dialogue between the free-associating, junkfood-loving Zippy and the cartoonist's alter ego Griffy:

© Bill Griffith. Reprinted with special permission of King Features Syndicate.

But among many professional women the existence of sex differences is still a source of discomfort. As one colleague said to me, "Look, I know that males and females are not identical. I see it in my kids, I see it in myself, I know about the research. I can't explain it, but when I read claims about sex differences, *steam comes out of my ears*." The most likely cause of her disquiet is captured in a recent editorial by Betty Friedan, the cofounder of the National Organization for Women and the author of the 1963 book *The Feminine Mystique*:

> Though the women's movement has begun to achieve equality for women on many economic and political measures, the victory remains incomplete. To take two of the simplest and most obvious indicators: women still earn no more than 72 cents for every dollar that men earn, and we are nowhere near equality in numbers at the very top of decision making in business, government, or the professions.[56]

Like Friedan, many people believe that the gender gap in wages and a "glass ceiling" that keeps women from rising to the uppermost levels of power are the two main injustices facing women in the West today. In his 1999 State of the Union address, Bill Clinton said, "We can be proud of this progress, but 75 cents on the dollar is still only three-quarters of the way there, and Americans can't be satisfied until we're all the way there." The gender gap and the glass ceiling have inspired lawsuits against companies that have too few women in the top positions, pressure on the government to regulate all salaries so men and women are paid according to the "comparable worth" of their jobs, and aggressive measures to change girls' attitudes to the professions, such as the annual Take Our Daughters to Work Day.

Scientists and engineers face the issue in the form of the "leaky pipeline." Though women make up almost 60 percent of university students and about half of the students majoring in many fields of science, the proportion advancing to the next career stage diminishes as they go from being undergraduates to graduate students to postdoctoral fellows to junior professors to tenured professors. Women make up less than 20 percent of the workforce in science, engineering, and technology development, and only 9 percent of the workforce in engineering.[57] Readers of the flagship journals *Science* and *Nature* have seen two decades of headlines such as "Diversity: Easier Said Than Done" and "Efforts to Boost Diversity Face Persistent Problems."[58] A typical story, commenting on the many national commissions set up to investigate the problem, said, "These activities are meant to continue chipping away at a problem that, experts say, begins with negative messages in elementary school, continues through undergraduate and graduate programs that erect barriers—financial, academic, and cultural—to all but the best candidates, and persists into the workplace."[59] A meeting in 2001 of the presidents of nine elite American universities called for "significant changes," such as setting aside grants and fellowships for women faculty, giving them the best parking spaces on campus, and ensuring that the percentage of women faculty equals the percentage of women students.[60]

But there is something odd in these stories about negative messages, hidden barriers, and gender prejudices. The way of science is to lay out every hypothesis that could account for a phenomenon and to eliminate all but the correct one. Scientists prize the ability to think up alternative explanations, and proponents of a hypothesis are expected to refute even the unlikely ones. Nonetheless, discussions of the leaky pipeline in science rarely even *mention* an alternative to the theory of barriers and bias. One of the rare exceptions was a sidebar to a 2000 story in *Science,* which quoted from a presentation at the National Academy of Engineering by the social scientist Patti Hausman:

The question of why more women don't choose careers in engineering has a rather obvious answer: Because they don't want to. Wherever you go, you will find females far less likely than males to see what is so fascinating about ohms, carburetors, or quarks. Reinventing the curriculum will not make me more interested in learning how my dishwasher works.[61]

An eminent woman engineer in the audience immediately denounced her analysis as "pseudoscience." But Linda Gottfredson, an expert in the literature on vocational preferences, pointed out that Hausman had the data on her side: "On average, women are more interested in dealing with people and men with things." Vocational tests also show that boys are more interested in "realistic," "theoretical," and "investigative" pursuits, and girls more interested in "artistic" and "social" pursuits.

Hausman and Gottfredson are lonely voices, because the gender gap is almost always analyzed in the following way. Any imbalance between men and women in their occupations or earnings is direct proof of gender bias—if not in the form of overt discrimination, then in the form of discouraging messages and hidden barriers. The possibility that men and women might differ from each other in ways that affect what jobs they hold or how much they get paid may never be mentioned in public, because it will set back the cause of equity in the workplace and harm the interests of women. It is this conviction that led Friedan and Clinton, for example, to say that we will not have attained gender equity until earnings and representation in the professions are identical for men and women. In a 1998 television interview, Gloria Steinem and the congresswoman Bella Abzug called the very idea of sex differences "poppycock" and "anti-American crazy thinking," and when Abzug was asked whether gender equality meant equal numbers in every field, she replied, "Fifty-fifty— absolutely."[62] This analysis of the gender gap has also become the official position of universities. That the presidents of the nation's elite universities are happy to accuse their colleagues of shameful prejudice without even considering alternative explanations (whether or not they would end up accepting them) shows how deeply rooted the taboo is.

The problem with this analysis is that inequality of *outcome* cannot be used as proof of inequality of *opportunity* unless the groups being compared are identical in all of their psychological traits, which is likely to be true only if we are blank slates. But the suggestion that the gender gap may arise, even in part, from differences between the sexes can be fightin' words. Anyone bringing it up is certain to be accused of "wanting to keep women in their place" or "justifying the status quo." This makes about as much sense as saying that a scientist who studies why women live longer than men "wants old men to die." And far from being a ploy by self-serving men, analyses exposing the flaws of the glass-ceiling theory have largely come from women, including Hausman, Gottfredson, Judith Kleinfeld, Karen Lehrman, Cathy Young, and Camilla Benbow, the economists Jennifer Roback, Felice Schwartz, Diana Furchtgott-Roth, and Christine Stolba, the legal scholar Jennifer Braceras, and, more guardedly, the economist Claudia Goldin and the legal scholar Susan Estrich.[63]

I believe these writers have given us a better understanding of the gender gap than the standard one, for a number of reasons. Their analysis is not afraid of the possibility that the sexes might differ, and therefore does not force us to choose between scientific findings on human nature and the fair treatment of women. It offers a more sophisticated understanding of the causes of the gender gap, one that is consistent with our best social science. It takes a more respectful view of women and their choices. And ultimately it promises more humane and effective remedies for gender inequities in the workplace.

Before presenting the new analysis of the gender gap from equity feminists, let me reiterate three points that are not in dispute. First, discouraging women from pursuing their ambitions, and discriminating against them on the basis of their sex, are injustices that should be stopped wherever they are discovered.

Second, there is no doubt that women faced widespread discrimination in the past and continue to face it in some sectors today. This cannot be proven by showing that men earn more than women or that the sex ratio departs from fifty-fifty, but it can be proven in other ways. Experimenters can send out fake résumés or grant proposals that are identical in all ways except the sex of the applicant and see whether they are treated differently. Economists can do a regression analysis that takes measures of people's qualifications and interests and determines whether the men and the women earn different amounts, or are promoted at different rates, *when their qualifications and interests are statistically held constant.* The point that differences in outcome don't show discrimination unless one has equated for other relevant traits is elementary social science (not to mention common sense), and is accepted by all economists when they analyze data sets looking for evidence of wage discrimination.[64]

Third, there is no question of whether women are "qualified" to be scientists, CEOs, leaders of nations, or elite professionals of any other kind. That was decisively answered years ago: some are and some aren't, just as some men are qualified and some aren't. The only question is whether the proportions of qualified men and women must be identical.

As in many other topics related to human nature, people's unwillingness to think in statistical terms has led to pointless false dichotomies. Here is how to think about gender distributions in the professions without having to choose between the extremes of "women are unqualified" and "fifty-fifty absolutely," or between "there is no discrimination" and "there is nothing but discrimination."

In a free and unprejudiced labor market, people will be hired and paid according to the match between their traits and the demands of the job. A given job requires some mixture of cognitive talents (such as mathematical or linguistic skill), personality traits (such as risk taking or cooperation), and tolerance of lifestyle demands (rigid schedules, relocations, updating job skills). And it offers some mixture of personal rewards: people, gadgets, ideas, the outdoors, pride in workmanship. The salary is influenced, among other things, by supply and demand: how many people want the job, how many can do it, and how many the employer can pay to do it. Readily filled jobs may pay less; difficult-to-fill jobs may pay more.

People vary in the traits relevant to employment. Most people can think logically, work with people, tolerate conflict or unpleasant surroundings, and so on, but not to an identical extent; everyone has a unique profile of strengths and tastes. Given all the evidence for sex differences (some biological, some cultural, some both), the statistical distributions for men and women in these strengths and tastes are unlikely to be identical. If one now matches the distribution of traits for men and for women with the distribution of the demands of the jobs in the economy, the chance that the proportion of men and of women in each profession will be identical, or that the mean salary of men and of women will be identical, is very close to zero—even if there were no barriers or discrimination.

None of this implies that women will end up with the short end of the stick. It depends on the menu of opportunities that a given society makes available. If there are more high-paying jobs that call for typical male strengths (say, willingness to put oneself in physical danger, or an interest in machines), men may do better on average; if there are more that call for typical female strengths (say, a proficiency with language, or an interest in people), women may do better on average. In either case, members of both sexes will be found in both kinds of jobs, just in different numbers. That is why some relatively prestigious professions are dominated by women. An example is my own field, the study of language development in children, in which women outnumber men by a large margin.[65] In her book *The First Sex: The Natural Talents of Women and How They Are Changing the World*, the anthropologist Helen Fisher speculates that the culture of business in our knowledge-driven, globalized economy will soon favor women. Women are more articulate and cooperative, are not as obsessed with rank, and are better able to negotiate win-win outcomes. The workplaces of the new century, she predicts, will increasingly demand these talents, and women may surpass men in status and earnings.

In today's world, of course, the gap favors men. Some of the gap is caused by discrimination. Employers may underestimate the skills of women, or assume that an all-male workplace is more efficient, or worry that their male employees will resent female supervisors, or fear resistance from prejudiced customers and clients. But the evidence suggests that not *all* sex differences in the professions are caused by these barriers.[66] It is unlikely, for example, that among academics the mathematicians are unusually biased against women, the developmental psycholinguists are unusually biased against men, and the evolutionary psychologists are unusually free of bias.

In a few professions, differences in ability may play some role. The fact that more men than women have exceptional abilities in mathematical reasoning and in mentally manipulating 3-D objects is enough to explain a departure from a fifty-fifty sex ratio among engineers, physicists, organic chemists, and professors in some branches of mathematics (though of course it does not mean that the proportion of women should be anywhere near zero).

In most professions, average differences in ability are irrelevant, but average differences in *preferences* may set the sexes on different paths. The most dramatic example comes from an analysis by David Lubinski and Camilla Benbow of a sample of mathematically precocious seventh-graders selected in a nationwide talent search.[67] The teenagers were born during the second wave of feminism, were encouraged by their parents to develop their talents (all were sent to summer programs in math and science), and were fully aware of their ability to achieve. But the gifted girls told the researchers that they were more interested in people, "social values," and humanitarian and altruistic goals, whereas the gifted boys said they were more interested in things, "theoretical values," and abstract intellectual inquiry. In college, the young women chose a broad range of courses in the humanities, arts, and sciences, whereas the boys were geeks who stuck to math and science. And sure enough, fewer than 1 percent of the young women pursued doctorates in math, physical sciences, or engineering, whereas 8 percent of the young men did. The women went into medicine, law, the humanities, and biology instead.

This asymmetry is writ large in massive surveys of job-related values and career choices, another kind of study in which men and women actually say what they want rather than having activists speak for them.[68] On average, men's self-esteem is more highly tied to their status, salary, and wealth, and so is their attractiveness as a sexual partner and marriage partner, as revealed in studies of what people look for in the opposite sex.[69] Not surprisingly, men say they are more keen to work longer hours and to sacrifice other parts of their lives—to live in a less attractive city, or to leave friends and family when they relocate—in order to climb the corporate ladder or achieve notoriety in their fields. Men, on average, are also more willing to undergo physical discomfort and danger, and thus are more likely to be found in grungy but relatively lucrative jobs such as repairing factory equipment, working on oil rigs, and jackhammering sludge from the inside of oil tanks. Women, on average, are more likely to choose administrative support jobs that offer lower pay in air-conditioned offices. Men are greater risk takers, and that is reflected in their career paths even when qualifications are held constant. Men prefer to work for corporations, women for government agencies and nonprofit organizations. Male doctors are more likely to specialize and to open up private practices; female doctors are more likely to be general practitioners on salary in hospitals and clinics. Men are more likely to be managers in factories, women more likely to be managers in human resources or corporate communications.

Mothers are more attached to their children, on average, than are fathers. That is true in societies all over the world and probably has been true of our lineage since the first mammals evolved some two hundred million years ago. As Susan Estrich puts it, "Waiting for the connection between gender and parenting to be broken is waiting for Godot." This does not mean that women in any society have ever been uninterested in work; among hunter-gatherers, women do most of the gathering and some of the hunting, especially when it involves nets rather than rocks and spears.[70] Nor does it mean that men in any society are indifferent to their children; male parental investment is a conspicuous and zoologically unusual feature of *Homo sapiens*. But it does mean that the biologically ubiquitous tradeoff between investing in a child and working to stay healthy (ultimately to beget or invest in other children) may be balanced at different points by males and females. Not only are women the sex who nurse, but women are more attentive to their babies' well-being and, in surveys, place a higher value on spending time with their children.[71]

So even if both sexes value work and both sexes value children, the different weightings may lead women, more often than men, to make career choices that allow them to spend more time with their children—shorter or more flexible hours, fewer relocations, skills that don't become obsolete as quickly—in exchange for lower wages or prestige. As the economist Jennifer Roback points out, "Once we observe that people sacrifice money income for other pleasurable things we can infer next to nothing by comparing the income of one person with another's."[72] The economist Gary Becker has shown that marriage can magnify the effects of sex differences, even if they are small to begin with, because of what economists call the law of comparative advantage. In couples where the husband can earn a bit more than the wife, but the wife is a somewhat better parent than the husband, they might rationally decide they are both better off if she works less than he does.[73]

To repeat: none of this means that sex discrimination has vanished, or that it is justified when it occurs. The point is only that gender gaps *by themselves* say nothing about discrimination unless the slates of men and women are blank, which they are not. The only way to establish discrimination is to compare their jobs or wages when choices and qualifications are equalized. And in fact a recent study of data from the National Longitudinal Survey of Youth found that childless women between the ages of twenty-seven and thirty-three earn 98 cents to men's dollar.[74] Even to people who are cynical about the motivations of American employers, this should come as no shock. In a cutthroat market, any company stupid enough to overlook qualified women or to overpay unqualified men would be driven out of business by a more meritocratic competitor.

Now, there is nothing in science or social science that would rule out policies implementing a fifty-fifty distribution of wages and jobs between the sexes, if a democracy decided that this was an inherently worthy goal. What the findings do say is that such policies will come with costs as well as benefits. The obvious benefit of equality-of-outcome policies is that they might neutralize the remaining discrimination against women. But if men and women are not interchangeable, the costs have to be considered as well.

Some costs would be borne by men or by both sexes. The two most obvious are the possibility of reverse discrimination against men and of a false presumption of sexism among the men and women who make decisions about hiring and salary today. Another cost borne by both sexes is the inefficiency that could result if employment decisions were based on factors other than the best match between the demands of a job and the traits of the person.

But many of the costs of equality-of-outcome policies would be borne by *women*. Many women scientists are opposed to hard gender preferences in science, such as designated faculty positions for women, or the policy (advocated by one activist) in which federal research grants would be awarded in exact proportion to the number of men and women who apply for them. The problem with these well-meaning policies is that they can plant seeds of doubt in people's minds about the excellence of the beneficiaries. As the astronomer Lynne Hillenbrand said, "If you're given an opportunity for the reason of being female, it doesn't do anyone any favors; it makes people question why you're there."[75]

Certainly there *are* institutional barriers to the advancement of women. People are mammals, and we should think through the ethical implications of the fact that it is women who bear, nurse, and disproportionately raise children. One ought not to assume that the default human being is a man and that children are an indulgence or an accident that strikes a deviant subset. Sex differences therefore can be used to justify, rather than endanger, woman-friendly policies such as parental leave, subsidized childcare, flexible hours, and stoppages of the tenure clock or the elimination of tenure altogether (a possibility recently broached by the biologist and Princeton University president Shirley Tilghman).

Of course, there is no such thing as a free lunch, and these policies are also decisions—perhaps justifiable ones—to penalize men and women who are childless, have grown children, or choose to stay at home with their children. But even when it comes to weighing these tradeoffs, thinking about human nature can raise deep new questions that could ultimately improve the lot of working women. Which of the onerous job demands that deter women really contribute to economic efficiency, and which are obstacle courses in which men compete for alpha status? In reasoning about fairness in the workplace, should we consider people as isolated individuals, or should we consider them as members of families who probably will have children at some point in their lives and who probably will care for aging parents at some point in their lives?

If we trade off some economic efficiency for more pleasant working conditions in all jobs, might there be a net increase in happiness? I don't have answers, but the questions are well worth asking.

There is one more reason that acknowledging sex differences can be more humane than denying them. It is men and women, not the male gender and the female gender, who prosper or suffer, and those men and women are endowed with brains—perhaps not identical brains—that give them values and an ability to make choices. Those choices should be respected. A regular feature of the lifestyle pages is the story about women who are made to feel ashamed about staying at home with their children. As they always say, "I thought feminism was supposed to be about choices." The same should apply to women who do choose to work but also to trade off some income in order to "have a life" (and, of course, to men who make that choice). It is not obviously progressive to insist that equal numbers of men and women work eighty-hour weeks in a corporate law firm or leave their families for months at a time to dodge steel pipes on a frigid oil platform. And it is grotesque to demand (as advocates of gender parity did in the pages of *Science*) that more young women "be conditioned to choose engineering," as if they were rats in a Skinner box.[76]

Gottfredson points out, "If you insist on using gender parity as your measure of social justice, it means you will have to keep many men and women out of the work they like best and push them into work they don't like."[77] She is echoed by Kleinfeld on the leaky pipeline in science: "We should not be sending [gifted] women the messages that they are less worthy human beings, less valuable to our civilization, lazy or low in status, if they choose to be teachers rather than mathematicians, journalists rather than physicists, lawyers rather than engineers."[78] These are not hypothetical worries: a recent survey by the National Science Foundation found that many more women than men say they majored in science, mathematics, or engineering under pressure from teachers or family members rather than to pursue their own aspirations—and that many eventually switched out for that reason.[79] I will give the final word to Margaret Mead, who, despite being wrong in her early career about the malleability of gender, was surely right when she said, "If we are to achieve a richer culture, rich in contrasting values, we must recognize the whole gamut of human potentialities, and so weave a less arbitrary social fabric, one in which each diverse human gift will find a fitting place."

～

OTHER THAN THE gender gap, the most combustible recent issue surrounding the sexes has been the nature and causes of rape. When the biologist Randy Thornhill and the anthropologist Craig Palmer published *A Natural History of Rape* in 2000, they threatened a consensus that had held firm in intellectual life for a quarter of a century, and they brought down more condemnation on evolutionary psychology than any issue had in years.[80] Rape is a painful issue to write about, but also an unavoidable one. Nowhere else in modern intellectual life is the denial of human nature more passionately insisted upon, and nowhere else is the alternative more deeply misunderstood. Clarifying these issues, I believe, would go a long way toward reconciling three ideals that have needlessly been put into conflict: women's rights, a biologically informed understanding of human nature, and common sense.

The horror of rape gives it a special gravity in our understanding of the psychology of men and women. There is an overriding moral imperative in the study of rape: to reduce its occurrence. Any scientist who illuminates the causes of rape deserves our admiration, like a medical researcher who illuminates the cause of a disease, because understanding an affliction is the first step toward eliminating it. And since no one acquires the truth by divine revelation, we must also respect those who explore theories that may turn out to be incorrect. Moral criticism would seem to be in order only for those who would enforce dogmas, ignore evidence, or shut down research, because they would be protecting their reputations at the expense of victims of rapes that might not have occurred if we understood the phenomenon better.

Current sensibilities, unfortunately, are very different. In modern intellectual life the overriding moral imperative in analyzing rape is to proclaim that rape has nothing to do with sex. The mantra must be repeated whenever the subject comes up. "Rape is an abuse of power and control in which the rapist seeks to humiliate, shame, embarrass, degrade, and terrify the victim," the United Nations declared in 1993. "The primary objective is to exercise power and control over another person."[81] This was echoed in a 2001 *Boston Globe* op-ed piece that said, "Rape is not about sex; it is about violence and the use of sex to exert power and control. . . . Domestic violence and sexual assault are manifestations of the same powerful social forces: sexism and the glorification of violence."[82] When an iconoclastic columnist wrote a dissenting article on rape and battering, a reader responded:

> As a man who has been actively engaged for more than a decade as an educator and a counselor to help men to stop their violence against women, I find Cathy Young's Oct. 15 column disturbing and discouraging. She confuses issues by failing to acknowledge that men are socialized in a patriarchal culture that still supports their violence against women if they choose it.[83]

So steeped in the prevailing ideology was this counselor that he didn't notice that Young was *arguing against* the dogma he took as self-evidently true, not "failing to acknowledge" it. And his wording—"men are socialized in a patriarchal culture"—reproduces a numbingly familiar slogan.

The official theory of rape originated in an important 1975 book, *Against Our Will,* by the gender feminist Susan Brownmiller. The book became an emblem of a revolution in our handling of rape that is one of second-wave feminism's greatest accomplishments. Until the 1970s, rape was often treated by the legal system and popular culture with scant attention to the interests of women. Victims had to prove they resisted their attackers to within an inch of their lives or else they were seen as having consented. Their style of dress was seen as a mitigating factor, as if men couldn't control themselves when an attractive woman walked by. Also mitigating was the woman's sexual history, as if choosing to have sex with one man on one occasion were the same as agreeing to have sex with any man on any occasion. Standards of proof that were not required for other violent crimes, such as eyewitness corroboration, were imposed on charges of rape. Women's consent was often treated lightly in the popular media. It was not uncommon in movies for a reluctant woman to be handled roughly by a man and then melt into his arms. The suffering of rape victims was treated lightly as well; I remember teenage girls, in the wake of the sexual revolution in the early 1970s, joking to one another, "If a rape is inevitable, you might as well lie back and enjoy it." Marital rape was not a crime, date rape was not a concept, and rape during wartime was left out of the history books. These affronts to humanity are gone or on the wane in Western democracies, and feminism deserves credit for this moral advance.

But Brownmiller's theory went well beyond the moral principle that women have a right not to be sexually assaulted. It said that rape had nothing to do with an individual man's desire for sex but was a tactic by which the entire male gender oppressed the entire female gender. In her famous words:

> Man's discovery that his genitalia could serve as a weapon to generate fear must rank as one of the most important discoveries of prehistoric times, along with the use of fire and the first crude stone axe. From prehistoric times to the present, I believe, rape has played a critical function . . . it is nothing more or less than a conscious process of intimidation by which *all* men keep *all* women in a state of fear.[84]

This grew into the modern catechism: rape is not about sex, our culture socializes men to rape, it glorifies violence against women. The analysis comes right out of the gender-feminist theory of human nature: people are blank slates (who must be trained or socialized to want things); the only significant human motive is power (so sexual desire is irrelevant); and all motives and interests must be located in groups (such as the male sex and the female sex) rather than in individual people.

The Brownmiller theory is appealing even to people who are not gender feminists because of the doctrine of the Noble Savage. Since the 1960s most educated people have come to believe that sex should be thought of as natural, not shameful or dirty. Sex is good because sex is natural and natural things are good. But rape is bad; therefore, rape is not about sex. The motive to rape must come from social institutions, not from anything in human nature.

The violence-not-sex slogan is right about two things. Both parts are absolutely true for the victim: a woman who is raped experiences it as a violent assault, not as a sexual act. And the part about violence is true for the perpetrator by definition: if there is no violence or coercion, we do not call it rape. But the fact that rape has something to do with violence does not mean it has nothing to do with sex, any more than the fact that armed robbery has something to do with violence means it has nothing to do with greed. Evil men may use violence to get sex, just as they use violence to get other things they want.

I believe that the rape-is-not-about-sex doctrine will go down in history as an example of extraordinary popular delusions and the madness of crowds. It is preposterous on the face of it, does not deserve its sanctity, is contradicted by a mass of evidence, and is getting in the way of the only morally relevant goal surrounding rape, the effort to stamp it out.

Think about it. First obvious fact: Men often want to have sex with women who don't want to have sex with them. They use every tactic that one human being uses to affect the behavior of another: wooing, seducing, flattering, deceiving, sulking, and paying. Second obvious fact: Some men use violence to get what they want, indifferent to the suffering they cause. Men have been known to kidnap children for ransom (sometimes sending their parents an ear or finger to show they mean business), blind the victim of a mugging so the victim can't identify them in court, shoot out the kneecaps of an associate as punishment for ratting to the police or invading their territory, and kill a stranger for his brand-name athletic footwear. It would be an extraordinary fact, contradicting everything else we know about people, if some men *didn't* use violence to get sex.

Let's also apply common sense to the doctrine that men rape to further the interests of their gender. A rapist always risks injury at the hands of the woman defending herself. In a traditional society, he risks torture, mutilation, and death at the hands of her relatives. In a modern society, he risks a long prison term. Are rapists really assuming these risks as an altruistic sacrifice to benefit the billions of strangers that make up the male gender? The idea becomes even less credible when we remember that rapists tend to be losers and nobodies, while presumably the main beneficiaries of the patriarchy are the rich and powerful. Men do sacrifice themselves for the greater good in wartime, of course, but they are either conscripted against their will or promised public adulation when their exploits are made public. But rapists usually commit their acts in private and try to keep them secret. And in most times and places, a man who rapes a woman in his community is treated as scum. The idea that all men are engaged in brutal warfare against all women clashes with the elementary fact that men have mothers, daughters, sisters, and wives, whom they care for more than they care for most other men. To put the same point in biological terms, every person's genes are carried in the bodies of other people, half of whom are of the opposite sex.

Yes, we must deplore the sometimes casual treatment of women's autonomy in popular culture. But can anyone believe that our culture literally "teaches men to rape" or "glorifies the rapist"? Even the callous treatment of rape victims in the judicial system of yesteryear has a simpler explanation than that all men benefit by rape. Until recently jurors in rape cases were given a warning from the seventeenth-century jurist Lord Matthew Hale that they should evaluate a woman's testimony with caution, because a rape charge is "easily made and difficult to defend against, even if the accused is innocent."[85] The principle is consistent with the presumption of innocence built into our judicial system and with its preference to let ten guilty people go free rather than jail one innocent. Even so, let's suppose that the men who applied this policy to rape did tilt it toward their own collective interests. Let's suppose that they leaned on the scales of justice to minimize their own chances of ever being falsely accused of rape (or accused under ambiguous circumstances) and that they placed insufficient value on the injustice endured by women who would not see their assailants put behind bars. That would indeed be unjust, but it is still not the same thing as *encouraging* rape as a conscious tactic to keep women down. If that were men's tactic, why would they have made rape a crime in the first place?

As for the morality of believing the not-sex theory, there is none. If we have to acknowledge that sexuality can be a source of conflict and not just wholesome mutual pleasure, we will have rediscovered a truth that observers of the human condition have noted throughout history. And if a man rapes for sex, that does not mean that he "just can't help it" or that we have to excuse him, any more than we have to excuse the man who shoots the owner of a liquor store to raid the cash register or who bashes a driver over the head to steal his BMW. The great contribution of feminism to the morality of rape is to put issues of *consent* and *coercion* at center stage. The ultimate motives of the rapist are irrelevant.

Finally, think about the humanity of the picture that the gender-feminist theory has painted. As the equity feminist Wendy McElroy points out, the theory holds that "even the most loving and gentle husband, father, and son is a beneficiary of the rape of women they love. No ideology that makes such vicious accusations against men as a class can heal any wounds. It can only provoke hostility in return."[86]

~

BROWNMILLER ASKED A revealing rhetorical question:

> Does one need scientific methodology in order to conclude that the anti-female propaganda that permeates our nation's cultural output promotes a climate in which acts of sexual hostility directed against women are not only tolerated but ideologically encouraged?

McElroy responded: "The answer is a clear and simple 'yes.' One needs scientific methodology to verify any empirical claim." And she called attention to the consequences of Brownmiller's attitude: "One of the casualties of the new dogma on rape has been research. It is no longer 'sexually correct' to conduct studies on the causes of rape, because—as any right-thinking person knows—there is only one cause: patriarchy. Decades ago, during the heyday of liberal feminism and sexual curiosity, the approach to research was more sophisticated."[87] McElroy's suspicions are borne out by a survey of published "studies" of rape that found that fewer than one in ten tested hypotheses or used scientific methods.[88]

Scientific research on rape and its connections to human nature was thrown into the spotlight in 2000 with the publication of *A Natural History of Rape*. Thornhill and Palmer began with a basic observation: a rape can result in a conception, which could propagate the genes of the rapist, including any genes that had made him likely to rape. Therefore, a male psychology that included a capacity to rape would not have been selected against, and could have been selected for. Thornhill and Palmer argued that rape is unlikely to be a *typical* mating strategy because of the risk of injury at the hands of the victim and her relatives and the risk of ostracism from the community. But it could be an *opportunistic* tactic, becoming more likely when the man is unable to win the consent of women, alienated from a community (and thus undeterred by ostracism), and safe from detection and punishment (such as in wartime or pogroms). Thornhill and Palmer then outlined two theories. Opportunistic rape could be a Darwinian adaptation that was specifically selected for, as in certain insects that have an appendage with no function other than restraining a female during forced copulation. Or rape could be a by-product of two other features of the male mind: a desire for sex and a capacity to engage in opportunistic violence in pursuit of a goal. The two authors disagreed on which hypothesis was better supported by the data, and they left that issue unresolved.

No honest reader could conclude that the authors think rape is "natural" in the vernacular sense of being welcome or unavoidable. The first words of the book are, "As scientists who would like to see rape eradicated from human life . . . ," which are certainly not the words of people who think it is inevitable. Thornhill and Palmer discuss the environmental circumstances that affect the likelihood of rape, and they offer suggestions on how to reduce it. The idea that most men have the capacity to rape works, if anything, in the interests of women, because it calls for vigilance against acquaintance rape, marital rape, and rape during societal breakdowns. Indeed, the analysis jibes with Brownmiller's own data that ordinary men, including "nice" American boys in Vietnam, may rape in wartime. For that matter, Thornhill and Palmer's hypothesis that rape is on a continuum with the rest of male sexuality makes them strange allies with the most radical gender feminists, such as Catharine MacKinnon and Andrea Dworkin, who said that "seduction is often difficult to distinguish from rape. In seduction, the rapist often bothers to buy a bottle of wine."[89]

Most important, the book focuses in equal part on the pain of the victims. (Its draft title was *Why Men Rape, Why Women Suffer.*) Thornhill and Palmer explain in Darwinian terms why females throughout the animal kingdom resist being forced into sex, and argue that the agony that rape victims feel is deeply rooted in women's nature. Rape subverts female choice, the core of the ubiquitous mechanism of sexual selection. By choosing the male and the circumstances for sex, a female can maximize the chances that her offspring will be fathered by a male with good genes, a willingness and ability to share the responsibility of rearing the offspring, or both. As John Tooby and Leda Cosmides have put it, this ultimate (evolutionary) calculus explains why women evolved "to exert control over their own sexuality, over the terms of their relationships, and over the choice of which men are to be the fathers of their children." They resist being raped, and they suffer when their resistance fails, because "control over their sexual choices and relationships was wrested from them."[90]

Thornhill and Palmer's theory reinforces many points of an equity-feminist analysis. It predicts that from the woman's point of view, rape and consensual sex are completely different. It affirms that women's repugnance toward rape is not a symptom of neurotic repression, nor is it a social construct that could easily be the reverse in a different culture. It predicts that the suffering caused by rape is deeper than the suffering caused by other physical traumas or body violations. That justifies our working harder to prevent rape, and punishing the perpetrators more severely, than we do for other kinds of assault. Compare this analysis with the dubious claim by two gender feminists that an aversion to rape has to be pounded into women by every social influence they can think of:

> Female fear . . . [results] not only from women's personal backgrounds but from what women as a group have imbibed from history, religion, culture, social institutions, and everyday social interactions. Learned early in life, female fear is continually reinforced by such

social institutions as the school, the church, the law, and the press. Much is also learned from parents, siblings, teachers, and friends.[91]

But despite the congeniality of their analysis to women's interests, Thornhill and Palmer had broken a taboo, and the response was familiar: there were demonstrations, disruptions of lectures, and invective that would curdle your hair, as the popular malaprop has it. "Latest nauseating scientific theory" was a typical reaction, and radical scientists applied their usual standards of accuracy to denounce it. Hilary Rose, discussing a presentation of the theory by another biologist, wrote, "The sociobiologist David Barash's appeal in defense of his misogynist claims that men are naturally predisposed to rape, 'If Nature is sexist don't blame her sons,' can no longer plug into the old deference to science as the view from nowhere."[92] Barash, of course, had said no such thing; he had referred to rapists as criminals who should be punished. The science writer Margaret Wertheim began her review of Thornhill and Palmer's book by calling attention to a recent epidemic of rape in South Africa.[93] Pitting the theory that rape is "a byproduct of social conditioning and chaos" against the theory that rape has evolutionary and genetic origins, she sarcastically wrote that if the latter were true, "South Africa must be a hothouse for such genes." Two slurs for the price of one: the statement puts Thornhill and Palmer on the simplistic side of a false dichotomy (in fact, they devote many pages to the social conditions fostering rape) and slips in the innuendo that their theory is racist, too. The psychologist Geoffrey Miller, in his own mixed review of the book, diagnosed the popular reaction:

> The Natural History of Rape has already suffered the worst possible fate for a popular science book. Like The Descent of Man and The Bell Curve, it has become an ideological touchstone. People who wish to demonstrate their sympathy for rape victims and women in general have already learned that they must dismiss this book as sexist, reactionary pseudo-science. News stories that treat the book as a symptom of chauvinist cultural decay have greatly outnumbered reviews that assess it as science. Viewed sociologically, turning books into ideological touchstones can be useful. People can efficiently sort themselves out into like-minded cliques without bothering to read or think. However, there can be more to human discourse than ideological self-advertisement.[94]

It's unfortunate that Thornhill and Palmer themselves set up a dichotomy between the theory that rape is an adaptation (a specifically selected sexual strategy) and the theory that it is a by-product (a consequence of using violence in general), because it diverted attention from the more basic claim that rape has something to do with sex. I think their dichotomy is drawn too sharply. Male sexuality may have evolved in a world in which women were more discriminating than men about partners and occasions for sex. That would have led men to treat female reluctance as an obstacle to be overcome. (Another way to put it is that one can imagine a species in which the male could become sexually interested only if he detected reciprocal signs of interest on the part of the female, but that humans do not appear to be such a species.) How the woman's reluctance is overcome depends on the rest of the man's psychology and on his assessment of the circumstances. His usual tactics may include being kind, persuading the woman of his good intentions, and offering the proverbial bottle of wine, but may become increasingly coercive as certain risk factors are multiplied in: the man is a psychopath (hence insensitive to the suffering of others), an outcast (hence immune to ostracism), a loser (with no other means to get sex), or a soldier or ethnic rioter who considers an enemy subhuman and thinks he can get away with it. Certainly most men in ordinary circumstances do *not* harbor a desire to rape. According to surveys, violent rape is unusual in pornography and sexual fantasies, and according to laboratory studies of men's sexual arousal, depictions of actual violence toward a woman or signs of her pain and humiliation are a turnoff.[95]

What about the more basic question of whether the motives of rapists include sex? The gender-feminist argument that they do not points to the rapists who target older, infertile women, those who suffer from sexual dysfunction during the rape, those who coerce nonreproductive sexual acts, and those who use a condom. The argument is unconvincing for two reasons. First, these examples make up a minority of rapes, so the argument could be turned around to show that most rapes do have a sexual motive. And all these phenomena occur with consensual sex, too, so the argument leads to the absurdity that sex itself has nothing to do with sex. And date rape is a particularly problematic case for the not-sex theory. Most people agree that women have the right to say no at any point during sexual activity, and that if the man persists he is a rapist—but should we also believe that his motive has instantaneously changed from wanting sex to oppressing women?

On the other side there is an impressive body of evidence (reviewed more thoroughly by the legal scholar Owen Jones than by Thornhill and Palmer) that the motives for rape overlap with the motives for sex:[96]

- Coerced copulation is widespread among species in the animal kingdom, suggesting that it is not selected against and may sometimes be selected for. It is found in many species of insects, birds, and mammals, including our relatives the orangutans, gorillas, and chimpanzees.
- Rape is found in all human societies.
- Rapists generally apply as much force as is needed to coerce the victim into sex. They rarely inflict a serious or fatal injury, which would preclude

conception and birth. Only 4 percent of rape victims sustain serious injuries, and fewer than one in five hundred is murdered.

- Victims of rape are mostly in the peak reproductive years for women, between thirteen and thirty-five, with a mean in most data sets of twenty-four. Though many rape victims are classified as children (under the age of sixteen), most of these are adolescents, with a median age of fourteen. The age distribution is very different from that of victims of other violent crimes, and is the opposite of what would happen if rape victims were picked for their physical vulnerability or by their likelihood of holding positions of power.
- Victims of rape are more traumatized when the rape can result in a conception. It is most psychologically painful for women in their fertile years, and for victims of forced intercourse as opposed to other forms of rape.
- Rapists are not demographically representative of the male gender. They are overwhelmingly young men, the age of the most intense sexual competitiveness. The young males who allegedly have been "socialized" to rape mysteriously lose that socialization as they get older.
- Though most rapes do not result in conception, many do. About 5 percent of rape victims of reproductive age become pregnant, resulting in more than 32,000 rape-related pregnancies in the United States each year. (That is why abortion in the case of rape is a significant issue.) The proportion would have been even higher in prehistory, when women did not use long-term contraception.[97] Brownmiller wrote that biological theories of rape are "fanciful" because "in terms of reproductive strategy, the hit or miss ejaculations of a single-strike rapist are a form of Russian roulette compared to ongoing consensual mating."[98] But ongoing consensual mating is not an option for every male, and dispositions that resulted in hit-or-miss sex could be evolutionarily more successful than dispositions that resulted in no sex at all. Natural selection can operate effectively with small reproductive advantages, as little as 1 percent.

THE PAYOFF FOR a reality-based understanding of rape is the hope of reducing or eliminating it. Given the theories on the table, the possible sites for levers of influence include violence, sexist attitudes, and sexual desire.

Everyone agrees that rape is a crime of violence. Probably the biggest amplifier of rape is lawlessness. The rape and abduction of women is often a goal of raiding in non-state societies, and rape is common in wars between states and riots between ethnic groups. In peacetime, the rates of rape tend to track rates of other violent crime. In the United States, for example, the rate of forcible rape went up in the 1960s and down in the 1990s, together with the rates of other violent crimes.[99] Gender feminists blame violence against women on civilization and social institutions, but this is exactly backwards.

Violence against women flourishes in societies that are outside the reach of civilization, and erupts whenever civilization breaks down.

Though I know of no quantitative studies, the targeting of sexist attitudes does not seem to be a particularly promising avenue for reducing rape, though of course it is desirable for other reasons. Countries with far more rigid gender roles than the United States, such as Japan, have far lower rates of rape, and within the United States the sexist 1950s were far safer for women than the more liberated 1970s and 1980s. If anything, the correlation might go in the opposite direction. As women gain greater freedom of movement because they are independent of men, they will more often find themselves in dangerous situations.

What about measures that focus on the sexual components of rape? Thornhill and Palmer suggested that teenage boys be forced to take a rape-prevention course as a condition for obtaining a driver's license, and that women should be reminded that dressing in a sexually attractive way may increase their risk of being raped. These untested prescriptions are an excellent illustration of why scientists should stay out of the policy business, but they don't deserve the outrage that followed. Mary Koss, described as an authority on rape, said, "The thinking is absolutely unacceptable in a democratic society." (Note the psychology of taboo—not only is their suggestion wrong, but merely *thinking* it is "absolutely unacceptable.") Koss continues, "Because rape is a gendered crime, such recommendations harm equality. They infringe more on women's liberties than men's."[100]

One can understand the repugnance at any suggestion that an attractively dressed woman excites an irresistible impulse to rape, or that culpability in any crime should be shifted from the perpetrator to the victim. But Thornhill and Palmer said neither of those things. They were offering a recommendation based on prudence, not an assignment of blame based on justice. Of course women have a right to dress in any way they please, but the issue is not what women have the right to do in a perfect world but how they can maximize their safety in this world. The suggestion that women in dangerous situations be mindful of reactions they may be eliciting or signals they may inadvertently be sending is just common sense, and it's hard to believe any grownup would think otherwise—unless she has been indoctrinated by the standard rape-prevention programs that tell women that "sexual assault is not an act of sexual gratification" and that "appearance and attractiveness are not relevant."[101] Equity feminists have called attention to the irresponsibility of such advice, in terms far harsher than anything by Thornhill and Palmer. Paglia, for example, wrote:

For a decade, feminists have drilled their disciples to say, "Rape is a crime of violence but not sex." This sugar-coated Shirley Temple nonsense has exposed young women to disaster. Misled by feminism, they do not expect rape from the nice boys from good homes who sit next to them in class. . . .

These girls say, "Well, I should be able to get drunk at a fraternity party and go upstairs to a guy's room without anything happening." And I say, "Oh, really? And when you drive your car to New York City, do you leave your keys on the hood?" My point is that if your car is stolen after you do something like that, yes, the police should pursue the thief and he should be punished. But at the same time, the police—and I—have the right to say to you, "You stupid idiot, what the hell were you think-ing?"[102]

Similarly, McElroy points out the illogic of arguments like Koss's that women should not be given practical advice that "infringes more on women's liberties than men's":

The fact that women are vulnerable to attack means we cannot have it all. We cannot walk at night across an unlit campus or down a back alley, without incurring real danger. These are things every woman *should* be able to do, but "shoulds" belong in a utopian world. They belong in a world where you drop your wallet in a crowd and have it returned, com-plete with credit cards and cash. A world in which unlocked Porsches are parked in the inner city. And children can be left unattended in the park. This is not the reality that confronts and confines us.[103]

The flight from reality of the rape-is-not-sex doctrine warps not just ad-vice to women but policies for deterring rapists. Some prison systems put sex offenders in group therapy and psychodrama sessions designed to uproot ex-periences of childhood abuse. The goal is to convince the offenders that ag-gression against women is a way of acting out anger at their mothers, fathers, and society. (A sympathetic story in the *Boston Globe* concedes that "there is no way to know what the success rate of [the] therapy is.")[104] Another program reeducates batterers and rapists with "pro-feminist therapy" consisting of lec-tures on patriarchy, heterosexism, and the connections between domestic vio-lence and racial oppression. In an article entitled "The Patriarchy Made Me Do It," the psychiatrist Sally Satel comments, "While it's tempting to conclude that perhaps pro-feminist 'therapy' is just what a violent man deserves, the tragic fact is that truly victimized women are put in even more danger when their husbands undergo a worthless treatment."[105] Savvy offenders who learn to mouth the right psychobabble or feminist slogans can be seen as successfully treated, which can win them earlier release and the opportunity to prey on women anew.

In his thoughtful review, Jones explores how the legal issues surrounding rape can be clarified by a more sophisticated understanding that does not rule the sexual component out of bounds. One example is "chemical castration," voluntary injections of the drug Depo-Provera, which inhibits the release of androgens and reduces the offender's sex drive. It is sometimes given to offenders who are morbidly obsessed with sex and compulsively commit crimes such as rape, indecent exposure, and child abuse. Chemical castration can cut recidivism rates dramatically—in one study, from 46 percent to 3 percent. Use of the drug certainly raises serious constitutional issues about privacy and punishment, which biology alone cannot decide. But the issues become cloudier, not clearer, when commentators declare a priori that "castration will not work because rape is not a crime about sex, but rather a crime about power and violence."

Jones is not advocating chemical castration (and neither am I). He is asking people to look at all the options for reducing rape and to evaluate them carefully and with an open mind. Anyone who is incensed by the very idea of mentioning rape and sex in the same breath should read the numbers again. If a policy is rejected out of hand that can reduce rape by a factor of fifteen, then many women will be raped who otherwise might not have been. People may have to decide which they value more, an ideology that claims to advance the interests of the female gender or what actually happens in the world to real women.

~

DESPITE ALL THE steam coming out of people's ears in the modern debate on the sexes, there are wide expanses of common ground. No one wants to accept sex discrimination or rape. No one wants to turn back the clock and empty the universities and professions of women, even if that were possible. No reasonable person can deny that the advances in the freedom of women during the past century are an incalculable enrichment of the human condition.

All the more reason not to get sidetracked by emotionally charged but morally irrelevant red herrings. The sciences of human nature can strengthen the interests of women by separating those herrings from the truly important goals. Feminism as a movement for political and social equity is important, but feminism as an academic clique committed to eccentric doctrines about human nature is not. Eliminating discrimination against women is important, but believing that women and men are born with indistinguishable minds is not. Freedom of choice is important, but ensuring that women make up exactly 50 percent of all professions is not. And eliminating sexual assaults is important, but advancing the theory that rapists are doing their part in a vast male conspiracy is not.

Chapter 19

Children

"THE NATURE-NURTURE DEBATE is over." So begins a recent article with a title—"Three Laws of Behavior Genetics and What They Mean"—as audacious as its opening sentence.[1] The nature-nurture debate is, of course, far from over when it comes to identifying the endowment shared by all human beings and understanding how it allows us to learn, which is the main topic of the preceding chapters. But when it comes to the question of what makes people within the mainstream of a society different from one another—whether they are smarter or duller, nicer or nastier, bolder or shyer—the nature-nurture debate, as it has been played out for millennia, really is over, or ought to be.

In announcing that the nature-nurture debate is over, the psychologist Eric Turkheimer was not just using the traditional mule-trainer's technique of getting his subjects' attention, namely whacking them over the head with a two-by-four. He was summarizing a body of empirical results that are unusually robust by the standards of psychology. They have been replicated in many studies, several countries, and over four decades. As the samples grew (often to many thousands), the tools were improved, and the objections were addressed, the results, like the Star-Spangled Banner, were still there.

The three laws of behavioral genetics may be the most important discoveries in the history of psychology. Yet most psychologists have not come to grips with them, and most intellectuals do not understand them, even when they have been explained in the cover stories of newsmagazines. It is not because the laws are abstruse: each can be stated in a sentence, without mathematical paraphernalia. Rather, it is because the laws run roughshod over the Blank Slate, and the Blank Slate is so entrenched that many intellectuals cannot *comprehend* an alternative to it, let alone argue about whether it is right or wrong.

Here are the three laws:

- The First Law: All human behavioral traits are heritable.
- The Second Law: The effect of being raised in the same family is smaller than the effect of the genes.
- The Third Law: A substantial portion of the variation in complex human behavioral traits is not accounted for by the effects of genes or families.

The laws are about what make us what we are (compared with our compatriots) and thus they are about the forces that impinge on us in childhood, the stage of life in which it is thought that our intellects and personalities are formed. "Just as the twig is bent, the tree's inclined," wrote Alexander Pope. "The child is father of the man," wrote Wordsworth, echoing Milton's "The childhood shows the man as morning shows the day." The Jesuits used to say, "Give me the child for the first seven years, and I'll give you the man," and the motto was used as the tag line of the documentary film series by Michael Apted that follows a cohort of British children every seven years (*Seven Up, Fourteen Up,* and so on). In this chapter I will walk you through the laws and explore what they mean for nature, nurture, and none of the above.

~

THE FIRST LAW: *All human behavioral traits are heritable.* Let's begin at the beginning. What is a "behavioral trait"? In many studies it is a stable property of a person that can be measured by standardized psychological tests. Intelligence tests ask people to recite a string of digits backwards, define words like *reluctant* and *remorse,* identify what an egg and a seed have in common, assemble four triangles into a square, and extrapolate sequences of geometric patterns. Personality tests ask people to agree or disagree with statements like "Often I cross the street in order not to meet someone I know," "I do not blame a person for taking advantage of someone who lays himself open to it," "Before I do something I try to consider how my friends will react to it," and "People say insulting and vulgar things about me." It sounds dodgy, but the tests have been amply validated: they give pretty much the same result each time a person is tested, and they statistically predict what they ought to predict reasonably well. IQ tests predict performance in school and on the job, and personality profiles correlate with other people's judgments of the person and with life outcomes such as psychiatric diagnoses, marriage stability, and brushes with the law.[2]

In other studies behavior is recorded more directly. Graduate students hang out in a schoolyard with a stopwatch and clipboard observing what the children do. Pupils are rated for aggressiveness by several teachers, and the ratings are averaged. People report how much television they watch or how many cigarettes they smoke. Researchers tally cut-and-dried outcomes such as high school graduation rates, criminal convictions, or divorces.

Once the measurements are made, the *variance* of the sample may be calculated: the average squared deviation of each person's score from the group mean. The variance is a number that captures the degree to which the members of a group differ from one another. For example, the variance in weight in a sample of Labrador retrievers will be smaller than the variance in weight in a sample that contains dogs of different breeds. Variance can be carved into pieces. It is mathematically meaningful to say that a certain percentage of the variance in a group overlaps with one factor (perhaps, though not necessarily, its cause), another percentage overlaps with a second factor, and so on, the percentages adding up to 100. The degree of overlap may be measured as a correlation coefficient, a number between −1 and +1 that captures the degree to which people who are high on one measurement are also high on another measurement. It is used in behavioral genetic research as an estimate of the proportion of variance accounted for by some factor.[3]

Heritability is the proportion of variance in a trait that correlates with genetic differences. It can be measured in several ways.[4] The simplest is to take the correlation between identical twins who were separated at birth and reared apart. They share all their genes and none of their environment (relative to the variation among environments in the sample), so any correlation between them must be an effect of their genes. Alternatively, one can compare identical twins reared together, who share all their genes and most of their environment, with fraternal twins reared together, who share half their genes and most of their environment (to be exact, they share half of the genes that vary among the people within the sample—obviously they share *all* the genes that are universal across the human species). If the correlation is higher for pairs of identical twins, it presumably reflects an effect of the extra genes they have in common. The bigger the difference between the two correlations, the higher the heritability estimate. Yet another technique is to compare biological siblings, who share half their genes and most of their environment, with adoptive siblings, who share none of their genes (among those that vary) and most of their environment.

The results come out roughly the same no matter what is measured or how it is measured. Identical twins reared apart are highly similar; identical twins reared together are more similar than fraternal twins reared together; biological siblings are far more similar than adoptive siblings.[5] All this translates into substantial heritability values, generally between .25 and .75. A conventional summary is that about half of the variation in intelligence, personality, and life outcomes is heritable—a correlate or an indirect product of the genes. It's hard to be much more precise than that, because heritability values vary within this range for a number of reasons.[6] One is whether measurement error (random noise) is included in the total variance to be explained or is estimated and pulled out of the equation. Another is whether *all* the effects of the genes are being estimated or only the *additive* effects: the ones that exert the same influence regardless of the person's other genes (in other words, the genes for traits that breed true). A third is how much variation there was in the sample to begin with: samples with homogeneous environments give large heritability estimates, those with varied environments give smaller ones. A fourth is when in the person's lifetime a trait is measured. The heritability of intelligence, for example, *increases* over the lifespan, and can be as high as .8 late in life.[7] Forget "As the twig is bent"; think "Omigod, I'm turning into my parents!"

"All traits are heritable" is a bit of an exaggeration, but not by much.[8] Concrete behavioral traits that patently depend on content provided by the home or culture are, of course, not heritable at all: which language you speak, which religion you worship in, which political party you belong to. But behavioral traits that reflect the underlying talents and temperaments *are* heritable: how proficient with language you are, how religious, how liberal or conservative. General intelligence is heritable, and so are the five major ways in which personality can vary (summarized by the acronym OCEAN): openness to experience, conscientiousness, extroversion-introversion, antagonism-agreeableness, and neuroticism. And traits that are surprisingly specific turn out to be heritable, too, such as dependence on nicotine or alcohol, number of hours of television watched, and likelihood of divorcing. Finally there are the Mallifert brothers in Chas Addams's patent office and their real-world counterparts: the identical twins separated at birth who both grew up to be captains of their volunteer fire departments, who both twirled their necklaces when answering questions, or who both told the researcher picking them up at the airport (separately) that a wheel bearing in his car needed to be replaced.

I once watched an interview in which Marlon Brando was asked about the childhood influences that made him an actor. He replied that identical twins separated at birth may both use the same hair tonic, smoke the same brand of cigarettes, vacation on the same beach, and so on. The interviewer, Connie Chung, pretended to snore as if she were sitting through a boring lecture, not realizing that he was answering her question—or, more accurately, explaining why he couldn't answer it. As long as the heritability of talents and tastes is not zero, none of us has any way of knowing whether a trait has been influenced by our genes, our childhood experiences, both, or neither. Chung is not alone in her failure to understand this point. The First Law implies that any study that measures something in parents and something in their biological children and then draws conclusions about the effects of parenting is worthless, because the correlations may simply reflect their shared genes (aggressive parents may breed aggressive children, talkative parents talkative children). But these expensive studies continue to be done and continue to be translated into parenting advice as if the heritability of all traits were zero. Perhaps Brando should be asked to serve on grant review panels.

Behavioral genetics does have its critics, who have tried to find alternative interpretations for the First Law. Perhaps children separated at birth are deliberately placed in similar adoptive families. Perhaps they have contact with each other during their separation. Perhaps parents expect identical twins to be more alike and so treat them more alike. Twins share a womb, not just their genes, and identical twins sometimes share a chorion (the membrane surrounding the fetus) and a placenta as well. Perhaps it is their shared prenatal experience, not their shared genes, that makes them more alike.

These possibilities have been tested, and though in some cases they may knock down a heritability estimate by a few points, they cannot reduce it by much.[9] The properties of adoptive parents and homes have been measured (their education, socioeconomic status, personalities, and so on), and they are not homogeneous enough to force identical twins into the same personalities and temperaments.[10] Identical twins are not earmarked for homes that both encourage twirling necklaces or sneezing in elevators. More important, the homes of identical twins who were separated at birth are no more similar than the homes of fraternal twins who were separated at birth, yet the identical twins are far more similar.[11] And most important of all, differences in home environments do not produce differences in grown children's intelligence and personality anyway (as we shall see in examining the Second Law), so the argument is moot.

As for contact between separated twins, it is unlikely that an occasional encounter between two people could revamp their personality and intelligence, but in any case the amount of contact turns out to have no correlation with the twins' degree of similarity.[12] What about the expectations of parents, friends, and peers? A neat test is provided by identical twins who are mistakenly thought to be fraternal until a genetic test shows otherwise. If it is expectations that make identical twins alike, these twins should not be alike; if it is the genes, they should be. In fact the twins are as alike as when the parents know they are identical.[13] And direct measures of how similarly twins are treated by their parents do not correlate with measures of how similar they are in intelligence or personality.[14] Finally, sharing a placenta can make identical twins more *different*, not just more similar (since one twin can crowd out the other), which is why studies have shown little or no consistent effect of sharing a placenta.[15] But even if it were to make them more similar, the inflation of heritability would be modest. As the behavioral geneticist Matt McGue noted of a recent mathematical model that tried to use prenatal effects to push down heritability estimates as much as possible, "That the IQ debate now centers on whether IQ is 50% or 70% heritable is a remarkable indication of how the nature-nurture debate has shifted over the past two decades."[16] In any case, studies comparing adoptees with biological siblings don't look at twins at all, and they come to the same conclusions as the twin studies, so no peculiarity of twinhood is likely to overturn the First Law.

Behavioral genetic methods do have three built-in limitations. First, studies of twins, siblings, and adoptees can help explain what makes people different, but they cannot explain what people have in common, that is, universal human nature. To say that the heritability of intelligence is .5, for example, does not imply that half of a person's intelligence is inherited (whatever that would mean); it implies only that half of the *variation* among people is inherited. Behavioral genetic studies of pathological conditions, such as those discussed in Chapters 3 and 4, *can* shed light on universal human nature, but they are not relevant to the topics of this chapter.

Second, behavioral genetic methods address variation within the group of people being examined, not variation *between* groups of people. If the twins or adoptees in a sample are all middle-class American whites, a heritability estimate can tell us about why middle-class American whites differ from other middle-class American whites, but not why the middle class differs from the lower or upper class, why Americans differ from non-Americans, or why whites differ from Asians or blacks.

Third, behavioral genetic methods can show only that traits *correlate* with genes, not that they are directly caused by them. The methods cannot distinguish traits that are relatively direct products of the genes—the result of genes that affect the wiring or metabolism of the brain—from traits that are highly indirect products, say, the result of having genes for a certain physical appearance. We know that tall men on average are promoted in their jobs more rapidly than short men, and that attractive people on average are more assertive than unattractive ones.[17] (In one experiment, subjects undergoing a fake interview had to cool their heels when the interviewer was called out of the room by a staged interruption. The plain-looking subjects waited nine minutes before complaining; the attractive ones waited three minutes and twenty seconds.)[18] Presumably people defer to tall and good-looking people, and that makes them more successful and entitled. Height and looks are obviously heritable, so if we didn't know about the effects of looks, we might think that these people's success comes directly from genes for ambition and assertiveness instead of coming indirectly from genes for long legs or a cute nose. The moral is that heritability always has to be interpreted in the light of all the evidence; it does not wear its meaning on its sleeve. That having been said, we know that the heritability of personality cannot, in fact, be reduced to genes for appearance. The effects of looks on personality are small and limited; blond jokes notwithstanding, not all attractive women are vain and entitled. The heritability of personality traits, in contrast, is large and pervasive, too large to be explained away as a by-product of looks.[19] And as we saw in Chapter 3, personality traits can in some cases be tied to actual genes with products in the nervous system. With the completion of the Human Genome Project, it is likely that geneticists soon will be discovering more of those linkages.

The First Law is a pain in the neck for radical scientists, who have tried unsuccessfully to discredit it. In 1974, Leon Kamin wrote that "there exist no data which should lead a prudent man to accept the hypothesis that IQ test scores are in any degree heritable," a conclusion he reiterated with Lewontin and Rose a decade later.[20] Even in the 1970s the argument was tortuous, but by the 1980s it was desperate and today it is a historical curiosity.[21] As usual, the attacks have not always come in dispassionate scholarly analyses. Thomas Bouchard, who directed the first large-scale study of twins reared apart, is one of the pioneers of the study of the genetics of personality. Campus activists at the University of Minnesota distributed handouts calling him a racist and linking him to "German fascism," spray-painted slogans calling him a Nazi, and demanded that he be fired. The psychologist Barry Mehler accused him of "rehabilitating" the work of Josef Mengele, the doctor who tormented twins in the Nazi death camps under the guise of research. As usual, the charges were unfair not just intellectually but personally: far from being a fascist, Bouchard was a participant in the Berkeley Free Speech Movement of the 1960s, was briefly jailed for his activism, and says he would do it again today.[22]

These attacks are transparently political and easy to discount. More pernicious is the way that the First Law is commonly interpreted: "So you're saying it's all in the genes," or, more angrily, "Genetic determinism!" I have already commented on this odd reflex in modern intellectual life: when it comes to genes, people suddenly lose their ability to distinguish 50 percent from 100 percent, "some" from "all," "affects" from "determines." The diagnosis for this intellectual crippling is clear: if the effects of the genes must, on theological grounds, be zero, then all nonzero values are equivalently heretical.

But the worst fallout from the Blank Slate is not that people misunderstand the effects of the genes. It is that they misunderstand the effects of the environment.

~

THE SECOND LAW: *The effect of being raised in the same family is smaller than the effect of the genes.* By now you appreciate that our genes play a role in making us different from our neighbors, and that our environments play an equally important role. At this point everyone draws the same conclusion. We are shaped both by our genes and by our family upbringing: how our parents treated us and what kind of home we grew up in.

Not so fast. Behavioral genetics allows us to distinguish two very different ways in which our environments might affect us.[23] The *shared* environment is what impinges on us and our siblings alike: our parents, our home life, and our neighborhood (as compared with other parents and neighborhoods in the sample). The *nonshared* or *unique* environment is everything else: anything that impinges on one sibling but not another, including parental favoritism (Mom always liked you best), the presence of the other siblings, unique experiences like falling off a bicycle or being infected by a virus, and for that matter anything that happens to us over the course of our lives that does not necessarily happen to our siblings.

The effects of the shared environment can be measured in twin studies by subtracting the heritability value from the correlation between the identical twins. The rationale is that identical twins are alike (measured by the correlation) because of their shared genes (measured by the heritability) and their shared environment, so the effects of the shared environment can be estimated by subtracting the heritability from the correlation. Alternatively, the effects can be estimated in adoption studies simply by looking at the correlation between two adoptive siblings: they do not share genes, so any similarities (relative to the sample) must come from the experiences they shared growing up in the same home. A third technique is to compare the correlation between siblings reared together (who share genes and a home environment) with the correlation between siblings reared apart (who share only genes).

The effects of the *unique* environment can be measured by subtracting the correlation between identical twins (who share genes and an environment) from 1 (which is the sum of the effects of the genes, the shared environment, and the unique environment). By the same reasoning, it can be measured in adoption studies by subtracting the heritability estimate and the shared-environment estimate from 1. In practice all these calculations are more complicated, because they may try to account for nonadditive effects, where the whole is not the sum of the parts, and for noise in the measurements. But you now have the basic logic behind them.

So what do we find? The effects of shared environment are small (less than 10 percent of the variance), often not statistically significant, often not replicated in other studies, and often a big fat zero.[24] Turkheimer was cautious in saying that the effects are smaller than those of the genes. Many behavioral geneticists go farther and say that they are negligible, particularly in adulthood. (IQ is affected by the shared environment in childhood, but over the years the effect peters out to nothing.)

Where do these conclusions come from? The actual findings are easy to understand. First, adult siblings are equally similar whether they grew up together or apart. Second, adoptive siblings are no more similar than two people plucked off the street at random. And third, identical twins are no more similar than one would expect from the effects of their shared genes. As with the First Law, the sheer consistency of the outcome across three completely different methods (comparisons of identical with fraternal twins, of siblings raised together with siblings raised apart, of adoptive siblings with biological siblings) emboldens one to conclude that the pattern is real. Whatever experiences siblings share by growing up in the same home makes little or no difference in the kind of people they turn out to be.

An important proviso: Differences among homes don't matter *within* the samples of homes netted by these studies, which tend to be more middle-class than the population as a whole. But differences between those samples and other kinds of homes *could* matter. The studies exclude cases of criminal neglect, physical and sexual abuse, and abandonment in a bleak orphanage, so they do not show that extreme cases fail to leave scars. Nor can they say anything about the differences between *cultures*—about what makes a child a middle-class American as opposed to a Yanomamö warrior or a Tibetan monk or even a member of an urban street gang. In general, if a sample comes from a restricted range of homes, it may underestimate effects of homes across a wider range.[25]

Despite these caveats, the Second Law is by no means trivial. The "middle class" (which includes most adoptive parents) can embrace a wide range of lifestyles, from fundamentalist Christians in the rural Midwest to Jewish doctors in Manhattan, with very different home environments and childrearing philosophies. Behavioral geneticists have found that their samples of parents in fact span a full range of personality types. And even if adoptive parents are unrepresentative in some other way, the Second Law would survive because it emerges from large studies of twins as well.[26] Though samples of adoptive parents span a narrower (and higher) range of IQs than the population at large, that cannot explain why the IQs of their adult children are uncorrelated, because they *were* correlated when the children were young.[27] Before exploring the revolutionary implications of these discoveries, let's turn to the Third Law.

THE THIRD LAW: *A substantial portion of the variation in complex human behavioral traits is not accounted for by the effects of genes or families.* This follows directly from the First Law, assuming that heritabilities are less than one, and the Second Law. If we carve up the variation among people into the effects of the genes, the shared environment, and the unique environment, and if the effects of the genes are greater than zero and less than one, and if the effects of the shared environment hover around zero, then the effects of the unique environment must be greater than zero. In fact, they are around 50 percent, depending as always on what is being measured and exactly how it is estimated. Concretely, this means that identical twins reared together (who share both their genes *and* a family environment) are far from identical in their intellects and personalities. There must be causes that are neither genetic *nor* common to the family that make identical twins different and, more generally, make people what they are.[28] As with Bob Dylan's Mister Jones, something is happening here but we don't know what it is.

A handy summary of the three laws is this: Genes 50 percent, Shared Environment 0 percent, Unique Environment 50 percent (or if you want to be charitable, Genes 40–50 percent, Shared Environment 0–10 percent, Unique

Environment 50 percent). A simple way of remembering what we are trying to explain is this: identical twins are 50 percent similar whether they grow up together or apart. Keep this in mind and watch what happens to your favorite ideas about the effects of upbringing in childhood.

~

THOUGH BEHAVIORAL GENETICISTS have known about the heritability of mental traits (First Law) for decades, it took a while for the absence of effects of the shared environment (Second Law) and the magnitude of the effects of the unique environment (Third Law) to sink in. Robert Plomin and Denise Daniels first sounded the alarm in a 1987 article called "Why Are Children in the Same Family So Different from One Another?" The enigma was noted by other behavioral geneticists such as Thomas Bouchard, Sandra Scarr, and David Lykken and spotlighted again by David Rowe in his 1994 book *The Limits of Family Influence*. It was also the springboard for the historian Frank Sulloway's widely discussed 1996 book on birth order and revolutionary temperament, *Born to Rebel*. Still, few people outside behavioral genetics really appreciated the importance of the Second and Third Laws.

It all hit the fan in 1998 when Judith Rich Harris, an unaffiliated scholar (whom the press quickly dubbed "a grandmother from New Jersey"), published *The Nurture Assumption*. A *Newsweek* cover story summed up the topic: "Do Parents Matter? A Heated Debate About How Kids Develop.", Harris brought the three laws out of the journals and tried to get people to recognize their implications: that the conventional wisdom about childrearing among experts and laypeople alike is wrong.

It was Rousseau who made parents and children the main actors in the human drama.[29] Children are noble savages, and their upbringing and education can either allow their essential nature to blossom or can saddle them with the corrupt baggage of civilization. Twentieth-century versions of the Noble Savage and the Blank Slate kept parents and children at center stage. The behaviorists claimed that children are shaped by contingencies of reinforcement, and advised parents not to respond to their children's distress because it would only reward them for crying and increase the frequency of crying behavior. Freudians theorized that we are shaped by our degree of success in weaning, toilet training, and identification with the parent of the same sex, and advised parents not to bring infants into their beds because it would arouse damaging sexual desires. Everyone theorized that psychological disorders could be blamed on mothers: autism on their coldness, schizophrenia on their "double binds," anorexia on their pressure on girls to be perfect. Low self-esteem was attributed to "toxic parents" and every other problem to "dysfunctional families." Patients in many forms of psychotherapy while away their fifty minutes reliving childhood conflicts, and most biographies scavenge through the subject's childhood for the roots of the grownup's tragedies and triumphs.

By now most well-educated parents believe that their children's fates are in their hands. They want their children to be popular and self-confident, to get good grades and stay in school, to avoid drugs, alcohol, and cigarettes, to avoid getting pregnant or fathering a child while a teenager, to stay on the right side of the law, and to become happily married and professionally successful. A parade of parenting experts has furnished them with advice, ever changing in content, never changing in certitude, on how to attain that outcome. The current recipe runs something like this. Parents should stimulate their babies with colorful toys and varied experiences. ("Take them outside. Let them feel tree bark," advised a pediatrician who shared a couch with me on a morning television show.) They should read and talk to their babies as much as possible to foster their language development. They should interact and communicate with their children at all ages, and no amount of time is too much. ("Quality time," the idea that working parents could spend an intense interlude with their children between dinner and bedtime to make up for their absence during the day, quickly became a national joke; it was seen as a rationalization by mothers who would not admit that their careers were compromising their children's welfare.) Parents should set firm but reasonable limits, neither bossing their children around nor giving them complete license. Physical punishment of any kind is out, because that perpetuates a cycle of violence. Nor should parents belittle their children or say that they are bad, because that will damage their self-esteem. On the contrary, they should shower them with hugs and unconditional affirmations of love and approval. And parents should communicate intensively with their adolescent children and take an interest in every aspect of their lives.

A few parents have begun to question the imperative to become round-the-clock parenting machines. A recent cover story in *Newsweek* entitled "The Parent Trap" reported on the frazzled mothers and fathers who devote every nonworking minute to entertaining and chauffeuring their children for fear that they will otherwise turn into ne'er-do-wells or cafeteria snipers. A similar story in the *Boston Globe Magazine* with the ironic title "How to Raise a Perfect Child . . ." elaborates:

> "I'm overwhelmed with parenting advice," says Alice Kelly of Newton. "I read all about how I'm supposed to be providing my children with enriching play experiences. I'm supposed to do lots of physical activity with them so I can instill in them a physical fitness habit so they'll grow up to be healthy, fit adults. And I'm supposed to do all kinds of intellectual play so they'll grow up smart. Also, there are all kinds of play, and I'm supposed to do each—clay for finger dexterity, word games for reading success, large-motor play, small-motor play. I feel like I could devote my life to figuring out what to play with my kids." . . .

Elizabeth Ward, a Stoneham dietician, has been puzzling over why parents are so "willing to be short-order cooks, preparing two or three meals at a time" in order to please the kids. . . . [One reason] is a belief that forcing a kid to choose between eating what's presented or skipping a meal will lead to eating disorders—a thought that probably never occurred to parents in earlier decades.[30]

The humorist Dave Barry comments on the experts' advice to parents of adolescents:

In addition to watching for warning signs, you must "keep the lines of communication open" between yourself and your child. Make a point of taking an interest in the things your child is interested in so that you can develop a rapport, as we see in this dialogue:

FATHER: What's that music you're listening to, son?
SON: It's a band called "Limp Bizkit," Dad.
FATHER: They suck.

. . . You should strive for this kind of closeness in your relationship with your child. And remember: If worse comes to worst, there is no parenting tool more powerful than a good hug. If you sense that your child is getting into trouble, you must give that child a great big fat hug in a public place with other young people around, while saying, in a loud, piercing voice, "You are MY LITTLE BABY and I love you NO MATTER WHAT!" That will embarrass your child so much that he or she may immediately run off and join a strict religious order whose entire diet consists of gravel. If one hug doesn't work, threaten to give your child another.[31]

Backlash aside, is it possible that the experts' advice might be sound? Perhaps the parent trap is the mixed blessing of scientists' knowing more and more about the effects of parenting. Parents can be forgiven for carving out some time for themselves, but if the experts are right they must realize that every such decision is a compromise.

So what do we really know about the long-term effects of parenting? Natural variation among parents, the raw material of behavioral genetics, offers one way of finding out. In any large sample of families, parents vary in how well they adhere to the ideals of parenting (if some didn't stray from the ideal, there would be no point in offering advice). Some mothers stay at home, others are workaholics. Some parents lose their tempers, others are infinitely patient. Some are garrulous, others taciturn; some unreserved in their affection, others more guarded. (As one academic said to me after pulling out a picture of her toddler, "We virtually adore her.") Some homes are filled with books, others with blaring

TV sets; some couples are lovey-dovey, others fight like Maggie and Jiggs. Some mothers are like June Cleaver, others are depressed or histrionic or disorganized. According to the conventional wisdom, these differences should make a difference. At a bare minimum, two children growing up in one of these homes— with the same mother, father, books, TVs, and everything else—should turn out more similar, on average, than two children growing up in different homes. Seeing whether they do is a remarkably direct and powerful test. It does not depend on any hypothesis about what parents have to do to change their children or how their children will respond. It does not depend on how well we measure the home environments. If *anything* that parents do affects their children in *any* systematic way, then children growing up with the same parents will turn out more similar than children growing up with different parents.

But they don't. Remember the discoveries behind the Second Law. Siblings reared together end up no more similar than siblings separated at birth. Adopted siblings are no more similar than strangers. And the similarities between siblings can be completely accounted for by their shared genes. All those differences among parents and homes have no predictable long-term effects on the personalities of their children. Not to put too fine a point on it, but much of the advice from the parenting experts is flapdoodle.

But surely the advice is grounded in research on children's development? Yes, from the many useless studies that show a correlation between the behavior of parents and the behavior of their biological children and conclude that the parenting shaped the child, as if there were no such thing as heredity. And in fact the studies are even worse than that. Even if there *were* no such thing as heredity, a correlation between parents and children would not imply that parenting practices shape children. It could imply that children shape parenting practices.[32] As any parent of more than one child knows, children are not indistinguishable lumps of raw material waiting to be shaped. They are little people, born with personalities. And people react to the personalities of other people, even if one is a parent and the other a child. The parents of an affectionate child may return that affection and thereby act differently from the parents of a child who squirms and wipes off his parents' kisses. The parents of a quiet, spacey child might feel they are talking to a wall and jabber at him less. The parents of a docile child can get away with setting firm but reasonable limits; the parents of a hellion might find themselves at their wits' end and either lay down the law or give up. In other words, correlation does not imply causation. A correlation between parents and children does not mean that parents affect children; it could mean that children affect parents, that genes affect both parents and children, or both.

It gets worse. In many studies, the same parties (in some studies the parents, in others the children) supply the data on both the parents' behavior and the child's. Parents tell the experimenter how they treat their children *and* what their children are like, or adolescents tell the experimenter what they are like *and* how their parents treat them. Those studies—suspiciously—show much stronger correlations than ones in which a third party assesses the parents and the child.[33] The problem is not just that people tend to look at themselves and at their families through the same rose-colored or jaundiced lenses, but also that the relationship between parents and adolescents is a two-way street. Harris sums up the problems when commenting on a widely publicized 1997 study. The authors claimed, solely on the basis of teenagers' responses to a questionnaire about themselves and their families, that "parent-family connectedness"—close bonds, high expectations, lots of affection—is "protective" against adolescent ills such as drugs, cigarettes, and unsafe sex. Harris notes:

> A happy person tends to check off upbeat answers to all the questions: Yes, my parents are good to me; yes, I'm doing fine. A person who cares about presenting a socially acceptable face to the world checks off socially acceptable responses: Yes, my parents are good to me; no, I haven't been in any fights or smoked anything illegal. A person who is angry or depressed checks off angry or depressed responses: My parents are jerks and I flunked the algebra test and to hell with your questionnaire. . . .
>
> . . . Perhaps what misled those eighteen federal agencies into thinking they were getting their 25 million dollars worth was the positive way the researchers phrased their findings: *good* relationships with parents exert a *protective* effect. Expressed in a different (but equally accurate) way, the results sound less interesting: adolescents who don't get along well with their parents are more likely to use drugs or engage in risky sex. The results sound still less interesting expressed this way: adolescents who use drugs or engage in risky sex don't get along well with their parents.[34]

Yet another problem crops up when researchers direct all their questions to the parents rather than to the offspring. People behave differently in different settings. That includes children, who tend to behave differently inside and outside the home. So even if parents' behavior does affect how their children behave *with them*, it may not affect how their children behave with other people. When parents describe their children's behavior, they describe the behavior they see in the home. To show that parents shape their children, then, a study would have to control for genes (by testing twins or adoptees), distinguish between parents affecting children and children affecting parents, measure the parents and the children independently, look at how children behave outside the home rather than inside, and test older children and young adults to see whether any effects are transient or permanent. No study that has claimed to show effects of parenting has met these standards.[35]

If behavioral genetic studies show no lasting effects of the home, and studies of parenting practices are uninformative, what about studies that compare radically different childhood milieus? The results, again, are bracing. Decades of studies have shown that, all things being equal, children turn out pretty much the same way whether their mothers work or stay at home, whether they are placed in daycare or not, whether they have siblings or are only children, whether their parents have a conventional or an open marriage, whether they grow up in an Ozzie-and-Harriet home or a hippie commune, whether their conceptions were planned, were accidental, or took place in a test tube, and whether they have two parents of the same sex or one of each.[36]

Even growing up without a father in the house, which does correlate with troubles such as dropping out of school, remaining idle, and having babies while a teenager, may not *cause* the troubles directly.[37] Children with experiences that should make up for the missing father, such as having a stepfather, a live-in grandmother, or frequent contact with the birth father, are no better off. The number of years that the father was in the house before leaving makes no difference. And children whose fathers died do not have the poor outcomes of children whose fathers walked out or were never there. The absence of a father may not be a cause of adolescent problems but a correlate of the true causes, which may include poverty, neighborhoods with lots of unattached men (who live in de facto polygyny and hence compete violently for status), frequent moves (which force children to start from the bottom of the pecking order in new peer groups), and genes that make both fathers and children more impulsive and quarrelsome.

The 1990s was the Decade of the Brain and the decade in which parents were told they were in charge of their babies' brains. The first three years of life was described as a critical window of opportunity in which the child's brain had to be constantly stimulated to keep it growing properly. Parents of late-talking children were blamed for not blanketing them in enough verbiage; the ills of the inner city were blamed on children's having to stare at empty walls. Bill and Hillary Clinton convened a conference at the White House to learn about the research, at which Mrs. Clinton said that the experiences of the first three years "can determine whether children will grow up to be peaceful or violent citizens, focused or undisciplined workers, attentive or detached parents themselves."[38] The governors of Georgia and Missouri asked their legislators for millions of dollars to issue every new mother with a Mozart CD. (They had confused experiments on infant brain development with experiments—since discredited—alleging that *adults* benefit from listening to a few minutes of Mozart.)[39] The pediatrician and childcare guru T. Berry Brazelton had the most hopeful suggestion of all: that nurturance during the first three years will protect children from the lure of tobacco when they become adolescents.[40]

In his book *The Myth of the First Three Years,* the cognitive neuroscience expert Jon Bruer showed that there was no science behind these astonishing claims.[41] No psychologist has ever documented a critical period for cognitive or language development that ends at three. And though *depriving* an animal of stimulation (by sewing an eye shut or keeping it in a barren cage) may hurt its brain growth, there is no evidence that providing *extra* stimulation (beyond what the organism would encounter in its normal habitat) *enhances* its brain growth.

So nothing in the research on family environments contradicts the behavioral geneticists' Second Law, which says that growing up in a particular family has little or no systematic effect on one's intellect and personality. And this leaves us with a maddening puzzle. No, it's not all in the genes; around half the variation in personality, intelligence, and behavior comes from something in the environment. But whatever that something is, it cannot be shared by two children growing up in the same home with the same parents. And that rules out all the obvious somethings. What is the elusive Mister Jones factor?

REFUSING TO GIVE up on parents, some developmental psychologists have trained their sights on the only remaining possibility that gives parents a starring role. The impotence of the shared environment says only that what parents do to *all* their children is powerless to shape them. But obviously parents don't treat their children alike. Perhaps the individualized parenting that mothers and fathers adapt to each child does have the power to shape them. It is the *interaction* between parents and children that affects them, not a one-size-fits-all parenting philosophy.[42]

At first this looks reasonable. But when you think it through, it does not restore a shaping role for parents, or for parenting advice, after all.[43]

What would individualized parenting look like? Presumably parents would tailor their parenting to the needs and talents of each child. A headstrong child would elicit firmer discipline than a compliant one; a fearful child would elicit more protectiveness than a bold one. The problem, as we saw in an earlier section, is that the differences in parenting cannot be separated from the preexisting differences in the children. If the fearful child turns into a fearful adult, we don't know whether it was an effect of the overprotective parent or a continuation of the fearfulness the child was born with.

And surprisingly, if children do elicit systematic differences in parenting it would show up as an effect of the *genes:* it would go into the heritability term, not the unique-environment term. The reason is that heritability is a measure of correlation and cannot distinguish direct effects of the genes (proteins that help wire the brain or trigger hormones) from indirect effects that operate many links away. Earlier I mentioned that attractive people are more assertive, presumably because they get accustomed to other people's kissing up to them. That is a highly indirect effect of the genes and would make assertiveness heritable even if there were no genes for assertive brains, just genes for violet eyes to die for. Similarly, if children with certain innate traits make their parents more patient, or encouraging, or strict, then parental patience, encouragement, and strictness would also count as "heritable." Now, if such individualized parenting does affect the way children turn out, a critic could legitimately say that the direct effects of the genes had been overestimated, because some of them would really be indirect effects of the children's genes on traits of the children that affect their parents' behavior, which in turn affects the children. (The hypothesis is baroque, and I will soon show why it is unlikely to be true, but let's assume it is true for argument's sake.) But at best, the effects of parenting would be fighting with other genetic effects (direct and indirect) for some portion of the 40 to 50 percent of the variation attributed to the genes. The 50 percent attributable to the unique environment would still be up for grabs.

Here is what would have to happen if the effects of the unique environment are to be explained by an interaction between parents and children (using the statistician's technical sense of the word "interaction," which is the one relevant to our puzzle). A given practice would have to affect some children one way, and other children another way, and the two effects would have to cancel out. For example, sparing the rod would have to spoil some children (making them more violent) and teach others that violence is not a solution (making them less violent). Displays of affection would have to make some children more affectionate (because they identify with their parents) and others less affectionate (because they react against their parents). The reason the effects have to go in opposite directions is that if a parenting practice had a consistent effect, on average, across all children, it would turn up as an effect of the shared environment. Adopted siblings would be similar, sibs growing up together would be more similar than sibs growing up apart—neither of which happens. And if it was applied successfully to some kinds of children and was avoided, or was ineffective, with other kinds, that would turn up as an effect of the genes.

The problems with the parent-child interaction idea now become obvious. It is implausible that any parenting process would have such radically different effects on different children that the sum of the effects (the shared environment) would add up to zero. If hugging merely makes some children more confident and has no effect on others, then the huggers should still have more confident children *on average* (some becoming more confident, others showing no change) than the cold fish. But, holding genes constant, they don't. (To put it in technical terms familiar to psychologists: it is rare to find a perfect crossover interaction, that is, an interaction with no main effects.) This is also, by the way, one of the reasons that heritability itself almost certainly cannot be reduced to child-specific parenting. Unless parents' behavior is *completely* determined by their child's inborn traits, some parents will behave somewhat differently from others across the board, and that would turn up in effects of the shared environment—which in fact are negligible.

But let's say that these parent-child interactions (in the technical sense) really do exist, and really do shape the child. The moral would be that across-the-board parenting advice is useless. Anything that parents do to make some children better will make an equal number of children worse.

In any case, the parent-child interaction theory can be tested directly. Psychologists can measure how parents treat the different children within a family, and see if the treatments correlate with how the children turn out, holding genes constant. The answer is that in almost every case they don't. Virtually all the differences in parenting within a family can be explained as reactions to genetic differences that the children were born with. And parental behavior that does differ among children for nongenetic reasons, such as marital conflict triggered by some siblings but not by others, or more parenting effort directed at one sibling than at another, has no effect.[44] The leader of a recent heroic study, who had hoped to prove that differences in parenting do affect how children turn out, confessed that he was "shocked" by his own results.[45]

There is another way that a home environment could differ among children in the same family for reasons having nothing to do with their genes: birth order. A firstborn usually has several years of undivided parental attention with no annoying siblings around. Laterborns have to compete with their siblings for parental attention and other family resources, and have to figure out how to hold their own against stronger and more entrenched competitors.

In *Born to Rebel*, Sulloway predicted that firstborns should parlay their advantages into a more assertive personality.[46] And because they identify with their parents, and by extension with the status quo, they should grow up to be more conservative and conscientious. Laterborns, in contrast, should be more conciliatory and open to new ideas and experiences. Though family therapists and laypeople have had these impressions for a long time, Sulloway tried to explain them in terms of Trivers's theory of parent-offspring conflict and its corollary, sibling rivalry. He found some support for these ideas in a meta-analysis (a quantitative literature review) of studies of birth order and personality.[47]

Sulloway's theory, however, also requires that children use the same strategies *outside* the home—with their peers and colleagues—as the ones that served them well *inside* the home. That does not follow from Trivers's theory; indeed, it contradicts the larger theory from evolutionary psychology that relationships with blood relatives should be very different from relationships with nonrelatives. Tactics that work on a sibling or parent may not work so well on a colleague or stranger. And in fact subsequent analyses have shown that any effects of birth order on personality turn up in the studies that ask siblings or parents to rate one another, or to rate themselves with respect to a sibling, which of course can assess only their family relationships. When personality is measured by neutral parties outside the family, birth-order effects diminish or disappear.[48] Any differences in the parenting of firstborns and laterborns—novice or experienced parents, divided or undivided attention, pressure to carry on the family legacy or indulgent babying—seem to have little or no effect on personality outside the home.

Similarities within a home don't shape children; differences within a home don't shape children. Perhaps, Harris says, we should look outside the home.

~

IF YOU GREW up in a different part of the world from where your parents grew up, consider this question: Do you sound like your parents, or like the people you grew up with? What about the way you dress, or the music you listen to, or the way you spend your free time? Consider the same question about your children if they grew up in a different part of the world from where you grew up—or for that matter, even if they didn't. In almost every case, people model themselves after their peers, not their parents.

This is Harris's explanation of the elusive environmental shaper of personality, which she calls Group Socialization theory. It's not all in the genes, but what isn't in the genes isn't from the parents either. Socialization—acquiring the norms and skills necessary to function in society—takes place in the peer group. Children have cultures, too, which absorb parts of the adult culture and also develop values and norms of their own. Children do not spend their waking hours trying to become better and better approximations of adults. They strive to be better and better children, ones that function well in their *own* society. It is in this crucible that our personalities are formed.

Multidecade, child-obsessed parenting, Harris points out, is an evolutionarily recent practice. In foraging societies, mothers carry their children on their hips or backs and nurse them on demand until the next child arrives two to four years later.[49] The child is then dumped into a play group with his older siblings and cousins, switching from being the beneficiary of almost all of the mother's attention to almost none of it. Children sink or swim in the milieu of other children.

Children are not just attracted to the norms of their peers; to some degree they are immune to the expectations of their parents. The theory of parent-offspring conflict predicts that parents do not always socialize a child in the child's best interests. So even if children acquiesce to their parents' rewards, punishments, examples, and naggings for the time being—because they are smaller and have no choice—they should not, according to the theory, allow their personalities to be shaped by these tactics. Children must learn what it takes to gain status among their peers, because status at one age gives them a leg up in the struggle for status at the next, including the young-adult stages in which they first compete for the attention of the opposite sex.[50]

What first attracted me to Harris's theory was its ability to explain a half-dozen puzzling facts in the part of psychology I work in the most, language.[51] Psycholinguists argue a lot about heredity and environment, but they all equate "the environment" with "parents." But many phenomena of children's language development just don't fit that equation. In traditional cultures, mothers don't say much to their children until they are old enough to hold up their end of the conversation; the children pick up language from other children. People's accents almost always resemble the accents of their childhood peers, not the accents of their parents. Children of immigrants acquire the language of their adopted homeland perfectly, without a foreign accent, as long as they have access to native speaking peers. They then try to force their parents to switch to the new language, and if they succeed, they may forget the mother tongue entirely. The same is true of hearing children of deaf parents, who learn the spoken language of their community without a hitch. Children thrown together without a common language from the grownups will quickly invent one; that is how creole languages, and the signed languages of the deaf, came into being. Now, a particular language like English or Japanese (as opposed to the instinct for language in general) is an example of learned social behavior par excellence. If children cultivate a fine ear for the nuances of their peers' speech, and if they cast their lot with their peers' language over their parents', it suggests that their social antennae are aimed peerward.

Children of immigrants soak up not just the language of their adopted homeland but the culture as well. For their entire lives, my shtetl-born grandparents were strangers in a strange land. Cars, banks, doctors, schools, and the urban concept of time left them baffled, and if the term "dysfunctional family" had been around in the 1930s and 1940s it would surely have applied to them. Nevertheless, my father, growing up in a community of immigrants who had arrived in different decades, gravitated to other children and families who knew the ropes, and ended up happy and successful. Such stories are common in chronicles of the immigrant experience.[52] So why do we insist that children's parents are the key to how they turn out?

Studies also confirm what every parent knows but what no one bothers to reconcile with theories of child development: that whether adolescents smoke, get into scrapes with the law, or commit serious crimes depends far more on what their peers do than on what their parents do.[53] Harris comments on a popular theory that children become delinquents to achieve "mature status," that is, adult power and privilege: "If teenagers wanted to be like adults they wouldn't be shoplifting nailpolish from drugstores or hanging off overpasses to spray I LOVE YOU LISA on the arch. If they really aspired to 'mature status' they would be doing boring adult things like sorting the laundry and figuring out their income taxes."[54]

Even the rare finding of an effect of the shared environment, and the equally elusive finding of an interaction between genes and the environment, emerge only when we substitute peers for parents in the "environment" part of the equation. Children who grow up in the same home tend to resemble each other in their vulnerability to delinquency, regardless of how closely related they are. But that similarity only holds if they are close in age and spend time together outside the home—which suggests they belong to the same peer group.[55] And in a large Danish adoption study, the biological children of convicts were somewhat more likely to get into trouble than the biological children of law-abiding citizens, which suggests a small across-the-board effect of the genes. But the susceptibility to crime was multiplied if they were adopted by parents who were criminals themselves *and* who lived in a large city, which suggests that the genetically at-risk children grew up in a high-crime neighborhood.[56]

It's not that parents "don't matter." In many ways parents matter a great deal. For most of human existence, the most important thing parents did for their children was keep them alive. Parents can certainly harm their children by abusing or neglecting them. Children appear to need some kind of nurturing figure in their early years, though it needn't be a parent, and possibly not even an adult: young orphans and refugees often turn out relatively well if they had the comfort of other children, even if they had no parents or other adults around them.[57] (This does not mean that the children were *happy*, but contrary to popular belief, unhappy children do not necessarily turn into dysfunctional adults.) Parents select an environment for their children and thereby select a peer group. They provide their children with skills and knowledge, such as reading and playing a musical instrument. And they certainly may affect their children's behavior in the home, just as any powerful people can affect behavior within their fiefdom. But parents' behavior does not seem to shape their children's intelligence or personality over the long term. Upon hearing this, many people ask, "So you're saying it doesn't matter how I treat my child?" It is a revealing question, and I will consider it at the end of the chapter. But first, the public reaction to Harris's theory, and my own assessment.

～

THE NURTURE ASSUMPTION was, by any standard, a major contribution to modern intellectual life. Though the main idea is at first counterintuitive, the book has the ring of truth, with real children running through it, not compliant little theoretical constructs that no one ever meets in real life. Harris backed up her hypothesis with voluminous data from many fields, interpreted with a keen analytical eye, and with a rarity in the social sciences: proposals for new empirical tests that might falsify it. The book also contains original policy suggestions on tough problems for which we sorely need new ideas, such as failing schools, teenage smoking, and juvenile delinquency. Even if major parts turn out to be wrong, the book forces one to think about childhood, and therefore what makes us what we are, in a fresh and insightful way.

So what was the public reaction? The first popular presentation of the theory was in a few pages of my book *How the Mind Works*, in which I presented the research behind the three laws of behavioral genetics and Harris's 1995 paper explaining them. Many reviews singled out those pages for discussion, such as the following analysis by Margaret Wertheim:

> Never in my fifteen years as a science writer have I seen the subject I love so dearly abused so greatly. . . . What is so appalling here—quite aside from the laughable grasp of family dynamics—is the misrepresentation of science. Science can never prove what percentage of personality is caused by upbringing. . . . By suggesting that it can and does, he invites us to see scientists as at best naïve and at worst fascistic. It is precisely this kind of claim that, in my opinion, is giving science a bad name and is helping to fuel a significant backlash against it.[58]

Wertheim, of course, confused "the percentage of personality that is caused by upbringing," which is indeed meaningless, with the percentage of *variance* in personality that is caused by variation in upbringing, which behavioral geneticists study all the time. And scientists can show, and have shown, that siblings are as similar when reared apart as when reared together and that adoptive siblings are not similar at all, which means that the conventional wisdom about "family dynamics" is simply wrong.

Wertheim is sympathetic to radical science and social constructionism. Her reaction is a sign of how behavioral genetics—and Harris's theory, which aims to explain its findings—touches a nerve on the political left, with its traditional emphasis on the malleability of children. The psychologist Oliver James wrote, "Harris's book can be safely ignored as yet another application of Friedmanite economics to the social realm" (an allusion to the economist who, according to James, stands for the idea that individuals should assume responsibility for their own lives). He suggested that Harris was downplaying research on parenting because it "would indirectly pose a real challenge to the theories of advanced consumer capitalism: if what parents do is critical, it calls into question the low priority given to it, compared with the pursuit of profit."[59] Actually, this fanciful diagnosis has it backwards. The most vehement propagandists for the importance of parents are the beer and tobacco companies, which sponsor ad campaigns such as "Family Talk About Drinking" and "Parents Should Talk to Kids About Not Smoking." (A sample ad: "Daughter speaks to the camera, as if it were her mother, reassuring her that her words about not smoking are with her, even when her mother is not with her.")[60] By putting the onus on parents to keep teens sober and smoke-free, these advanced consumer capitalists can divert attention from their own massive influence on adolescent peer culture.

In any case, Harris drew even more venom from the political right. The columnist John Leo called her theory "stupid," ridiculed her lack of a Ph.D. and a university affiliation, and compared her to deniers of the Holocaust. He ended his column, "It's not time to celebrate a foolish book that justifies self-absorption and makes non-parenting a respectable, mainstream activity."[61]

Why do conservatives hate the theory too? An axiom of the contemporary American right is that the traditional family is under assault from feminists, a licentious popular culture, and left-wing social analysts. The root of social ills, conservatives believe, is the failure of parents to teach their children discipline and values, a failure that can be traced to working mothers, absent fathers, easy divorce, and a welfare system that rewards young women for having babies out of wedlock. When the unmarried sitcom character Murphy Brown had a child, Vice President Dan Quayle denounced her for setting a bad example for American women (a headline of the time: "Murphy Has a Baby; Quayle Has a Cow"). Harris's review showing that Murphy's baby would probably have turned out fine was not welcome. (To be fair, concerns about fatherlessness may not be ill founded, but the problem may be the absence of fathers from all the families in a neighborhood rather than the absence of a father from an individual family. These fatherless children lack access to *other* families in which an adult male is present, and worse, they have access to packs of single men, whose values trickle down to their own peer groups.) Also, the Great Satan, Hillary Clinton, had written a book on childhood called *It Takes a Village*, based on the African saying "It takes a village to raise a child." Conservatives despised it because they thought the whole idea was a pretext for social engineers to take childrearing out of the hands of parents and give it to the government. But Harris quoted the saying too, and her theory implies there is some truth to it.

And then there were the experts. Brazelton called the thesis "absurd."[62] Jerome Kagan, one of the deans of scholarly research on children, said, "I'm embarrassed for psychology."[63] Another developmental psychologist, Frank Farley, told *Newsweek*:

She's all wrong. She's taking an extreme position based on a limited set of data. Her thesis is absurd on its face, but consider what might happen if parents believe this stuff! Will it free some to mistreat their kids, since "it doesn't matter"? Will it tell parents who are tired after a long day that they needn't bother even paying any attention to their kid since "it doesn't matter"?[64]

Kagan and other developmentalists told reporters about the "many, many good studies that show parents can affect how children turn out."

What were these "many, many good studies"? In the *Boston Globe*, Kagan laid out what he called the "ample evidence."[65] He mentioned the usual see-no-genetics studies showing that smart parents have smart children, verbal parents have verbal children, and so on. He observed that "a 6-year-old raised in New England will be very different from a 6-year-old raised in Malaysia, Uganda, or the southern tip of Argentina. The reason is that they experience different child-rearing practices by their parents." But of course a child growing up in Malaysia has both Malaysian parents *and* Malaysian peers. If Kagan had considered what would happen to a six-year-old child of Malaysian parents who grew up in a New England town, he might have thought twice before using the example to illustrate the power of parenting. The other "evidence" was that when authors write their memoirs, they credit their parents, never their childhood friends, with making them what they are. An irony in these feeble arguments is that Kagan himself, in the course of a distinguished career, often chided his fellow psychologists for overlooking genetics and for accepting their culture's folk theories on childhood instead of holding them up to scientific scrutiny. I can only imagine that on this occasion he felt compelled to defend his field against an exposé by a grandmother from New Jersey. In any case, the other "good studies" produced by defensive psychologists were no more informative.[66]

~

So HAS HARRIS solved the mystery of the Third Law, the unique environment that comes neither from the genes nor from the family? Not exactly. I am convinced that children are socialized—that they acquire the values and skills of the culture—in their peer groups, not their families. But I am not convinced, at least not yet, that peer groups explain how children develop their *personalities:* why they turn out shy or bold, anxious or confident, open-minded or old-school. Socialization and the development of personality are not the same thing, and peers may explain the first without necessarily explaining the second.

One way that peers could explain personality is that children in the same family may join different peer groups—the jocks, the brains, the preppies, the punks, the Goths—and assimilate their values. But then how do children get sorted into peer groups? If it is by their inborn traits—smart kids join the brains, aggressive kids join the punks, and so on—then effects of the peer group would show up as indirect effects of the genes, not as effects of the unique environment. If it is their parents' choice of neighborhoods, it would turn up as effects of the shared environment, because siblings growing up together share a neighborhood as well as a set of parents. In some cases, as with delinquency and smoking, the missing variance might be explained as an interaction between genes and peers: violence-prone adolescents become violent only in dangerous neighborhoods, addiction-prone children become smokers only in the company of peers who think smoking is cool. But those interactions are unlikely to explain most of the differences among children. Let's return to our touchstone: identical twins growing up together. They share their genes, they share their family environments, *and they share their peer groups,* at least on average. But the correlations between them are only around 50 percent. Ergo, neither genes nor families nor peer groups can explain what makes them different.

Harris is forthcoming about this limitation, and suggests that children differentiate themselves *within* a peer group, not by their *choice* of a peer group. Within each group, some become leaders, others foot soldiers, still others jesters, loose cannons, punching bags, or peacemakers, depending on what niche is available, how suited a child is to filling it, and chance. Once a child acquires a role, it is hard to shake it off, both because other children force the child to stay in the niche and because the child specializes in the skills necessary to prosper in it. This part of the theory, Harris notes, is untested, and difficult to test, because the crucial first step—which child fills which niche in which group—is so capricious.

The filling of niches in peer groups, then, is largely a matter of chance. But once we allow Lady Luck into the picture, she can act at other stages in life. When reminiscing on how we got to where we are, we all can think of forks in the road where we could have gone on very different life paths. If I hadn't gone to that party, I wouldn't have met my spouse. If I hadn't picked up that brochure, I wouldn't have known about the field that would become my life's calling. If I hadn't answered the phone, if I hadn't missed that flight, if only I had caught that ball. Life is a pinball game in which we bounce and graze through a gantlet of chutes and bumpers. Perhaps our history of collisions and near misses explains what made us what we are. One twin was once beaten up by a bully, the other was home sick that day. One inhaled a virus, the other didn't. One twin got the top bunk bed, the other got the bottom bunk bed.

We still don't know whether these unique experiences leave their fingerprints on our intellects and personalities. But an even earlier pinball game certainly could do so, the one that wires up our brain in the womb and early childhood. As I have mentioned, the human genome cannot possibly specify every last connection among neurons. But the "environment," in the sense of information encoded by the sense organs, isn't the only other option. Chance is another. One twin lies one way in the womb and stakes out her share of the placenta, the other has to squeeze around her. A cosmic ray mutates a stretch of DNA, a neurotransmitter zigs instead of zags, the growth cone of an axon goes left instead of right, and one identical twin's brain might gel into a slightly different configuration from the other's.[67]

We know this happens in the development of other organisms. Even genetically homogeneous strains of flies, mice, and worms, raised in monotonously controlled laboratories, can differ from one another. A fruit fly may have more or fewer bristles under one wing than its bottlemates. One mouse may have three times as many oocytes (cells destined to become eggs) as her genetically identical sister reared in the same lab. One roundworm may live three times as long as its virtual clone in the next dish. The biologist Steven Austad commented on the roundworms' lifespans: "Astonishingly, the degree of variability they exhibit in longevity is not much less than that of a genetically mixed population of humans, who eat a variety of diets, attend to or abuse their health, and are subject to all the vagaries of circumstance—car crashes, tainted beef, enraged postal workers—of modern industrialized life."[68] And a roundworm is composed of only 959 cells! A human brain, with its hundred billion neurons, has even more opportunities to be buffeted by the outcomes of molecular coin flips.

If chance in development is to explain the less-than-perfect similarity of identical twins, it says something interesting about development in general. One can imagine a developmental process in which millions of small chance events cancel one another out, leaving no difference in the end product. One can imagine a different process in which a chance event could derail development entirely, or send it on a chaotic developmental path resulting in a freak or a monster. Neither of these happens to identical twins. They are distinct enough that our crude instruments can pick up the differences, yet both are healthy instances of that staggeringly improbable, exquisitely engineered system we call a human being. The development of organisms must use complex feedback loops rather than prespecified blueprints. Random events can divert the trajectory of growth, but the trajectories are confined within an envelope of functioning designs for the species. Biologists refer to such developmental dynamics as robustness, buffering, or canalization.[69]

If the nongenetic component of personality is the outcome of neurodevelopmental roulette, it would present us with two surprises. One is that just as the "genetic" term in the behavioral geneticist's equation is not necessarily genetic, the "environmental" term is not necessarily environmental. If the unexplained variance is a product of chance events in brain assembly, yet another chunk of our personalities would be "biologically determined" (though not genetic) and beyond the scope of the best-laid plans of parents and society.

The other surprise is that we may have to make room for a pre-scientific explanatory concept in our view of human nature—not free will, as many people have suggested to me, but fate. It is not free will because among the traits that may differ between identical twins reared together are ones that are stubbornly involuntary. No one chooses to become schizophrenic, homosexual, musically gifted, or, for that matter, anxious or self-confident or open to experience. But the old idea of fate—in the sense of uncontrollable fortune, not strict predestination—can be reconciled with modern biology once we remember the many openings for chance to operate in development. Harris, noting how recent and parochial is the belief that we can shape our children, quotes a woman living in a remote village of India in the 1950s. When asked what kind of man she hoped her child would grow into, she shrugged and replied, "It is in his fate, no matter what I want."[70]

NOT EVERYONE IS so accepting of fate, or of the other forces beyond a parent's control, like genes and peers. "I hope to God this isn't true," one mother said to the *Chicago Tribune*. "The thought that all this love that I'm pouring into him counts for nothing is too terrible to contemplate."[71] As with other discoveries about human nature, people hope to God it isn't true. But the truth doesn't care about our hopes, and sometimes it can force us to revisit those hopes in a liberating way.

Yes, it is disappointing that there is no algorithm for growing a happy and successful child. But would we really want to specify the traits of our children in advance, and never be delighted by the unpredictable gifts and quirks that every child brings into the world? People are appalled by human cloning and its dubious promise that parents can design their children by genetic engineering. But how different is that from the fantasy that parents can design their children by how they bring them up? Realistic parents would be less anxious parents. They could enjoy their time with their children rather than constantly trying to stimulate them, socialize them, and improve their characters. They could read stories to their children for the pleasure of it, not because it's good for their neurons.

Many critics accuse Harris of trying to absolve parents of responsibility for their children's lives: if the kids turn out badly, parents can say it's not their fault. But by the same token she is assigning adults responsibility for their *own* lives: if your life is not going well, stop moaning that it's all your parents' fault. She is rescuing mothers from fatuous theories that blame them for every misfortune that befalls their children, and from the censorious know-it-alls who make them feel like ogres if they slip out of the house to work or skip a reading of *Goodnight Moon*. And the theory assigns us all a collective responsibility for the health of the neighborhoods and culture in which peer groups are embedded.

Finally: "So you're saying it doesn't matter how I treat my children?" What a question! Yes, of course it matters. Harris reminds her readers of the reasons.

First, parents wield enormous power over their children, and their actions can make a big difference to their happiness. Childrearing is above all an ethical responsibility. It is not OK for parents to beat, humiliate, deprive, or neglect their children, because those are awful things for a big strong person to do to a small helpless one. As Harris writes, "We may not hold their tomorrows in our hands but we surely hold their todays, and we have the power to make their todays very miserable."[72]

Second, a parent and a child have a human relationship. No one ever asks, "So you're saying it doesn't matter how I treat my husband or wife?" even though no one but a newlywed believes that one can change the personality of one's spouse. Husbands and wives are nice to each other (or should be) not to pound the other's personality into a desired shape but to build a deep and satisfying relationship. Imagine being told that one cannot revamp the personality of a husband or wife and replying, "The thought that all this love I'm pouring into him (or her) counts for nothing is too terrible to contemplate." So it is with parents and children: one person's behavior toward another has consequences for the quality of the relationship between them. Over the course of a lifetime the balance of power shifts, and children, complete with memories of how they were treated, have a growing say in their dealings with their parents. As Harris puts it, "If you don't think the moral imperative is a good enough reason to be nice to your kid, try this one: Be nice to your kid when he's young so that he will be nice to you when you're old."[73] There are well-functioning adults who still shake with rage when recounting the cruelties their parents inflicted on them as children. There are others who moisten up in private moments when recalling a kindness or sacrifice made for their happiness, perhaps one that the mother or father has long forgotten. If for no other reason, parents should treat their children well to allow them to grow up with such memories.

I have found that when people hear these explanations they lower their eyes and say, somewhat embarrassedly, "Yes. I knew that." The fact that people can forget these simple truths when intellectualizing about children shows how far modern doctrines have taken us. They make it easy to think of children as lumps of putty to be shaped instead of partners in a human relationship. Even the theory that children adapt to their peer group becomes less surprising when we think of them as human beings like ourselves. "Peer group" is a patronizing term we use in connection with children for what we call "friends and colleagues and associates" when we talk about ourselves. We groan when children obsess over wearing the right kind of cargo pants, but we would be just as mortified if a very large person forced us to wear pink overalls to a corporate board meeting or a polyester disco suit to an academic conference. "Being socialized by a peer group" is another way of saying "living successfully within a society," which for a social organism means "living." It is children, above all, who are alleged to be blank slates, and that can make us forget they are people.

Chapter 20

The Arts

THE ARTS ARE in trouble. I didn't say it; they did: the critics, scholars, and (as we now say) content providers who make their living in the arts and humanities. According to the theater director and critic Robert Brustein:

> The possibility of sustaining high culture in our time is becoming increasingly problematical. Serious book stores are losing their franchise; small publishing houses are closing shop; little magazines are going out of business; nonprofit theaters are surviving primarily by commercializing their repertory; symphony orchestras are diluting their programs; public television is increasing its dependence on reruns of British sitcoms; classical radio stations are dwindling; museums are resorting to blockbuster shows; dance is dying.[1]

In recent years the higher-brow magazines and presses have been filled with similar laments. Here is a sample of titles:

> The Death of Literature[2] • The Decline and Fall of Literature[3] • The Decline of High Culture[4] • Have the Humanities Disciplines Collapsed?[5] • The Humanities—At Twilight?[6] • Humanities in the Age of Money[7] • The Humanities' Plight[8] • Literature: An Embattled Profession[9] • Literature Lost[10] • Music's Dying Fall[11] • The Rise and Fall of English[12] • What's Happened to the Humanities?[13] • Who Killed Culture?[14]

If we are to believe the pessimists, the decline has been going on for some time. In 1948 T. S. Eliot wrote, "We can assert with some confidence that our own period is one of decline; that the standards of culture are lower than they were fifty years ago; and that the evidences of this decline are visible in every department of human activity."[15]

Some of the vital signs of the arts and humanities are indeed poor. In 1997 the U.S. House of Representatives voted to kill the National Endowment for the Arts, and the Senate was able to save it only by cutting its budget nearly in half. Universities have disinvested in the humanities: since 1960, the proportion of faculty in liberal arts has fallen by half, salaries and working conditions have stagnated, and more and more teaching is done by graduate students and part-time faculty.[16] New Ph.D.s are often unemployed or resigned to a life of one-year appointments. In many liberal arts colleges, humanities departments have been downsized, merged, or eliminated altogether.

One cause of the decline in academia is competition from the efflorescence of science and engineering. Another may be a surfeit of Ph.D.s pumped out by graduate programs that failed to practice academic birth control. But the problem is as much a reduction in the demand by students as an increase in the supply of professors. While the total number of bachelor's degrees rose by almost 40 percent between 1970 and 1994, the number of degrees in English *declined* by 40 percent. It may get worse: only 9 percent of high school students today indicate an interest in majoring in the humanities.[17] One university was so desperate to restore enrollment in its College of Arts and Sciences that it hired an advertising firm to come up with a "Think for a Living" campaign. Here are some of the slogans they came up with:

> Do what you want when you graduate or wait 20 years for your mid-
> life crisis.
> Insurance for when the robots take over all the boring jobs.
> Okay then. Follow your dreams in your next life.
> Yeah, like your parents are so happy.

Careerism may explain the disenchantment some students feel with liberal arts, but not all of it. The economy is in better shape today than it was in periods in which the humanities were more popular, and many young people still do not shoot themselves from cannons into their careers but use their college years to enrich themselves in various ways. There is no good reason that the arts and humanities should not be able to compete for students' attention during this interlude. A knowledge of culture, history, and ideas is still an asset in most professions, as it is in everyday life. But students stay away from the humanities anyway.

In this chapter I will diagnose the malaise of the arts and humanities and offer some suggestions for revitalizing them. They didn't ask me, but by their own accounts they need all the help they can get, and I believe that part of the answer lies within the theme of this book. I will begin by circumscribing the problem.

~

As a matter of fact, the arts and humanities are *not* in trouble. According to recent assessments based on data from the National Endowment for the Arts and the Statistical Abstract of the United States, they have never been in better shape.[18] In the past two decades, symphony orchestras, booksellers, libraries, and new independent films have all increased in number. Attendance is up, in some cases at record levels, at classical music concerts, live theater, opera performances, and art museums, as we see in blockbuster shows with long lines and scarce tickets. The number of books in print (including books of art, poetry, and drama) has exploded, as have book sales. Nor have people become passive consumers of art. The year 1997 broke records for the proportion of adults drawing, taking art photographs, buying art, and doing creative writing.

Advances in technology have made art more accessible than ever before. A couple of hours of minimum-wage income can buy any of tens of thousands of audiophile-quality musical recordings, including many versions of any classical work performed by the world's great orchestras. Video stores allow people in the boondocks to arrange cheap private screenings of the great classics of cinema. Instead of the three television networks with their sitcoms, variety shows, and soaps, most Americans can now choose from a menu of fifty to a hundred stations, including ones that specialize in history, science, politics, and the arts. Inexpensive video equipment and streaming video on the World Wide Web are allowing independent filmmaking to flourish. Virtually any book in print is available within days to anyone with a credit card and a modem. On the Web one can find the text of all the major novels, poems, plays, and works of philosophy and scholarship that have fallen out of copyright, as well as virtual tours of the world's great art museums. New intellectual e-zines and web sites have proliferated, and back issues are instantly available.

We are swimming in culture, drowning in it. So why all the lamentations about its plight, decline, fall, collapse, twilight, and death?

One response from the doomsayers is that the current frenzy of consumption involves past classics and current mediocrities but that few new works of quality are coming into the world. That is doubtful.[19] As historians of the arts repeatedly tell us, all the supposed sins of contemporary culture—mass appeal, the profit motive, themes of sex and violence, and adaptations to popular formats (such as serialization in newspapers)—may be found in the great artists of past centuries. Even in recent decades, many artists were seen in their time as commercial hacks and only later attained artistic respectability. Examples include the Marx Brothers, Alfred Hitchcock, the Beatles, and, if we are to judge by recent museum shows and critical appreciations, even Norman Rockwell. There are dozens of excellent novelists from countries all over the world, and though most television and cinema is dreadful, the best can be very good indeed: Carla on *Cheers* was wittier than Dorothy Parker, and the plot of *Tootsie* is cleverer than the plots of any of Shakespeare's cross-dressing comedies.

As for music, though it may be hard for anyone to compete against the best composers from the eighteenth and nineteenth centuries, the past century has been anything but barren. Jazz, Broadway, country, blues, folk, rock, soul, samba, reggae, world music, and contemporary composition have blossomed. Each has produced gifted artists and has introduced new complexities of rhythm, instrumentation, vocal style, and studio production into our total musical experience. Then there are genres that are flourishing as never before, such as animation and industrial design, and still others that have only recently come into existence but have already achieved moments of high accomplishment, such as computer graphics and rock videos (for instance, Peter Gabriel's *Sledgehammer*).

In every era for thousands of years critics have bemoaned the decline of culture, and the economist Tyler Cowen suggests they are the victims of a cognitive illusion. The best works of art are more likely to appear in a past decade than in the present decade for the same reason that another line in the supermarket always moves faster than the one you are in: there are more of them. We get to enjoy the greatest hits winnowed from all those decades, listening to the Mozarts and forgetting the Salieris. Also, genres of art (opera, Impressionist painting, Broadway musicals, film noir) usually blossom and fade in a finite span of time. It's hard to recognize nascent art forms when they are on the rise, and by the time they are widely appreciated their best days are behind them. Cowen also notes, citing Hobbes, that putting down the present is a backhanded way of putting down one's rivals: "Competition of praise inclineth to a reverence of antiquity. For men contend with the living, not with the dead."[20]

But in three circumscribed areas the arts really do have something to be depressed about. One is the traditions of elite art that descended from prestigious European genres, such as the music performed by symphony orchestras, the art shown in major galleries and museums, and the ballet performed by major companies. Here there really may be a drought of compelling new material. For example, 90 percent of "classical music" was composed before 1900, and the most influential composers in the twentieth century were active before 1940.[21]

The second is the guild of critics and cultural gatekeepers, who have seen their influence dwindle. The 1939 comedy *The Man Who Came to Dinner* is about a literary critic who achieved such celebrity that we can believe that the burghers of a small Ohio town would coo and fawn over him. It is hard to think of a contemporary critic who could plausibly inspire such a character.

And the third, of course, is the groves of academe, where the foibles of the humanities departments have been fodder for satirical novels and the subject of endless fretting and analyzing.

After nineteen chapters, you can probably guess where I will seek a diagnosis for these three ailing endeavors. The giveaway lies in a statement (attributed to Virginia Woolf) that can be found in countless English course outlines: "On or about December, 1910, human nature changed." She was referring to the new philosophy of modernism that would dominate the elite arts and criticism for much of the twentieth century, and whose denial of human nature was carried over with a vengeance to postmodernism, which seized control in its later decades. The point of this chapter is that the elite arts, criticism, and scholarship are in trouble because the statement is wrong. Human nature did not change in 1910, or in any year thereafter.[22]

~

ART IS IN our nature—in the blood and in the bone, as people used to say; in the brain and in the genes, as we might say today. In all societies people dance, sing, decorate surfaces, and tell and act out stories. Children begin to take part in these activities in their twos and threes, and the arts may even be reflected in the organization of the adult brain: neurological damage may leave a person able to hear and see but unable to appreciate music or visual beauty.[23] Paintings, jewelry, sculpture, and musical instruments go back at least 35,000 years in Europe, and probably far longer in other parts of the world where the archaeological record is scanty. The Australian aborigines have been painting on rocks for 50,000 years, and red ochre has been used as body makeup for at least twice that long.[24]

Though the exact forms of art vary widely across cultures, the activities of making and appreciating art are recognizable everywhere. The philosopher Denis Dutton has identified seven universal signatures:[25]

1. Expertise or virtuosity. Technical artistic skills are cultivated, recognized, and admired.
2. Nonutilitarian pleasure. People enjoy art for art's sake, and don't demand that it keep them warm or put food on the table.
3. Style. Artistic objects and performances satisfy rules of composition that place them in a recognizable style.
4. Criticism. People make a point of judging, appreciating, and interpreting works of art.
5. Imitation. With a few important exceptions like music and abstract painting, works of art simulate experiences of the world.
6. Special focus. Art is set aside from ordinary life and made a dramatic focus of experience.
7. Imagination. Artists and their audiences entertain hypothetical worlds in the theater of the imagination.

The psychological roots of these activities have become a topic of recent research and debate. Some researchers, such as the scholar Ellen Dissanayake, believe that art is an evolutionary adaptation like the emotion of fear or the ability to see in depth.[26] Others, such as myself, believe that art (other than narrative) is a by-product of three other adaptations: the hunger for status, the aesthetic pleasure of experiencing adaptive objects and environments, and the ability to design artifacts to achieve desired ends.[27] On this view art is a pleasure technology, like drugs, erotica, or fine cuisine—a way to purify and concentrate pleasurable stimuli and deliver them to our senses. For the discussion in this chapter it does not matter which view is correct. Whether art is an adaptation or a by-product or a mixture of the two, it is deeply rooted in our mental faculties. Here are some of those roots.

Organisms get pleasure from things that promoted the fitness of their ancestors, such as the taste of food, the experience of sex, the presence of children, and the attainment of know-how. Some forms of visual pleasure in natural environments may promote fitness, too. As people explore an environment, they seek patterns that help them negotiate it and take advantage of its contents. The patterns include well-delineated regions, improbable but informative features like parallel and perpendicular lines, and axes of symmetry and elongation. All are used by the brain to carve the visual field into surfaces, group the surfaces into objects, and organize the objects so people can recognize them the next time they see them. Vision researchers such as David Marr, Roger Shepard, and V. S. Ramachandran have suggested that the pleasing visual motifs used in art and decoration exaggerate these patterns, which tell the brain that the visual system is functioning properly and analyzing the world accurately.[28] By the same logic, tonal and rhythmic patterns in music may tap into mechanisms used by the auditory system to organize the world of sound.[29]

As the visual system converts raw colors and forms to interpretable objects and scenes, the aesthetic coloring of its products gets even richer. Surveys of art, photography, and landscape design, together with experiments on people's visual tastes, have found recurring motifs in the sights that give people pleasure.[30] Some of the motifs may belong to a search image for the optimal human habitat, a savanna: open grassland dotted with trees and bodies of water and inhabited by animals and flowering and fruiting plants. The enjoyment of the forms of living things has been dubbed *biophilia* by E. O. Wilson, and it appears to be a human universal.[31] Other patterns in a landscape may be pleasing because they are signals of safety, such as protected but panoramic views. Still others may be compelling because they are geographic features that make a terrain easy to explore and remember, such as landmarks, boundaries, and paths. The study of evolutionary aesthetics is also documenting the features that make a face or body beautiful.[32] The prized lineaments are those that signal health, vigor, and fertility.

People are imaginative animals who constantly recombine events in their mind's eye. That ability is one of the engines of human intelligence, allowing us to envision new technologies (such as snaring an animal or purifying a plant extract) and new social skills (such as exchanging promises or finding common enemies).[33] Narrative fiction engages this ability to explore hypothetical worlds, whether for edification—expanding the number of scenarios whose outcomes can be predicted—or for pleasure—vicariously experiencing love, adulation, exploration, or victory.[34] Hence Horace's definition of the purpose of literature: to instruct and to delight.

In good works of art, these aesthetic elements are layered so that the whole is more than the sum of its parts.[35] A good landscape painting or photograph will simultaneously evoke an inviting environment and be composed of geometric shapes with pleasing balance and contrast. A compelling story may simulate juicy gossip about desirable or powerful people, put us in an exciting time or place, tickle our language instincts with well-chosen words, and teach us something new about the entanglements of families, politics, or love. Many kinds of art are contrived to induce a buildup and release of psychological tension, mimicking other forms of pleasure. And a work of art is often embedded in a social happening in which the emotions are evoked in many members of a community at the same time, which can multiply the pleasure and grant a sense of solidarity. Dissanayake emphasizes this spiritual part of the art experience, which she calls "making special."[36]

A final bit of psychology engaged by the arts is the drive for status. One of the items on Dutton's list of the universal signatures of art is impracticality. But useless things, paradoxically, can be highly useful for a certain purpose: appraising the assets of the bearer. Thorstein Veblen first made the point in his theory of social status.[37] Since we cannot easily peer into the bank books or Palm Pilots of our neighbors, a good way to size up their means is to see whether they can afford to waste them on luxuries and leisure. Veblen wrote that the psychology of taste is driven by three "pecuniary canons": conspicuous consumption, conspicuous leisure, and conspicuous waste. They explain why status symbols are typically objects made by arduous and specialized labor out of rare materials, or else signs that the person is not bound to a life of manual toil, such as delicate and restrictive clothing or expensive and time-consuming hobbies. In a beautiful convergence, the biologist Amotz Zahavi used the same principle to explain the evolution of outlandish ornamentation in animals, such as the tail of the peacock.[38] Only the healthiest peacocks can afford to divert nutrients to expensive and cumbersome plumage. The peahen sizes up mates by the splendor of their tails, and evolution selects for males who muster the best ones.

Though most aficionados are aghast at the suggestion, art—especially elite art—is a textbook example of conspicuous consumption. Almost by definition, art has no practical function, and as Dutton points out in his list, it universally entails virtuosity (a sign of genetic quality, the free time to hone skills, or both) and criticism (which sizes up the worth of the art and the artist). Through most of European history, fine art and sumptuosity went hand in hand, as in the ostentatious decorations of opera and theater halls, the ornate frames around paintings, the formal dress of musicians, and the covers and bindings of old books. Art and artists were under the patronage of aristocrats or of the nouveau riche seeking instant respectability. Today, paintings, sculptures, and manuscripts continue to be sold at exorbitant and much-discussed prices (such as the $82.5 million paid for van Gogh's *Portrait of Dr. Gachet* in 1990).

In *The Mating Mind,* the psychologist Geoffrey Miller argues that the impulse to create art is a mating tactic: a way to impress prospective sexual and marriage partners with the quality of one's brain and thus, indirectly, one's genes. Artistic virtuosity, he notes, is unevenly distributed, neurally demanding, hard to fake, and widely prized. Artists, in other words, are sexy. Nature even gives us a precedent, the bowerbirds of Australia and New Guinea. The males construct elaborate nests and fastidiously decorate them with colorful objects such as orchids, snail shells, berries, and bark. Some of them literally paint their bowers with regurgitated fruit residue using leaves or bark as a brush. The females appraise the bowers and mate with the creators of the most symmetrical and well-ornamented ones. Miller argues that the analogy is exact:

> If you could interview a male Satin Bowerbird for *Artforum* magazine, he might say something like "I find this implacable urge for self-expression, for playing with color and form for their own sake, quite inexplicable. I cannot remember when I first developed this raging thirst to present richly saturated color-fields within a monumental yet minimalist stage-set, but I feel connected to something beyond myself when I indulge these passions. When I see a beautiful orchid high in a tree, I simply must have it for my own. When I see a single shell out of place in my creation, I must put it right. . . . It is a happy coincidence that females sometimes come to my gallery openings and appreciate my work, but it would be an insult to suggest that I create in order to procreate." Fortunately, bowerbirds cannot talk, so we are free to use sexual selection to explain their work, without them begging to differ.[39]

I am partial to a weaker version of the theory, in which one of the functions (not the only function) of creating and owning art is to impress other people (not just prospective mates) with one's social status (not just one's genetic quality). The idea goes back to Veblen and has been amplified by the art historian Quentin Bell and by Tom Wolfe in his fiction and nonfiction.[40] Perhaps its greatest champion today is the sociologist Pierre Bourdieu, who argues that connoisseurship of difficult and inaccessible works of culture serves as a membership badge in society's upper strata.[41] Remember that in all these theories, proximate and ultimate causes may be different. As with Miller's bowerbird, status and fitness need not enter the minds of people who create or appreciate art; they may simply explain how an urge for self-expression and an eye for beauty and skill evolved.

Regardless of what lies behind our instincts for art, those instincts bestow it with a transcendence of time, place, and culture. Hume noted that "the general principles of taste are uniform in human nature. . . . the same Homer who pleased at Athens and Rome two thousand years ago, is still admired at Paris and London."[42] Though people can argue about whether the glass is half full or half empty, a universal human aesthetic really can be discerned beneath the variation across cultures. Dutton comments:

> It is important to note how remarkably well the arts travel outside their home cultures: Beethoven and Shakespeare are beloved in Japan, Japanese prints are adored by Brazilians, Greek tragedy is performed worldwide, while, much to the regret of many local movie industries, Hollywood films have wide cross-cultural appeal. . . . Even Indian music . . . , while it sounds initially strange to the Western ear, can be shown to rely on rhythmic pulse and acceleration, repetition, variation and surprise, as well as modulation and divinely sweet melody: in fact, all the same devices found in Western music.[43]

One can extend the range of the human aesthetic even further. The Lascaux cave paintings, crafted in the late old Stone Age, continue to dazzle viewers in the age of the Internet. The faces of Nefertiti and Botticelli's Venus could appear on the cover of a twenty-first-century fashion magazine. The plot of the hero myth found in countless traditional cultures was transplanted effectively into the *Star Wars* saga. Western museum collectors plundered the prehistoric treasures of Africa, Asia, and the Americas not to add to the ethnographic record but because their patrons found the works beautiful to gaze at.

A wry demonstration of the universality of basic visual tastes came from a 1993 stunt by two artists, Vitaly Komar and Alexander Melamid, who used marketing research polls to assess Americans' taste in art.[44] They asked respondents about their preferences in color, subject matter, composition, and style, and found considerable uniformity. People said they liked realistic, smoothly painted landscapes in green and blue containing animals, women, children, and heroic figures. To satisfy this consumer demand, Komar and Melamid painted a composite of the responses: a lakeside landscape in a nineteenth-century realist style featuring children, deer, and George Washington. That's mildly amusing, but no one was prepared for what came next. When the painters replicated the polling in nine other countries, including Ukraine, Turkey, China, and Kenya, they found pretty much the same preferences: an idealized landscape, like the ones on calendars, and only minor substitutions from the American standard (hippos instead of deer, for example). What is even more interesting is that these McPaintings exemplify the kind of landscape that had been characterized as optimal for our species by researchers in evolutionary aesthetics.[45]

The art critic Arthur Danto had a different explanation: Western calendars are marketed all over the world, just like the rest of Western culture and art.[46] To many intellectuals, the globalization of Western styles is proof that tastes in art are arbitrary. People show similar aesthetic preferences, they claim, only because Western ideals have been exported to the world by imperialism, global business, and electronic media. There may be some truth to this, and for many people it is the morally correct position because it implies that there is nothing superior about Western culture or inferior about the indigenous ones it is replacing.

But there is another side to the story. Western societies are good at providing things that people want: clean water, effective medicine, varied and abundant food, rapid transportation and communication. They perfect these goods and services not from benevolence but from self-interest, namely the profits to be made in selling them. Perhaps the aesthetics industry also perfected ways of giving people what they like—in this case, art forms that appeal to basic human tastes, such as calendar landscapes, popular songs, and Hollywood romances and adventures. So even if an art form matured in the West, it may be not an arbitrary practice spread by a powerful navy but a successful product that engages a universal human aesthetic. This all sounds very parochial and Eurocentric, and I wouldn't push it too far, but it must have an element of truth: if there is a profit to be made in appealing to global human tastes, it would be surprising if entrepreneurs hadn't taken advantage of it. And it isn't as Eurocentric as one might think. Western culture, like Western technology and Western cuisine, is voraciously eclectic, appropriating any trick that pleases people from any culture it encounters. An example is one of America's most important culture exports, popular music. Ragtime, jazz, rock, blues, soul, and rap grew out of African American musical forms, which originally incorporated African rhythms and vocal styles.

∼

So what happened in 1910 that supposedly changed human nature? The event that stood out in Virginia Woolf's recollection was a London exhibition of the paintings of the post-Impressionists, including Cézanne, Gauguin, Picasso, and van Gogh. It was an unveiling of the movement called modernism, and when Woolf wrote her declaration in the 1920s, the movement was taking over the arts.

Modernism certainly proceeded *as if* human nature had changed. All the tricks that artists had used for millennia to please the human palate were cast aside. In painting, realistic depiction gave way to freakish distortions of shape and color and then to abstract grids, shapes, dribbles, splashes, and, in the $200,000 painting featured in the recent comedy *Art,* a blank white canvas. In literature, omniscient narration, structured plots, the orderly introduction of characters, and general readability were replaced by a stream of consciousness, events presented out of order, baffling characters and causal sequences, subjective and disjointed narration, and difficult prose. In poetry, the use of rhyme, meter, verse structure, and clarity were frequently abandoned. In music, conventional rhythm and melody were set aside in favor of atonal, serial, dissonant, and twelve-tone compositions. In architecture, ornamentation, human scale, garden space, and traditional craftsmanship went out the window (or would have if the windows could have been opened), and buildings were "machines for living" made of industrial materials in boxy shapes. Modernist architecture culminated both in the glass-and-steel towers of multinational corporations and in the dreary high-rises of American housing projects, postwar British council flats, and Soviet apartment blocks.

Why did the artistic elite spearhead a movement that called for such masochism? In part it was touted as a reaction to the complacency of the Victorian era and to the naïve bourgeois belief in certain knowledge, inevitable progress, and the justice of the social order. Weird and disturbing art was supposed to remind people that the world was a weird and disturbing place. And science, supposedly, was offering the same message. According to the version that trickled into the humanities, Freud showed that behavior springs from unconscious and irrational impulses, Einstein showed that time and space can be defined only relative to an observer, and Heisenberg showed that the position and momentum of an object were inherently uncertain because they were affected by the act of observation. Much later, this embroidery of physics inspired the famous hoax in which the physicist Alan Sokal successfully published a paper filled with gibberish in the journal *Social Text.*[47]

But modernism wanted to do more than just afflict the comfortable. Its glorification of pure form and its disdain for easy beauty and bourgeois pleasure had an explicit rationale and a political and spiritual agenda. In a review of a book defending the mission of modernism, the critic Frederick Turner explains them:

> The great project of modern art was to diagnose, and cure, the sickness unto death of modern humankind. . . . [Its artistic mission] is to identify and strip away the false sense of routine experience and interpretive framing provided by conformist mass commercial society, and to make us experience nakedly and anew the immediacy of reality through our peeled and rejuvenated senses. This therapeutic work is also a spiritual

mission, in that a community of such transformed human beings would, in theory, be able to construct a better kind of society. The enemies of the process are cooptation, commercial exploitation and reproduction, and kitsch. . . . Fresh, raw experience—to which artists have an unmediated and childlike access—is routinized, compartmentalized, and dulled into insensibility by society.[48]

Beginning in the 1970s, the mission of modernism was extended by the set of styles and philosophies called postmodernism. Postmodernism was even more aggressively relativistic, insisting that there are many perspectives on the world, none of them privileged. It denied even more vehemently the possibility of meaning, knowledge, progress, and shared cultural values. It was more Marxist and far more paranoid, asserting that claims to truth and progress were tactics of political domination which privileged the interests of straight white males. According to the doctrine, mass-produced commodities and media-disseminated images and stories were designed to make authentic experience impossible.

The goal of postmodernist art is to help us break out of this prison. The artists try to preempt cultural motifs and representational techniques by taking capitalist icons (such as ads, package designs, and pinup photos) and defacing them, exaggerating them, or presenting them in odd contexts. The earliest examples were Andy Warhol's paintings of soup can labels and his repetitive false-color images of Marilyn Monroe. More recent ones include the Whitney Museum's "Black Male" exhibit described in Chapter 12 and Cindy Sherman's photographs of grotesquely assembled bi-gendered mannequins. (I saw them as part of an MIT exhibit that explored "the female body as a site of conflicting desires, and femininity as a taut web of social expectations, historical assumptions, and ideological constructions.") In postmodernist literature, authors comment on what they are writing while they are writing it. In postmodernist architecture, materials and details from different kinds of buildings and historical periods are thrown together in incongruous ways, such as an awning made of chain-link fencing in a fancy shopping mall or Corinthian columns holding up nothing on the top of a sleek skyscraper. Postmodernist films contain sly references to the filmmaking process or to earlier films. In all these forms, irony, self-referential allusions, and the pretense of not taking the work seriously are meant to draw attention to the representations themselves, which (according to the doctrine) we are ordinarily in danger of mistaking for reality.

∼

ONCE WE RECOGNIZE what modernism and postmodernism have done to the elite arts and humanities, the reasons for their decline and fall become all too obvious. The movements are based on a false theory of human psychology, the Blank Slate. They fail to apply their most vaunted ability—stripping away pretense—to themselves. And they take all the fun out of art!

Modernism and postmodernism cling to a theory of perception that was rejected long ago: that the sense organs present the brain with a tableau of raw colors and sounds and that everything else in perceptual experience is a learned social construction. As we saw in preceding chapters, the visual system of the brain comprises some fifty regions that take raw pixels and effortlessly organize them into surfaces, colors, motions, and three-dimensional objects. We can no more turn the system off and get immediate access to pure sensory experience than we can override our stomachs and tell them when to release their digestive enzymes. The visual system, moreover, does not drug us into a hallucinatory fantasy disconnected from the real world. It evolved to feed us information about the consequential things out there, like rocks, cliffs, animals, and other people and their intentions.

Nor does innate organization stop at apprehending the physical structure of the world. It also colors our visual experience with universal emotions and aesthetic pleasures. Young children prefer calendar landscapes to pictures of deserts and forests, and babies as young as three months old gaze longer at a pretty face than at a plain one.[49] Babies prefer consonant musical intervals over dissonant ones, and two-year-olds embark on a lifetime of composing and appreciating narrative fiction when they engage in pretend play.[50]

When we perceive the products of other people's behavior, we evaluate them through our intuitive psychology, our theory of mind. We do not take a stretch of language or an artifact like a product or work of art at face value, but try to guess why the producers came out with them and what effect they hope to have on us (as we saw in Chapter 12). Of course, people can be taken in by a clever liar, but they are not trapped in a false world of words and images and in need of rescue by postmodernist artists.

Modernist and postmodernist artists and critics fail to acknowledge another feature of human nature that drives the arts: the hunger for status, especially their *own* hunger for status. As we saw, the psychology of art is entangled with the psychology of esteem, with its appreciation of the rare, the sumptuous, the virtuosic, and the dazzling. The problem is that whenever people seek rare things, entrepreneurs make them less rare, and whenever a dazzling performance is imitated, it can become commonplace. The result is the perennial turnover of styles in the arts. The psychologist Colin Martindale has documented that every art form increases in complexity, ornamentation, and emotional charge until the evocative potential of the style is fully exploited.[51] Attention then turns to the style itself, at which point the style gives way to a new one. Martindale attributes this cycle to habituation on the part of the audience, but it also comes from the desire for attention on the part of the artists.

In twentieth-century art, the search for the new new thing became desperate because of the economies of mass production and the affluence of the middle class. As cameras, art reproductions, radios, records, magazines, movies, and paperbacks became affordable, ordinary people could buy art by the carload. It is hard to distinguish oneself as a good artist or discerning connoisseur if people are up to their ears in the stuff, much of it of reasonable artistic merit. The problem for artists is not that popular culture is so bad but that it is so good, at least some of the time. Art could no longer confer prestige by the rarity or excellence of the works themselves, so it had to confer it by the rarity of the powers of appreciation. As Bourdieu points out, only a special elite of initiates could get the point of the new works of art. And with beautiful things spewing out of printing presses and record plants, distinctive works need not be beautiful. Indeed, they had better not be, because now any schmo could have beautiful things.

One result is that modernist art stopped trying to appeal to the senses. On the contrary, it *disdained* beauty as saccharine and lightweight.[52] In his 1913 book *Art,* the critic Clive Bell (Virginia Woolf's brother-in-law and Quentin's father) argued that beauty had no place in good art because it was rooted in crass experiences.[53] People use *beautiful* in phrases like "beautiful huntin' and shootin'," he wrote, or worse, to refer to beautiful women. Bell assimilated the behaviorist psychology of his day and argued that ordinary people come to enjoy art by a process of Pavlovian conditioning. They appreciate a painting only if it depicts a beautiful woman, music only if it evokes "emotions similar to those provoked by young ladies in musical farces," and poetry only if it arouses feelings like the ones once felt for the vicar's daughter. Thirty-five years later, the abstract painter Barnett Newman approvingly declared that the impulse of modern art was "the desire to destroy beauty."[54] Postmodernists were even more dismissive. Beauty, they said, consists of arbitrary standards dictated by an elite. It enslaves women by forcing them to conform to unrealistic ideals, and it panders to market-oriented art collectors.[55]

To be fair, modernism comprises many styles and artists, and not all of them rejected beauty and other human sensibilities. At its best, modernist design perfected a visual elegance and an aesthetic of form-following-function that were welcome alternatives to Victorian bric-a-brac and ostentatious displays of wealth. The art movements opened up new stylistic possibilities, including motifs from Africa and Oceania. The fiction and poetry offered invigorating intellectual workouts, and countered a sentimental romanticism that saw art as a spontaneous overflow of the artist's personality and emotion. The problem with modernism was that its philosophy did not acknowledge the ways in which it was appealing to human pleasure. As its denial of beauty became an orthodoxy, and as its aesthetic successes were appropriated into commercial culture (such as minimalism in graphic design), modernism left nowhere for artists to go.

Quentin Bell suggested that when the variations within a genre are exhausted, people avail themselves of a different canon of status, which he added to Veblen's list. In "conspicuous outrage," bad boys (and girls) flaunt their ability to get away with shocking the bourgeoisie.[56] The never-ending campaign by postmodernist artists to attract the attention of a jaded public progressed from puzzling audiences to doing everything they could to offend them. Everyone has heard of the notorious cases: Robert Mapplethorpe's photographs of sadomasochistic acts, Andres Serrano's *Piss Christ* (a photo of a crucifix in a jar of the artist's urine), Chris Ofili's painting of the Virgin Mary smeared in elephant dung, and the nine-hour performance piece "Flag Fuck (w/Beef) #17B," in which Ivan Hubiak danced on stage wearing an American flag as a diaper while draping himself with raw meat. Actually, this last one never happened; it was invented by writers for the satirical newspaper *The Onion* in an article entitled "Performance Artist Shocks U.S. Out of Apathetic Slumber."[57] But I bet I had you fooled.

Another result is that elite art could no longer be appreciated without a support team of critics and theoreticians. They did not simply evaluate and interpret art, like movie critics or book reviewers, but supplied the art with its rationale. Tom Wolfe wrote *The Painted Word* after reading an art review in the *New York Times* that criticized realist painting because it lacked "something crucial," namely, "a persuasive theory." Wolfe explains:

Then and there I experienced a flash known as the *Aha!* phenomenon, and the buried life of contemporary art was revealed to me for the first time. . . . All these years I, like so many others, had stood in front of a thousand, two thousand, God-knows-how-many thousand Pollocks, de Koonings, Newmans, Nolands, Rothkos, Rauschenbergs, Judds, Johnses, Olitskis, Louises, Stills, Franz Klines, Frankenthalers, Kellys, and Frank Stellas, now squinting, now popping the eye sockets open, now drawing back, now moving closer—waiting, waiting, forever waiting for . . . *it* . . . for *it* to come into focus, namely, the visual reward (for so much effort) which must be there, which everyone (*tout le monde*) knew to be there—waiting for something to radiate directly from the paintings on these invariably pure white walls, in this room, in this moment, into my own optic chiasma. All these years, in short, I had assumed that in art, if nowhere else, seeing is believing. Well—how very shortsighted! Now, at last, on April 28, 1974, I could see. I had gotten it backward all along. Not "seeing is believing," you ninny, but "believing is seeing," for *Modern Art has become completely literary: the paintings and other works exist only to illustrate the text.*[58]

Once again, postmodernism took this extreme to an even greater extreme in which the theory upstaged the subject matter and became a genre of performance art in itself. Postmodernist scholars, taking off from the critical theorists Theodor Adorno and Michel Foucault, distrust the demand for "linguistic transparency" because it hobbles the ability "to think the world more radically" and puts a text in danger of being turned into a mass-market commodity.[59] This attitude has made them regular winners of the annual Bad Writing Contest, which "celebrates the most stylistically lamentable passages found in scholarly books and articles."[60] In 1998, first prize went to the lauded professor of rhetoric at Berkeley, Judith Butler, for the following sentence:

> The move from a structuralist account in which capital is understood to structure social relations in relatively homologous ways to a view of hegemony in which power relations are subject to repetition, convergence, and rearticulation brought the question of temporality into the thinking of structure, and marked a shift from a form of Althusserian theory that takes structural totalities as theoretical objects to one in which the insights into the contingent possibility of structure inaugurate a renewed conception of hegemony as bound up with the contingent sites and strategies of the rearticulation of power.

Dutton, whose journal *Philosophy and Literature* sponsors the contest, assures us that this is not a satire. The rules of the contest forbid it: "Deliberate parody cannot be allowed in a field where unintended self-parody is so widespread."

A final blind spot to human nature is the failure of contemporary artists and theorists to deconstruct their own moral pretensions. Artists and critics have long believed that an appreciation of elite art is ennobling and have spoken of cultural philistines in tones ordinarily reserved for child molesters (as we see in the two meanings of the word *barbarian*). The affectation of social reform that surrounds modernism and postmodernism is part of this tradition.

Though moral sophistication requires an appreciation of history and cultural diversity, there is no reason to think that the elite arts are a particularly good way to instill it compared with middlebrow realistic fiction or traditional education. The plain fact is that there are no obvious moral consequences to how people entertain themselves in their leisure time. The conviction that artists and connoisseurs are morally advanced is a cognitive illusion, arising from the fact that our circuitry for morality is cross-wired with our circuitry for status (see Chapter 15). As the critic George Steiner has pointed out, "We know that a man can read Goethe or Rilke in the evening, that he can play Bach and Schubert, and go to his day's work at Auschwitz in the morning."[61] Conversely there must be many unlettered people who give blood, risk their lives as volunteer firefighters, or adopt handicapped children, but whose opinion of modern art is "My four-year-old daughter could have done that."

The moral and political track record of modernist artists is nothing to be proud of. Some were despicable in the conduct of their personal lives, and many embraced fascism or Stalinism. The modernist composer Karlheinz Stockhausen described the September 11, 2001, terrorist attacks as "the greatest work of art imaginable for the whole cosmos" and added, enviously, that "artists, too, sometimes go beyond the limits of what is feasible and conceivable, so that we wake up, so that we open ourselves to another world."[62] Nor is the theory of postmodernism especially progressive. A denial of objective reality is no friend to moral progress, because it prevents one from saying, for example, that slavery or the Holocaust really took place. And as Adam Gopnik has pointed out, the political messages of most postmodernist pieces are utterly banal, like "racism is bad." But they are stated so obliquely that viewers are made to feel morally superior for being able to figure them out.

As for sneering at the bourgeoisie, it is a sophomoric grab at status with no claim to moral or political virtue. The fact is that the values of the middle class—personal responsibility, devotion to family and neighborhood, avoidance of macho violence, respect for liberal democracy—are good things, not bad things. Most of the world wants to join the bourgeoisie, and most artists are members in good standing who adopted a few bohemian affectations. Given the history of the twentieth century, the reluctance of the bourgeoisie to join mass utopian uprisings can hardly be held against them. And if they want to hang a painting of a red barn or a weeping clown above their couch, it's none of our damn business.

The dominant theories of elite art and criticism in the twentieth century grew out of a militant denial of human nature. One legacy is ugly, baffling, and insulting art. The other is pretentious and unintelligible scholarship. And they're surprised that people are staying away in droves?

~

A REVOLT HAS begun. Museum-goers have become bored with the umpteenth exhibit on the female body featuring dismembered torsos or hundreds of pounds of lard chewed up and spat out by the artist.[63] Graduate students in the humanities are grumbling in emails and conference hallways about being locked out of the job market unless they write in gibberish while randomly dropping the names of authorities like Foucault and Butler. Maverick scholars are doffing the blinders that prevented them from looking at exciting developments in the sciences of human nature. And younger artists are wondering how the art world got itself into the bizarre place in which beauty is a dirty word.

These currents of discontent are coming together in a new philosophy of the arts, one that is consilient with the sciences and respectful of the minds and senses of human beings. It is taking shape both in the community of artists and in the community of critics and scholars.

In the year 2000, the composer Stefania de Kenessey puckishly announced a new "movement" in the arts, Derrière Guard, which celebrates beauty, technique, and narrative.[64] If that sounds too innocuous to count as a movement, consider the response of the director of the Whitney, the shrine of the dismembered-torso establishment, who called the members of the movement "a bunch of crypto-Nazi conservative bullshitters."[65] Ideas similar to Derrière Guard's have sprung up in movements called the Radical Center, Natural Classicism, the New Formalism, the New Narrativism, Stuckism, the Return of Beauty, and No Mo Po Mo.[66] The movements combine high and low culture and are opposed equally to the postmodernist left, with its disdain for beauty and artistry, and to the cultural right, with its narrow canons of "great works" and fire-and-brimstone sermons on the decline of civilization. It includes classically trained musicians who mix classical and popular compositions, realist painters and sculptors, verse poets, journalistic novelists, and dance directors and performance artists who use rhythm and melody in their work.

Within the academy, a growing number of mavericks are looking to evolutionary psychology and cognitive science in an effort to reestablish human nature at the center of any understanding of the arts. They include Brian Boyd, Joseph Carroll, Denis Dutton, Nancy Easterlin, David Evans, Jonathan Gottschall, Paul Hernadi, Patrick Hogan, Elaine Scarry, Wendy Steiner, Robert Storey, Frederick Turner, and Mark Turner.[67] A good grasp of how the mind works is indispensable to the arts and humanities for at least two reasons.

One is that the real medium of artists, whatever their genre, is human mental representations. Oil paint, moving limbs, and printed words cannot penetrate the brain directly. They trigger a cascade of neural events that begin with the sense organs and culminate in thoughts, emotions, and memories. Cognitive science and cognitive neuroscience, which map out the cascade, offer a wealth of information to anyone who wants to understand how artists achieve their effects. Vision research can illuminate painting and sculpture.[68] Psychoacoustics and linguistics can enrich the study of music.[69] Linguistics can give insight on poetry, metaphor, and literary style.[70] Mental imagery research helps to explain the techniques of narrative prose.[71] The theory of mind (intuitive psychology) can shed light on our ability to entertain fictional worlds.[72] The study of visual attention and short-term memory can help explain the experience of cinema.[73] And evolutionary aesthetics can help explain the feelings of beauty and pleasure that can accompany all of these acts of perception.[74]

Ironically, the early modernist painters were avid consumers of perception research. It may have been introduced to them by Gertrude Stein, who studied psychology with William James at Harvard and conducted research on visual attention under his supervision.[75] The Bauhaus designers and artists, too, were appreciators of perceptual psychology, particularly the contemporary Gestalt school.[76] But the consilience was lost as the two cultures drifted apart, and only recently have they begun to come back together. I predict that the application of cognitive science and evolutionary psychology to the arts will become a growth area in criticism and scholarship.

The other point of contact may be more important still. Ultimately what draws us to a work of art is not just the sensory experience of the medium but its emotional content and insight into the human condition. And these tap into the timeless tragedies of our biological predicament: our mortality, our finite knowledge and wisdom, the differences among us, and our conflicts of interest with friends, neighbors, relatives, and lovers. All are topics of the sciences of human nature.

The idea that art should reflect the perennial and universal qualities of the human species is not new. Samuel Johnson, in the preface to his edition of Shakespeare's plays, comments on the lasting appeal of that great intuitive psychologist:

Nothing can please many, and please long, but just representations of general nature. Particular manners can be known to few, and therefore few only can judge how nearly they are copied. The irregular combinations of fanciful invention may delight a-while, by that novelty of which the common satiety of life sends us all in quest; but the pleasures of sudden wonder are soon exhausted, and the mind can only repose on the stability of truth.

Today we may be seeing a new convergence of explorations of the human condition by artists and scientists—not because scientists are trying to take over the humanities, but because artists and humanists are beginning to look to the sciences, or at least to the scientific mindset that sees us as a species with a complex psychological endowment. In explaining this connection I cannot hope to compete with the words of the artists themselves, and I will conclude with the overtures of three fine novelists.

Iris Murdoch, haunted by the origins of the moral sense, comments on its endurance in fiction:

We make, in many respects though not in all, the same kinds of moral judgments as the Greeks did, and we recognize good or decent people in times and literatures remote from our own. Patroclus, Antigone, Cordelia, Mr. Knightley, Alyosha. Patroclus' invariable kindness. Cordelia's truthfulness. Alyosha telling his father not to be afraid of hell. It is just as important that Patroclus should be kind to the captive women as that Emma should be kind to Miss Bates, and we feel this importance in an immediate and natural way in both cases in spite of the fact that nearly three thousand years divide the writers. And this, when one reflects on it, is a remarkable testimony to the existence of a single durable human nature.[77]

A. S. Byatt, asked by the editors of the *New York Times Magazine* for the best narrative of the millennium, picked the story of Scheherazade:

The stories in "The Thousand and One Nights" . . . are stories about storytelling without ever ceasing to be stories about love and life and death and money and food and other human necessities. Narration is as much a part of human nature as breath and the circulation of the blood. Modernist literature tried to do away with storytelling, which it thought vulgar, replacing it with flashbacks, epiphanies, streams of consciousness. But storytelling is intrinsic to biological time, which we cannot escape. Life, Pascal said, is like living in a prison from which every day fellow prisoners are taken away to be executed. We are all, like Scheherazade, under sentences of death, and we all think of our lives as narratives, with beginnings, middles, and ends.[78]

John Updike, also asked for reflections at the turn of the millennium, commented on the future of his own profession. "A writer of fiction, a professional liar, is paradoxically obsessed with what is true," he wrote, and "the unit of truth, at least for a fiction writer, is the human animal, belonging to the species *Homo sapiens,* unchanged for at least 100,000 years."

Evolution moves more slowly than history, and much slower than the technology of recent centuries; surely sociobiology, surprisingly maligned in some scientific quarters, performs a useful service in investigating what traits are innate and which are acquired. What kind of cultural software can our evolved hard-wiring support? Fiction, in its groping way, is drawn to those moments of discomfort when society asks more than its individual members can, or wish to, provide. Ordinary people experiencing friction on the page is what warms our hands and hearts as we write. . . .

To be human is to be in the tense condition of a death-foreseeing, consciously libidinous animal. No other earthly creature suffers such a capacity for thought, such a complexity of envisioned but frustrated possibilities, such a troubling ability to question the tribal and biological imperatives.

So conflicted and ingenious a creature makes an endlessly interesting focus for the meditations of fiction. It seems to me true that *Homo sapiens* will never settle into any utopia so complacently as to relax all its conflicts and erase all its perversity-breeding neediness.[79]

Literature has three voices, wrote the scholar Robert Storey: those of the author, the audience, and the species.[80] These novelists are reminding us of the voice of the species, an essential constituent of all the arts, and a fitting theme with which to wrap up my own story.

PART VI

THE VOICE OF
THE SPECIES

The Blank Slate was an attractive vision. It promised to make racism, sexism, and class prejudice factually untenable. It appeared to be a bulwark against the kind of thinking that led to ethnic genocide. It aimed to prevent people from slipping into a premature fatalism about preventable social ills. It put a spotlight on the treatment of children, indigenous peoples, and the underclass. The Blank Slate thus became part of a secular faith and appeared to constitute the common decency of our age.

But the Blank Slate had, and has, a dark side. The vacuum that it posited in human nature was eagerly filled by totalitarian regimes, and it did nothing to prevent their genocides. It perverts education, childrearing, and the arts into forms of social engineering. It torments mothers who work outside the home and parents whose children did not turn out as they would have liked. It threatens to outlaw biomedical research that could alleviate human suffering. Its corollary, the Noble Savage, invites contempt for the principles of democracy and of "a government of laws and not of men." It blinds us to our cognitive and moral shortcomings. And in matters of policy it has elevated sappy dogmas above the search for workable solutions.

The Blank Slate is not some ideal that we should all hope and pray is true. No, it is an anti-life, anti-human theoretical abstraction that denies our common humanity, our inherent interests, and our individual preferences. Though it has pretensions of celebrating our potential, it does the opposite, because our potential comes from the combinatorial interplay of wonderfully complex faculties, not from the passive blankness of an empty tablet.

Regardless of its good and bad effects, the Blank Slate is an empirical hypothesis about the functioning of the brain and must be evaluated in terms of whether or not it is true. The modern sciences of mind, brain, genes, and evolution are increasingly showing that it is not true. The result is a rearguard effort to salvage the Blank Slate by disfiguring science and intellectual life:

denying the possibility of objectivity and truth, dumbing down issues into dichotomies, replacing facts and logic with political posturing.

The Blank Slate became so entrenched in intellectual life that the prospect of doing without it can be deeply unsettling. In topics from childrearing to sexuality, from natural foods to violence, ideas that seemed immoral even to question turn out to be not just questionable but probably wrong. Even people with no ideological ax to grind can feel a sense of vertigo when they learn of such taboos being broken: "O brave new world that has such people in it!" Is science leading to a place where prejudice is all right, where children may be neglected, where Machiavellianism is accepted, where inequality and violence are met with resignation, where people are treated like machines?

Not at all! By unhandcuffing widely shared values from moribund factual dogmas, the rationale for those values can only become clearer. We understand *why* we condemn prejudice, cruelty to children, and violence against women, and can focus our efforts on how to implement the goals we value most. We thereby protect those goals against the upheavals of factual understanding that science perennially delivers.

Abandoning the Blank Slate, in any case, is not as radical as it might first appear. True, it is a revolution in many sectors of modern intellectual life. But except for a few intellectuals who have let their theories get the better of them, it is not a revolution in the world views of most people. I suspect that few people really believe, deep down, that boys and girls are interchangeable, that all differences in intelligence come from the environment, that parents can micromanage the personalities of their children, that humans are born free of selfish tendencies, or that appealing stories, melodies, and faces are arbitrary social constructions. Margaret Mead, an icon of twentieth-century egalitarianism, told her daughter that she credited her own intellectual talent to her genes, and I can confirm that such split personalities are common among academics.[1] Scholars who publicly deny that intelligence is a meaningful concept treat it as anything but meaningless in their professional lives. Those who argue that gender differences are a reversible social construction do not treat them that way in their advice to their daughters, their dealings with the opposite sex, and their unguarded gossip, humor, and reflections on their lives.

Acknowledging human nature does not mean overturning our personal world views, and I would have nothing to suggest as a replacement if it did. It means only taking intellectual life out of its parallel universe and reuniting it with science and, when it is borne out by science, with common sense. The alternative is to make intellectual life increasingly irrelevant to human affairs, to turn intellectuals into hypocrites, and to turn everyone else into anti-intellectuals.

Scientists and public intellectuals are not the only people who have pondered how the mind works. We are all psychologists, and some people, without the benefit of credentials, are great psychologists. Among them are poets and novelists, whose business, as we saw in the preceding chapter, is to create "just representations of general nature." Paradoxically, in today's intellectual climate novelists may have a clearer mandate than scientists to speak the truth about human nature. Sophisticated people sneer at feel-good comedies and saccharine romances in which all loose ends are tied and everyone lives happily ever after. Life is nothing like that, we note, and we look to the arts for edification about the painful dilemmas of the human condition.

Yet when it comes to the *science* of human beings, this same audience says: Give us schmaltz! "Pessimism" is considered a legitimate criticism of observations of human nature, and people expect theories to be a source of sentimental uplift. "Shakespeare had no conscience; neither do I," said George Bernard Shaw. This was not a confession of psychopathy but an affirmation of a good playwright's obligation to take every character's point of view seriously. Scientists of human behavior have the same obligation, and it does not require them to turn off their consciences in the spheres in which they must be exercised.

Poets and novelists have made many of the points of this book with greater wit and power than any academic scribbler could hope to do. They allow me to conclude the book by revisiting some of its main themes without merely repeating them. What follows are five vignettes from literature that capture, for me, some of the morals of the sciences of human nature. They underscore that the discoveries of those sciences should be faced not with fear and loathing but with the balance and discernment we use when we reflect on human nature in the rest of our lives.

~

The Brain—is wider than the Sky—
For—put them side by side—
The one the other will contain
With ease—and you—beside—

The Brain is deeper than the sea—
For—hold them—Blue to Blue—
The one the other will absorb—
As Sponges—Buckets—do—

The Brain is just the weight of God—
For—Heft them—Pound for Pound—
And they will differ—if they do—
As Syllable from Sound—

The first two verses of Emily Dickinson's "The Brain Is Wider Than the Sky" express the grandeur in the view of the mind as consisting in the activity of the brain.[2] Here and in her other poems, Dickinson refers to "the brain," not "the soul" or even "the mind," as if to remind her readers that the seat of our thought and experience is a hunk of matter. Yes, science is, in a sense, "reducing" us to the physiological processes of a not-very-attractive three-pound organ. But what an organ! In its staggering complexity, its explosive combinatorial computation, and its limitless ability to imagine real and hypothetical worlds, the brain, truly, is wider than the sky. The poem itself proves it. Simply to understand the comparison in each verse, the brain of the reader must contain the sky and absorb the sea and visualize each one at the same scale as the brain itself.

The enigmatic final verse, with its startling image of God and the brain being hefted like cabbages, has puzzled readers since the poem was published. Some read it as creationism (God made the brain), others as atheism (the brain thought up God). The simile with phonology—sound is a seamless continuum, a syllable is a demarcated unit of it—suggests a kind of pantheism: God is everywhere and nowhere, and every brain incarnates a finite measure of divinity. The loophole "if they do" suggests mysticism—the brain and God may somehow be the same thing—and, of course, agnosticism. The ambiguity is surely intentional, and I doubt that anyone could defend a single interpretation as the correct one.

I like to read the verse as suggesting that the mind, in contemplating its place in the cosmos, at some point reaches its own limitations and runs into puzzles that seem to belong in a separate, divine realm. Free will and subjective experience, for example, are alien to our concept of causation and feel like a divine spark inside us. Morality and meaning seem to inhere in a reality that exists independent of our judgments. But that separateness may be the illusion of a brain that makes it impossible for us *not* to think they are separate from us. Ultimately we have no way of knowing, because we *are* our brains and have no way of stepping outside them to check. But if we are thereby trapped, it is a trap that we can hardly bemoan, for it is wider than the sky, deeper than the sea, and perhaps as weighty as God.

~

KURT VONNEGUT'S STORY "Harrison Bergeron" is as transparent as Dickinson's poem is cryptic. Here is how it begins:

> The year was 2081, and everybody was finally equal. They weren't only equal before God and the law. They were equal every which way. Nobody was smarter than anybody else. Nobody was better looking than anybody else. Nobody was stronger or quicker than anybody else. All this equality was due to the 211th, 212th, and 213th Amendments to the Constitution, and to the unceasing vigilance of agents of the United States Handicapper General.[3]

The Handicapper General enforces equality by neutralizing any inherited (hence undeserved) asset. Intelligent people have to wear radios in their ears tuned to a government transmitter that sends out a sharp noise every twenty seconds (such as the sound of a milk bottle struck with a ball-peen hammer) to prevent them from taking unfair advantage of their brains. Ballerinas are laden with bags of birdshot and their faces are hidden by masks so that no one can feel bad at seeing someone prettier or more graceful than they. Newscasters are selected for their speech impediments. The hero of the story is a multiply gifted teenager forced to wear headphones, thick wavy glasses, three hundred pounds of scrap iron, and black caps on half his teeth. The story is about his ill-fated rebellion.

Subtle it is not, but "Harrison Bergeron" is a witty reductio of an all too common fallacy. The ideal of political equality is not a guarantee that people are innately indistinguishable. It is a policy to treat people in certain spheres (justice, education, politics) on the basis of their individual merits rather than the statistics of any group they belong to. And it is a policy to recognize inalienable rights in all people by virtue of the fact that they are sentient human beings. Policies that insist that people be identical in their outcomes must impose costs on humans who, like all living things, vary in their biological endowment. Since talents by definition are rare, and can be fully realized only in rare circumstances, it is easier to achieve forced equality by lowering the top (and thereby depriving everyone of the fruits of people's talents) than by raising the bottom. In Vonnegut's America of 2081 the desire for equality of outcome is played out as a farce, but in the twentieth century it frequently led to real crimes against humanity, and in our own society the entire issue is often a taboo.

Vonnegut is a beloved author who has never been called a racist, sexist, elitist, or Social Darwinist. Imagine the reaction if he had stated his message in declarative sentences rather than in a satirical story. Every generation has its designated jokers, from Shakespearean fools to Lenny Bruce, who give voice to truths that are unmentionable in polite society. Today part-time humorists like Vonnegut, and full-time ones like Richard Pryor, Dave Barry, and the writers of *The Onion,* are continuing that tradition.

~

VONNEGUT'S DYSTOPIAN FANTASY was played out as a story-length farce, but the most famous of such fantasies was played out as a novel-length nightmare. George Orwell's *1984* is a vivid depiction of what life would look like if the repressive strands of society and government were extrapolated into the future. In the half-century since the novel was published, many developments have been condemned because of their associations to Orwell's world: government euphemism, national identity cards, surveillance cameras, personal data on the Internet, and even, in the first television commercial for the

Macintosh computer, the IBM PC. No other work of fiction has had such an impact on people's opinions of real-world issues.

Nineteen Eighty-four was unforgettable literature, not just a political screed, because of the way Orwell thought through the details of how his society would work. Every component of the nightmare interlocked with the others to form a rich and credible whole: the omnipresent government, the eternal war with shifting enemies, the totalitarian control of the media and private life, the Newspeak language, the constant threat of personal betrayal.

Less widely known is that the regime had a well-articulated philosophy. It is explained to Winston Smith in the harrowing sequence in which he is strapped to a table and alternately tortured and lectured by the government agent O'Brien. The philosophy of the regime is thoroughly postmodernist, O'Brien explains (without, of course, using the word). When Winston objects that the Party cannot realize its slogan, "Who controls the past controls the future; who controls the present controls the past," O'Brien replies:

> You believe that reality is something objective, external, existing in its own right. You also believe that the nature of reality is self-evident. When you delude yourself into thinking that you see something, you assume that everyone else sees the same thing as you. But I tell you, Winston, that reality is not external. Reality exists in the human mind, and nowhere else. Not in the individual mind, which can make mistakes, and in any case soon perishes; only in the mind of the Party, which is collective and immortal.[4]

O'Brien admits that for certain purposes, such as navigating the ocean, it is useful to assume that the Earth goes around the sun and that there are stars in distant galaxies. But, he continues, the Party could also use alternative astronomies in which the sun goes around the Earth and the stars are bits of fire a few kilometers away. And though O'Brien does not explain it in this scene, Newspeak is the ultimate "prisonhouse of language," a "language that thinks man and his 'world.'"

O'Brien's lecture should give pause to the advocates of postmodernism. It is ironic that a philosophy that prides itself on deconstructing the accoutrements of power should embrace a relativism that makes challenges to power impossible, because it denies that there are objective benchmarks against which the deceptions of the powerful can be evaluated. For the same reason, the passages should give pause to radical scientists who insist that other scientists' aspirations to theories with objective reality (including theories about human nature) are really weapons to preserve the interests of the dominant class, gender, and race.[5] Without a notion of objective truth, intellectual life degenerates into a struggle of who can best exercise the raw force to "control the past."

A second precept of the Party's philosophy is the doctrine of the super-organism:

> Can you not understand, Winston, that the individual is only a cell? The weariness of the cell is the vigor of the organism. Do you die when you cut your fingernails?[6]

The doctrine that a collectivity (a culture, a society, a class, a gender) is a living thing with its own interests and belief system lies behind Marxist political philosophies and the social science tradition begun by Durkheim. Orwell is showing its dark side: the dismissal of the individual—the only entity that literally feels pleasure and pain—as a mere component that exists to further the interests of the whole. The sedition of Winston and his lover Julia began in the pursuit of simple human pleasures—sugar and coffee, white writing paper, private conversation, affectionate lovemaking. O'Brien makes it clear that such individualism will not be tolerated: "There will be no loyalty, except loyalty to the Party. There will be no love, except the love of Big Brother."[7]

The Party also believes that emotional ties to family and friends are "habits" that get in the way of a smoothly functioning society:

> Already we are breaking down the habits of thought that have survived from before the Revolution. We have cut the links between child and parent, and between man and man, and between man and woman. No one dares trust a wife or a child or a friend any longer. But in the future there will be no wives and no friends. Children will be taken from their mothers at birth, as one takes eggs from a hen. The sex instinct will be eradicated. . . . There will be no distinction between beauty and ugliness.[8]

It is hard to read the passage and not think of the current enthusiasm for proposals in which enlightened mandarins would reengineer childrearing, the arts, and the relationship between the sexes in an effort to build a better society.

Dystopian novels, of course, work by grotesque exaggeration. Any idea can be made to look terrifying in caricature, even if it is reasonable in moderation. I do not mean to imply that a concern with the interests of society or in improving human relationships is a step toward totalitarianism. But satire can show how popular ideologies may have forgotten downsides—in this case, how the notion that language, thought, and emotions are social conventions creates an opening for social engineers to try to reform them. Once we become aware of the downsides, we no longer have to treat the ideologies as sacred cows to which factual discoveries must be subordinated.

And finally we get to the core of the Party's philosophy. O'Brien has refuted every one of Winston's arguments, dashed every one of his hopes. He has informed him, "If you want a picture of the future, imagine a boot stamping on a human face—forever." Toward the end of this dialogue, O'Brien reveals the proposition that makes the whole nightmare possible (and whose falsehood, we may surmise, will make it impossible).

> As usual, the voice had battered Winston into helplessness. Moreover he was in dread that if he persisted in his disagreement O'Brien would twist the dial again. And yet he could not keep silent. Feebly, without arguments, with nothing to support him except his inarticulate horror of what O'Brien had said, he returned to the attack.
>
> "I don't know—I don't care. Somehow you will fail. Something will defeat you. Life will defeat you."
>
> "We control life, Winston, at all its levels. You are imagining that there is something called human nature which will be outraged by what we do and will turn against us. But we create human nature. Men are infinitely malleable."[9]

THE THREE WORKS I have discussed are didactic and unanchored in any existing time and place. The remaining two are different. Both are rooted in a culture, a locale, and an era. Both savor their characters' language, milieu, and philosophies of life. And both authors warned their readers not to generalize from the stories. Yet both authors are famous for their insight into human nature, and I believe I am doing them no injustice by presenting episodes from their works in that light.

Mark Twain's *Adventures of Huckleberry Finn* is an especially perilous source for lessons because it begins with the following order of the author: "Persons attempting to find a motive in this narrative will be prosecuted; persons attempting to find a moral in it will be banished; persons attempting to find a plot in it will be shot." That has not deterred a century of critics from noting its dual power. *Huckleberry Finn* shows us both the foibles of the antebellum South and the foibles of human nature, as seen through the eyes of two noble savages who sample them as they float down the Mississippi River.

Huckleberry Finn revels in many human imperfections, but perhaps the most tragicomic is the origin of violence in a culture of honor. The culture of honor is really a psychology of honor: a package of emotions that includes a loyalty to kin, a hunger for revenge, and a drive to maintain a reputation for toughness and valor. When sparked by other human sins—envy, lust, self-deception—they can fuel a vicious cycle of violence, as each side finds itself unable to abjure revenge against the other. The cycle can become amplified in certain places, among them the American South.

Huck met up with the culture of honor on two occasions in quick succession. The first was when he stowed away on a barge manned by a "rough-looking lot" of hard-drinking men. After one of them was about to belt out the fifteenth verse of a raunchy song, an altercation of relatively trivial origin broke out, and two men squared off to fight.

> [Bob, the biggest man on the boat] jumped up in the air and cracked his heels together again and shouted out: "Whoo-oop! I'm the original iron-jawed, brass-mounted, copper-bellied corpse-maker from the wilds of Arkansaw! Look at me! I'm the man they call Sudden Death and General Desolation! Sired by a hurricane, dam'd by an earthquake, half-brother to the cholera, nearly related to the smallpox on the mother's side! Look at me! I take nineteen alligators and a bar'l of whisky for breakfast when I'm in robust health, and a bushel of rattlesnakes and a dead body when I'm ailing. I split the everlasting rocks with my glance, and I squench the thunder when I speak! Whoo-oop! Stand back and give me room according to my strength! Blood's my natural drink and the wails of the dying is music to my ear. Cast your eye on me, gentle-men! and lay low and hold your breath, for I'm 'bout to turn myself loose!" . . .
>
> Then the man that had started the row . . . jumped up and cracked his heels together three times before he lit again . . . , and he began to shout like this: "Whoo-oop! bow your neck and spread, for the kingdom of sorrow's a coming! Hold me down to the earth, for I feel my powers a-working! . . . I put my hand on the sun's face and make it night in the earth; I bite a piece out of the moon and hurry the seasons; I shake my-self and crumble the mountains! Contemplate me through leather—*don't* use the naked eye! I'm the man with a petrified heart and biler-iron bowels! The massacre of isolated communities is the pastime of my idle moments, the destruction of nationalities the serious business of my life! The boundless vastness of the great American desert is my inclosed property, and I bury the dead on my own premises! . . . Whoo-oop! bow your neck and spread, for the Pet Child of Calamity's a'coming!"[10]

They circled and flailed at each other and knocked each other's hats off, until Bob said, as Huck describes it,

> . . . never mind, this warn't going to be the last of this thing, because he was a man that never forgot and never forgive, and so The Child better

look out for there was a time a-coming, just as sure as he was a living man, that he would have to answer to him with the best blood in his body. The Child said no man was willinger than he for that time to come, and he would give Bob fair warning, *now,* never to cross his path again, for he could never rest till he had waded in his blood, for such was his nature, though he was sparing him now on account of his family, if he had one.[11]

And then a "little black-whiskered chap" sent them both sprawling. With black eyes and red noses, they shook hands, said they had always respected each other, and agreed to let bygones be bygones.

Later in the chapter Huck swims ashore and stumbles onto the cabin of a family called the Grangerfords. Huck is frozen in his tracks by menacing dogs, until a voice from the window beckons him to enter the cabin slowly. He opens the door and finds himself staring down the barrels of three shotguns. When the Grangerfords see that Huck is not a Shepherdson, the family with whom they are feuding, they welcome him to live with them. Huck is captivated by their genteel life: their lovely furnishings, their elegant dress, and their refined manners, especially the patriarch, Col. Grangerford. "He was a gentleman all over, and so was his family. He was well born, as the saying is, and that's worth as much in a man as it is in a horse."

Three of the six Grangerford sons had been killed in the feud, and the youngest survivor, Buck, has befriended Huck. When the two boys go for a walk and Buck shoots at a Shepherdson boy, Huck asks why he wants to kill someone who has done nothing to hurt him. Buck explains the concept of a feud:

"Well," says Buck, "a feud is this way: A man has a quarrel with another man, and kills him; then that other man's brother kills *him;* then the other brothers on both sides goes for one another; then the *cousins* chip in—and by and by everybody's killed off and there ain't no more feud. But it's kind of slow and takes a long time."

"Has this one been going on long, Buck?"

"Well, I should *reckon*! It started thirty years ago, or som'ers along there. There was trouble 'bout something and then a lawsuit to settle it, and the suit went agin one of the men and so he up and shot the man that won the suit—which he would naturally do, of course. Anybody would."

"What was the trouble about, Buck?—land?"

"I reckon maybe—I don't know."

"Well, who done the shooting? Was it a Grangerford or a Shepherdson?"

"Laws, how do *I* know? It was so long ago."

"Don't anybody know?"

"Oh, yes, pa knows, I reckon, and some of the other old people; but they don't know now what the row was about in the first place."[12]

Buck adds that the feud is carried along by the two families' sense of honor: "There ain't a coward amongst them Shepherdsons—not a one. And there ain't no cowards amongst the Grangerfords either."[13] The reader anticipates trouble, and it comes soon enough. A Grangerford girl runs off with a Shepherdson boy, the Grangerfords head off in hot pursuit, and all the Grangerford males are killed in an ambush. "I ain't a'going to tell *all* that happened," says Huck; "it would make me sick again if I was to do that. I wished I hadn't ever come ashore that night to see such things."[14]

In the course of the chapter Huck has met up with two instances of the Southern culture of honor. Among the low-lifes it amounted to hollow bluster and was played for laughs; among the aristocrats it led to the devastation of two families and played out as tragedy. I think Twain was commenting on the twisted logic of violence and how it cuts across our stereotypes of refined and coarse classes of people. Indeed, the moral reckoning does not just cut across the classes but inverts them: the riffraff resolve their pointless dispute with face-saving verbiage; the gentlemen pursue their equally pointless one to a dreadful conclusion.

Though thoroughly Southern, the perverse psychology of the Grangerford-Shepherdson feud is familiar from the history and ethnography of just about any region of the world. (In particular, Huck's introduction to the Grangerfords was hilariously replayed in Napoleon Chagnon's famous account of his baptism into anthropological fieldwork, in which he stumbled into a feuding Yanomamö village and found himself trapped by dogs and staring down the shafts of poison arrows.) And it is familiar in the cycles of violence that continue to be played out by gangs, militias, ethnic groups, and respectable nation-states. Twain's depiction of the origins of endemic violence in an entrapping psychology of honor has a timelessness that will, I predict, make it outlast fashionable theories of the causes and cures of violence.

~

THE FINAL THEME I wish to reprise is that the human tragedy lies in the partial conflicts of interest that are inherent to all human relationships. I suppose I could illustrate it with just about any great work of fiction. An immortal literary text expresses "all the principal constants of conflict in the condition of man," wrote George Steiner about *Antigone;* "Ordinary people experiencing friction on the page is what warms our hands and hearts as we write," observed John Updike. But one novel caught my eye by flaunting the idea in its title: Isaac Bashevis Singer's *Enemies, A Love Story.*[15]

Singer, like Twain, protests too much against the possibility that his readers might draw morals from the slice of life he presents. "Although I did not have the privilege of going through the Hitler holocaust, I have lived for years in New York with refugees from this ordeal. I therefore hasten to say that this novel is by no means the story of the typical refugee, his life, and struggle. . . . The characters are not only Nazi victims but victims of their own personalities and fates." In literature the exception is the rule, Singer writes, but only after noting that the exception is rooted in the rule. Singer has been praised as a keen observer of human nature, not least because he imagines what happens when fate puts ordinary characters in extraordinary dilemmas. This is the conceit behind his book and the superb 1989 film adaptation, directed by Paul Mazursky and featuring Anjelica Huston and Ron Silver.

Herman Broder lives in Brooklyn in 1949 with his second wife, Yadwiga, a peasant girl who worked for his parents as a servant when they lived in Poland. A decade earlier his first wife, Tamara, had taken their two children to visit her parents, and while they were separated the Nazis invaded Poland. Tamara and the children were shot; Herman survived because Yadwiga hid him in her family's hayloft. At the end of the war he learned of his family's fate and married Yadwiga, and they found their way to New York.

While in the refugee camps, Herman had fallen in love with Masha, whom he meets again in New York and with whom he carries on a consuming affair (later in the book he will marry her, too). Yadwiga and Masha are, in part, male fantasies: the first pure but simple, the second ravishing but histrionic. Herman's conscience prevents him from leaving Yadwiga; his passion prevents him from leaving Masha. This brings much misery all around, but Singer does not let us hate Herman too much because we see how the capricious horror of the Holocaust has left him a fatalist with no confidence that his decisions can affect the course of his life. Moreover, Herman is amply punished for his duplicity by a life of high anxiety, which Singer portrays with comic, at times sadistic, relish.

The cruel joke continues when Herman learns that he has even more of too much of a good thing. It turns out that his first wife survived the Nazi bullet and escaped to Russia; she has moved to New York and is staying with her pious elderly uncle and aunt. Every Jew in the postwar period knows of emotional reunions of the survivors of Holocaust-ravaged families, but the reunion of a husband and a wife whom he had given up for dead is a scene of almost unimaginable poignancy. Herman enters the apartment of Reb Abraham:

ABRAHAM: A miracle from heaven, Broder, a miracle . . . Your *wife* has
   returned.
[Abraham leaves. Tamara enters.]
TAMARA: Hello, Herman.

HERMAN: I didn't know that you were alive.

TAMARA: That's something you never knew.

HERMAN: It's as if you've risen from the dead.

TAMARA: We were dumped in an open pit. They thought we were all dead. But I crawled over some corpses and escaped at night. How is it my uncle didn't know where you were—we had to put an advertisement in the paper?

HERMAN: I don't have my own apartment. I live with someone else.

TAMARA: What do you do? Where do you live?

HERMAN: I didn't know you were alive and—

TAMARA [smiles]: Who is the lucky woman who has taken my place?

HERMAN [stunned; then replies]: She was our servant. You knew her . . . Yadwiga.

TAMARA [about to laugh]: You married *her?* Forgive me, but wasn't she simple-minded? She didn't even know how to put on a pair of shoes. I remember your mother telling me how she tried to put the left shoe on the right foot. If she was given money to buy something, she would lose it.

HERMAN: She saved my life.

TAMARA: Was there no other way to repay her? Well, I'd better not ask. Do you have any children by her?

HERMAN: No.

TAMARA: It wouldn't shock me if you did. I assumed you crawled into bed with her even when you were with me.

HERMAN: That's nonsense. I never crawled into bed with her—

TAMARA: Oh, really. Well we never really did have a marriage. All we ever did was argue. You never had any respect for me, for my ideas—

HERMAN: That's not true. You know that—

ABRAHAM [enters the room, addresses Herman]: You may stay with us until you find an apartment. Hospitality is an act of charity, and besides, you are relatives. As the Holy Book says, "And thou shalt not hide thyself from thine own flesh."

TAMARA [interrupting]: Uncle, he has another wife.[16]

Yes, within seconds of the miraculous reunion they are bickering, picking up from where they left off when they were separated a decade before. What a wealth of psychology is folded into that scene! Men's inclination to polygamy and the frustrations it inevitably brings. Women's keener social intelligence and their preference for verbal over physical aggression against romantic rivals. The stability of personality over the lifespan. The way that social behavior is elicited by the specifics of a situation, especially the specifics of other people, so that two people play out the same dynamic whenever they are together.

Though it is a scene of considerable sadness, it has a streak of sly humor, as we watch these pathetic souls forgo their chance to savor a moment of rare good fortune and slip instead into petty quarreling. And Singer's biggest joke is on us. Dramatic conventions, and a belief in cosmic justice, lead us to expect that suffering has ennobled these characters and that we are about to witness a scene of great drama and pathos. Instead we are shown what we ought to have expected all along: real human beings with all their follies. Nor is the episode a display of cynicism or misanthropy: we are not surprised when later in the story Herman and Tamara share moments of tenderness, or that a wise Tamara will offer him his only chance at redemption. It is a scene that has the voice of the species in it: that infuriating, endearing, mysterious, predictable, and eternally fascinating thing we call human nature.

APPENDIX

Donald E. Brown's List of Human Universals

THIS LIST, COMPILED in 1989 and published in 1991, consists primarily of "surface" universals of behavior and overt language noted by ethnographers. It does not list deeper universals of mental structure that are revealed by theory and experiments. It also omits near-universals (traits that most, but not all, cultures show) and conditional universals ("If a culture has trait A, it always has trait B"). A list of items added since 1989 is provided at the end. For discussion and references, see Brown's *Human Universals* (1991) and his entry for "Human Universals" in *The MIT Encyclopedia of the Cognitive Sciences* (Wilson & Keil, 1999).

abstraction in speech and thought
actions under self-control distinguished from those not under control
aesthetics
affection expressed and felt
age grades
age statuses
age terms
ambivalence
anthropomorphization
antonyms
baby talk
belief in supernatural/religion
beliefs, false
beliefs about death

beliefs about disease
beliefs about fortune and misfortune
binary cognitive distinctions
biological mother and social mother normally the same person
black (color term)
body adornment
childbirth customs
childcare
childhood fears
childhood fear of loud noises
childhood fear of strangers
choice making (choosing alternatives)

classification
classification of age
classification of behavioral propensities
classification of body parts
classification of colors
classification of fauna
classification of flora
classification of inner states
classification of kin
classification of sex
classification of space
classification of tools
classification of weather conditions
coalitions
collective identities
conflict

conflict, consultation to deal with
conflict, means of dealing with
conflict, mediation of
conjectural reasoning
containers
continua (ordering as cognitive pattern)
contrasting marked and nonmarked sememes (meaningful elements in language)
cooking
cooperation
cooperative labor
copulation normally conducted in privacy
corporate (perpetual) statuses
coyness display
crying
cultural variability
culture
culture/nature distinction
customary greetings
daily routines
dance
death rituals
decision making
decision making, collective
directions, giving of
discrepancies between speech, thought, and action
dispersed groups
distinguishing right and wrong
diurnality
divination
division of labor
division of labor by age
division of labor by sex
dreams

dream interpretation
economic inequalities
economic inequalities, consciousness of
emotions
empathy
entification (treating patterns and relations as things)
environment, adjustments to
envy
envy, symbolic means of coping with
ethnocentrism
etiquette
explanation
face (word for)
facial communication
facial expression of anger
facial expression of contempt
facial expression of disgust
facial expression of fear
facial expression of happiness
facial expression of sadness
facial expression of surprise
facial expressions, masking/modifying of
family (or household)
father and mother, separate kin terms for
fears
fears, ability to overcome some
feasting
females do more direct childcare
figurative speech
fire
folklore
food preferences
food sharing

future, attempts to predict
generosity admired
gestures
gift giving
good and bad distinguished
gossip
government
grammar
group living
groups that are not based on family
hairstyles
hand (word for)
healing the sick (or attempting to)
hospitality
hygienic care
identity, collective
incest between mother and son unthinkable or tabooed
incest, prevention or avoidance
in-group distinguished from out-group(s)
in-group, biases in favor of
inheritance rules
insulting
intention
interest in bioforms (living things or things that resemble them)
interpreting behavior
intertwining (e.g., weaving)
jokes
kin, close distinguished from distant
kin groups
kin terms translatable by basic relations of procreation
kinship statuses
language

language employed to manipulate others
language employed to misinform or mislead
language is translatable
language not a simple reflection of reality
language, prestige from proficient use of
law (rights and obligations)
law (rules of membership)
leaders
lever
linguistic redundancy
logical notions
logical notion of "and"
logical notion of "equivalent"
logical notion of "general/particular"
logical notion of "not"
logical notion of "opposite"
logical notion of "part/whole"
logical notion of "same"
magic
magic to increase life
magic to sustain life
magic to win love
male and female and adult and child seen as having different natures
males dominate public/political realm
males more aggressive
males more prone to lethal violence
males more prone to theft
manipulate social relations
marking at phonemic, syntactic, and lexical levels

marriage
materialism
meal times
meaning, most units of are non-universal
measuring
medicine
melody
memory
metaphor
metonym
mood- or consciousness-altering techniques and/or substances
morphemes
mother normally has consort during child-rearing years
mourning
murder proscribed
music
music, children's
music related in part to dance
music related in part to religious activity
music seen as art (a creation)
music, vocal
music, vocal, includes speech forms
musical redundancy
musical repetition
musical variation
myths
narrative
nomenclature (perhaps the same as classification)
nonbodily decorative art
normal distinguished from abnormal states
nouns
numerals (counting)
Oedipus complex
oligarchy (de facto)
one (numeral)

onomatopoeia
overestimating objectivity of thought
pain
past/present/future
person, concept of
personal names
phonemes
phonemes defined by sets of minimally contrasting features
phonemes, merging of
phonemes, range from 10 to 70 in number
phonemic change, inevitability of
phonemic change, rules of
phonemic system
planning
planning for future
play
play to perfect skills
poetry/rhetoric
poetic line, uniform length range
poetic lines characterized by repetition and variation
poetic lines demarcated by pauses
polysemy (one word has several related meanings)
possessive, intimate
possessive, loose
practice to improve skills
preference for own children and close kin (nepotism)
prestige inequalities
private inner life
promise
pronouns
pronouns, minimum two numbers
pronouns, minimum three persons

proper names
property
psychological defense mechanisms
rape
rape proscribed
reciprocal exchanges (of labor, goods, or services)
reciprocity, negative (revenge, retaliation)
reciprocity, positive
recognition of individuals by face
redress of wrongs
rhythm
right-handedness as population norm
rites of passage
rituals
role and personality seen in dynamic interrelationship (i.e., departures from role can be explained in terms of individual personality)
sanctions
sanctions for crimes against the collectivity
sanctions include removal from the social unit
self distinguished from other
self as neither wholly passive nor wholly autonomous
self as subject and object
self is responsible
semantics
semantic category of affecting things and people
semantic category of dimension
semantic category of giving
semantic category of location
semantic category of motion
semantic category of speed
semantic category of other physical properties
semantic components
semantic components, generation
semantic components, sex
sememes, commonly used ones are short, infrequently used ones are longer
senses unified
sex (gender) terminology is fundamentally binary
sex statuses
sexual attraction
sexual attractiveness
sexual jealousy
sexual modesty
sexual regulation
sexual regulation includes incest prevention
sexuality as focus of interest
shelter
sickness and death seen as related
snakes, wariness around
social structure
socialization
socialization expected from senior kin
socialization includes toilet training
spear
special speech for special occasions
statuses and roles
statuses, ascribed and achieved
statuses distinguished from individuals
statuses on other than sex, age, or kinship bases
stop/nonstop contrasts (in speech sounds)
succession
sweets preferred
symbolism
symbolic speech
synonyms
taboos
tabooed foods
tabooed utterances
taxonomy
territoriality
time
time, cyclicity of
tools
tool dependency
tool making
tools for cutting
tools to make tools
tools patterned culturally
tools, permanent
tools for pounding
trade
triangular awareness (assessing relationships among the self and two other people)
true and false distinguished
turn-taking
two (numeral)
tying material (i.e., something like string)
units of time
verbs
violence, some forms of proscribed
visiting

vocalic/nonvocalic
contrasts in
phonemes
vowel contrasts

weaning
weapons
weather control (at-
tempts to)

white (color term)
world view

Additions Since 1989

anticipation
attachment
critical learning periods
differential valuations
dominance/submission
fairness (equity),
concept of
fear of death
habituation
hope
husband older than wife
on average
imagery
institutions (organized
co-activities)
intention
interpolation
judging others
likes and dislikes
making comparisons

males, on average, travel
greater distances over
lifetime
males engage in more
coalitional violence
mental maps
mentalese
moral sentiments
moral sentiments,
limited effective
range of
precedence, concept of
(that's how the leop-
ard got its spots)
pretend play
pride
proverbs, sayings
proverbs, sayings—in
mutually contradic-
tory forms

resistance to abuse of
power, to dominance
risk taking
self-control
self-image, awareness of
(concern for what
others think)
self-image, manipulation
of
self-image, wanted to be
positive
sex differences in spatial
cognition and behavior
shame
stinginess, disapproval of
sucking wounds
synesthetic metaphors
thumb sucking
tickling
toys, playthings

NOTES

PREFACE

1. Herrnstein & Murray, 1994, p. 311.
2. Harris, 1998a, p. 2.
3. Thornhill & Palmer, 2000, p. 176; quotation modified to make it gender-neutral.
4. Hunt, 1999; Jensen, 1972; Kors & Silverglate, 1998; J. P. Rushton, "The new enemies of evolutionary science," *Liberty*, March 1998, pp. 31–35; "Psychologist Hans Eysenck, Freudian critic, dead at 81," Associated Press, September 8, 1997.

PART I: THE BLANK SLATE, THE NOBLE SAVAGE, AND THE GHOST IN THE MACHINE

1. Macnamara, 1999; Passmore, 1970; Stevenson & Haberman, 1998; Ward, 1998.
2. Genesis 1:26.
3. Genesis 3:16.
4. This is according to interpretations postdating the Bible, which did not clearly distinguish mind from body.
5. Creation: Opinion Dynamics, August 30, 1999; miracles: Princeton Survey Research Associates, April 15, 2000; angels: Opinion Dynamics, December 5, 1997; devil: Princeton Survey Research Associates, April 20, 2000; afterlife: Gallup Organization, April 1, 1998; evolution: Opinion Dynamics, August 30, 1999. Available through the Roper Center at the University of Connecticut Public Opinion Online: www.ropercenter.uconn.edu.

Chapter 1: The Official Theory

1. Locke, 1690/1947, bk. II, chap. 1, p. 26.
2. Hacking, 1999.

3. Rousseau, 1755/1994, pp. 61–62.
4. Hobbes, 1651/1957, pp. 185–186.
5. Descartes, 1641/1967, Meditation VI, p. 177.
6. Ryle, 1949, pp. 13–17.
7. Descartes, 1637/2001, part V, p. 10.
8. Ryle, 1949, p. 20.
9. Cohen, 1997.
10. Rousseau, 1755/1986, p. 208.
11. Rousseau, 1762/1979, p. 92.
12. Quoted in Sowell, 1987, p. 63.
13. Originally in *Red Flag* (Beijing), June 1, 1958; quoted in Courtois et al., 1999.
14. J. Kalb, "The downtown gospel according to Reverend Billy," *New York Times,* February 27, 2000.
15. D. R. Vickery, "And who speaks for our earth?" *Boston Globe*, December 1, 1997.
16. Green, 2001; R. Mishra, "What can stem cells really do?" *Boston Globe,* August 21, 2001.

Chapter 2: Silly Putty

1. Jespersen, 1938/1982, pp. 2–3.
2. Degler, 1991; Fox, 1989; Gould, 1981; Richards, 1987.
3. Degler, 1991; Fox, 1989; Gould, 1981; Rachels, 1990; Richards, 1987; Ridley, 2000.
4. Degler, 1991; Gould, 1981; Kevles, 1985; Richards, 1987; Ridley, 2000.
5. The term "Standard Social Science Model" was introduced by John Tooby and Leda Cosmides (1992). The philosophers Ron Mallon and Stephen Stich (2000) use "social constructionism" because it is close in meaning but shorter. "Social construction" was coined by one of the founders of sociology, Emile Durkheim, and is analyzed by Hacking, 1999.

6. See Curti, 1980; Degler, 1991; Fox, 1989; Freeman, 1999; Richards, 1987; Shipman, 1994; Tooby & Cosmides, 1992.

7. Degler, 1991, p. viii.

8. White, 1996.

9. Quoted in Fox, 1989, p. 68.

10. Watson, 1924/1998.

11. Quoted in Degler, 1991, p. 139.

12. Quoted in Degler, 1991, pp. 158–159.

13. Breland & Breland, 1961.

14. Skinner, 1974.

15. Skinner, 1971.

16. Fodor & Pylyshyn, 1988; Gallistel, 1990; Pinker & Mehler, 1988.

17. Gallistel, 2000.

18. Preuss, 1995; Preuss, 2001.

19. Hirschfeld & Gelman, 1994.

20. Ekman & Davidson, 1994; Haidt, in press.

21. Daly, Salmon, & Wilson, 1997.

22. McClelland, Rumelhart, & the PDP Research Group, 1986; Rumelhart, McClelland, & the PDP Research Group, 1986.

23. Rumelhart & McClelland, 1986, p. 143.

24. Quoted in Degler, 1991, p. 148.

25. Boas, 1911. My thanks to David Kemmerer for the examples.

26. Degler, 1991; Fox, 1989; Freeman, 1999.

27. Quoted in Degler, 1991, p. 84.

28. Quoted in Degler, 1991, p. 95.

29. Quoted in Degler, 1991, p. 96.

30. Durkheim, 1895/1962, pp. 103–106.

31. Durkheim, 1895/1962, p. 110.

32. Quoted in Degler, 1991, p. 161.

33. Quoted in Tooby & Cosmides, 1992, p. 26.

34. Ortega y Gasset, 1935/2001.

35. Montagu, 1973a, p. 9. The portion before the ellipsis is from an earlier edition, quoted in Degler, 1991, p. 209.

36. Benedict, 1934/1959, p. 278.

37. Mead, 1935/1963, p. 280.

38. Quoted in Degler, 1991, p. 209.

39. Mead, 1928.

40. Geertz, 1973, p. 50.

41. Geertz, 1973, p. 44.

42. Shweder, 1990.

43. Quoted in Tooby & Cosmides, 1992, p. 22.

44. Quoted in Degler, 1991, p. 208.

45. Quoted in Degler, 1991, p. 204.

46. Degler, 1991; Shipman, 1994.

47. Quoted in Degler, 1991, p. 188.

48. Quoted in Degler, 1991, pp. 103–104.

49. Quoted in Degler, 1991, p. 210.

50. Cowie, 1999; Elman et al., 1996, pp. 390–391.

51. Quoted in Degler, 1991, p. 330.

52. Quoted in Degler, 1991, p. 95.

53. Quoted in Degler, 1991, p. 100.

54. Charles Singer, *A short history of biology;* quoted in Dawkins, 1998, p. 90.

Chapter 3: The Last Wall to Fall

1. Wilson, 1998. The idea was first developed by John Tooby and Leda Cosmides, 1992.

2. Anderson, 1995; Crevier, 1993; Gardner, 1985; Pinker, 1997.

3. Fodor, 1994; Haugeland, 1981; Newell, 1980; Pinker, 1997, chap. 2.

4. Brutus. 1, by Selmer Bringsjord. S. Bringsjord, "Chess is too easy," *Technology Review,* March/April 1998, pp. 23–28.

5. EMI (Experiments in Musical Intelligence), by David Cope. G. Johnson, "The artist's angst is all in your head," *New York Times,* November 16, 1997, p. 16.

6. Aaron, by Harold Cohen. G. Johnson, "The artist's angst is all in your head," *New York Times,* November 16, 1997, p. 16.

7. Goldenberg, Mazursky, & Solomon, 1999.

8. Leibniz, 1768/1996, bk. II, chap. i, p. 111.

9. Leibniz, 1768/1996, preface, p. 68.

10. Chomsky, 1975; Chomsky, 1988b; Fodor, 1981.

11. Elman et al., 1996; Rumelhart & McClelland, 1986.

12. Dennett, 1986.

13. Elman et al., 1996, p. 82.

14. Elman et al., 1996, pp. 99–100.

15. Chomsky, 1975; Chomsky, 1993; Chomsky, 2000; Pinker, 1994.

16. See also Miller, Galanter, & Pribram, 1960; Pinker, 1997, chap. 2; Pinker, 1999, chaps. 1, 10.

17. Baker, 2001.

18. Baker, 2001.

19. Shweder, 1994; see Ekman & Davidson, 1994, and Lazarus, 1991, for discussion.

20. See Lazarus, 1991, for a review of theories of emotion.

21. Mallon & Stich, 2000.

22. Ekman & Davidson, 1994; Lazarus, 1991.

23. Ekman & Davidson, 1994.

24. Fodor, 1983; Gardner, 1983; Hirschfeld & Gelman, 1994; Pinker, 1994; Pinker, 1997.

25. Elman et al., 1996; Karmiloff-Smith, 1992.

26. Anderson, 1995; Gazzaniga, Ivry, & Mangun, 1998.

27. Calvin, 1996a; Calvin, 1996b; Calvin & Ojemann, 2001; Crick, 1994; Damasio, 1994; Gazzaniga, 2000a; Gazzaniga, 2000b; Gazzaniga, Ivry, & Mangun, 1998; Kandel, Schwartz, & Jessell, 2000.

28. Crick, 1994.

29. 1948, translated by C. B. Garnett (New York: Macmillan), p. 664.

30. Damasio, 1994.

31. Damasio, 1994; Dennett, 1991; Gazzaniga, 1998.

32. Gazzaniga, 1992; Gazzaniga, 1998.

33. Anderson et al., 1999; Blair & Cipolotti, 2000; Lykken, 1995.

34. Monaghan & Glickman, 1992.

35. Bourgeois, Goldman-Rakic, & Rakic, 2000; Chalupa, 2000; Geary & Huffman, 2002; Katz, Weliky, & Crowley, 2000; Rakic, 2000; Rakic, 2001. See also Chapter 5.

36. Thompson et al., 2001.

37. Thompson et al., 2001.

38. Witelson, Kigar, & Harvey, 1999.

39. LeVay, 1993.

40. Davidson, Putnam, & Larson, 2000; Raine et al., 2000.

41. Bouchard, 1994; Hamer & Copeland, 1998; Lykken, 1995; Plomin, 1994; Plomin et al., 2001; Ridley, 2000.

42. Hyman, 1999; Plomin, 1994.

43. Bouchard, 1994; Bouchard, 1998; Damasio, 2000; Lykken et al., 1992; Plomin, 1994; Thompson et al., 2001; Tramo et al., 1995; Wright, 1995.

44. Segal, 2000.

45. Lai et al., 2001; Pinker, 2001b.

46. Frangiskakis et al., 1996.

47. Chorney et al., 1998.

48. Benjamin et al., 1996.

49. Lesch et al., 1996.

50. Lai et al., 2001; Pinker, 2001b.

51. Charlesworth, 1987; Miller, 2000b; Mousseau & Roff, 1987; Tooby & Cosmides, 1990.

52. Bock & Goode, 1996; Lykken, 1995; Mealey, 1995.

53. Blair & Cipolotti, 2000; Hare, 1993; Kirwin, 1997; Lykken, 1995; Mealey, 1995.

54. Anderson et al., 1999; Blair & Cipolotti, 2000; Lalumière, Harris, & Rice, 2001; Lykken, 2000; Mealey, 1995; Rice, 1997.

55. Barkow, Cosmides, & Tooby, 1992; Betzig, 1997; Buss, 1999; Cartwright, 2000; Crawford & Krebs, 1998; Evans & Zarate, 1999; Gaulin & McBurney, 2000; Pinker, 1997; Pope, 2000; Wright, 1994.

56. Dawkins, 1983; Dawkins, 1986; Gould, 1980; Maynard Smith, 1975/1993; Ridley, 1986; Williams, 1966.

57. Dawkins, 1983; Dawkins, 1986; Maynard Smith, 1975/1993; Ridley, 1986; Williams, 1966.

58. The improved metaphor "megalomaniacal gene" was suggested by the philosopher Colin McGinn.

59. Etcoff, 1999.

60. Frank, 1988; Haidt, in press; Trivers, 1971.

61. Daly & Wilson, 1988; Frank, 1988.

62. McGuinness, 1997; Pinker, 1994.

63. Brown, 1991; Brown, 2000.

64. Baron-Cohen, 1995; Hirschfeld & Gelman, 1994; Spelke, 1995.

65. Boyd & Silk, 1996; Calvin & Bickerton, 2000; Kingdon, 1993; Klein, 1989; Mithen, 1996.

66. Gallistel, 1992; Hauser, 1996; Hauser, 2000; Trivers, 1985.

67. James, 1890/1950, vol. 2, chap. 24.

68. Freeman, 1983; Freeman, 1999.

69. Wrangham & Peterson, 1996.

70. Wrangham & Peterson, 1996.

71. Keeley, 1996, graph adapted by Ed Hagen from fig. 6.2 on p. 90.

72. Ghiglieri, 1999; Keeley, 1996; Wrangham & Peterson, 1996.

73. Ember, 1978. See also Ghiglieri, 1999; Keeley, 1996; Knauft, 1987; Wrangham & Peterson, 1996.

74. Divale, 1972; see Eibl-Eibesfeldt, 1989, p. 323, for discussion.

75. Bamforth, 1994; Chagnon, 1996; Daly & Wilson, 1988; Divale, 1972; Edgerton, 1992; Ember, 1978; Ghiglieri, 1999; Gibbons, 1997; Keeley, 1996; Kingdon, 1993; Knauft, 1987; Krech, 1994; Krech, 1999; Wrangham & Peterson, 1996.

76. Axelrod, 1984; Brown, 1991; Ridley, 1997; Wright, 2000.

77. Brown, 1991.

Chapter 4: Culture Vultures

1. Borges, 1964, p. 30.

2. Pinker, 1984a.

3. Boyer, 1994; Hirschfeld & Gelman, 1994; Norenzayan & Atran, in press; Schaller & Crandall, in press; Sperber, 1994; Talmy, 2000; Tooby & Cosmides, 1992.

4. Adams et al., 2000.

5. Tomasello, 1999.

6. Baron-Cohen, 1995; Karmiloff-Smith et al., 1995.

7. Rapin, 2001.

8. Baldwin, 1991.

9. Carpenter, Akhtar, & Tomasello, 1998.

10. Meltzoff, 1995.

11. Pinker, 1994; Pinker, 1996; Pinker, 1999.

12. Campbell & Fairey, 1989; Frank, 1985; Kelman, 1958; Latané & Nida, 1981.

13. Deutsch & Gerard, 1955.

14. Harris, 1985.

15. Cronk, 1999; Cronk, Chagnon, & Irons, 2000.

16. Pinker, 1999, chap. 10.

17. Searle, 1995.

18. Sperber, 1985; Sperber, 1994.

19. Boyd & Richerson, 1985; Cavalli-Sforza & Feldman, 1981; Durham, 1982; Lumsden & Wilson, 1981.

20. Cavalli-Sforza, 1991; Cavalli-Sforza & Feldman, 1981.

21. Toussaint-Samat, 1992.

22. Degler, 1991.

23. Sowell, 1996, p. 378. See also Sowell, 1994, and Sowell, 1998.

24. Diamond, 1992; Diamond, 1998.

25. Diamond, 1997.

26. Putnam, 1973.

27. Chomsky, 1980, p. 227; Marr, 1982; Tinbergen, 1952.

28. Pinker, 1999.

Chapter 5: The Slate's Last Stand

1. Venter et al., 2001.

2. See, e.g., the contributors to Rose & Rose, 2000.

3. R. McKie, in *The Guardian,* February 11, 2001. See also S. J. Gould, "Humbled by the genome's mysteries," *New York Times,* February 19, 2001.

4. *The Observer,* February 11, 2001.

5. E. Pennisi, "The human genome," *Science, 291,* 2001, 1177–1180; see pp. 1178–1179.

6. "Gene count," *Science, 295,* 2002, p. 29; R. Mishar, "Biotech CEO says map missed much of genome," *Boston Globe,* April 9, 2001; Wright et al., 2001.

7. Claverie, 2001; Szathmáry, Jordán, & Pál, 2001; Venter et al., 2001.

8. Szathmáry, Jordán, & Pál, 2001.

9. Claverie, 2001.

10. Venter et al., 2001.

11. Evan Eichler, quoted by G. Vogel, "Objection #2: Why sequence the junk?" *Science, 291,* 2001, p. 1184.

12. Elman et al., 1996; McClelland, Rumelhart, & the PDP Research Group, 1986; McLeod, Plunkett, & Rolls, 1998; Pinker, 1997, pp. 98–111; Rumelhart, McClelland, & the PDP Research Group, 1986.

13. Anderson, 1993; Fodor & Pylyshyn, 1988; Hadley, 1994a; Hadley, 1994b; Hummel & Holyoak, 1997; Lachter & Bever, 1988; Marcus, 1998; Marcus, 2001a; McCloskey & Cohen, 1989; Minsky & Papert, 1988; Shastri & Ajjanagadde, 1993; Smolensky, 1995; Sougné, 1998.

14. Berent, Pinker, & Shimron, 1999; Marcus et al., 1995; Pinker, 1997; Pinker, 1999; Pinker, 2001a; Pinker & Prince, 1988.

15. Pinker, 1997, pp. 112–131.

16. Pinker, 1999. See also Clahsen, 1999; Marcus, 2001a; Marslen-Wilson & Tyler, 1998; Pinker, 1991.

17. See Marcus et al., 1995, and Marcus, 2001a, for examples.

18. Hinton & Nowlan, 1987; Nolfi, Elman, & Parisi, 1994.

19. For examples, see Hummel & Biederman, 1992; Marcus, 2001a; Shastri, 1999; Smolensky, 1990.

20. Deacon, 1997; Elman et al., 1996; Hardcastle & Buller, 2000; Panskepp & Panskepp, 2000; Quartz & Sejnowski, 1997.

21. Elman et al., 1996, p. 108.

22. Quartz & Sejnowski, 1997, pp. 552, 555.

23. Maguire et al., 2000.

24. E. K. Miller, 2000.

25. Sadato et al., 1996.

26. Neville & Bavelier, 2000; Petitto et al., 2000.

27. Pons et al., 1991; Ramachandran & Blakeslee, 1998.

28. Curtiss, de Bode, & Shields, 2000; Stromswold, 2000.

29. Catalano & Shatz, 1998; Crair, Gillespie, & Stryker, 1998; Katz & Shatz, 1996; Miller, Keller, & Stryker, 1989.

30. Sharma, Angelucci, & Sur, 2000; Sur, 1988; Sur, Angelucci, & Sharma, 1999.

31. For related arguments, see Geary & Huffman, 2002; Katz & Crowley, 2002; Katz & Shatz, 1996; Katz, Weliky, & Crowley, 2000; Marcus, 2001b.

32. R. Restak, "Rewiring" (Review of *The talking cure* by S. C. Vaughan), *New York Times Book Review,* June 22, 1997, pp. 14–15.

33. D. Milmore, " 'Wiring' the brain for life," *Boston Globe,* November 2, 1997, pp. N5–N8.

34. William Jenkins, quoted in A. Ellin, "Can 'neurobics' do for the brain what aerobics do for the lungs?" *New York Times,* October 3, 1999.

35. Quotations from A. Ellin, "Can 'neurobics' do for the brain what aerobics do for the lungs?" *New York Times,* October 3, 1999.

36. G. Kolata, "Muddling fact and fiction in policy," *New York Times,* August 8, 1999.

37. Bruer, 1997; Bruer, 1999.

38. R. Saltus, "Study shows brain adaptable," *Boston Globe,* April 20, 2000.

39. Van Essen & Deyoe, 1995, p. 388.

40. Crick & Koch, 1995.

41. Bishop, Coudreau, & O'Leary, 2000; Bourgeois, Goldman-Rakic, & Rakic, 2000; Chalupa, 2000; Katz, Weliky, & Crowley, 2000; Levitt, 2000; Miyashita-Lin et al., 1999; Rakic, 2000; Rakic, 2001; Verhage et al., 2000; Zhou & Black, 2000.

42. See the references cited in the preceding note, and also Geary & Huffman, 2002; Krubitzer & Huffman, 2000; Preuss, 2000; Preuss, 2001; Tessier-Lavigne & Goodman, 1996.

43. Geary & Huffman, 2002; Krubitzer & Huffman, 2000; Preuss, 2000; Preuss, 2001.

44. D. Normile, "Gene expression differs in human and chimp brains," *Science, 292,* 2001, pp. 44–45.

45. Kaas, 2000, p. 224.

46. Hardcastle & Buller, 2000; Panksepp & Panksepp, 2000.

47. Gu & Spitzer, 1995.

48. Catalano & Shatz, 1998; Crair, Gillespie, & Stryker, 1998; Katz & Shatz, 1996.

49. Catalano & Shatz, 1998; Crair, Gillespie, & Stryker, 1998; Katz & Shatz, 1996; Stryker, 1994.

50. Catalano & Shatz, 1998; Stryker, 1994.

51. Wang et al., 1998.

52. Brown, 1985; Hamer & Copeland, 1994.

53. J. R. Skoyles, June 7, 1999, on an email discussion list for evolutionary psychology.

54. Recanzone, 2000, p. 245.

55. Van Essen & Deyoe, 1995.

56. Kosslyn, 1994.

57. Kennedy, 1993; Kosslyn, 1994, pp. 334–335; Zimler & Keenan, 1983; though see also Arditi, Holtzman, & Kosslyn, 1988.

58. Petitto et al., 2000.

59. Klima & Bellugi, 1979; Padden & Perlmutter, 1987; Siple & Fischer, 1990.

60. Cramer & Sur, 1995; Sharma, Angelucci, & Sur, 2000; Sur, 1988; Sur, Angelucci, & Sharma, 1999.

61. Sur, 1988, pp. 44, 45.

62. Bregman, 1990; Bregman & Pinker, 1978; Kubovy, 1981.

63. Hubel, 1988.

64. Bishop, Coudreau, & O'Leary, 2000; Bourgeois, Goldman-Rakic, & Rakic, 2000; Chalupa, 2000; Geary & Huffman, 2002; Katz, Weliky, & Crowley, 2000; Krubitzer & Huffman, 2000; Levitt, 2000; Miyashita-Lin et al., 1999; Preuss, 2000; Preuss, 2001; Rakic, 2000; Rakic, 2001; Tessier-Lavigne & Goodman, 1996; Verhage et al., 2000; Zhou & Black, 2000.

65. Katz, Weliky, & Crowley, 2000, p. 209.

66. Crowley & Katz, 2000.

67. Verhage et al., 2000.

68. Miyashita-Lin et al., 1999.

69. Bishop, Coudreau, & O'Leary, 2000. See also Rakic, 2001.

70. Thompson et al., 2001.

71. Brugger et al., 2000; Melzack, 1990; Melzack et al., 1997; Ramachandran, 1993.

72. Curtiss, de Bode, & Shields, 2000; Stromswold, 2000.

73. Described in Stromswold, 2000.

74. Farah et al., 2000.

75. Anderson et al., 1999.

76. Anderson, 1976; Pinker, 1979; Pinker, 1984a; Quine, 1969.

77. Adams et al., 2000.

78. Tooby & Cosmides, 1992; Williams, 1966.

79. Gallistel, 2000; Hauser, 2000.

80. Barkow, Cosmides, & Tooby, 1992; Burnham & Phelan, 2000; Wright, 1994.

81. Brown, 1991.

82. Hirschfeld & Gelman, 1994; Pinker, 1997, chap. 5.

83. Baron-Cohen, 1995; Gopnik, Meltzoff, & Kuhl, 1999; Hirschfeld & Gelman, 1994; Leslie, 1994; Spelke, 1995; Spelke et al., 1992.

84. Baron-Cohen, 1995; Fisher et al., 1998; Frangiskakis et al., 1996; Hamer & Copeland, 1998; Lai et al., 2001; Rossen et al., 1996.

85. Bouchard, 1994; Plomin et al., 2001.

86. Caspi, 2000; McCrae et al., 2000.

87. Bouchard, 1994; Harris, 1998a; Plomin et al., 2001; Turkheimer, 2000.

88. See the references cited in this chapter.

PART II: FEAR AND LOATHING

Chapter 6: Political Scientists

1. Weizenbaum, 1976.

2. Lewontin, Rose, & Kamin, 1984, p. x.

3. Herrnstein, 1971.

4. Jensen, 1969; Jensen, 1972.

5. Herrnstein, 1973.

6. Darwin, 1872/1998; Pinker, 1998.

7. Ekman, 1987; Ekman, 1998.

8. Wilson, 1975/2000.

9. Sahlins, 1976, p. 3.

10. Sahlins, 1976, p. x.

11. Allen et al., 1975, p. 43.

12. Chorover, 1979, pp. 108–109.

13. Wilson, 1975/2000, p. 548.

14. Wilson, 1975/2000, p. 555.

15. Wilson, 1975/2000, p. 550.

16. Wilson, 1975/2000, p. 554.

17. Wilson, 1975/2000, p. 569.

18. Segerstråle, 2000; Wilson, 1994.

19. Wright, 1994.

20. Trivers & Newton, 1982.

21. Trivers, 1981.

22. Trivers, 1981, p. 37.

23. Gould, 1976a; Gould, 1981; Gould, 1998a; Lewontin, 1992; Lewontin, Rose, & Kamin, 1984; Rose & Rose, 2000; Rose, 1997.

24. In titles alone, we find "determinism" in Gould, 1976a; Rose, 1997; Rose & the Dialectics of Biology Group, 1982; and four of the nine chapters in Lewontin, Rose, & Kamin, 1984.

25. Lewontin, Rose, & Kamin, 1984, p. 236.
26. Lewontin, Rose, & Kamin, 1984, p. 5.
27. Dawkins, 1976/1989, p. 164.
28. Lewontin, Rose, & Kamin, 1984, p. 11.
29. Dawkins, 1985.
30. Lewontin, Rose, & Kamin, 1984, p. 287.
31. Dawkins, 1976/1989, p. 20, emphasis added.
32. Levins & Lewontin, 1985, pp. 88, 128; Lewontin, 1983, p. 68; Lewontin, Rose, & Kamin, 1984, p. 287. In Lewontin, 1982, p. 18, the quotation is paraphrased as "ruled by our genes."
33. Lewontin, Rose, & Kamin, 1984, p. 149.
34. Lewontin, Rose, & Kamin, 1984, p. 260.
35. Rose, 1997, p. 211.
36. Freeman, 1999.
37. Turner and Sponsel's letter may be found at chief.anth.uconn.edu/gradstudents/dhume/darkness_in_el_dorado.
38. Chagnon, 1988; Chagnon, 1992.
39. Tierney, 2000.
40. University of Michigan Report on the Ongoing Investigation of the Neel-Chagnon Allegations (www.umich.edu/~urel/darkness.html); John J. Miller, "The Fierce People: The wages of anthropological incorrectness," *National Review,* November 20, 2000.
41. John Tooby, "Jungle fever: Did two U.S. scientists start a genocidal epidemic in the Amazon, or was The New Yorker duped?" *Slate,* October 24, 2000; University of Michigan Report on the Ongoing Investigation of the Neel-Chagnon Allegations (www.umich.edu/~urel/darkness.html); John J. Miller, "The Fierce People: The wages of anthropological incorrectness," *National Review,* November 20, 2000; "A statement from Bruce Alberts," National Academy of Sciences, November 9, 2000, www.nas. org; John Tooby, "Preliminary Report," Department of Anthropology, University of California, Santa Barbara, December 10, 2000 (www.anth.ucsb.edu/ucsbprelimnaryreport.pdf; see also www.anth.ucsb.edu/chagnon.html); Lou Marano, "Darkness in anthropology," UPI, October 20, 2000; Michael Shermer, "Spin-doctoring the Yanomamö," *Skeptic,* 2001; Virgilio Bosh & eight other signatories, "Venezuelan response to Yanomamö book," *Science, 291,* 2001, pp. 985–986; "The Yanomamö and the 1960s measles epidemic": letters from J. V. Neel, Jr., K. Hill, and S. L. Katz, *Science, 292,* June 8, 2001, pp. 1836–1837; "Yanomamö wars continue," *Science, 295,* January 4, 2002, p. 41; yahoo. com/group/evolutionary-psychology/files/aaa.html. November 2001. An extensive collection of documents related to the Tierney affair may be found on the web site www.anth.uconn.edu/gradstudents/dhume/index4.htm.
42. Edward Hagen, "Chagnon and Neel saved hundreds of lives," The Fray, *Slate,* December 8, 2000 (www.anth.uconn.edu/gradstudents/dhume/dark/darkness.0250.html); S. L. Katz, "The Yanomamö and the 1960s measles epidemic" (letter), *Science, 292,* June 8, 2001, p. 1837.
43. In the *Pittsburgh Post-Gazette,* quoted in John J. Miller, "The Fierce People: The wages of anthropological incorrectness," *National Review,* November 20, 2000.
44. Chagnon, 1992, chaps. 5–6.
45. Valero & Biocca, 1965/1996.
46. Ember, 1978; Keeley, 1996; Knauft, 1987.
47. Tierney, 2000, p. 178.
48. Redmond, 1994, p. 125; quoted in John Tooby, *Slate,* October 24, 2000.
49. Sponsel, 1996, p. 115.
50. Sponsel, 1996, pp. 99, 103.
51. Sponsel, 1998, p. 114.
52. Tierney, 2000, p. 38.
53. Neel, 1994.
54. John J. Miller, "The Fierce People: The wages of anthropological incorrectness," *National Review,* November 20, 2000.
55. Tierney, 2000, p. xxiv.

Chapter 7: The Holy Trinity

1. Hunt, 1999.
2. Halpern, Gilbert, & Coren, 1996.
3. Allen et al., 1975.
4. Gould, 1976a.
5. Lewontin, Rose, & Kamin, 1984, p. 267.
6. Lewontin, Rose, & Kamin, 1984, p. 267.
7. Lewontin, Rose, & Kamin, 1984, p. 14.
8. Lewontin, 1992, p. 123.
9. Précis of Lewontin, 1982, on the book jacket.
10. Lewontin, 1992, p. 123.
11. Montagu, 1973a.
12. S. Gould, "A time of gifts," *New York Times,* September 26, 2001.
13. Gould, 1998b.
14. Mealey, 1995.
15. Gould, 1998a, p. 262.
16. Bamforth, 1994; Chagnon, 1996; Daly & Wilson, 1988; Divale, 1972; Edgerton, 1992; Ember, 1978; Ghiglieri, 1999; Gibbons, 1997; Keeley, 1996; Kingdon, 1993; Knauft, 1987; Krech, 1994; Krech, 1999; Wrangham & Peterson, 1996.
17. Gould, 1998a, p. 262.
18. Gould, 1998a, p. 265.
19. Levins & Lewontin, 1985, p. 165.
20. Lewontin, Rose, & Kamin, 1984, p. ix.

21. Lewontin, Rose, & Kamin, 1984, p. 76.

22. Lewontin, Rose, & Kamin, 1984, p. 270.

23. Rose, 1997, pp. 7, 309.

24. Gould, 1992.

25. Hunt, 1999.

26. Quoted in J. Salamon, "A stark explanation for mankind from an unlikely rebel" (Review of the PBS series "Evolution"), *New York Times,* September 24, 2001.

27. D. Wald, "Intelligent design meets congressional designers," *Skeptic, 8,* 2000, p. 13. Lyrics from "Bad Touch" by the Bloodhound Gang.

28. Quoted in D. Falk, "Design or chance?" *Boston Globe Magazine,* October 21, 2001, pp. 14–23, quotation on p. 21.

29. National Center for Science Education, www.ncseweb.org/pressroom.asp?branch= statement. See also Berra, 1990; Kitcher, 1982; Miller, 1999; Pennock, 2000; Pennock, 2001.

30. Quoted in L. Arnhart, M. J. Behe, & W. A. Dembski, "Conservatives, Darwin, and design: An exchange," *First Things, 107,* November 2000, pp. 23–31.

31. Behe, 1996.

32. Behe, 1996; Crews, 2001; Dorit, 1997; Miller, 1999; Pennock, 2000; Pennock, 2001; Ruse, 1998.

33. R. Bailey, "Origin of the specious," *Reason,* July 1997.

34. D. Berlinski, "The deniable Darwin," *Commentary,* June 1996. See R. Bailey, "Origin of the specious," *Reason,* July 1997. The Pope's views on evolution are discussed in Chapter 11.

35. A 1991 essay, quoted in R. Bailey, "Origin of the specious," *Reason,* July 1997.

36. Quoted in R. Bailey, "Origin of the specious," *Reason,* July 1997.

37. R. Bailey, "Origin of the specious," *Reason,* July 1997.

38. L. Kass, "The end of courtship," *Public Interest, 126,* Winter 1997.

39. A. Ferguson, "The end of nature and the next man" (Review of F. Fukuyama's *The great disruption*), *Weekly Standard,* June 28, 1999.

40. A. Ferguson, "How Steven Pinker's mind works" (Review of S. Pinker's *How the mind works*), *Weekly Standard,* January 12, 1998.

41. T. Wolfe, "Sorry, but your soul just died," *Forbes ASAP,* December 2, 1996; reprinted in slightly different form in Wolfe, 2000. Ellipses in original.

42. T. Wolfe, "Sorry, but your soul just died," *Forbes ASAP,* December 2, 1996; reprinted in slightly different form in Wolfe, 2000.

43. C. Holden, "Darwin's brush with racism," *Science, 292,* 2001, p. 1295. Resolution HLS 01–2652, Regular Session, 2001, House Concurrent Resolution No. 74 by Representative Broome.

44. R. Wright, "The accidental creationist," *New Yorker,* December 13, 1999. Similarly, the creationist Discovery Institute used Lewontin's attacks on evolutionary psychology to help criticize the 2001 PBS television documentary series "Evolution," www.reviewevolution.com.

45. Rose, 1978.

46. T. Wolfe, "Sorry, but your soul just died," *Forbes ASAP,* December 2, 1996; reprinted in slightly different form in Wolfe, 2000.

47. Gould, 1976b.

48. A. Ferguson, "The end of nature and the next man" (Review of F. Fukuyama's *The great disruption*), *Weekly Standard,* 1999.

49. See Dennett, 1995, p. 263, for a similar report.

50. E. Smith, "Look who's stalking," *New York,* February 14, 2000.

51. Alcock, 1998.

52. For example, the articles entitled "Eugenics revisited" (Horgan, 1993), "The new Social Darwinists" (Horgan, 1995), and "Is a new eugenics afoot?" (Allen, 2001).

53. *New Republic,* April 27, 1998, p. 33.

54. *New York Times,* February 18, 2001, Week in Review, p. 3.

55. Tooby & Cosmides, 1992, p. 49.

56. Chimps: Montagu, 1973b, p. 4. Heritability of IQ: Kamin, 1974; Lewontin, Rose, & Kamin, 1984, p. 116. IQ as reification: Gould, 1981. Personality and social behavior: Lewontin, Rose, & Kamin, 1984, chap. 9. Sex differences: Lewontin, Rose, & Kamin, 1984, p. 156. Pacific clans: Gould, 1998a, p. 262.

57. Daly, 1991.

58. Alcock, 2001.

59. Buss, 1995; Daly & Wilson, 1988; Daly & Wilson, 1999; Etcoff, 1999; Harris, 1998a; Hrdy, 1999; Ridley, 1993; Ridley, 1997; Symons, 1979; Wright, 1994.

60. Plomin et al., 2001.

PART III: HUMAN NATURE WITH A HUMAN FACE

1. Drake, 1970; Koestler, 1959.

2. Galileo, 1632/1967, pp. 58–59.

Chapter 8: The Fear of Inequality

1. From *The Rambler,* no. 60.

2. From the *Analects.*

3. Charlesworth, 1987; Lewontin, 1982; Miller, 2000b; Mousseau & Roff, 1987; Tooby & Cosmides, 1990.

4. Tooby & Cosmides, 1990.

5. Lander et al., 2001.

6. Bodmer & Cavalli-Sforza, 1970.

7. Tooby & Cosmides, 1990.

8. Patai & Patai, 1989.

9. Sowell, 1994; Sowell, 1995a.

10. Patterson, 1995; Patterson, 2000.

11. Cappon, 1959, pp. 387–392.

12. Seventh Lincoln-Douglas debate, October 15, 1858.

13. Mayr, 1963, p. 649. For a more recent statement of this argument from an evolutionary geneticist, see Crow, 2002.

14. Chomsky, 1973, pp. 362–363. See also Segerstråle, 2000.

15. For further discussion, see Tribe, 1971.

16. *Los Angeles Times* poll, December 21, 2001.

17. Nozick, 1974.

18. Gould, 1981, pp. 24–25. For reviews, see Blinkhorn, 1982; Davis, 1983; Jensen, 1982; Rushton, 1996; Samelson, 1982.

19. Putnam, 1973, p. 142.

20. See the consensus statements by Neisser et al., 1996; Snyderman & Rothman, 1988; and Gottfredson, 1997; and also Andreasen et al., 1993; Caryl, 1994; Deary, 2000; Haier et al., 1992; Reed & Jensen, 1992; Thompson et al., 2001; Van Valen, 1974; Willerman et al., 1991.

21. Moore & Baldwin, 1903/1996; Rachels, 1990.

22. Rawls, 1976.

23. Hayek, 1960/1978.

24. Chirot, 1994; Courtois et al., 1999; Glover, 1999.

25. Horowitz, 2001; Sowell, 1994; Sowell, 1996.

26. Lykken et al., 1992.

27. Interview in *Boston Phoenix* in the late 1970s, quotation reproduced from memory. Ironically, Wald's son Elijah became a radical science writer, like his father and his mother, the biologist Ruth Hubbard.

28. Degler, 1991; Kevles, 1985; Ridley, 2000.

29. Bullock, 1991; Chirot, 1994; Glover, 1999; Gould, 1981.

30. Richards, 1987, p. 533.

31. Glover, 1999; Murphy, 1999.

32. Proctor, 1999.

33. Laubichler, 1999.

34. For discussions of the Marxist genocides of the twentieth century and comparisons to the Nazi Holocaust, see Besançon, 1998; Bullock, 1991; Chandler, 1999; Chirot, 1994; Conquest, 2000; Courtois et al., 1999; Getty, 2000; Minogue, 1999; Shatz, 1999; Short, 1999.

35. For discussions of the intellectual roots of Marxism and comparisons with the intellectual roots of Nazism, see Berlin, 1996; Besançon, 1981; Besançon, 1998; Bullock, 1991; Chirot, 1994; Glover, 1999; Minogue, 1985; Minogue, 1999; Scott, 1998; Sowell, 1985. For discussions of the Marxist theory of human nature, see Archibald, 1989; Bauer, 1952; Plamenatz, 1963; Plamenatz, 1975; Singer, 1999; Stevenson & Haberman, 1998; Venable, 1945.

36. See, e.g., Venable, 1945, p. 3.

37. Marx, 1847/1995, chap. 2.

38. Marx & Engels, 1846/1963, part I.

39. Marx, 1859/1979, preface.

40. Marx, 1845/1989; Marx & Engels, 1846/1963.

41. Marx, 1867/1993, vol. 1, p. 10.

42. Marx & Engels, 1844/1988.

43. Glover, 1999, p. 254.

44. Minogue, 1999.

45. Glover, 1999, p. 275.

46. Glover, 1999, pp. 297–298.

47. Courtois et al., 1999, p. 620.

48. See the references cited in notes 34 and 35.

49. Marx quotation from Stevenson & Haberman, 1998, p. 146; Hitler quotation from Glover, 1999, p. 315.

50. Besançon, 1998.

51. Watson, 1985.

52. Tajfel, 1981.

53. Originally in *Red Flag* (Beijing), June 1, 1958; quoted in Courtois et al., 1999.

Chapter 9: The Fear of Imperfectibility

1. *The Prelude,* Book Sixth, "Cambridge and the Alps," I. Published 1799–1805.

2. Passmore, 1970, epigraph.

3. For example, the Seville Statement on Violence, 1990.

4. "Study says rape has its roots in evolution," *Boston Herald,* January 11, 2000, p. 3.

5. Thornhill & Palmer, 2001.

6. Brownmiller & Merhof, 1992.

7. Gould, 1995, p. 433.

8. Well, almost. The cartoonist, Jim Johnson, told me that he may have slandered walruses: he subsequently learned that it is leopard seals that kill penguins for fun.

9. Williams, 1988.

10. Jones, 1999; Williams, 1988.

11. Williams, 1966, p. 255.

12. On the relevance of human nature to morality, see McGinn, 1997; Petrinovich, 1995; Rachels, 1990; Richards, 1987; Singer, 1981; Wilson, 1993.

13. Masters, 1989, p. 240.

14. Daly & Wilson, 1988; Daly & Wilson, 1999.

15. Jones, 1997.

16. Daly & Wilson, 1999, pp. 58–66.

17. *Science Friday,* National Public Radio, May 7, 1999.

18. Singer, 1981.

19. Maynard Smith & Szathmáry, 1997; Wright, 2000.

20. De Waal, 1998; Fry, 2000.

21. Axelrod, 1984; Brown, 1991; Fry, 2000; Ridley, 1997; Wright, 2000.

22. Singer, 1981.

23. Skinner, 1948/1976; Skinner, 1971; Skinner, 1974.

24. Chomsky, 1973.

25. Berlin, 1996; Chirot, 1994; Conquest, 2000; Glover, 1999; Minogue, 1985; Minogue, 1999; Scott, 1998.

26. Scott, 1998.

27. Quoted in Scott, 1998, pp. 114–115.

28. Perry, 1997.

29. Harris, 1998a.

30. From a dialogue with Betty Friedan in *Saturday Review,* June 14, 1975, p. 18, quoted in Sommers, 1994, p. 18.

31. Quoted by Elizabeth Powers, *Commentary,* January 1, 1997.

32. From a talk at the Cornell University Institute on Women and Work, quoted by C. Young, "The mommy wars," *Reason,* July 2000.

33. Liza Mundy, "The New Critics," *Lingua Franca, 3,* September/October 1993, p. 27.

34. "From Carol Gilligan's chair," interview by Michael Norman, *New York Times Magazine,* November 7, 1997.

35. Letter by Bruce Bodner, *New York Times Magazine,* November 30, 1997.

36. C. Young, "Where the boys are," *Reason,* February 2, 2001.

37. Sommers, 2000.

Chapter 10: The Fear of Determinism

1. Kaplan, 1973, p. 10.

2. E. Felsenthal, "Man's genes made him kill, his lawyers claim," *Wall Street Journal,* November 15, 1994. The defense was unsuccessful: see "Mobley v. The State," Supreme Court of Georgia, March 17, 1995, 265 Ga. 292, 455 S.E.2d 61.

3. "Lawyers may use genetics study in rape defense," *National Post* (Canada), January 22, 2000, p. A8.

4. Jones, 2000; Jones, 1999.

5. Dennett, 1984. See also Kane, 1998; Nozick, 1981, pp. 317–362; Ridley, 2000; Staddon, 1999.

6. Dershowitz, 1994; J. Ellement, "Alleged con man's defense: 'Different' mores," *Boston Globe,* February 25, 1999; N. Hall, "Metis woman avoids jail term for killing her husband," *National Post* (Canada), January 20, 1999.

7. B. English, "David Lisak seeks out a dialogue with murderers," *Boston Globe,* July 27, 2000.

8. M. Williams, "Social work in the city: Rewards and risks," *New York Times,* July 30, 2000.

9. S. Morse, Review of C. Sandford's *Springsteen point blank, Boston Globe,* November 19, 1999.

10. M. Udovich, Review of M. Meade's *The unruly life of Woody Allen, New York Times,* March 5, 2000.

11. L. Franks, Interview with Hillary Clinton, *Talk,* August 1999.

12. K. Q. Seelye, "Clintons try to quell debate over interview," *New York Times,* August 5, 1999.

13. Dennett, 1984; Kane, 1998; Nozick, 1981, pp. 317–362; Ridley, 2000; Staddon, 1999.

14. Quoted in Kaplan, 1973, p. 16.

15. Daly & Wilson, 1988; Frank, 1988; Pinker, 1997; Schelling, 1960.

16. Quoted in Kaplan, 1973, p. 29.

17. Daly & Wilson, 1988, p. 256.

18. Dershowitz, 1994; Faigman, 1999; Kaplan, 1973; Kirwin, 1997.

19. Rice, 1997.

Chapter 11: The Fear of Nihilism

1. October 22, 1996; reprinted in the English edition of *L'Osservatore Romano,* October 30, 1996.

2. Macnamara, 1999; Miller, 1999; Newsome, 2001; Ruse, 2000.

3. See Nagel, 1970; Singer, 1981.

4. Cummins, 1996; Trivers, 1971; Wright, 1994.

5. Zahn-Wexler et al., 1992.

6. Brown, 1991.

7. Hare, 1993; Lykken, 1995; Mealey, 1995; Rice, 1997.

8. Rachels, 1990.

9. Murphy, 1999.

10. Damewood, 2001.

11. Ron Rosenbaum, "Staring into the heart of darkness," *New York Times Magazine,* June 4, 1995; Daly & Wilson, 1988, p. 79.

12. Antonaccio & Schweiker, 1996; Brink, 1989; Murdoch, 1993; Nozick, 1981; Sayre-McCord, 1988.

13. Singer, 1981.

PART IV: KNOW THYSELF

1. Alexander, 1987, p. 40.

Chapter 12: In Touch with Reality

1. Quotation from Cartmill, 1998.
2. Shepard, 1990.
3. www-bcs.mit.edu/persci/high/gallery/ checkershadow illusion.html.
4. www-bcs.mit.edu/persci/high/gallery/ checkershadow illusion.html.
5. From the computer scientist Oliver Selfridge; reproduced in Neisser, 1967.
6. Brown, 1991.
7. Brown, 1985; Lee, Jussim, & McCauley, 1995.
8. "Phony science wars" (Review of Ian Hacking's *The social construction of what?*), *Atlantic Monthly,* November 1999.
9. Hacking, 1999.
10. Searle, 1995.
11. Anderson, 1990; Pinker, 1997, chaps. 2, 5; Pinker, 1999, chap. 10; Pinker & Prince, 1996.
12. Armstrong, Gleitman, & Gleitman, 1983; Erikson & Kruschke, 1998; Marcus, 2001a; Pinker, 1997, chaps. 2, 5; Pinker, 1999, chap. 10; Sloman, 1996.
13. Ahn et al., 2001.
14. Lee, Jussim, & McCauley, 1995.
15. McCauley, 1995; Swim, 1994.
16. Jussim, McCauley, & Lee, 1995; McCauley, 1995.
17. Jussim & Eccles, 1995.
18. Brown, 1985; Jussim, McCauley, & Lee, 1995; McCauley, 1995.
19. Gilbert & Hixon, 1991; Pratto & Bargh, 1991.
20. Brown, 1985, p. 595.
21. Jussim & Eccles, 1995; Smith, Jussim, & Eccles, 1999.
22. Flynn, 1999; Loury, 2002; Valian, 1998.
23. Galileo, 1632/1967, p. 105.
24. Whorf, 1956.
25. Geertz, 1973, p. 45.
26. Quotations from Lehman, 1992.
27. Barthes, 1972, p. 135.
28. Pinker, 1994, chap. 3.
29. Pinker, 1984a.
30. Lakoff & Johnson, 1980.
31. Jackendoff, 1996.
32. Baddeley, 1986.
33. Dehaene et al., 1999.
34. Pinker, 1994, chap. 3; Siegal, Varley, & Want, 2001; Weiskrantz, 1988.
35. Gallistel, 1992; Gopnik, Meltzoff, & Kuhl, 1999; Hauser, 2000.
36. Anderson, 1983.
37. Pinker, 1994.
38. "'Minority' a bad word in San Diego," *Boston Globe,* April 4, 2001; S. Schweitzer, "Council mulls another word for 'minority,'" *Boston Globe,* August 9, 2001.
39. Brooker, 1999, pp. 115–116.
40. Leslie, 1995.
41. Abbott, 2001; Leslie, 1995.
42. Frith, 1992.
43. Kosslyn, 1980; Kosslyn, 1994; Pinker, 1984b; Pinker, 1997, chap. 4.
44. Kosslyn, 1980; Pinker, 1997, chap. 5.
45. Chase & Simon, 1973.
46. Dennett, 1991, pp. 56–57.
47. A. Gopnik, "Black studies," *New Yorker,* December 5, 1994, pp. 138–139.

Chapter 13: Out of Our Depths

1. Caramazza & Shelton, 1998; Gallistel, 2000; Gardner, 1983; Hirschfeld & Gelman, 1994; Keil, 1989; Pinker, 1997, chap. 5; Tooby & Cosmides, 1992.
2. Spelke, 1995.
3. Atran, 1995; Atran, 1998; Gelman, Coley, & Gottfried, 1994; Keil, 1995.
4. Bloom, 1996; Keil, 1989.
5. Gallistel, 1990; Kosslyn, 1994.
6. Butterworth, 1999; Dehaene, 1997; Devlin, 2000; Geary, 1994; Lakoff & Nunez, 2000.
7. Cosmides & Tooby, 1996; Gigerenzer, 1997; Kahneman & Tversky, 1982.
8. Braine, 1994; Jackendoff, 1990; Macnamara & Reyes, 1994; Pinker, 1989.
9. Pinker, 1994; Pinker, 1999.
10. Quoted in Ravitch, 2000, p. 388.
11. McGuinness, 1997.
12. Geary, 1994; Geary, 1995.
13. Carey, 1986; Carey & Spelke, 1994; Gardner, 1983; Gardner, 1999; Geary, 1994; Geary, 1995; Geary, in press.
14. Carey, 1986; McCloskey, 1983.
15. Gardner, 1999.
16. McGuinness, 1997.
17. Dehaene et al., 1999.
18. Bloom, 1994.
19. Pinker, 1990.
20. Carey & Spelke, 1994.
21. Geary, 1995; Geary, in press; Harris, 1998a.
22. Green, 2001, chap. 2.
23. S. G. Stolberg, "Reconsidering embryo research," *New York Times,* July 1, 2001.
24. Brock, 1993, p. 372, n. 14 p. 385; Glover, 1977; Tooley, 1972; Warren, 1984.
25. Green, 2001.
26. R. Bailey, "Dr. Strangelunch, or: Why we

should learn to stop worrying and love genetically modified food," *Reason,* January 2001.

27. "EC-sponsored research on safety of genetically modified organisms—A review of results." Report EUR 19884, October 2001, European Union Office for Publications.

28. Ames, Profet, & Gold, 1990.

29. Ames, Profet, & Gold, 1990.

30. E. Schlosser, "Why McDonald's fries taste so good," *Atlantic Monthly,* January 2001.

31. Ahn et al., 2001; Frazer, 1890/1996; Rozin, 1996; Rozin, Markwith, & Stoess, 1997; P. Stevens, 2001 (but see also M. Stevens, 2001).

32. Rozin & Fallon, 1987.

33. Ahn et al., 2001.

34. Rozin, 1996; Rozin & Fallon, 1987; Rozin, Markwith, & Stoess, 1997.

35. Rozin, 1996.

36. Mayr, 1982.

37. Ames, Profet, & Gold, 1990; Lewis, 1990; G. Gray & D. Ropeik, "What, me worry?" *Boston Globe,* November 11, 2001, p. E8.

38. Marks & Nesse, 1994; Seligman, 1971.

39. Slovic, Fischof, & Lichtenstein, 1982.

40. Sharpe, 1994.

41. Cosmides & Tooby, 1996; Gigerenzer, 1991; Gigerenzer, 1997; Pinker, 1997, chap. 5.

42. Hoffrage et al., 2000; Tversky & Kahneman, 1973.

43. Slovic, Fischof, & Lichtenstein, 1982.

44. Tooby & DeVore, 1987.

45. Fiske, 1992.

46. Cosmides & Tooby, 1992.

47. Sowell, 1980.

48. Sowell, 1980; Sowell, 1996.

49. Sowell, 1994; Sowell, 1996.

50. R. Radford (writing in 1945), quoted in Sowell, 1994, p. 57.

51. From "The figure of the youth as virile poet"; Stevens, 1965.

52. Jackendoff, 1987; Pinker, 1997; Pinker, 1999.

53. Bailey, 2000.

54. Sen, 1984.

55. Simon, 1996.

56. Bailey, 2000; Romer, 1991; Romer & Nelson, 1996; P. Romer, "Ideas and things," *Economist,* September 11, 1993.

57. Romer & Nelson, 1996.

58. Quoted in M. Kumar, "Quantum reality," *Prometheus, 2,* pp. 20–21, 1999.

59. Quoted in M. Kumar, "Quantum reality," *Prometheus, 2,* pp. 20–21, 1999.

60. Quoted in Dawkins, 1998, p. 50.

61. McGinn, 1993; McGinn, 1999; Pinker, 1997, chap. 8.

Chapter 14: The Many Roots of Our Suffering

1. Trivers, 1976.

2. Trivers, 1971; Trivers, 1972; Trivers, 1974; Trivers, 1976; Trivers, 1985.

3. Alexander, 1987; Cronin, 1992; Dawkins, 1976/1989; Ridley, 1997; Wright, 1994.

4. Hamilton, 1964; Trivers, 1971; Trivers, 1972; Trivers, 1974; Williams, 1966.

5. "Renewing American Civilization," a talk presented at Reinhardt College, January 7, 1995.

6. Chagnon, 1988; Daly, Salmon, & Wilson, 1997; Fox, 1984; Mount, 1992; Shoumatoff, 1985.

7. Chagnon, 1992; Daly, Salmon, & Wilson, 1997; Daly & Wilson, 1988; Gaulin & McBurney, 2001, pp. 321–329.

8. Burnstein, Crandall, & Kitayama, 1994; Petrinovich, O'Neill, & Jorgensen, 1993.

9. Petrinovich, O'Neill, & Jorgensen, 1993; Singer, 1981.

10. Masters, 1989, pp. 207–208.

11. Quoted in J. Muravchick, "Socialism's last stand," *Commentary*, March 2002, pp. 47–53, quotation from p. 51.

12. Broadcast on Radio Free LA, January 1997, www.radiofreela.com. Transcript available at www.zmag.org/chomsky/rage or as a cached page on www.google.com.

13. Daly, Salmon, & Wilson, 1997; Mount, 1992.

14. Johnson, Ratwik, & Sawyer, 1987; Salmon, 1998.

15. Fiske, 1992.

16. Fiske, 1992, p. 698.

17. Trivers, 1974; Trivers, 1985.

18. Agrawal, Brodie, & Brown, 2001; Godfray, 1995; Trivers, 1985.

19. Haig, 1993.

20. Daly & Wilson, 1988; Hrdy, 1999.

21. Hrdy, 1999.

22. Trivers, 1976; Trivers, 1981.

23. Trivers, 1985.

24. Harris, 1998a; Plomin & Daniels, 1987; Rowe, 1994; Sulloway, 1996; Turkheimer, 2000.

25. Trivers, 1985, p. 159.

26. Used as the epigraph to Judith Rich Harris's *The nurture assumption.*

27. Dunn & Plomin, 1990.

28. Hrdy, 1999.

29. Daly & Wilson, 1988; Wilson, 1993.

30. Wilson, 1993.

31. Trivers, 1972; Trivers, 1985.

32. Blum, 1997; Buss, 1994; Geary, 1998; Ridley, 1993; Symons, 1979.

33. Buss, 1994; Kenrick et al., 1993; Salmon & Symons, 2001; Symons, 1979.

34. Buss, 2000.

35. Alexander, 1987.

36. Brown, 1991; Symons, 1979.

37. K. Kelleher, "When students 'hook up,' someone inevitably gets let down," *Los Angeles Times,* August 13, 2001.

38. Symons, 1979.

39. Daly, Salmon, & Wilson, 1997.

40. Wilson & Daly, 1992.

41. Ridley, 1997. See also Lewontin, 1990.

42. Rose & Rose, 2000.

43. Fiske, 1992.

44. Axelrod, 1984; Dawkins, 1976/1989; Ridley, 1997; Trivers, 1971.

45. Cosmides & Tooby, 1992; Frank, Gilovich, & Regan, 1993; Gigerenzer & Hug, 1992; Kanwisher & Moscovitch, 2000; Mealey, Daood, & Krage, 1996.

46. Yinon & Dovrat, 1987.

47. Gaulin & McBurney, 2001, pp. 329–338; Haidt, in press; Trivers, 1971, pp. 49–54.

48. Fehr & Gächter, 2000; Gintis, 2000; Price, Cosmides, & Tooby, 2002.

49. Ridley, 1997, p. 84.

50. Fehr & Gächter, 2000; Gaulin & McBurney, 2001, pp. 333–335.

51. Fehr & Gächter, 2000; Ridley, 1997.

52. Williams, Harkins, & Latané, 1981.

53. Klaw, 1993; McCord, 1989; Muravchik, 2002; Spann, 1989.

54. J. Muravchik, "Socialism's last stand," *Commentary,* March 2002, pp. 47–53, quotation from p. 53.

55. Fiske, 1992.

56. Cashdan, 1989; Cosmides & Tooby, 1992; Eibl-Eibesfeldt, 1989; Fiske, 1992; Hawkes, O'Connell, & Rogers, 1997; Kaplan, Hill, & Hurtado, 1990; Ridley, 1997.

57. Ridley, 1997, p. 111.

58. Junger, 1997, p. 76.

59. Cited in Williams, 1966, p. 116.

60. Williams, 1966.

61. Fehr, Fischbacher, & Gächter, in press; Gintis, 2000.

62. Nunney, 1998; Reeve, 2000; Trivers, 1998; Wilson & Sober, 1994.

63. Williams, 1988, pp. 391–392.

64. Frank, 1988; Hirshleifer, 1987; Trivers, 1971.

65. Hare, 1993; Lykken, 1995; Mealey, 1995.

66. On the heritability of antisocial traits, see Bock & Goode, 1996; Deater-Deckard & Plomin, 1999; Krueger, Hicks, & McGue, 2001; Lykken, 1995; Mealey, 1995; Rushton et al., 1986. Regarding altruism, one study failed to find that it is heritable (Krueger, Hicks, &

McGue, 2001); another study, with twice as many subjects, found it to be substantially heritable (Rushton et al., 1986).

67. Miller, 2000b.

68. Tooby & Cosmides, 1990.

69. Axelrod, 1984; Dawkins, 1976/1989; Nowak, May, & Sigmund, 1995; Ridley, 1997.

70. Dugatkin, 1992; Harpending & Sobus, 1987; Mealey, 1995; Rice, 1997.

71. Rice, 1997.

72. Lalumière, Harris, & Rice, 2001.

73. M. Kakutani, "The strange case of the writer and the criminal," *New York Times Book Review,* September 20, 1981.

74. S. McGraw, "Some used their second chance at life; others squandered it," *The Record* (Bergen County, N.J.), October 12, 1998.

75. Rice, 1997.

76. Trivers, 1976.

77. Goleman, 1985; Greenwald, 1988; Krebs & Denton, 1997; Lockard & Paulhaus, 1988; Rue, 1994; Taylor, 1989; Trivers, 1985; Wright, 1994.

78. Nesse & Lloyd, 1992.

79. Gazzaniga, 1998.

80. Damasio, 1994, p. 68.

81. Babcock & Loewenstein, 1997; Rue, 1994; Taylor, 1989.

82. Aronson, 1980; Festinger, 1957; Greenwald, 1988.

83. Haidt, 2001.

84. Dutton, 2001, p. 209; Fox, 1989; Hogan, 1997; Polti, 1921/1977; Storey, 1996, pp. 110, 142.

85. Steiner, 1984, p. 1.

86. Steiner, 1984, p. 231.

87. Steiner, 1984, pp. 300–301.

88. Symons, 1979, p. 271.

89. D. Symons, personal communication, July 30, 2001.

Chapter 15: The Sanctimonious Animal

1. Alexander, 1987; Haidt, in press; Krebs, 1998; Trivers, 1971; Wilson, 1993; Wright, 1994.

2. Haidt, Koller, & Dias, 1993.

3. Haidt, 2001.

4. Haidt, in press.

5. Shweder et al., 1997.

6. Haidt, in press; Rozin, 1997; Rozin, Markwith, & Stoess, 1997.

7. Glendon, 2001; Sen, 2000.

8. Cronk, 1999; Sommers, 1998; Wilson, 1993; C. Sommers, 1998, "Why Johnny can't tell right from wrong," *American Outlook,* Summer 98, pp. 45–47.

9. D. Symons, personal communication, July 26, 2001.

10. Etcoff, 1999.

11. Glover, 1999.

12. L. Kass, "The wisdom of repugnance," *New Republic,* June 2, 1997.

13. Rozin, 1997; Rozin, Markwith, & Stoess, 1997.

14. Tetlock, 1999; Tetlock et al., 2000.

15. Tetlock, 1999.

16. Tetlock et al., 2000.

17. Hume, 1739/2000.

18. I. Buruma, Review of Ian Kershaw's *Hitler 1936–45: Nemesis, New York Times Book Review,* December 10, 2000, p. 13.

PART V: HOT BUTTONS

1. Haidt & Hersh, 2001; Tetlock, 1999; Tetlock et al., 2000.

2. Haidt & Hersh, 2001; Tetlock, 1999; Tetlock et al., 2000.

Chapter 16: Politics

1. From *Iolanthe.*

2. Personal communication, D. Lykken, April 11, 2001. Other estimates of the heritability of conservative attitudes are typically in the range of .4 to .5: Bouchard et al., 1990; Eaves, Eysenck, & Martin, 1989; Holden, 1987; Martin et al., 1986; Plomin et al., 1997, p. 206; Scarr & Weinberg, 1981.

3. Tesser, 1993.

4. Wilson, 1994, pp. 338–339.

5. Masters, 1982; Masters, 1989.

6. Dawkins, 1976/1989; Williams, 1966.

7. Boyd & Silk, 1996; Ridley, 1997; Trivers, 1985.

8. Sowell, 1987.

9. Sowell, 1995b.

10. From the preface to *On the rocks: A political fantasy in two acts.*

11. Smith, 1759/1976, pp. 233–234.

12. Burke, 1790/1967, p. 93.

13. Quoted in E. M. Kennedy, "Tribute to Senator Robert F. Kennedy," June 8, 1968, www.jfklibrary.org/e060868.htm.

14. Hayek, 1976, pp. 64, 33.

15. Quoted in Sowell, 1995, pp. 227, 112.

16. "If the law supposes that . . . the law is a ass—a idiot" (from *Oliver Twist*).

17. Quoted in Sowell, 1995, p. 11.

18. Hayek, 1976.

19. This is a point of contact with an alternative theory of the psychological underpinnings of the left-right divide proposed by the linguist George Lakoff: that the left believes that government should act like a nurturant parent, whereas the right believes it should act like a strict parent; see Lakoff, 1996.

20. See Chapter 14, and also Burnstein, Crandall, & Kitayama, 1994; Chagnon, 1992; Daly, Salmon, & Wilson, 1997; Daly & Wilson, 1988; Fox, 1984; Gaulin & McBurney, 2001, pp. 321–329; Mount, 1992; Petrinovich, O'Neill, & Jorgensen, 1993; Shoumatoff, 1985.

21. See Chapter 14, and also Bowles & Gintis, 1999; Cosmides & Tooby, 1992; Fehr, Fischbacher, & Gächter, in press; Fehr & Gächter, 2000; Fiske, 1992; Gaulin & McBurney, 2001, pp. 333–335; Gintis, 2000; Klaw, 1993; McCord, 1989; Muravchik, 2002; Price, Cosmides, & Tooby, 2002; Ridley, 1997; Spann, 1989; Williams, Harkins, & Latané, 1981.

22. See Chapters 3 and 17, especially the references in notes 39, 52, 53, 72, 73, and 74 in Chapter 3, and notes 42, 43, and 45 in Chapter 17.

23. Brown, 1991; Brown, 1985; Sherif, 1966; Tajfel, 1981.

24. See Chapters 3 and 19, and also Bouchard, 1994; Neisser et al., 1996; Plomin et al., 2001.

25. See Chapter 14, and also Aronson, 1980; Festinger, 1957; Gazzaniga, 1998; Greenwald, 1988; Nesse & Lloyd, 1992; Wright, 1994.

26. See Chapter 15, and also Haidt, in press; Haidt, Koller, & Dias, 1993; Petrinovich, O'Neill, & Jorgensen, 1993; Rozin, Markwith, & Stoess, 1997; Shweder et al., 1997; Singer, 1981; Tetlock, 1999; Tetlock et al., 2000.

27. Sowell, 1987.

28. Marx & Engels, 1844/1988.

29. Quoted in Singer, 1999, p. 4.

30. Bullock, 1991; Chirot, 1994; Conquest, 2000; Courtois et al., 1999; Glover, 1999.

31. Quoted in J. Getlin, "Natural wonder: At heart, Edward Wilson's an ant man," *Los Angeles Times,* October 21, 1994, p. E1.

32. Federalist Papers No. 51, Rossiter, 1961, p. 322.

33. Bailyn, 1967/1992; Maier, 1997.

34. Lutz, 1984.

35. McGinnis, 1996; McGinnis, 1997.

36. Federalist Papers No. 10, Rossiter, 1961, p. 78.

37. Quoted in McGinnis, 1997, p. 236.

38. Federalist Papers No. 72, Rossiter, 1961, p. 437.

39. Federalist Papers No. 51, Rossiter, 1961, p. 322.

40. Federalist Papers No. 51, Rossiter, 1961, pp. 331–332.

41. From *Helvedius No. 4,* quoted in McGinnis, 1997, p. 130.

42. Boehm, 1999; de Waal, 1998; Dunbar, 1998.

43. Singer, 1999, p. 5.

44. L. Arnhart, M. J. Behe, & W. A. Dembski, "Conservatives, Darwin, and design: An exchange," *First Things, 107,* November 2000, pp. 23–31.

45. For arguments similar to Singer's, see Brociner, 2001.

46. Singer, 1999, p. 6.

47. Singer, 1999, pp. 8–9.

48. Chomsky, 1970, p. 22.

49. See Barsky, 1997; Chomsky, 1988a.

50. Chomsky, 1975, p. 131.

51. Trivers, 1981.

52. A. Wooldridge, "Bell Curve liberals," *New Republic,* February 27, 1995.

53. Herrnstein & Murray, 1994, chap. 22. See also Murray's afterword in the 1996 paperback edition.

54. Gigerenzer & Selten, 2001; Jones, 2001; Kahneman & Tversky, 1984; Thaler, 1994; Tversky & Kahneman, 1974.

55. Akerlof, 1984; Daly & Wilson, 1994; Jones, 2001; Rogers, 1994.

56. Frank, 1999; Frank, 1985.

57. Bowles & Gintis, 1998; Bowles & Gintis, 1999.

58. Gintis, 2000.

59. Wilkinson, 2000.

60. Daly & Wilson, 1988; Daly, Wilson, & Vasdev, 2001; Wilson & Daly, 1997.

Chapter 17: Violence

1. Quoted by R. Cooper in "The long peace," *Prospect,* April 1999.

2. National Defense Council Foundation, Alexandria, Va., www.ndcf.org/index.htm.

3. Bamforth, 1994; Chagnon, 1996; Daly & Wilson, 1988; Ember, 1978; Ghiglieri, 1999; Gibbons, 1997; Keeley, 1996; Kingdon, 1993; Knauft, 1987; Krech, 1994; Krech, 1999; Wrangham & Peterson, 1996.

4. Keeley, 1996; Walker, 2001.

5. Gibbons, 1997; Holden, 2000.

6. Fernández-Jalvo et al., 1996.

7. *FBI Uniform Crime Reports 1999:* www.fbi.gov/ucr/99cius.htm.

8. Seville, 1990.

9. Ortega y Gasset, 1932/1985, epilogue.

10. *New York Times,* June 13, 1999.

11. Paul Billings, quoted in B. H. Kevles & D. J. Kevles, "Scapegoat biology," *Discover,* October 1997, pp. 59–62, quotation from p. 62.

12. B. H. Kevles & D. J. Kevles, "Scapegoat biology," *Discover,* October 1997, pp. 59–62, quotation from p. 62.

13. Daphne White, quoted in M. Wilkinson, "Parent group lists 'dirty dozen' toys," *Boston Globe,* December 5, 2000, p. A5.

14. H. Spivak & D. Prothrow-Stith, "The next tragedy of Jonesboro," *Boston Globe,* April 5, 1998.

15. C. Burrell, "Study of inmates cites abuse factor," Associated Press, April 27, 1998.

16. G. Kane, "Violence as a cultural imperative," *Boston Sunday Globe,* October 6, 1996.

17. Quoted in A. Flint, "Some see bombing's roots in a US culture of conflict," *Boston Globe,* June 1, 1995.

18. A. Flint, "Some see bombing's roots in a US culture of conflict," *Boston Globe,* June 1, 1995.

19. M. Zuckoff, "More murders, more debate," *Boston Globe,* July 31, 1999.

20. A. Diamant, "What's the matter with men?" *Boston Globe Magazine,* March 14, 1993.

21. Mesquida & Wiener, 1996.

22. Freedman, 2002.

23. Fischoff, 1999; Freedman, 1984; Freedman, 1996; Freedman, 2002; Renfrew, 1997.

24. Charlton, 1997.

25. J. Q. Wilson, "Hostility in America," *New Republic,* August 25, 1997, pp. 38–41.

26. Nisbett & Cohen, 1996.

27. E. Marshal, "The shots heard 'round the world," *Science, 289,* 2000, pp. 570–574.

28. Wakefield, 1992.

29. M. Enserink, "Searching for the mark of Cain," *Science, 289,* 2000, pp. 575–579; quotation from p. 579.

30. Clark, 1970, p. 220.

31. Daly & Wilson, 1988, p. ix.

32. Shipman, 1994, p. 252.

33. E. Marshal, "A sinister plot or victim of politics?" *Science, 289,* 2000, p. 571.

34. Shipman, 1994, p. 243.

35. Quoted in R. Wright, "The biology of violence," *New Yorker,* March 13, 1995, pp. 68–77; quotation from p. 69.

36. Daly & Wilson, 1988.

37. Daly & Wilson, 1988; Rogers, 1994; Wilson & Herrnstein, 1985.

38. Quoted by Frederick Goodwin in R. Wright, "The biology of violence," *New Yorker,* March 13, 1995, p. 70.

39. C. Holden, "The violence of the lambs," *Science, 289,* 2000, pp. 580–581.

40. Hare, 1993; Lykken, 1995; Rice, 1997.

41. Ghiglieri, 1999; Wrangham & Peterson, 1996.

42. Davidson, Putnam, & Larson, 2000; Renfrew, 1997.

43. Geary, 1998, pp. 226–227; Sherif, 1966.

44. R. Tremblay, quoted in C. Holden, "The violence of the lambs," *Science, 289,* 2000, pp. 580–581.

45. Buss & Duntley, in press; Kenrick & Sheets, 1994.

46. Hobbes, 1651/1957, p. 185.

47. Dawkins, 1976/1989, p. 66.

48. Bueno de Mesquita, 1981.

49. Trivers, 1972.

50. Chagnon, 1992; Daly & Wilson, 1988; Keeley, 1996.

51. Daly & Wilson, 1988, p. 163.

52. Rogers, 1994; Wilson & Daly, 1997.

53. Wilson & Herrnstein, 1985.

54. Mesquida & Wiener, 1996.

55. Singer, 1981.

56. Wright, 2000.

57. Glover, 1999.

58. Zimbardo, Maslach, & Haney, 2000.

59. Quoted in Glover, 1999, p. 53.

60. Quoted in Glover, 1999, pp. 37–38.

61. Bourke, 1999, pp. 63–64; Graves, 1992; Spiller, 1988.

62. Bourke, 1999; Glover, 1999; Horowitz, 2001.

63. Daly & Wilson, 1988; Glover, 1999; Schelling, 1960.

64. Chagnon, 1992; Daly & Wilson, 1988; Wrangham & Peterson, 1996.

65. Van den Berghe, 1981.

66. Epstein, 1994; Epstein & Axtell, 1996; Richardson, 1960; Saperstein, 1995.

67. Chagnon, 1988; Chagnon, 1992.

68. Glover, 1999.

69. Vasquez, 1992.

70. Rosen, 1992.

71. Wrangham, 1999.

72. Daly & Wilson, 1988.

73. Daly & Wilson, 1988, pp. 225–226.

74. Daly & Wilson, 1988; Frank, 1988; Schelling, 1960.

75. Brown, 1985; Horowitz, 2001.

76. Daly & Wilson, 1988.

77. Daly & Wilson, 1988; Fox & Zawitz, 2000; Nisbett & Cohen, 1996.

78. Daly & Wilson, 1988, p. 127.

79. Daly & Wilson, 1988, p. 229.

80. Chagnon, 1992; Daly & Wilson, 1988; Frank, 1988.

81. Nisbett & Cohen, 1996.

82. Nisbett & Cohen, 1996.

83. E. Anderson, "The code of the streets," *Atlantic Monthly,* May 1994, pp. 81–94.

84. See also Patterson, 1997.

85. E. Anderson, "The code of the streets," *Atlantic Monthly,* May 1994, pp. 81–94, quotation from p. 82.

86. Quoted in L. Helmuth, "Has America's tide of violence receded for good?" *Science, 289,* 2000, pp. 582–585, quotation from p. 582.

87. L. Helmuth, "Has America's tide of violence receded for good?" *Science, 289,* 2000, pp. 582–585, quotation from p. 583.

88. Wilkinson, 2000; Wilson & Daly, 1997.

89. Harris, 1998a, pp. 212–213.

90. Hobbes, 1651/1957, p. 190.

91. Hobbes, 1651/1957, p. 223.

92. Fry, 2000.

93. Daly & Wilson, 1988; Keeley, 1996.

94. Daly & Wilson, 1988; Nisbett & Cohen, 1996.

95. Daly & Wilson, 1988.

96. Daly & Wilson, 1988.

97. Wilson & Herrnstein, 1985.

98. L. Helmuth, "Has America's tide of violence receded for good?" *Science, 289,* 2000; Kelling & Sousa, 2001.

99. *Time,* October 17, 1969, p. 47.

100. Kennedy, 1997.

101. National Defense Council Foundation, Alexandria, Va., www.ndcf.org/index.htm.

102. Quoted by Glover, 1999, p. 227.

103. Horowitz, 2001; Keegan, 1976.

104. C. Nickerson, "Canadians remain gun-shy of Americans," *Boston Globe,* February 11, 2001.

105. Quoted in Wright, 2000, p. 61.

106. Chagnon, 1988; Chagnon, 1992.

107. Axelrod, 1984.

108. Glover, 1999, p. 159.

109. Glover, 1999, p. 202.

110. Axelrod, 1984; Ridley, 1997.

111. Glover, 1999, pp. 231–232.

112. M. J. Wilkinson, personal communication, October 29, 2001; Wilkinson, in press.

113. See Chapters 3 and 13, and also Fodor & Pylyshyn, 1988; Miller, Galanter, & Pribram, 1960; Pinker, 1997, chap. 2; Pinker, 1999, chap. 1.

Chapter 18: Gender

1. Jaggar, 1983.

2. Quoted in Jaggar, 1983, p. 27.

3. J. N. Wilford, "Sexes equal on South Sea isle," *New York Times,* March 29, 1994.

4. L. Tye, "Girls appear to be closing aggression gap with boys," *Boston Globe,* March 26, 1998.

5. M. Zoll, "What about the boys?" *Boston Globe,* April 23, 1998.

6. Quoted in Young, 1999, p. 247.

7. Crittenden, 1999; Shalit, 1999.

8. L. Kass, "The end of courtship," *Public Interest, 126,* Winter 1997.

9. Patai, 1998.

10. Grant, 1993; Jaggar, 1983; Tong, 1998.

11. Sommers, 1994. See also Jaggar, 1983.

12. Quoted in Sommers, 1994, p. 22.

13. Gilligan, 1982.

14. Jaffe & Hyde, 2000; Sommers, 1994, chap. 7; Walker, 1984.

15. Belenky et al., 1986.

16. Denfeld, 1995; Kaminer, 1990; Lehrman, 1997; McElroy, 1996; Paglia, 1992; Patai, 1998; Patai & Koertge, 1994; Sommers, 1994; Taylor, 1992; Young, 1999.

17. Sommers, 1994.

18. Denfeld, 1995; Lehrman, 1997; Roiphe, 1993; Walker, 1995.

19. S. Boxer, "One casualty of the women's movement: Feminism," *New York Times,* December 14, 1997.

20. C. Paglia, "Crying wolf," *Salon,* February 7, 2001.

21. Patai, 1998; Sommers, 1994.

22. Trivers, 1976; Trivers, 1981; Trivers, 1985.

23. Trivers & Willard, 1973.

24. Jensen, 1998, chap. 13.

25. Blum, 1997; Eagly, 1995; Geary, 1998; Halpern, 2000; Kimura, 1999.

26. Salmon & Symons, 2001; Symons, 1979.

27. Daly & Wilson, 1988. Surgery anecdote from Barry, 1995.

28. Geary, 1998; Maccoby & Jacklin, 1987.

29. Geary, 1998; Halpern, 2000; Kimura, 1999.

30. Blum, 1997; Geary, 1998; Halpern, 2000; Hedges & Nowell, 1995; Lubinski & Benbow, 1992.

31. Hedges & Nowell, 1995; Lubinski & Benbow, 1992.

32. Blum, 1997; Geary, 1998; Halpern, 2000; Kimura, 1999.

33. Blum, 1997; Geary, 1998; Halpern, 2000; Kimura, 1999.

34. Provine, 1993.

35. Hrdy, 1999.

36. Fausto-Sterling, 1985, pp. 152–153.

37. Brown, 1991.

38. Buss, 1999; Geary, 1998; Ridley, 1993; Symons, 1979; Trivers, 1972.

39. Daly & Wilson, 1983; Geary, 1998; Hauser, 2000.

40. Geary, 1998; Silverman & Eals, 1992.

41. Gibbons, 2000.

42. Blum, 1997; Geary, 1998; Halpern, 2000; Kimura, 1999.

43. Blum, 1997; Geary, 1998; Gur & Gur, in press; Gur et al., 1999; Halpern, 2000; Jensen, 1998; Kimura, 1999; Neisser et al., 1996.

44. Dabbs & Dabbs, 2000; Geary, 1998; Halpern, 2000; Kimura, 1999; Sapolsky, 1997.

45. A. Sullivan, "Testosterone power," *Women's Quarterly,* Summer 2000.

46. Kimura, 1999.

47. Blum, 1997; Gangestad & Thornhill, 1998.

48. Blum, 1997; Geary, 1998; Halpern, 2000; Kimura, 1999.

49. Symons, 1979, chap. 9.

50. Reiner, 2000.

51. Quoted in Halpern, 2000, p. 9.

52. Quoted in Colapinto, 2000.

53. Colapinto, 2000; Diamond & Sigmundson, 1997.

54. Skuse et al., 1997.

55. Barkley et al., 1977; Harris, 1998a; Lytton & Romney, 1991; Maccoby & Jacklin, 1987.

56. B. Friedan, "The future of feminism," *Free Inquiry,* Summer 1999.

57. "Land of plenty: Diversity as America's competitive edge in science, engineering, and technology," Report of the Congressional Commission on the Advancement of Women and Minorities in Science, Engineering, and Technology Development, September 2000.

58. J. Alper, "The pipeline is leaking women all the way along," *Science, 260,* April 16, 1993; J. Mervis, "Efforts to boost diversity face persistent problems," *Science, 284,* June 11, 1999; J. Mervis, "Diversity: Easier said than done," *Science, 289,* March 16, 2000; J. Mervis, "NSF searches for right way to help women," *Science, 289,* July 21, 2000; J. Mervis, "Gender equity: NSF program targets institutional change," *Science, 291,* July 21, 2001.

59. J. Mervis, "Efforts to boost diversity face persistent problems," *Science, 284,* June 11, 1999, p. 1757.

60. P. Healy, "Faculty shortage: Women in sciences," *Boston Globe,* January 31, 2001.

61. C. Holden, "Parity as a goal sparks bitter battle," *Science, 289,* July 21, 2000, p. 380.

62. Quoted in Young, 1999, pp. 22, 34–35.

63. Estrich, 2000; Furchtgott-Roth & Stolba, 1999; Goldin, 1990; Gottfredson, 1988; Hausman, 1999; Kleinfeld, 1999; Lehrman, 1997; Lubinski & Benbow, 1992; Roback, 1993; Schwartz, 1992; Young, 1999.

64. Browne, 1998; Furchtgott-Roth & Stolba, 1999; Goldin, 1990.

65. In a random sample of 100 members of the International Association for the Study of Child Language, I counted 75 women and 25 men. The Stanford Child Language Research Forum lists 18 past keynote speakers on its web site (csli.stanford.edu/~clrf/history.html): 15 women and 3 men.

66. Browne, 1998; Furchtgott-Roth & Stolba, 1999; Goldin, 1990; Gottfredson, 1988; Kleinfeld, 1999; Roback, 1993; Young, 1999.

67. Lubinski & Benbow, 1992.

68. See Browne, 1998, and the references in note 63.

69. Buss, 1992; Ellis, 1992.

70. Hrdy, 1999.

71. Browne, 1998; Hrdy, 1999.

72. Roback, 1993.

73. Becker, 1991.

74. Furchtgott-Roth & Stolba, 1999.

75. Quoted in C. Young, "Sex and science," *Salon,* April 12, 2001.

76. Quoted in C. Holden, "Parity as a goal sparks bitter battle," *Science, 289,* July 21, 2000.

77. Quoted in C. Holden, "Parity as a goal sparks bitter battle," *Science, 289,* July 21, 2000.

78. Kleinfeld, 1999.

79. National Science Foundation, *Women, Minorities, and Persons with Disabilities in Science and Engineering: 1998,* www.nsf.gov/sbe/srs/nsf99338.

80. Thornhill & Palmer, 2000.

81. "Report on the situation of human rights in the territory of the former Yugoslavia," 1993. United Nations Document E/CN.4/1993/50.

82. J. E. Beals, "Ending the silence on sexual violence," *Boston Globe,* April 10, 2000.

83. R. Haynor, "Violence against women," *Boston Globe,* October 22, 2000.

84. Brownmiller, 1975, p. 14.

85. Young, 1999, p. 139.

86. McElroy, 1996.

87. McElroy, 1996.

88. Thiessen & Young, 1994.

89. Dworkin, 1993.

90. J. Tooby & L. Cosmides, "Reply to Jerry Coyne," www.psych.ucsb.edu/research/cep/tnr.html.

91. Gordon & Riger, 1991, p. 47.

92. Rose & Rose, 2000, p. 139.

93. M. Wertheim, "Born to rape?" *Salon,* February 29, 2000.

94. G. Miller, "Why men rape," *Evening Standard,* March 6, 2000, p. 53.

95. Symons, 1979; Thornhill & Palmer, 2000.

96. Jones, 1999. See also Check & Malamuth, 1985; Ellis & Beattie, 1983; Symons, 1979; Thornhill & Palmer, 2000.

97. Gottschall & Gottschall, 2001.

98. Jones, 1999, p. 890.

99. Bureau of Justice Statistics, www.ojp.usdoj.gov/bjs.

100. Quoted in A. Humphreys, "Lawyers may use genetics study in rape defense," *National Post* (Canada), January 22, 2000, p. A8.

101. Quoted in Jones, 1999.

102. Paglia, 1990, pp. 51, 57.

103. McElroy, 1996.

104. J. Phillips, "Exploring inside to live on the outside," *Boston Globe,* March 21, 1999.

105. S. Satel, "The patriarchy made me do it," *Women's Freedom Newsletter, 5,* September/October 1998.

Chapter 19: Children

1. Turkheimer, 2000.

2. Goldberg, 1968; Janda, 1998; Neisser et al., 1996.

3. Jensen, 1971.

4. Plomin et al., 2001.

5. Bouchard, 1994; Bouchard et al., 1990; Bouchard, 1998; Loehlin, 1992; Plomin, 1994; Plomin et al., 2001.

6. Plomin et al., 2001.

7. McLearn et al., 1997; Plomin, Owen, & McGuffin, 1994.

8. Bouchard, 1994; Bouchard et al., 1990; Bouchard, 1998; Loehlin, 1992; Lykken et al., 1992; Plomin, 1990; Plomin, 1994; Stromswold, 1998.

9. Plomin et al., 2001.

10. Bouchard et al., 1990; Plomin, 1991; Plomin, 1994; Plomin & Daniels, 1987.

11. Bouchard et al., 1990; Pedersen et al., 1992.

12. Bouchard et al., 1990; Bouchard, 1998.

13. Scarr & Carter-Saltzman, 1979.

14. Loehlin & Nichols, 1976.

15. Bouchard, 1998; Gutknecht, Spitz, & Carlier, 1999.

16. McGue, 1997.

17. Etcoff, 1999; Persico, Postlewaite, & Silverman, 2001.

18. Jackson & Huston, 1975.

19. Bouchard, 1994; Bouchard et al., 1990.

20. Kamin, 1974; Lewontin, Rose, & Kamin, 1984, p. 116.

21. Neisser et al., 1996; Snyderman & Rothman, 1988.

22. Hunt, 1999, pp. 50–51.

23. Plomin & Daniels, 1987; Plomin et al., 2001.

24. Bouchard, 1994; Harris, 1998a; Plomin & Daniels, 1987; Rowe, 1994; Turkheimer & Waldron, 2000. An example of a nonreplicated finding is the recent claim by Krueger, Hicks, & McGue, 2001, that altruism is affected by the shared environment, which is contradicted by a study by Rushton et al., 1986, which used similar methods and a larger sample.

25. Stoolmiller, 2000.

26. Bouchard et al., 1990; Plomin & Daniels, 1987; Reiss et al., 2000; Rowe, 1994.

27. Plomin, 1991; Plomin & Daniels, 1987, p. 6; Plomin et al., 2001.

28. Bouchard, 1994; Plomin & Daniels, 1987; Rowe, 1994; Turkheimer, 2000; Turkheimer & Waldron, 2000.

29. Schütze, 1987.

30. B. Singer, "How to raise a perfect child . . . ," *Boston Globe Magazine,* March 26, 2000, pp. 12–36.

31. D. Barry, "Is your kid's new best friend named 'Bessie'? Be very afraid," *Miami Herald,* October 31, 1999.

32. Harris, 1998a, chap. 2; Lytton, 1990.

33. Harris, 1998a, chap. 4; Harris, 2000b.

34. Harris, 1998a, pp. 319–320, 323.

35. Harris, 1998a; Harris, 1998b; Harris, 2000a; Harris, 2000b.

36. Harris, 1998a, chaps. 2, 3; Maccoby & Martin, 1983.

37. Harris, 1998a, pp. 300–311.

38. Bruer, 1999, p. 5.

39. Chabris, 1999.

40. T. B. Brazelton, "To curb teenage smoking, nurture children in their earliest years," *Boston Globe,* May 21, 1998.

41. Bruer, 1999.

42. Collins et al., 2000; Vandell, 2000.

43. Harris, 1995; Harris, 1998b; Harris, 2000b; Loehlin, 2001; Rowe, 2001.

44. Plomin, DeFries, & Fulker, 1988; Reiss et al., 2000; Turkheimer & Waldron, 2000.

45. D. Reiss, quoted in A. M. Paul, "Kid stuff: Do parents really matter?" *Psychology Today,* January/February 1998, pp. 46–49, 78.

46. Sulloway, 1996.

47. Sulloway, 1995.

48. Harris, 1998a, appendix 1; Harris, in press.

49. Hrdy, 1999.

50. Dunphy, 1963.

51. Pinker, 1994, chaps. 2, 9.

52. Kosof, 1996.

53. Harris, 1998a, chaps. 9, 12, 13.

54. Harris, 1998a, p. 264.

55. Harris, 1998a, chap. 13; Rowe, 1994; Rutter, 1997.

56. Gottfredson & Hirschi, 1990; Harris, 1998a, chap. 13.

57. Harris, 1998a, chap. 8.

58. M. Wertheim, "Mindfield" (Review of S. Pinker's *How the mind works*), *The Australian's Review of Books,* 1998.

59. O. James, "It's a free market on the nature of nurture," *The Independent,* October 20, 1998.

60. www.philipmorrisusa.com/DisplayPage WithTopic.asp?ID=189. See also Anheuser-Busch's www.beeresponsible.com/ftad/review. html.

61. J. Leo, "Parenting without a care," *US News and World Report,* September 21, 1998.

62. Quoted in J. Leo, "Parenting without a care," *US News and World Report,* September 21, 1998.

63. S. Begley, "The parent trap," *Newsweek,* September 7, 1998, p. 54.

64. S. Begley, "The parent trap," *Newsweek,* September 7, 1998, p. 54.

65. J. Kagan, "A parent's influence is peerless," *Boston Globe,* September 13, 1998, p. E3.

66. Harris, 1998b; Harris, 2000a; Harris, 2000b; Loehlin, 2001; Rowe, 2001.

67. See also Miller, 1997.

68. Austad, 2000; Finch & Kirkwood, 2000.

69. Hartman, Garvik, & Hartwell, 2001; Waddington, 1957.

70. Harris, 1998a, pp. 78–79.

71. Quoted in B. M. Rubin, "Raising a ruckus being a parent is difficult, but is it necessary?" *Chicago Tribune,* August 31, 1998.

72. Harris, 1998a, p. 291.

73. Harris, 1998a, p. 342.

Chapter 20: The Arts

1. R. Brustein, "The decline of high culture," *New Republic,* November 3, 1997.

2. A. Kernan, Yale University Press, 1992.

3. A. Delbanco, *New York Review of Books,* November 4, 1999.

4. R. Brustein, *New Republic,* November 3, 1997.

5. Conference at the Stanford University Humanities Center, April 23, 1999.

6. G. Steiner, *PN Review, 25,* March–April 1999.

7. J. Engell & A. Dangerfield, *Harvard Magazine,* May–June 1998, pp. 48–55, 111.

8. A. Louch, *Philosophy and Literature, 22,* April 1998, pp. 231–241.

9. C. Woodring, Columbia University Press, 1999.

10. J. M. Ellis, Yale University Press, 1997.

11. G. Wheatcroft, *Prospect,* August/September 1998.

12. R. E. Scholes, Yale University Press, 1998.

13. A. Kernan (Ed.), Princeton University Press, 1997.

14. C. P. Freund, *Reason,* March 1998, pp. 33–38.

15. Quoted in Cowen, 1998, pp. 9–10.

16. J. Engell & A. Dangerfield, "Humanities in the age of money," *Harvard Magazine,* May–June 1998, pp. 48–55, 111.

17. J. Engell & A. Dangerfield, "Humanities in the age of money," *Harvard Magazine,* May–June 1998, pp. 48–55, 111.

18. Cowen, 1998; N. Gillespie, "All culture, all the time," *Reason,* April 1999, pp. 24–35.

19. Cowen, 1998.

20. Quoted in Cowen, 1998, p. 188.

21. Cowen, 1998.

22. Actually, "human character changed," from her essay "Character in Fiction."

23. Crick, 1994; Gardner, 1983; Peretz, Gagnon, & Bouchard, 1998.

24. Miller, 2000a.

25. Dutton, 2001.

26. Dissanayake, 1992; Dissanayake, 2000.

27. Pinker, 1997, chap. 8.

28. Marr, 1982; Pinker, 1997, chap. 8; Ramachandran & Hirstein, 1999; Shepard, 1990. See also Gombrich, 1982/1995; Miller, 2001.

29. Pinker, 1997, chap. 8.

30. Kaplan, 1992; Orians, 1998; Orians & Heerwgen, 1992; Wilson, 1984.

31. Wilson, 1984.

32. Etcoff, 1999; Symons, 1995; Thornhill, 1998.

33. Tooby & DeVore, 1987.

34. Abbott, 2001; Pinker, 1997.

35. Dissanayake, 1998.

36. Dissanayake, 1992.

37. Frank, 1999; Veblen, 1899/1994.

38. Zahavi & Zahavi, 1997.

39. Miller, 2000a, p. 270.

40. Bell, 1992; Wolfe, 1975; Wolfe, 1981.

41. Bourdieu, 1984.

42. From his 1757 essay "Of the standard of taste," quoted in Dutton, 2001, p. 206.

43. Dutton, 2001, p. 213.

44. Dutton, 1998; Komar, Melamid, & Wypijewski, 1997.

45. Dissanayake, 1998.

46. Dutton, 1998.

47. *Lingua Franca,* 2000.

48. Turner, 1997, pp. 170, 174–175.

49. Etcoff, 1999; Kaplan, 1992; Orians & Heerwgen, 1992.

50. Leslie, 1994; Schellenberg & Trehub, 1996; Storey, 1996; Zentner & Kagan, 1996.

51. Martindale, 1990.

52. Steiner, 2001.

53. Quoted in Dutton, 2000.

54. C. Darwent, "Art of staying pretty," *New Statesman,* February 13, 2000.

55. Steiner, 2001.

56. Bell, 1992.

57. *The Onion, 36,* September 21–27, 2000, p. 1.

58. Wolfe, 1975, pp. 2–4.

59. J. Miller, "Is bad writing necessary? George Orwell, Theodor Adorno, and the politics of language," *Lingua Franca*, December/January, 2000.

60. www.cybereditions.com/aldaily/bwc.htm.

61. Steiner, 1967, preface.

62. *New York Times,* September 19, 2001.

63. By the sculptor Janine Antoni; G. Beauchamp, "Dissing the middle class: The view from Burns Park," *American Scholar,* Summer 1995, pp. 335–349.

64. K. Limaye, "Adieu to the Avant-Garde," *Reason,* July 1997.

65. K. Limaye, "Adieu to the Avant-Garde," *Reason,* July 1997.

66. C. Darwent, "Art of staying pretty," *New Statesman,* February 13, 2000; C. Lambert, "The stirring of sleeping beauty," *Harvard Magazine,* September–October 1999, pp. 46–53; K. Limaye, "Adieu to the Avant-Garde," *Reason,* July 1997; A. Delbanco, "The decline and fall of literature," *New York Review of Books,* November 4, 1999; Perloff, 1999; Turner, 1985; Turner, 1995.

67. Abbott, 2001; Boyd, 1998; Carroll, 1995; Dutton, 2001; Easterlin, Riebling, & Crews, 1993; Evans, 1998; Gottschall & Jobling, in preparation; Hernadi, 2001; Hogan, 1997; Steiner, 2001; Turner, 1985; Turner, 1995; Turner, 1996.

68. Goguen, 1999; Gombrich, 1982/1995; Kubovy, 1986.

69. Aiello & Sloboda, 1994; Lerdahl & Jackendoff, 1983.

70. Keyser, 1999; Keyser & Halle, 1998; Turner, 1991; Turner, 1996; Williams, 1990.

71. Scarry, 1999.

72. Abbott, 2001.

73. A. Quart, "David Bordwell blows the whistle on film studies," *Lingua Franca,* March 2000, pp. 35–43.

74. Abbott, 2001; Aiken, 1998; Cooke & Turner, 1999; Dissanayake, 1992; Etcoff, 1999; Kaplan, 1992; Orians & Heerwgen, 1992; Thornhill, 1998.

75. Teuber, 1997.

76. Behrens, 1998.

77. Quoted in Storey, 1996, p. 182.

78. A. S. Byatt, "Narrate or die," *New York Times Magazine,* April 18, 1999, pp. 105–107.

79. John Updike, "The tried and the trëowe," *Forbes ASAP,* October 2, 2000, pp. 201, 215.

80. Storey, 1996, p. 114.

PART VI: THE VOICE OF THE SPECIES

1. Degler, 1991, p. 135.

2. Dickinson, 1976.

3. Vonnegut, 1968/1998.

4. Orwell, 1949/1983, p. 205.

5. For example, Gould, 1981; Lewontin, Rose, & Kamin, 1984, pp. ix–x.

6. Orwell, 1949/1983, p. 217.

7. Orwell, 1949/1983, p. 220.

8. Orwell, 1949/1983, p. 220.

9. Orwell, 1949/1983, p. 222.

10. Twain, 1884/1983, pp. 293–295.

11. Twain, 1884/1983, p. 295.

12. Twain, 1884/1983, pp. 330–331.

13. Twain, 1884/1983, p. 332.

14. Twain, 1884/1983, p. 339.

15. Singer, 1972.

16. The dialogue is condensed from Singer, 1972, pp. 68–78, and from the film adaptation.

REFERENCES

Abbott, H. P. E. 2001. Imagination and the adapted mind: A special double issue. *SubStance*, 30.

Adams, B., Breazeal, C., Brooks, R. A., & Scassellati, B. 2000. Humanoid robots: A new kind of tool. *IEEE Intelligent Systems*, 25–31.

Agrawal, A. F., Brodie, E. D. I., & Brown, J. 2001. Parent-offspring coadaptation and the dual genetic control of maternal care. *Science, 292,* 1710–1712.

Ahn, W.-K., Kalish, C., Gelman, S. A., Medin, D. L., Luhmann, C., Atran, S., Coley, J. D., & Shafto, P. 2001. Why essences are essential in the psychology of concepts. *Cognition, 82,* 59–69.

Aiello, R., & Sloboda, J. A. (Eds.) 1994. *Musical perceptions.* New York: Oxford University Press.

Aiken, N. E. 1998. *The biological origins of art.* Westport, Conn.: Praeger.

Akerlof, G. A. 1984. *An economic theorist's book of tales: Essays that entertain the consequences of new assumptions in economic theory.* New York: Cambridge University Press.

Alcock, J. 1998. Unpunctuated equilibrium in the *Natural History* essays of Stephen Jay Gould. *Evolution and Human Behavior, 19,* 321–336.

Alcock, J. 2001. *The triumph of sociobiology.* New York: Oxford University Press.

Alexander, R. D. 1987. *The biology of moral systems.* Hawthorne, N.Y.: Aldine de Gruyter

Allen, E., Beckwith, B., Beckwith, J., Chorover, S., Culver, D., Duncan, M., Gould, S. J., Hubbard, R., Inouye, H., Leeds, A., Lewontin, R., Madansky, C., Miller, L., Pyeritz, R., Rosenthal, M., & Schreier, H. 1975. Against "sociobiology." *New York Review of Books, 22,* 43–44.

Allen, G. E. 2001. Is a new eugenics afoot? *Science, 294,* 59–61.

Ames, B., Profet, M., & Gold, L. S. 1990. Dietary pesticides (99.9% all natural). *Proceedings of the National Academy of Sciences, 87,* 7777–7781.

Anderson, J. R. 1976. *Language, memory, and thought.* Mahwah, N.J.: Erlbaum.

Anderson, J. R. 1983. *The architecture of cognition.* Cambridge, Mass.: Harvard University Press.

Anderson, J. R. 1990. *The adaptive character of thought.* Mahwah, N.J.: Erlbaum.

Anderson, J. R. 1993. *Rules of the mind.* Mahwah, N.J.: Erlbaum.

Anderson, J. R. 1995. *Cognitive psychology and its implications* (4th ed.). New York: W. H. Freeman.

Anderson, S. W., Bechara, A., Damasio, H., Tranel, D., & Damasio, A. R. 1999. Impairment of social and moral behavior related to early damage in human prefrontal cortex. *Nature Neuroscience, 2,* 1032–1037.

Andreasen, N. C., Flaum, M., Swayze, V., O'Leary, D. S., Alliger, R., Cohen, G., Ehrhardt, J., & Yuh, W. T. C. 1993. Intelligence and brain structure in normal individuals. *American Journal of Psychiatry, 150,* 130–134.

Antonaccio, M., & Schweiker, W. (Eds.) 1996. *Iris Murdoch and the search for human goodness.* Chicago: University of Chicago Press.

Archibald, W. P. 1989. *Marx and the missing link: Human nature.* Atlantic Highlands, N.J.: Humanities Press International.

Arditi, A., Holtzman, J. D., & Kosslyn, S. M. 1988. Mental imagery and sensory experience in congenital blindness. *Neuropsychologia, 26,* 1–12.

Armstrong, S. L., Gleitman, L. R., & Gleitman, H. 1983. What some concepts might not be. *Cognition, 13,* 263–308.

Aronson, E. 1980. *The social animal.* San Francisco: W. H. Freeman.

Atran, S. 1995. Causal constraints on categories and categorical constraints on biological reasoning across cultures. In D. Sperber, D. Premack, & A. J. Premack (Eds.), *Causal cognition.* New York: Oxford University Press.

Atran, S. 1998. Folk biology and the anthropology of science: Cognitive universals and cultural particulars. *Behavioral and Brain Sciences, 21,* 547–609.

Austad, S. 2000. Varied fates from similar states. *Science, 290,* 944.

Axelrod, R. 1984. *The evolution of cooperation.* New York: Basic Books.

Babcock, L., & Loewenstein, G. 1997. Explaining bargaining impasse: The role of self-serving biases. *Journal of Economic Perspectives, 11,* 109–126.

Baddeley, A. D. 1986. *Working memory.* New York: Oxford University Press.

Bailey, R. 2000. The law of increasing returns. *Public Interest, 59,* 113–121.

Bailyn, B. 1967/1992. *The ideological origins of the American revolution.* Cambridge, Mass.: Harvard University Press.

Baker, M. 2001. *The atoms of language.* New York: Basic Books.

Baldwin, D. A. 1991. Infants' contribution to the achievement of joint reference. *Child Development, 62,* 875–890.

Bamforth, D. B. 1994. Indigenous people, indigenous violence: Precontact warfare on the North American Great Plains. *Man, 29,* 95–115.

Barkley, R. A., Ullman, D. G., Otto, L., & Brecht, J. M. 1977. The effects of sex typing and sex appropriateness of modeled behavior on children's imitation. *Child Development, 48,* 721–725.

Barkow, J. H., Cosmides, L., & Tooby, J. 1992. *The adapted mind: Evolutionary psychology and the generation of culture.* New York: Oxford University Press.

Baron-Cohen, S. 1995. *Mindblindness: An essay on autism and theory of mind.* Cambridge, Mass.: MIT Press.

Barry, D. 1995. *Dave Barry's complete guide to guys.* New York: Ballantine.

Barsky, R. F. 1997. *Noam Chomsky: A life of dissent.* Cambridge, Mass.: MIT Press.

Barthes, R. 1972. To write: An intransitive verb? In R. Macksey & E. Donato (Eds.), *The languages of criticism and the science of man: The structuralist controversy.* Baltimore: Johns Hopkins University Press.

Bauer, R. A. 1952. *The new man in Soviet psychology.* Cambridge, Mass.: Harvard University Press.

Becker, G. S. 1991. *A treatise on the family* (enlarged ed.). Cambridge, Mass.: Harvard University Press.

Behe, M. J. 1996. *Darwin's black box: The biochemical challenge to evolution.* New York: Free Press.

Behrens, R. R. 1998. Art, design, and gestalt theory. *Leonardo, 31,* 299–304.

Belenky, M. F., Clinchy, B. M., Goldberger, N. R., & Tarule, J. M. 1986. *Women's ways of knowing.* New York: Basic Books.

Bell, Q. 1992. *On human finery.* London: Allison & Busby.

Benedict, R. 1934/1959. Anthropology and the abnormal. In M. Mead (Ed.), *An anthropologist at work: Writings of Ruth Benedict.* Boston: Houghton Mifflin.

Benjamin, J., Li, L., Patterson, C., Greenberg, B. D., Murphy, D. L., & Hamer, D. H. 1996. Population and familial association between the D4 dopamine receptor gene and measures of novelty seeking. *Nature Genetics, 12,* 81–84.

Berent, I., Pinker, S., & Shimron, J. 1999. Default nominal inflection in Hebrew: Evidence for mental variables. *Cognition, 72,* 1–44.

Berlin, I. 1996. *The sense of reality: Studies in ideas and their history.* New York: Farrar, Straus, & Giroux.

Berra, T. M. 1990. *Evolution and the myth of creationism.* Stanford, Calif.: Stanford University Press.

Besançon, A. 1981. *The intellectual origins of Leninism.* Oxford: Basil Blackwell.

Besançon, A. 1998. Forgotten communism. *Commentary,* 24–27.

Betzig, L. L. 1997. *Human nature: A critical reader.* New York: Oxford University Press.

Bishop, K. M., Coudreau, G., & O'Leary, D. D. M. 2000. Regulation of area identity in the mammalian neocortex by *Emx2* and *Pax6. Science, 288,* 344–349.

Blair, J., & Cipolotti, L. 2000. Impaired social response reversal: A case of "acquired sociopathy." *Brain, 123,* 1122–1141.

Blinkhorn, S. 1982. Review of S. J. Gould's "The mismeasure of man." *Nature, 296,* 506.

Bloom, P. 1994. Generativity within language and other cognitive domains. *Cognition, 51,* 177–189.

Bloom, P. 1996. Intention, history, and artifact concepts. *Cognition, 60,* 1–29.

Blum, D. 1997. *Sex on the brain: The biological differences between men and women*. New York: Viking.

Boas, F. 1911. Language and thought. In *Handbook of American Indian languages*. Lincoln, Nebr.: Bison Books.

Bock, G. R., & Goode, J. A. (Eds.) 1996. *The genetics of criminal and antisocial behavior*. New York: Wiley.

Bodmer, W. F., & Cavalli-Sforza, L. L. 1970. Intelligence and race. *Scientific American*.

Boehm, C. 1999. *Hierarchy in the forest: The evolution of egalitarian behavior*. Cambridge, Mass.: Harvard University Press.

Borges, J. L. 1964. The lottery in Babylon. In *Labyrinths: Selected stories and other writings*. New York: New Directions.

Bouchard, T. J., Jr. 1994. Genes, environment, and personality. *Science, 264,* 1700–1701.

Bouchard, T. J., Jr. 1998. Genetic and environmental influences on intelligence and special mental abilities. *Human Biology, 70,* 257–259.

Bouchard, T. J., Jr., Lykken, D. T., McGue, M., Segal, N. L., & Tellegen, A. 1990. Sources of human psychological differences: The Minnesota Study of Twins Reared Apart. *Science, 250,* 223–228.

Bourdieu, P. 1984. *Distinction: A social critique of the judgment of taste*. Cambridge, Mass.: Harvard University Press.

Bourgeois, J.-P., Goldman-Rakic, P. S., & Rakic, P. 2000. Formation, elimination, and stabilization of synapses in the primate cerebral cortex. In M. S. Gazzaniga (Ed.), *The new cognitive neurosciences*. Cambridge, Mass.: MIT Press.

Bourke, J. 1999. *An intimate history of killing: Face-to-face killing in 20th-century warfare*. New York: Basic Books.

Bowles, S., & Gintis, H. 1998. Is equality passé? Homo reciprocans and the future of egalitarian politics. *Boston Review*.

Bowles, S., & Gintis, H. 1999. *Recasting egalitarianism: New rules for communities, states, and markets*. New York: Verso.

Boyd, B. 1998. Jane, meet Charles: Literature, evolution, and human nature. *Philosophy and Literature, 22,* 1–30.

Boyd, R., & Richerson, P. 1985. *Culture and the evolutionary process*. Chicago: University of Chicago Press.

Boyd, R., & Silk, J. R. 1996. *How humans evolved*. New York: Norton.

Boyer, P. 1994. Cognitive constraints on cultural representations: Natural ontologies and religious ideas. In L. A. Hirschfeld & S. A. Gelman (Eds.), *Mapping the mind: Domain specificity in cognition and culture*. New York: Cambridge University Press.

Braine, M. D. S. 1994. Mental logic and how to discover it. In J. Macnamara & G. Reyes (Eds.), *The logical foundations of cognition*. New York: Oxford University Press.

Bregman, A. S. 1990. *Auditory scene analysis: The perceptual organization of sound*. Cambridge, Mass.: MIT Press.

Bregman, A. S., & Pinker, S. 1978. Auditory streaming and the building of timbre. *Canadian Journal of Psychology, 32,* 19–31.

Breland, K., & Breland, M. 1961. The misbehavior of organisms. *American Psychologist, 16,* 681–684.

Brink, D. O. 1989. *Moral realism and the foundations of ethics*. New York: Cambridge University Press.

Brociner, K. 2001. Utopianism, human nature, and the left. *Dissent,* 89–92.

Brock, D. W. 1993. *Life and death: Philosophical essays in biomedical ethics*. New York: Cambridge University Press.

Brooker, P. 1999. *A concise glossary of cultural theory*. New York: Oxford University Press.

Brown, D. E. 1991. *Human universals*. New York: McGraw-Hill.

Brown, D. E. 2000. Human universals and their implications. In N. Roughley (Ed.), *Being humans: Anthropological universality and particularity in transdisciplinary perspectives*. New York: Walter de Gruyter.

Brown, R. 1985. *Social psychology: The second edition*. New York: Free Press.

Browne, K. 1998. *Divided labors: An evolutionary view of women at work*. London: Weidenfeld and Nicholson.

Brownmiller, S. 1975. *Against our will: Men, women, and rape*. New York: Fawcett Columbine.

Brownmiller, S., & Merhof, B. 1992. A feminist response to rape as an adaptation in men. *Behavioral and Brain Sciences, 15,* 381–382.

Bruer, J. 1997. Education and the brain: A bridge too far. *Educational Researcher, 26,* 4–16.

Bruer, J. 1999. *The myth of the first three years: A new understanding of brain development and life-long learning.* New York: Free Press.

Brugger, P., Kollias, S. S., Müri, R. M., Crelier, G., Hepp-Reymond, M.-C., & Regard, M. 2000. Beyond re-membering: Phantom sensations of congenitally absent limbs. *Proceedings of the National Academy of Science, 97,* 6167–6172.

Bueno de Mesquita, B. 1981. *The war trap.* New Haven, Conn.: Yale University Press.

Bullock, A. 1991. *Hitler and Stalin: Parallel lives.* London: HarperCollins.

Burke, E. 1790/1967. *Reflections on the revolution in France.* London: J. M. Dent & Sons.

Burnham, R., & Phelan, J. 2000. *Mean genes: From sex to money to food: Taming our primal instincts.* Cambridge, Mass.: Perseus.

Burnstein, E., Crandall, C., & Kitayama, S. 1994. Some neo-Darwinian decision rules for altruism: Weighing cues for inclusive fitness as a function of the biological importance of the decision. *Journal of Personality and Social Psychology, 67,* 773–789.

Buss, D., & Duntley, J. D. In press. Why the mind is designed for murder: The coevolution of killing and death prevention strategies. *Behavioral and Brain Sciences.*

Buss, D. M. 1992. Mate preference mechanisms: Consequences for partner choice and intrasexual competition. In J. Barkow, L. Cosmides, & J. Tooby (Eds.), *The adapted mind: Evolutionary psychology and the generation of culture.* New York: Oxford University Press.

Buss, D. M. 1994. *The evolution of desire.* New York: Basic Books.

Buss, D. M. 1995. Evolutionary psychology: A new paradigm for psychological science. *Psychological Inquiry, 6,* 1–30.

Buss, D. M. 1999. *Evolutionary psychology: The new science of the mind.* Boston: Allyn and Bacon.

Buss, D. M. 2000. *The dangerous passion: Why jealousy is as necessary as love and sex.* New York: Free Press.

Butterworth, B. 1999. *The mathematical brain.* London: Macmillan.

Calvin, W. H. 1996a. *The cerebral code.* Cambridge, Mass.: MIT Press.

Calvin, W. H. 1996b. *How brains think.* New York: Basic Books.

Calvin, W. H., & Bickerton, D. 2000. *Lingua ex machina: Reconciling Darwin and Chomsky with the human brain.* Cambridge, Mass.: MIT Press.

Calvin, W. H., & Ojemann, G. A. 2001. *Inside the brain: Mapping the cortex, exploring the neuron,* www.iuniverse.com.

Campbell, J. D., & Fairey, P. J. 1989. Informational and normative routes to conformity: The effect of faction size as a function of norm extremity and attention to the stimulus. *Journal of Personality and Social Psychology, 51,* 315–324.

Cappon, L. J. (Ed.) 1959. *The Adams-Jefferson letters.* New York: Simon & Schuster.

Caramazza, A., & Shelton, J. A. 1998. Domain-specific knowledge systems in the brain: The animate-inanimate distinction. *Journal of Cognitive Neuroscience, 10,* 1–34.

Carey, S. 1986. Cognitive science and science education. *American Psychologist, 41,* 1123–1130.

Carey, S., & Spelke, E. 1994. Domain-specific knowledge and conceptual change. In L. A. Hirschfeld & S. A. Gelman (Eds.), *Mapping the mind: Domain specificity in cognition and culture.* New York: Cambridge University Press.

Carpenter, M., Akhtar, N., & Tomasello, M. 1998. Fourteen- through eighteen-month-old infants differentially imitate intentional and accidental actions. *Infant Behavior and Development, 21,* 315–330.

Carroll, J. 1995. *Evolution and literary theory.* Columbia: University of Missouri Press.

Cartmill, M. 1998. Oppressed by evolution. *Discover, 19,* 78–83.

Cartwright, J. 2000. *Evolution and human behavior.* Cambridge, Mass.: MIT Press.

Caryl, P. G. 1994. Early event-related potentials correlate with inspection time and intelligence. *Intelligence, 18,* 15–46.

Cashdan, E. 1989. Hunters and gatherers: Economic behavior in bands. In S. Plattner (Ed.), *Economic anthropology.* Stanford, Calif.: Stanford University Press.

Caspi, A. 2000. The child is father of the man: Personality continuities from childhood to adulthood. *Journal of Personality and Social Psychology, 78,* 158–172.

Catalano, S. M., & Shatz, C. J. 1998. Activity-dependent cortical target selection by thalamic axons. *Science, 24,* 559–562.

Cavalli-Sforza, L. L. 1991. Genes, people, and languages. *Scientific American, 223,* 104–110.

Cavalli-Sforza, L. L., & Feldman, M. W. 1981. *Cultural transmission and evolution: A quantitative approach.* Princeton, N.J.: Princeton University Press.

Chabris, C. F. 1999. Prelude or requiem for the "Mozart effect"? *Nature, 400,* 826–828.

Chagnon, N. A. 1988. Life histories, blood revenge, and warfare in a tribal population. *Science, 239,* 985–992.

Chagnon, N. A. 1992. *Yanomamö: The last days of Eden.* New York: Harcourt Brace.

Chagnon, N. A. 1996. Chronic problems in understanding tribal violence and warfare. In G. Bock & J. Goode (Eds.), *The genetics of criminal and antisocial behavior.* New York: Wiley.

Chalupa, L. M. 2000. A comparative perspective on the formation of retinal connections in the mammalian brain. In M. S. Gazzaniga (Ed.), *The new cognitive neurosciences.* Cambridge, Mass.: MIT Press.

Chandler, D. P. 1999. *Brother number one: A political biography of Pol Pot.* Boulder, Colo.: Westview Press.

Charlesworth, B. 1987. The heritability of fitness. In J. W. Bradbury & M. B. Andersson (Eds.), *Sexual selection: Testing the hypotheses.* New York: Wiley.

Charlton, T. 1997. The inception of broadcast television: A naturalistic study of television's effects in St. Helena, South Atlantic. In T. Charlton & K. David (Eds.), *Elusive links: Television, video games, and children's behavior.* Cheltenham, U.K.: Park Published Papers.

Chase, W. G., & Simon, H. A. 1973. Perception in chess. *Cognitive Psychology, 4,* 55–81.

Check, J. V. P., & Malamuth, N. 1985. An empirical assessment of some feminist hypotheses about rape. *International Journal of Women's Studies, 8,* 414–423.

Chirot, D. 1994. *Modern tyrants.* Princeton, N.J.: Princeton University Press.

Chomsky, N. 1970. Language and freedom. *Abraxas, 1,* 9–24.

Chomsky, N. 1973. Psychology and ideology. In N. Chomsky (Ed.), *For reasons of state.* New York: Vintage.

Chomsky, N. 1975. *Reflections on language.* New York: Pantheon.

Chomsky, N. 1980. *Rules and representations.* New York: Columbia University Press.

Chomsky, N. 1988a. *Language and politics.* Montreal: Black Rose Books.

Chomsky, N. 1988b. *Language and problems of knowledge: The Managua lectures.* Cambridge, Mass.: MIT Press.

Chomsky, N. 1993. *Language and thought.* Wakefield, R.I.: Moyer Bell.

Chomsky, N. 2000. *New horizons in the study of language and mind.* New York: Cambridge University Press.

Chorney, M. J., Chorney, K., Seese, N., Owen, M. J., McGuffin, P., Daniels, J., Thompson, L. A., Detterman, D. K., Benbow, C. P., Lubinski, D., Eley, T. C., & Plomin, R. 1998. A quantitative trait locus (QTL) associated with cognitive ability in children. *Psychological Science, 9,* 159–166.

Chorover, S. L., 1979. *From genesis to genocide: The meaning of human nature and the power of behavior control.* Cambridge, Mass.: MIT Press.

Clahsen, H. 1999. Lexical entries and rules of language: A multidisciplinary study of German inflection. *Behavioral and Brain Sciences, 22,* 991–1013.

Clark, R. 1970. *Crime in America: Observations on its nature, causes, prevention, and control.* New York: Simon & Schuster.

Claverie, J.-M. 2001. What if there are only 30,000 human genes? *Science, 291,* 1255–1257.

Cohen, J. 1997. The natural goodness of humanity. In A. Reath, B. Herman, & C. Korsgaard (Eds.), *Reclaiming the history of ethics: Essays for John Rawls.* New York: Cambridge University Press.

Colapinto, J. 2000. *As nature made him: The boy who was raised as a girl.* New York: HarperCollins.

Collins, W. A., Maccoby, E. E., Steinberg, L., Hetherington, E. M., & Bornstein, M. H. 2000. Contemporary research on parenting: The case for nature *and* nurture. *American Psychologist, 55,* 218–232.

Conquest, R. 2000. *Reflections on a ravaged century.* New York: Norton.

Cooke, B., & Turner, F. (Eds.) 1999. *Biopoetics: Evolutionary explorations in the arts.* St. Paul, Minn.: Paragon House.

Cosmides, L., & Tooby, J. 1992. Cognitive adaptations for social exchange. In J. H. Barkow, L. Cosmides, & J. Tooby (Eds.), *The adapted mind: Evolutionary psychology and the generation of culture.* New York: Oxford University Press.

Cosmides, L., & Tooby, J. 1996. Are humans good intuitive statisticians after all? Rethinking some conclusions from the literature on judgment under uncertainty. *Cognition, 58,* 1–73.

Courtois, S., Werth, N., Panné, J.-L., Paczkowski, A., Bartošek, K., & Margolin, J.-L. 1999. *The black book of communism: Crimes, terror, repression.* Cambridge, Mass.: Harvard University Press.

Cowen, T. 1998. *In praise of commercial culture.* Cambridge, Mass.: Harvard University Press.

Cowie, F. 1999. *What's within? Nativism reconsidered.* New York: Oxford University Press.

Crair, M. C., Gillespie, D. C., & Stryker, M. P. 1998. The role of visual experience in the development of columns in cat visual cortex. *Science, 279,* 566–570.

Cramer, K. S., & Sur, M. 1995. Activity-dependent remodeling of connections in the mammalian visual system. *Current Opinion in Neurobiology, 5,* 106–111.

Crawford, C., & Krebs, D. L. (Eds.) 1998. *Handbook of evolutionary psychology: Ideas, issues, and applications.* Mahwah, N.J.: Erlbaum.

Crevier, D. 1993. *AI: The tumultuous history of the search for artificial intelligence.* New York: Basic Books.

Crews, F. 2001. Saving us from Darwin. *New York Review of Books,* October 4 and October 18.

Crick, F. 1994. *The astonishing hypothesis: The scientific search for the soul.* New York: Simon & Schuster.

Crick, F., & Koch, C. 1995. Are we aware of neural activity in primary visual cortex? *Nature, 375,* 121–123.

Crittenden, D. 1999. *What our mothers didn't tell us: Why happiness eludes the modern woman.* New York: Simon & Schuster.

Cronin, H. 1992. *The ant and the peacock.* New York: Cambridge University Press.

Cronk, L. 1999. *That complex whole: Culture and the evolution of human behavior.* Boulder, Colo.: Westview Press.

Cronk, L., Chagnon, N., & Irons, W. (Eds.) 2000. *Adaptation and human behavior.* Hawthorne, N.Y.: Aldine de Gruyter.

Crow, J. F. 2002. Unequal by nature: A geneticist's perspective on human differences. *Daedalus,* Winter, 81–88.

Crowley, J. C., & Katz, L. C. 2000. Early development of ocular dominance columns. *Science, 290,* 1321–1324.

Cummins, D. D. 1996. Evidence for the innateness of deontic reasoning. *Mind and Language, 11,* 160–190.

Curti, M. 1980. *Human nature in American thought: A history.* Madison: University of Wisconsin Press.

Curtiss, S., de Bode, S., & Shields, S. 2000. Language after hemispherectomy. In J. Gilkerson, M. Becker, & N. Hyams (Eds.), *UCLA Working Papers in Linguistics* (Vol. 5, pp. 91–112). Los Angeles: UCLA Department of Linguistics.

Dabbs, J. M., & Dabbs, M. G. 2000. *Heroes, rogues, and lovers: Testosterone and behavior.* New York: McGraw-Hill.

Daly, M. 1991. Natural selection doesn't have goals, but it's the reason organisms do (Commentary on P. J. H. Shoemaker, "The quest for optimality: A positive heuristic of science?"). *Behavioral and Brain Sciences, 14,* 219–220.

Daly, M., Salmon, C., & Wilson, M. 1997. Kinship: The conceptual hole in psychological studies of social cognition and close relationships. In J. Simpson & D. Kenrick (Eds.), *Evolutionary social psychology.* Mahwah, N.J.: Erlbaum.

Daly, M., & Wilson, M. 1983. *Sex, evolution, and behavior* (2nd ed.). Belmont, Calif.: Wadsworth.

Daly, M., & Wilson, M. 1988. *Homicide.* Hawthorne, N.Y.: Aldine de Gruyter.

Daly, M., & Wilson, M. 1994. Evolutionary psychology of male violence. In J. Archer (Ed.), *Male violence.* London: Routledge.

Daly, M., & Wilson, M. 1999. *The truth about Cinderella: A Darwinian view of parental love.* New Haven, Conn.: Yale University Press.

Daly, M., Wilson, M., & Vasdev, S. 2001. Income inequality and homicide rates in Canada and the United States. *Canadian Journal of Criminology, 43,* 219–236.

Damasio, A. R. 1994. *Descartes' error: Emotion, reason, and the human brain.* New York: Putnam.

Damasio, H. 2000. The lesion method in cognitive neuroscience. In F. Boller & J. Grafman (Eds.), *Handbook of neuropsychology* (2nd ed.), Vol. 1. New York: Elsevier.

Damewood, M. D. 2001. Ethical implications of a new application of preimplantation diagnosis. *Journal of the American Medical Association, 285,* 3143–3144.

Darwin, C. 1872/1998. *The expression of the emotions in man and animals: Definitive edition.* New York: Oxford University Press.

Davidson, R. J., Putnam, K. M., & Larson, C. L. 2000. Dysfunction in the neural circuitry of emotion regulation: A possible prelude to violence. *Science, 289,* 591–594.

Davis, B. D. 1983. Neo-Lysenkoism, IQ, and the press. *Public Interest, 73,* 41–59.

Dawkins, R. 1976/1989. *The selfish gene* (new ed.). New York: Oxford University Press.

Dawkins, R. 1983. Universal Darwinism. In D. S. Bendall (Ed.), *Evolution from molecules to man.* New York: Cambridge University Press.

Dawkins, R. 1985. Sociobiology: The debate continues (Review of Lewontin, Rose, & Kamin's "Not in our genes"). *New Scientist, 24,* 59–60.

Dawkins, R. 1986. *The blind watchmaker: Why the evidence of evolution reveals a universe without design.* New York: Norton.

Dawkins, R. 1998. *Unweaving the rainbow: Science, delusion and the appetite for wonder.* Boston: Houghton Mifflin.

de Waal, F. 1998. *Chimpanzee politics: Power and sex among the apes.* Baltimore: Johns Hopkins University Press.

Deacon, T. 1997. *The symbolic species: The coevolution of language and the brain.* New York: Norton.

Deary, J. J. 2000. *Looking down on human intelligence: From psychometrics to the brain.* New York: Oxford University Press.

Deater-Deckard, K., & Plomin, R. 1999. An adoption study of the etiology of teacher and parent reports of externalising behavior problems in middle childhood. *Child Development, 70,* 144–154.

Degler, C. N. 1991. *In search of human nature: The decline and revival of Darwinism in American social thought.* New York: Oxford University Press.

Dehaene, S. 1997. *The number sense: How the mind creates mathematics.* New York: Oxford University Press.

Dehaene, S., Spelke, L., Pinel, P., Stanescu, R., & Tsivkin, S. 1999. Sources of mathematical thinking: Behavioral and brain-imaging evidence. *Science, 284,* 970–974.

Denfeld, R. 1995. *The new Victorians: A young woman's challenge to the old feminist order.* New York: Warner Books.

Dennett, D. C. 1984. *Elbow room: The varieties of free will worth wanting.* Cambridge, Mass.: MIT Press.

Dennett, D. C. 1986. The logical geography of computational approaches: A view from the East Pole. In M. Harnish & M. Brand (Eds.), *The representation of knowledge and belief.* Tucson: University of Arizona Press.

Dennett, D. C. 1991. *Consciousness explained.* Boston: Little, Brown.

Dennett, D. C. 1995. *Darwin's dangerous idea: Evolution and the meanings of life.* New York: Simon & Schuster.

Dershowitz, A. M. 1994. *The abuse excuse.* Boston: Little, Brown.

Descartes, R. 1637/2001. *Discourse on method.* New York: Bartleby.com.

Descartes, R. 1641/1967. Meditations on first philosophy. In R. Popkin (Ed.), *The philosophy of the 16th and 17th centuries.* New York: Free Press.

Deutsch, M., & Gerard, G. B. 1955. A study of normative and informational social influence upon individual judgment. *Journal of Abnormal and Social Psychology, 51,* 629–636.

Devlin, K. 2000. *The math gene: How mathematical thinking evolved and why numbers are like gossip.* New York: Basic Books.

Diamond, J. 1992. *The third chimpanzee: The evolution and future of the human animal.* New York: HarperCollins.

Diamond, J. 1997. *Guns, germs, and steel: The fates of human societies.* New York: Norton.

Diamond, J. 1998. *Why is sex fun? The evolution of human sexuality.* New York: Basic Books.

Diamond, M., & Sigmundson, K. 1997. Sex reassignment at birth: Long-term review and clinical implications. *Archives of Pediatric and Adolescent Medicine, 151,* 298–304.

Dickinson, E. 1976. *The complete poems of Emily Dickinson.* New York: Little, Brown.

Dissanayake, E. 1992. *Homo aestheticus: Where art comes from and why.* New York: Free Press.

Dissanayake, E. 1998. Komar and Melamid discover Pleistocene taste. *Philosophy and Literature, 22,* 486–496.

Dissanayake, E. 2000. *Art and intimacy: How the arts began.* Seattle: University of Washington Press.

Divale, W. T. 1972. System population control in the middle and upper Paleolithic: Inferences based on contemporary hunter-gatherers. *World Archaeology, 4,* 222–243.

Dorit, R. 1997. Review of Michael Behe's "Darwin's black box." *American Scientist, 85,* 474–475.

Drake, S. 1970. *Galileo studies: Personality, tradition, and revolution.* Ann Arbor: University of Michigan Press.

Dugatkin, L. 1992. The evolution of the con artist. *Ethology and Sociobiology, 13,* 3–18.

Dunbar, R. 1998. *Grooming, gossip, and the evolution of language.* Cambridge, Mass.: Harvard University Press.

Dunn, J., & Plomin, R. 1990. *Separate lives: Why siblings are so different.* New York: Basic Books.

Dunphy, D. 1963. The social structure of early adolescent peer groups. *Sociometry, 26,* 230–246.

Durham, W. H. 1982. Interactions of genetic and cultural evolution: Models and examples. *Human Ecology, 10,* 299–334.

Durkheim, E. 1895/1962. *The rules of the sociological method.* Glencoe, Ill.: Free Press.

Dutton, D. 1998. America's most wanted, and why no one wants it. *Philosophy and Literature, 22,* 530–543.

Dutton, D. 2000. Mad about flowers. *Philosophy and Literature, 24,* 249–260.

Dutton, D. 2001. Aesthetic universals. In B. Gaut & D. M. Lopes (Eds.), *The Routledge companion to aesthetics.* New York: Routledge.

Dworkin, A. 1993. Sexual economics: The terrible truth. In *Letters from a war-zone.* New York: Lawrence Hill.

Eagly, A. H. 1995. The science and politics of comparing women and men. *American Psychologist, 50,* 145–158.

Easterlin, N., Riebling, B., & Crews, F. 1993. *After poststructuralism: Interdisciplinarity and literary theory (rethinking theory).* Evanston, Ill.: Northwestern University Press.

Eaves, L. J., Eysenck, H. J., & Martin, N. G. 1989. *Genes, culture, and personality: An empirical approach.* San Diego: Academic Press.

Edgerton, R. B. 1992. *Sick societies: Challenging the myth of primitive harmony.* New York: Free Press.

Eibl-Eibesfeldt, I. 1989. *Human ethology.* Hawthorne, N.Y.: Aldine de Gruyter.

Ekman, P. 1987. A life's pursuit. In T. A. Sebeok & J. Umiker-Sebeok (Eds.), *The semiotic web 86: An international yearbook.* Berlin: Mouton de Gruyter.

Ekman, P. 1998. Afterword: Universality of emotional expression? A personal history of the dispute. In C. Darwin, *The expression of the emotions in man and animals: Definitive edition.* New York: Oxford University Press.

Ekman, P., & Davidson, R. J. 1994. *The nature of emotion.* New York: Oxford University Press.

Ellis, B. J. 1992. The evolution of sexual attraction: Evaluative mechanisms in women. In J. H. Barkow, L. Cosmides, & J. Tooby (Eds.), *The adapted mind: Evolutionary psychology and the generation of culture.* New York: Oxford University Press.

Ellis, L., & Beattie, C. 1983. The feminist explanation for rape: An empirical test. *Journal of Sex Research, 19,* 74–91.

Elman, J. L., Bates, E. A., Johnson, M. H., Karmiloff-Smith, A., Parisi, D., & Plunkett, K. 1996. *Rethinking innateness: A connectionist perspective on development.* Cambridge, Mass.: MIT Press.

Ember, C. 1978. Myths about hunter-gatherers. *Ethnology, 27,* 239–248.

Epstein, J. 1994. On the mathematical biology of arms races, wars, and revolutions. In L. Nadel & D. Stein (Eds.), *1992 lectures in complex systems,* Vol. 5. Reading, Mass.: Addison Wesley.

Epstein, J., & Axtell, R. L. 1996. *Growing artificial societies: Social science from the bottom up.* Cambridge, Mass.: MIT Press.

Erikson, M. A., & Kruschke, J. K., 1998. Rules and exemplars in category learning. *Journal of Experimental Psychology: General, 127,* 107–140.

Estrich, S. 2000. *Sex and power.* New York: Riverhead Press.

Etcoff, N. L. 1999. *Survival of the prettiest: The science of beauty.* New York: Doubleday.

Evans, D., & Zarate, O. 1999. *Introducing evolutionary psychology.* New York: Totem Books.

Evans, D. A. 1998. Evolution and literature. *South Dakota Review, 36,* 33–46.

Faigman, D. L. 1999. *Legal alchemy: The use and misuse of science in the law.* New York: W. H. Freeman.

Farah, M. J., Rabinowitz, C., Quinn, G. E., & Liu, G. T. 2000. Early commitment of neural substrates for face recognition. *Cognitive Neuropsychology, 17,* 117–123.

Fausto-Sterling, A. 1985. *Myths of gender: Biological theories about women and men.* New York: Basic Books.

Fehr, E., Fischbacher, U., & Gächter, S. In press. Strong reciprocity, human cooperation and the enforcement of social norms. *Human Nature.*

Fehr, E., & Gächter, S. 2000. Fairness and retaliation: The economics of reciprocity. *Journal of Economic Perspectives, 14,* 159–181.

Fernández-Jalvo, Y., Diez, J. C., Bermúdez de Castro, J. M., Carbonell, E., & Arsuaga, J. L. 1996. Evidence of early cannibalism. *Science, 271,* 277–278.

Festinger, L. 1957. *A theory of cognitive dissonance.* Stanford, Calif.: Stanford University Press.

Finch, C. E., & Kirkwood, T. B. L. 2000. *Chance, development, and aging.* New York: Oxford University Press.

Fischoff, S. 1999. Psychology's quixotic quest for the media-violence connection. *Journal of Media Psychology, 4.*

Fisher, S. E., Vargha-Khadem, F., Watkins, K. E., Monaco, A. P., & Pembrey, M. E. 1998. Localisation of a gene implicated in a severe speech and language disorder. *Nature Genetics, 18,* 168–170.

Fiske, A. P. 1992. The four elementary forms of sociality: Framework for a unified theory of social relations. *Psychological Review, 99,* 689–723.

Flynn, J. R. 1999. Searching for justice: The discovery of IQ gains over time. *American Psychologist, 54,* 5–20.

Fodor, J. A. 1981. The present status of the innateness controversy. In J. A. Fodor (Ed.), *RePresentations.* Cambridge, Mass.: MIT Press.

Fodor, J. A. 1983. *The modularity of mind.* Cambridge, Mass.: MIT Press.

Fodor, J. A. 1994. *The elm and the expert: Mentalese and its semantics.* Cambridge, Mass.: MIT Press.

Fodor, J. A., & Pylyshyn, Z. 1988. Connectionism and cognitive architecture: A critical analysis. *Cognition, 28,* 3–71.

Fox, J. A., & Zawitz, M. W. 2000. *Homicide trends in the United States.* Washington, D.C.: U.S. Department of Justice. Available: www.ojp.usdoj.gov/bjs/homicide/homtrnd.htm.

Fox, R. 1984. *Kinship and marriage: An anthropological perspective.* New York: Cambridge University Press.

Fox, R. 1989. *The search for society: Quest for a biosocial science and morality.* New Brunswick, N.J.: Rutgers University Press.

Frangiskakis, J. M., Ewart, A. K., Morris, A. C., Mervis, C. B., Bertrand, J., Robinson, B. F., Klein, B. P., Ensing, G. J., Everett, L. A., Green, E. D., Proschel, C., Gutowski, N. J., Noble, M., Atkinson, D. L., Odelberg, S. J., & Keating, M. T. 1996. LIM-Kinase1 hemizygosity implicated in impaired visuospatial constructive cognition. *Cell, 86,* 59–69.

Frank, R. 1999. *Luxury fever: When money fails to satisfy in an era of excess.* New York: Free Press.

Frank, R. H. 1985. *Choosing the right pond: Human behavior and the quest for status.* New York: Oxford University Press.

Frank, R. H. 1988. *Passions within reason: The strategic role of the emotions.* New York: Norton.

Frank, R. H., Gilovich, T., & Regan, D. 1993. The evolution of one-shot cooperation: An experiment. *Ethology and Sociobiology, 14,* 247–256.

Frazer, J. G. 1890/1996. *The golden bough.* New York: Simon & Schuster.

Freedman, J. L. 1984. Effect of television violence on aggressiveness. *Psychological Bulletin, 96,* 227–246.

Freedman, J. L. 1996. Violence in the mass media and violence in society: The link is unproven. *Harvard Mental Health Letter, 12,* 4–6.

Freedman, J. L. 2002. *Media violence and aggression: No evidence for a connection.* Toronto: University of Toronto Press.

Freeman, D. 1983. *Margaret Mead and Samoa: The making and unmaking of an anthropological myth.* Cambridge, Mass.: Harvard University Press.

Freeman, D. 1999. *The fateful hoaxing of Margaret Mead: A historical analysis of her Samoan research.* Boulder, Colo.: Westview Press.

Frith, C. 1992. *The cognitive neuropsychology of schizophrenia.* New York: Psychology Press.

Fry, D. 2000. Conflict management in cross-cultural perspective. In F. Aureli & F. B. M. de Waal (Eds.), *Natural conflict resolution.* Berkeley: University of California Press.

Furchtgott-Roth, D., & Stolba, C. 1999. *Women's figures: An illustrated guide to the economic progress of women in America.* Washington, D.C.: American Enterprise Institute Press.

Galileo, G. 1632/1967. *Dialogue concerning the two chief world systems.* Berkeley: University of California Press.

Gallistel, C. R. 1990. *The organization of learning.* Cambridge, Mass.: MIT Press.

Gallistel, C. R. (Ed.) 1992. *Animal cognition*. Cambridge, Mass.: MIT Press.

Gallistel, C. R. 2000. The replacement of general-purpose theories with adaptive specializations. In M. S. Gazzaniga (Ed.), *The new cognitive neurosciences*. Cambridge, Mass.: MIT Press.

Gangestad, S., & Thornhill, R. 1998. Menstrual cycle variation in women's preferences for the scent of symmetrical men. *Proceedings of the Royal Society of London B, 265*, 927–933.

Gardner, H. 1983. *Frames of mind: The theory of multiple intelligences*. New York: Basic Books.

Gardner, H. 1985. *The mind's new science: A history of the cognitive revolution*. New York: Basic Books.

Gardner, H. 1999. *Intelligence reframed: Multiple intelligences for the 21st century*. New York: Basic Books.

Gaulin, S., & McBurney, D. 2000. *Evolutionary psychology*. Englewood Cliffs, N.J.: Prentice Hall.

Gaulin, S. J. C., & McBurney, D. H. 2001. *Psychology: An evolutionary approach*. Upper Saddle River, N.J.: Prentice Hall.

Gazzaniga, M. S. 1992. *Nature's mind: The biological roots of thinking, emotion, sexuality, language, and intelligence*. New York: Basic Books.

Gazzaniga, M. S. 1998. *The mind's past*. Berkeley: University of California Press.

Gazzaniga, M. S. 2000a. *Cognitive neuroscience: A reader*. Malden, Mass.: Blackwell.

Gazzaniga, M. S. (Ed.) 2000b. *The new cognitive neurosciences*. Cambridge, Mass.: MIT Press.

Gazzaniga, M. S., Ivry, R. B., & Mangun, G. R. 1998. *Cognitive neuroscience: The biology of the mind*. New York: Norton.

Geary, D. C. 1994. *Children's mathematical development*. Washington, D.C.: American Psychological Association.

Geary, D. C. 1995. Reflections on evolution and culture in children's cognition. *American Psychologist, 50*, 24–37.

Geary, D. C. 1998. *Male, female: The evolution of human sex differences*. Washington, D.C.: American Psychological Association.

Geary, D. C. In press. Principles of evolutionary educational psychology. *Learning and Individual Differences*.

Geary, D. C., & Huffman, K. J. 2002. Brain and cognitive evolution: Forms of modularity and functions of mind. *Psychological Bulletin*.

Geertz, C. 1973. *The interpretation of cultures: Selected essays*. New York: Basic Books.

Gelman, S. A., Coley, J. D., & Gottfried, G. M. 1994. Essentialist beliefs in children: The acquisition of concepts and theories. In L. A. Hirschfeld & S. A. Gelman (Eds.), *Mapping the mind: Domain specificity in cognition and culture*. New York: Cambridge University Press.

Getty, J. A. 2000. The future did not work (Reviews of Furet's "The passing of an illusion" and Courtois et al.'s "The black book of communism"). *Atlantic Monthly, 285*, 113–116.

Ghiglieri, M. P. 1999. *The dark side of man: Tracing the origins of male violence*. Reading, Mass.: Perseus Books.

Gibbons, A. 1997. Archaeologists rediscover cannibals. *Science, 277*, 635–637.

Gibbons, A. 2000. Europeans trace ancestry to Paleolithic people. *Science, 290*, 1080–1081.

Gigerenzer, G. 1991. How to make cognitive illusions disappear: Beyond heuristics and biases. *European Review of Social Psychology, 2*, 83–115.

Gigerenzer, G. 1997. Ecological intelligence: An adaptation for frequencies. In D. Cummins & C. Allen (Eds.), *The evolution of mind*. New York: Oxford University Press.

Gigerenzer, G., & Hug, K. 1992. Domain specific reasoning: Social contracts, cheating and perspective change. *Cognition, 43*, 127–171.

Gigerenzer, G., & Selten, R. (Eds.) 2001. *Bounded rationality: The adaptive toolbox*. Cambridge, Mass.: MIT Press.

Gilbert, D. T., & Hixon, J. G. 1991. The trouble of thinking: Activation and application of stereotypic beliefs. *Journal of Personality and Social Psychology, 60*, 509–517.

Gilligan, C. 1982. *In a different voice: Psychological theory and women's development*. Cambridge, Mass.: Harvard University Press.

Gintis, H. 2000. Strong reciprocity and human sociality. *Journal of Theoretical Biology, 206*, 169–179.

Glendon, M. A. 2001. *A world made new: Eleanor Roosevelt and the Universal Declaration of Human Rights*. New York: Random House.

Glover, J. 1977. *Causing death and saving lives*. London: Penguin.

Glover, J. 1999. *Humanity: A moral history of the twentieth century*. London: Jonathan Cape.

Godfray, H. C. 1995. Evolutionary theory of parent-offspring conflict. *Nature, 376,* 133–138.

Goguen, J. A. E. 1999. Special Issue on Art and the Brain. *Journal of Consciousness Studies, 6.*

Goldberg, L. R. 1968. Simple models or simple processes? Some research on clinical judgments. *American Psychologist, 23,* 483–496.

Goldenberg, J., Mazursky, D., & Solomon, S. 1999. Creative sparks. *Science, 285,* 1495–1496.

Goldin, C. 1990. *Understanding the gender gap: An economic history of American workers.* New York: Oxford University Press.

Goleman, D. 1985. *Vital lies, simple truths: The psychology of self-deception.* New York: Simon & Schuster.

Gombrich, E. 1982/1995. *The sense of order: A study in the psychology of decorative art* (2nd ed.). London: Phaidon Press.

Gopnik, A., Meltzoff, A. N., & Kuhl, P. K. 1999. *The scientist in the crib: Minds, brains, and how children learn.* New York: William Morrow.

Gordon, M. T., & Riger, S. 1991. *The female fear: The social cost of rape.* Urbana: University of Illinois Press.

Gottfredson, L. S. 1988. Reconsidering fairness: A matter of social and ethical priorities. *Journal of Vocational Behavior, 29,* 379–410.

Gottfredson, L. S. 1997. Mainstream science on intelligence: An editorial with 52 signatories, history, and bibliography. *Intelligence, 24,* 13–23.

Gottfredson, M. R., & Hirschi, T. 1990. *A general theory of crime.* Stanford, Calif.: Stanford University Press.

Gottschall, J., & Gottschall, R. 2001. The reproductive success of rapists: An exploration of the per-incident rape-pregnancy rate. Paper presented at the Annual Meeting of the Human Behavior and Evolution Society, London.

Gottschall, J., & Jobling, I. (Eds.) In preparation. *Evolutionary psychology and literary studies: Toward integration.*

Gould, S. J. 1976a. Biological potential vs. biological determinism. In S. J. Gould (Ed.), *Ever since Darwin: Reflections in natural history.* New York: Norton.

Gould, S. J. 1976b. Criminal man revived. *Natural History, 85,* 16–18.

Gould, S. J. 1980. *The panda's thumb.* New York: Norton.

Gould, S. J. 1981. *The mismeasure of man.* New York: Norton.

Gould, S. J. 1992. Life in a punctuation. *Natural History, 101,* 10–21.

Gould, S. J. 1995. Ordering nature by budding and full-breasted sexuality. In *Dinosaur in a haystack.* New York: Harmony Books.

Gould, S. J. 1998a. The Diet of Worms and the defenestration of Prague. In *Leonardo's mountain of clams and the Diet of Worms: Essays in natural history.* New York: Harmony Books.

Gould, S. J. 1998b. The great asymmetry. *Science, 279,* 812–813.

Grant, J. 1993. *Fundamental feminism: Contesting the core concepts of feminist theory.* New York: Routledge.

Graves, D. E. 1992. "Naked truths for the asking": Twentieth-century military historians and the battlefield narrative. In D. A. Charters, M. Milner, & J. B. Wilson (Eds.), *Military history and the military profession.* Westport, Conn.: Greenwood Publishing Group.

Green, R. M. 2001. *The human embryo research debates: Bioethics in the vortex of controversy.* New York: Oxford University Press.

Greenwald, A. 1988. Self-knowledge and self-deception. In J. S. Lockard & D. L. Paulhaus (Eds.), *Self-deception: An adaptive mechanism.* Englewood Cliffs, N.J.: Prentice Hall.

Gu, X., & Spitzer, N. C. 1995. Distinct aspects of neuronal differentiation encoded by frequency of spontaneous Ca^{2+} transients. *Nature, 375,* 784–787.

Gur, R. C., & Gur, R. E. In press. Gender differences in neuropsychological functions. In L. J. Dickstein & B. L. Kennedy (Eds.), *Gender differences in the brain: Linking biology to psychiatry.* New York: Guilford Publications.

Gur, R. C., Turetsky, B. I., Matsui, M., Yan, M., Bilker, W., Hughett, P., & Gur, R. E. 1999. Sex differences in brain gray and white matter in healthy young adults: Correlations with cognitive performance. *Journal of Neuroscience, 19,* 4065–4072.

Gutknecht, L., Spitz, E., & Carlier, M. 1999. Long-term effect of placental type on anthropometrical and psychological traits among monozygotic twins: A follow-up study. *Twin Research, 2,* 212–217.

Hacking, I. 1999. *The social construction of what?* Cambridge, Mass.: Harvard University Press.

Hadley, R. F. 1994a. Systematicity in connectionist language learning. *Mind and Language, 9,* 247–272.

Hadley, R. F. 1994b. Systematicity revisited: Reply to Christiansen and Chater and Niklasson and Van Gelder. *Mind and Language, 9,* 431–444.

Haidt, J. 2001. The emotional dog and its rational tail: A social intuitionist approach to moral judgment. *Psychological Review, 108,* 813–834.

Haidt, J. In press. The moral emotions. In R. J. Davidson (Ed.), *Handbook of affective sciences.* New York: Oxford University Press.

Haidt, J., & Hersh, M. A. 2001. Sexual morality: The cultures and emotions of conservatives and liberals. *Journal of Applied Social Psychology, 31,* 191–221.

Haidt, J., Koller, H., & Dias, M. G. 1993. Affect, culture, and morality, or Is it wrong to eat your dog? *Journal of Personality and Social Psychology, 65,* 613–628.

Haier, R. J., Siegel, B., Tang, C., Abel, L., & Buchsbaum, M. S. 1992. Intelligence and changes in regional cerebral glucose metabolic rate following learning. *Intelligence, 16,* 415–426.

Haig, D. 1993. Genetic conflicts in human pregnancy. *Quarterly Review of Biology, 68,* 495–532.

Halpern, D. 2000. *Sex differences in cognitive abilities* (3rd ed.). Mahwah, N.J.: Erlbaum.

Halpern, D. F., Gilbert, R., & Coren, S. 1996. PC or not PC? Contemporary challenges to unpopular research findings. *Journal of Social Distress and the Homeless, 5,* 251–271.

Hamer, D., & Copeland, P. 1994. *The science of desire: The search for the gay gene and the biology of behavior.* New York: Simon & Schuster.

Hamer, D., & Copeland, P. 1998. *Living with our genes: Why they matter more than you think.* New York: Doubleday.

Hamilton, W. D. 1964. The genetical evolution of social behaviour (I and II). *Journal of Theoretical Biology, 7,* 1–16, 17–52.

Hardcastle, V. G., & Buller, D. J. 2000. Evolutionary psychology, meet developmental neurobiology: Against promiscuous modularity. *Brain and Mind, 1,* 307–325.

Hare, R. D. 1993. *Without conscience: The disturbing world of the psychopaths around us.* New York: Guilford Press.

Harpending, H., & Sobus, J. 1987. Sociopathy as an adaptation. *Ethology and Sociobiology, 8,* 63–72.

Harris, J. R. 1995. Where is the child's environment? A group socialization theory of development. *Psychological Review, 102,* 458–489.

Harris, J. R. 1998a. *The nurture assumption: Why children turn out the way they do.* New York: Free Press.

Harris, J. R. 1998b. The trouble with assumptions (Commentary on "Parental socialization of emotion" by Eisenberg, Cumberland, and Spinrad). *Psychological Inquiry, 9,* 294–297.

Harris, J. R. 2000a. Research on child development: What we can learn from medical research. Paper presented at the Children's Roundtable, Brookings Institution, Washington, D.C., September 28.

Harris, J. R. 2000b. Socialization, personality development, and the child's environments: Comment on Vandell (2000). *Developmental Psychology, 36,* 711–723.

Harris, J. R. In press. Personality and birth order: Explaining the differences between siblings. *Politics and the Life Sciences.*

Harris, M. 1985. *Good to eat: Riddles of food and culture.* New York: Simon & Schuster.

Hartman, J. L., Garvik, B., & Hartwell, L. 2001. Principles for the buffering of genetic variation. *Science, 291,* 1001–1004.

Haugeland, J. 1981. Semantic engines: An introduction to mind design. In J. Haugeland (Ed.), *Mind design: Philosophy, psychology, artificial intelligence.* Cambridge, Mass.: MIT Press.

Hauser, M. D. 1996. *The evolution of communication.* Cambridge, Mass.: MIT Press.

Hauser, M. D. 2000. *Wild minds: What animals really think.* New York: Henry Holt.

Hausman, P. 1999. *On the rarity of mathematically and mechanically gifted females.* The Fielding Institute, Santa Barbara, Calif.

Hawkes, K., O'Connell, J., & Rogers, L. 1997. The behavioral ecology of modern hunter-gatherers, and human evolution. *Trends in Evolution and Ecology, 12,* 29–32.

Hayek, F. A. 1960/1978. *The constitution of liberty.* Chicago: University of Chicago Press.

Hayek, F. A. 1976. *Law, legislation, and liberty* (Vol. 2: *The mirage of social justice*). Chicago: University of Chicago Press.

Hedges, L. V., & Nowell, A. 1995. Sex differences in mental test scores, variability, and numbers of high-scoring individuals. *Science, 269,* 41–45.

Hernadi, P. 2001. Literature and evolution. *SubStance, 30,* 55–71.

Herrnstein, R. 1971. I.Q. *Atlantic Monthly,* 43–64.

Herrnstein, R. J. 1973. On challenging an orthodoxy. *Commentary,* 52–62.

Herrnstein, R. J., & Murray, C. 1994. *The bell curve: Intelligence and class structure in American life.* New York: Free Press.

Hinton, G. E., & Nowlan, S. J. 1987. How learning can guide evolution. *Complex Systems, 1,* 495–502.

Hirschfeld, L. A., & Gelman, S. A. 1994. *Mapping the mind: Domain specificity in cognition and culture.* New York: Cambridge University Press.

Hirshleifer, J. 1987. On the emotions as guarantors of threats and promises. In J. Dupré (Ed.), *The latest on the best: Essays on evolution and optimality.* Cambridge, Mass.: MIT Press.

Hobbes, T. 1651/1957. *Leviathan.* New York: Oxford University Press.

Hoffrage, U., Lindsey, S., Hertwig, R., & Gigerenzer, G. 2000. Communicating statistical information. *Science, 290,* 2261–2262.

Hogan, P. C. 1997. Literary universals. *Poetics Today, 18,* 224–249.

Holden, C. 1987. The genetics of personality. *Science, 237,* 598–601.

Holden, C. 2000. Molecule shows Anasazi ate their enemies. *Science, 289,* 1663.

Horgan, J. 1993. Eugenics revisited: Trends in behavioral genetics. *Scientific American, 268,* 122–131.

Horgan, J. 1995. The new Social Darwinists. *Scientific American, 273,* 174–181.

Horowitz, D. L. 2001. *The deadly ethnic riot.* Berkeley: University of California Press.

Hrdy, S. B. 1999. *Mother nature: A history of mothers, infants, and natural selection.* New York: Pantheon Books.

Hubel, D. H. 1988. *Eye, brain, and vision.* New York: Scientific American.

Hume, D. 1739/2000. *A treatise of human nature.* New York: Oxford University Press.

Hummel, J. E., & Biederman, I. 1992. Dynamic binding in a neural network for shape recognition. *Psychological Review, 99,* 480–517.

Hummel, J. E., & Holyoak, K. J. 1997. Distributed representations of structure: A theory of analogical access and mapping. *Psychological Review, 104,* 427–466.

Hunt, M. 1999. *The new know-nothings: The political foes of the scientific study of human nature.* New Brunswick, N.J.: Transaction Publishers.

Hyman, S. E. 1999. Introduction to the complex genetics of mental disorders. *Biological Psychiatry, 45,* 518–521.

Jackendoff, R. 1990. *Semantic structures.* Cambridge, Mass.: MIT Press.

Jackendoff, R. 1996. How language helps us think. *Pragmatics and Cognition, 4,* 1–34.

Jackendoff, R. S. 1987. *Consciousness and the computational mind.* Cambridge, Mass.: MIT Press.

Jackson, D. J., & Huston, T. L. 1975. Physical attractiveness and assertiveness. *Journal of Social Psychology, 96,* 79–84.

Jaffe, S., & Hyde, J. S. 2000. Gender differences in moral orientation. *Psychological Bulletin, 126,* 703–726.

Jaggar, A. M. 1983. *Feminist politics and human nature.* Lanham, Md.: Rowman & Littlefield.

James, W. 1890/1950. *The principles of psychology.* New York: Dover.

Janda, L. H. 1998. *Psychological testing: Theory and applications.* Boston: Allyn & Bacon.

Jensen, A. 1969. How much can we boost IQ and scholastic achievement? *Harvard Educational Review, 39,* 1–123.

Jensen, A. 1971. A note on why genetic correlations are not squared. *Psychological Bulletin, 75,* 223–224.

Jensen, A. R. 1972. *Genetics and education.* New York: Harper and Row.

Jensen, A. R. 1982. The debunking of scientific fossils and straw persons: Review of "The mismeasure of man." *Contemporary Education Review, 1,* 121–135.

Jensen, A. R. 1998. *The g factor: The science of mental ability.* Westport, Conn.: Praeger.

Jespersen, O. 1938/1982. *Growth and structure of the English language.* Chicago: University of Chicago Press.

Johnson, G. R., Ratwik, S. H., & Sawyer, T. J. 1987. The evocative significance of kin terms in patriotic speech. In V. Reynolds, V. Falger, & I. Vine (Eds.), *The sociobiology of ethnocentrism.* London: Croon Helm.

Jones, O. 2000. Reconsidering rape. *National Law Journal,* February 21, A21.

Jones, O. 2001. Time-shifted rationality and the Law of Law's Leverage: Behavioral economics meets behavioral biology. *Northwestern University Law Review, 95,* 1141–1205.

Jones, O. D. 1997. Evolutionary analysis in law: An introduction and application to child abuse. *North Carolina Law Review, 75,* 1117–1242.

Jones, O. D. 1999. Sex, culture, and the biology of rape: Toward explanation and prevention. *California Law Review, 87,* 827–942.

Junger, S. 1997. *The perfect storm: A true story of men against the sea.* New York: Norton.

Jussim, L. J., & Eccles, J. 1995. Are teacher expectations biased by students' gender, social class, or ethnicity? In Y.-T. Lee, L. J. Jussim, & C. R. McCauley (Eds.), *Stereotype accuracy: Toward appreciating group differences.* Washington, D.C.: American Psychological Association.

Jussim, L. J., McCauley, C. R., & Lee, Y.-T. 1995. Why study stereotype accuracy and inaccuracy? In Y.-T. Lee, L. J. Jussim, & C. R. McCauley (Eds.), *Stereotype accuracy: Toward appreciating group differences.* Washington, D.C.: American Psychological Association.

Kaas, J. H. 2000. The reorganization of sensory and motor maps after injury in adult mammals. In M. S. Gazzaniga (Ed.), *The new cognitive neurosciences.* Cambridge, Mass.: MIT Press.

Kahneman, D., & Tversky, A. 1982. On the study of statistical intuitions. *Cognition, 11,* 123–141.

Kahneman, D., & Tversky, A. 1984. Choices, values, and frames. *American Psychologist, 39,* 341–350.

Kamin, L. 1974. *The science and politics of IQ.* Mahwah, N.J.: Erlbaum.

Kaminer, W. 1990. *A fearful freedom: Women's flight from equality.* Reading, Mass.: Addison Wesley.

Kandel, E. R., Schwartz, J. H., & Jessell, T. M. 2000. *Principles of neural science* (4th ed.). New York: McGraw-Hill.

Kane, R. 1998. *The significance of free will.* New York: Oxford University Press.

Kanwisher, N., & Moscovitch, M. 2000. The cognitive neuroscience of face processing: An introduction. *Cognitive Neuropsychology, 17,* 1–13.

Kaplan, H., Hill, K., & Hurtado, A. M. 1990. Risk, foraging, and food sharing among the Ache. In E. Cashdan (Ed.), *Risk and uncertainty in tribal and peasant economies.* Boulder, Colo.: Westview Press.

Kaplan, J. 1973. *Criminal justice: Introductory cases and materials.* Mineola, N.Y.: The Foundation Press.

Kaplan, S. 1992. Environmental preference in a knowledge-seeking, knowledge-using organism. In J. H. Barkow, L. Cosmides, & J. Tooby (Eds.), *The adapted mind: Evolutionary psychology and the generation of culture.* New York: Oxford University Press.

Karmiloff-Smith, A. 1992. *Beyond modularity: A developmental perspective on cognitive science.* Cambridge, Mass.: MIT Press.

Karmiloff-Smith, A., Klima, E. S., Bellugi, U., Grant, J., & Baron-Cohen, S. 1995. Is there a social module? Language, face processing, and Theory of Mind in individuals with Williams syndrome. *Journal of Cognitive Neuroscience, 7,* 196–208.

Katz, L. C., & Crowley, J. C. 2002. Development of cortical circuits: Lessons from ocular dominance columns. *Nature Neuroscience Reviews, 3,* 34–42.

Katz, L. C., & Shatz, C. J. 1996. Synaptic activity and the construction of cortical circuits. *Science, 274,* 1133–1137.

Katz, L. C., Weliky, M., & Crowley, J. C. 2000. Activity and the development of the visual cortex: New perspectives. In M. S. Gazzaniga (Ed.), *The new cognitive neurosciences.* Cambridge, Mass.: MIT Press.

Keegan, J. 1976. *The face of battle.* New York: Penguin.

Keeley, L. H. 1996. *War before civilization: The myth of the peaceful savage.* New York: Oxford University Press.

Keil, F. C. 1989. *Concepts, kinds, and cognitive development.* Cambridge, Mass.: MIT Press.

Keil, F. C. 1995. The growth of casual understandings of natural kinds. In D. Sperber, D. Premack, & A. J. Premack (Eds.), *Causal cognition.* New York: Oxford University Press.

Kelling, G. L., & Sousa, W. H. 2001. *Do police matter? An analysis of the impact of New York City's police reforms* (Civic Report 22). New York: Manhattan Institute for Policy Research.

Kelman, H. 1958. Compliance, identification, and internalization: Three processes of attitude change. *Journal of Conflict Resolution, 2,* 51–60.

Kennedy, J. 1993. *Drawings in the blind.* New Haven, Conn.: Yale University Press.

Kennedy, R. 1997. *Race, crime, and the law.* New York: Vintage.

Kenrick, D., Groth, G., Trost, M., & Sadalla, E. 1993. Integrating evolutionary and social exchange perspectives on relationships: Effects of gender, self-appraisal, and involvement level on mate selection criteria. *Journal of Personality and Social Psychology, 64,* 951–969.

Kenrick, D., & Sheets, V. 1994. Homicide fantasies. *Ethology and Sociobiology, 14,* 231–246.

Kevles, D. J. 1985. *In the name of eugenics: Genetics and the uses of human heredity*. Cambridge, Mass.: Harvard University Press.

Keyser, S. J. 1999. Meter and poetry. In R. A. Wilson & F. C. Keil (Eds.), *The MIT Encyclopedia of the Cognitive Sciences*. Cambridge, Mass.: MIT Press.

Keyser, S. J., & Halle, M. 1998. On meter in general and on Robert Frost's loose iambics in particular. In E. Iwamoto (Ed.), *Festschrift for Professor K. Inoue*. Tokyo: Kanda University of International Studies.

Kimura, D. 1999. *Sex and cognition*. Cambridge, Mass.: MIT Press.

Kingdon, J. 1993. *Self-made man: Human evolution from Eden to extinction?* New York: Wiley.

Kirwin, B. R. 1997. *The mad, the bad, and the innocent: The criminal mind on trial*. Boston: Little, Brown.

Kitcher, P. 1982. *Abusing science: The case against creationism.* Cambridge, Mass.: MIT Press.

Klaw, S. 1993. *Without sin: The life and death of the Oneida community*. New York: Penguin.

Klein, R. G. 1989. *The human career: Human biological and cultural origins*. Chicago: University of Chicago Press.

Kleinfeld, J. 1999. *MIT tarnishes its reputation with gender junk science* (Special report www.uaf.edu/northern/mitstudy). Arlington, Va.: Independent Women's Forum.

Klima, E., & Bellugi, U. 1979. *The signs of language*. Cambridge, Mass.: Harvard University Press.

Knauft, B. 1987. Reconsidering violence in simple human societies. *Current Anthropology, 28,* 457–500.

Koestler, A. 1959. *The sleepwalkers: A history of man's changing vision of the universe*. London: Penguin.

Komar, B., Melamid, A., & Wypijewski, J. 1997. *Painting by numbers: Komar and Melamid's scientific guide to art*. New York: Farrar, Straus & Giroux.

Kors, A. C., & Silverglate, H. A. 1998. *The shadow university: The betrayal of liberty on America's campuses*. New York: Free Press.

Kosof, A. 1996. *Living in two worlds: The immigrant children's experience*. New York: Twenty-First Century Books.

Kosslyn, S. M. 1980. *Image and mind*. Cambridge, Mass.: Harvard University Press.

Kosslyn, S. M. 1994. *Image and brain: The resolution of the imagery debate*. Cambridge, Mass.: MIT Press.

Krebs, D., & Denton, K. 1997. Social illusions and self-deception: The evolution of biases in person perception. In J. A. Simpson & D. T. Kenrick (Eds.), *Evolutionary social psychology*. Mahwah, N.J.: Erlbaum.

Krebs, D. L. 1998. The evolution of moral behaviors. In C. Crawford & D. L. Krebs (Eds.), *Handbook of evolutionary psychology: Ideas, issues, and applications*. Mahwah, N.J.: Erlbaum.

Krech, S. 1994. Genocide in tribal society. *Nature, 371,* 14–15.

Krech, S. 1999. *The ecological Indian: Myth and history*. New York: Norton.

Krubitzer, L., & Huffman, K. J. 2000. A realization of the neocortex in mammals: Genetic and epigenetic contributions to the phenotype. *Brain, Behavior, and Evolution, 55,* 322–335.

Krueger, R. F., Hicks, B. M., & McGue, M. 2001. Altruism and antisocial behavior: Independent tendencies, unique personality correlates, distinct etiologies. *Psychological Science, 12,* 397–402.

Kubovy, M. 1981. Concurrent pitch segregation and the theory of indispensable attributes. In M. Kubovy & J. Pomerantz (Eds.), *Perceptual organization*. Mahwah, N.J.: Erlbaum.

Kubovy, M. 1986. *The psychology of perspective and Renaissance art*. New York: Cambridge University Press.

Lachter, J., & Bever, T. G. 1988. The relation between linguistic structure and associative theories of language learning—A constructive critique of some connectionist learning models. *Cognition, 28,* 195–247.

Lai, C. S. L., Fisher, S. E., Hurst, J. A., Vargha-Khadem, F., & Monaco, A. P. 2001. A novel forkhead-domain gene is mutated in a severe speech and language disorder. *Nature, 413,* 519–523.

Lakoff, G. 1996. *Moral politics: What conservatives know that liberals don't*. Chicago: University of Chicago Press.

Lakoff, G., & Johnson, M. 1980. *Metaphors we live by*. Chicago: University of Chicago Press.

Lakoff, G., & Nunez, R. E. 2000. *Where mathematics comes from: How the embodied mind brings mathematics into being*. New York: Basic Books.

Lalumière, M. L., Harris, G. T., & Rice, M. E. 2001. Psychopathy and developmental instability. *Evolution and Human Behavior, 22,* 75–92.

Lander, E. S., Patrinos, A., Morgan, J. J., & International Human Genome Sequencing Consortium. 2001. Intitial sequencing and analysis of the human genome. *Nature, 409,* 813–958.

Latané, B., & Nida, S. 1981. Ten years of research on group size and helping. *Psychological Bulletin, 89,* 308–324.

Laubichler, M. D. 1999. Frankenstein in the land of *Dichter* and *Denker. Science, 286,* 1859–1860.

Lazarus, R. S. 1991. *Emotion and adaptation.* New York: Oxford University Press.

Lee, Y.-T., Jussim, L. J., & McCauley, C. R. (Eds.) 1995. *Stereotype accuracy: Toward appreciating group differences.* Washington, D.C.: American Psychological Association.

Lehman, D. 1992. *Signs of the times: Deconstructionism and the fall of Paul de Man.* New York: Simon & Schuster.

Lehrman, K. 1997. *The lipstick proviso: Women, sex, and power in the real world.* New York: Doubleday.

Leibniz, G. W. 1768/1996. *New essays on human understanding.* New York: Cambridge University Press.

Lerdahl, F., & Jackendoff, R. 1983. *A generative theory of tonal music.* Cambridge, Mass.: MIT Press.

Lesch, K.-P., Bengel, D., Heils, A., Sabol, S. Z., Greenberg, B. D., Petri, S., Benjamin, J., Muller, C. R., Hamer, D. H., & Murphy, D. L. 1996. Association of anxiety-related traits with a polymorphism in the serotonin transporter gene regulatory region. *Science, 274,* 1527–1531.

Leslie, A. M. 1994. ToMM, ToBY, and agency: Core architecture and domain specificity. In L. A. Hirschfeld & S. A. Gelman (Eds.), *Mapping the mind: Domain specificity in cognition and culture.* New York: Cambridge University Press.

Leslie, A. M. 1995. Pretending and believing: Issues in the theory of ToMM. *Cognition, 50,* 193–220.

LeVay, S. 1993. *The sexual brain.* Cambridge, Mass.: MIT Press.

Levins, R., & Lewontin, R. C. 1985. *The dialectical biologist.* Cambridge, Mass.: Harvard University Press.

Levitt, P. 2000. Molecular determinants of regionalization of the forebrain and cerebral cortex. In M. S. Gazzaniga (Ed.), *The new cognitive neurosciences.* Cambridge, Mass.: MIT Press.

Lewis, H. W. 1990. *Technological risk.* New York: Norton.

Lewontin, R. 1990. How much did the brain have to change for speech? (Commentary on Pinker & Bloom's "Natural language and natural selection"). *Behavioral and Brain Sciences, 13,* 740–741.

Lewontin, R. 1992. *Biology as ideology: The doctrine of DNA.* New York: HarperCollins.

Lewontin, R. C. 1982. *Human diversity.* San Francisco: Scientific American.

Lewontin, R. C. 1983. The organism as the subject and object of evolution. *Scientia, 118,* 65–82.

Lewontin, R. C., Rose, S., & Kamin, L. J. 1984. *Not in our genes.* New York: Pantheon.

Lingua Franca, Editors of. 2000. *The Sokal hoax: The sham that shook the academy.* Lincoln: University of Nebraska Press.

Lockard, J. S., & Paulhaus, D. L. (Eds.) 1988. *Self-deception: An adaptive mechanism.* Englewood Cliffs, N.J.: Prentice Hall.

Locke, J. 1690/1947. *An essay concerning human understanding.* New York: E. P. Dutton.

Loehlin, J. C. 1992. *Genes and environment in personality development.* Newbury Park, Calif.: Sage.

Loehlin, J. C. 2001. Behavior genetics and parenting theory. *American Psychologist, 56,* 169–170.

Loehlin, J. C., & Nichols, R. C. 1976. *Heredity, environment, and personality: A study of 850 sets of twins.* Austin: University of Texas Press.

Loury, G. 2002. *The anatomy of racial inequality: Stereotypes, stigma, and the elusive quest for racial justice in the United States.* Cambridge, Mass.: Harvard University Press.

Lubinski, D., & Benbow, C. 1992. Gender differences in abilities and preferences among the gifted: Implications for the math-science pipeline. *Current Directions in Psychological Science, 1,* 61–66.

Lumsden, C., & Wilson, E. O. 1981. *Genes, mind, and culture.* Cambridge, Mass.: Harvard University Press.

Lutz, D. 1984. The relative influence of European writers on late eighteenth-century American political thought. *American Political Science Review, 78,* 189–197.

Lykken, D. T. 1995. *The antisocial personalities.* Mahwah, N.J.: Erlbaum.

Lykken, D. T. 2000. The causes and costs of crime and a controversial cure. *Journal of Personality, 68,* 559–605.

Lykken, D. T., McGue, M., Tellegen, Á., & Bouchard, T. J., Jr. 1992. Emergenesis: Genetic traits that may not run in families. *American Psychologist, 47,* 1565–1577.

Lytton, H. 1990. Child effects—Still unwelcome? Response to Dogge and Wahler. *Developmental Psychology, 26,* 705–709.

Lytton, H., & Romney, D. M. 1991. Parents' differential socialization of boys and girls: A meta-analysis. *Psychological Bulletin, 109,* 267–296.

Maccoby, E. E., & Jacklin, C. N. 1987. *The psychology of sex differences.* Stanford, Calif.: Stanford University Press.

Maccoby, E. E., & Martin, J. A. 1983. Socialization in the context of the family: Parent-child interaction. In P. H. Mussen & E. M. Hetherington (Eds.), *Handbook of child psychology: Socialization, personality, and social development* (4 ed., Vol. 4). New York: Wiley.

Macnamara, J. 1999. *Through the rearview mirror: Historical reflections on psychology.* Cambridge, Mass.: MIT Press.

Macnamara, J., & Reyes, G. E. (Eds.) 1994. *The logical foundations of cognition.* New York: Oxford University Press.

Maguire, E. A., Gadian, D. G., Johnsrude, I. S., Good, C. D., Ashburner, J., Frackowiak, R. S. J., & Frith, C. D. 2000. Navigation-related structural change in the hippocampi of taxi drivers. *PNAS, 97,* 4398–4403.

Maier, P. 1997. *American scripture: Making the Declaration of Independence.* New York: Knopf.

Mallon, R., & Stich, S. 2000. The odd couple: The compatibility of social construction and evolutionary psychology. *Philosophy of Science, 67,* 133–154.

Marcus, G. F. 1998. Rethinking eliminative connectionism. *Cognitive Psychology, 37,* 243–282.

Marcus, G. F. 2001a. *The algebraic mind: Reflections on connectionism and cognitive science.* Cambridge, Mass.: MIT Press.

Marcus, G. F. 2001b. Plasticity and nativism: Towards a resolution of an apparent paradox. In S. Wermter, J. Austin, & D. Willshaw (Eds.), *Emergent neural computational architectures based on neuroscience.* New York: Springer-Verlag.

Marcus, G. F., Brinkmann, U., Clahsen, H., Wiese, R., & Pinker, S. 1995. German inflection: The exception that proves the rule. *Cognitive Psychology, 29,* 189–256.

Marks, I. M., & Nesse, R. M. 1994. Fear and fitness: An evolutionary analysis of anxiety disorders. *Ethology and Sociobiology, 15,* 247–261.

Marr, D. 1982. *Vision.* San Francisco: W. H. Freeman.

Marslen-Wilson, W. D., & Tyler, L. K. 1998. Rules, representations, and the English past tense. *Trends in Cognitive Science, 2,* 428–435.

Martin, N. G., Eaves, L. J., Heath, A. C., Jardine, R., Feingold, L. M., & Eysenck, H. J. 1986. Transmission of social attitudes. *Proceedings of the National Academy of Science, 83,* 4364–4368.

Martindale, C. 1990. *The clockwork muse: The predictability of artistic change.* New York: Basic Books.

Marx, K. 1845/1989. Theses on Feuerbach. In K. Marx & F. Engels, *Basic writings on politics and philosophy.* New York: Anchor Books.

Marx, K. 1847/1995. *The poverty of philosophy.* Amherst, N.Y.: Prometheus Books.

Marx, K. 1859/1979. *Contribution to the critique of political economy.* New York: International Publishers.

Marx, K. 1867/1993. *Capital: A critique of political economy.* London: Penguin.

Marx, K., & Engels, F. 1844/1988. *The economic and philosophic manuscripts of 1844.* Amherst, N.Y.: Prometheus Books.

Marx, K., & Engels, F. 1846/1963. *The German ideology: Parts I & III.* New York: New World Paperbacks/International Publishers.

Masters, R. D. 1982. Is sociobiology reactionary? The political implications of inclusive-fitness theory. *Quarterly Review of Biology, 57,* 275–292.

Masters, R. D. 1989. *The nature of politics.* New Haven, Conn.: Yale University Press.

Maynard Smith, J. 1975/1993. *The theory of evolution.* New York: Cambridge University Press.

Maynard Smith, J., & Szathmáry, E. 1997. *The major transitions in evolution.* New York: Oxford University Press.

Mayr, E. 1963. *Animal species and evolution.* Cambridge, Mass.: Harvard University Press.

Mayr, E. 1982. *The growth of biological thought.* Cambridge, Mass.: Harvard University Press.

McCauley, C. R. 1995. Are stereotypes exaggerated? A sampling of racial, gender, academic, occupational, and political stereotypes. In Y.-T. Lee, L. J. Jussim, & C. R. McCauley (Eds.), *Stereotype accuracy: Toward appreciating group differences.* Washington, D.C.: American Psychological Association.

McClelland, J. L., Rumelhart, D. E., & the PDP Research Group. 1986. *Parallel distributed processing: Explorations in the microstructure of cognition* (Vol. 2: *Psychological and biological models*). Cambridge, Mass.: MIT Press.

McCloskey, M. 1983. Intuitive physics. *Scientific American, 248,* 122–130.

McCloskey, M., & Cohen, N. J. 1989. Catastrophic interference in connectionist networks: The sequential learning problem. In G. H. Bower (Ed.), *The psychology of learning and motivation* (Vol. 23). New York: Academic Press.

McCord, W. M. 1989. *Voyages to Utopia: From monastery to commune: The search for the perfect society in modern times.* New York: Norton.

McCrae, R. R., Costa, P. T., Ostendorf, F., Angleitner, A., Hrebickova, M., Avia, M. D., Sanz, J., Sanchez-Bernardos, M. L., Kusdil, M. E., Woodfield, R., Saunders, P. R., & Smith, P. B. 2000. Nature over nurture: Temperament, personality, and life span development. *Journal of Personality and Social Psychology, 78,* 173–186.

McElroy, W. 1996. *Sexual correctness: The gender-feminist attack on women.* Jefferson, N.C.: McFarland.

McGinn, C. 1993. *Problems in philosophy: The limits of inquiry.* Cambridge, Mass.: Blackwell.

McGinn, C. 1997. *Evil, ethics, and fiction.* New York: Oxford University Press.

McGinn, C. 1999. *The mysterious flame: Conscious minds in a material world.* New York: Basic Books.

McGinnis, J. O. 1996. The original constitution and our origins. *Harvard Journal of Law and Public Policy, 19,* 251–261.

McGinnis, J. O. 1997. The human constitution and constitutive law: A prolegomenon. *Journal of Contemporary Legal Issues, 8,* 211–239.

McGue, M. 1997. The democracy of the genes. *Nature, 388,* 417–418.

McGuinness, D. 1997. *Why our children can't read.* New York: Free Press.

McLearn, G. E., Johansson, B., Berg, S., Pedersen, N. L., Ahern, F., Petrill, S. A., & Plomin, R. 1997. Substantial genetic influence on cognitive abilities in twins 80 or more years old. *Science, 276,* 1560–1563.

McLeod, P., Plunkett, K., & Rolls, E. T. 1998. *Introduction to connectionist modeling of cognitive processes.* New York: Oxford University Press.

Mead, M. 1928. *Coming of age in Samoa: A psychological study of primitive youth for Western Civilisation.* New York: Blue Ribbon Books.

Mead, M. 1935/1963. *Sex and temperament in three primitive societies.* New York: William Morrow.

Mealey, L. 1995. The sociobiology of sociopathy: An integrated evolutionary model. *Behavioral and Brain Sciences, 18,* 523–541.

Mealey, L., Daood, C., & Krage, M. 1996. Enhanced memory for faces of cheaters. *Ethology and Sociobiology, 17,* 119–128.

Meltzoff, A. N. 1995. Understanding the intentions of others: Re-enactment of intended acts by 18-month-old children. *Developmental Psychology, 31,* 838–850.

Melzack, R. 1990. Phantom limbs and the concept of a neuromatrix. *Trends in Neurosciences, 13,* 88–92.

Melzack, R., Israel, R., Lacroix, R., & Schultz, G. 1997. Phantom limbs in people with congenital limb deficiency or amputation in early childhood. *Brain, 120,* 1603–1620.

Mesquida, C. G., & Wiener, N. I. 1996. Human collective aggression: A behavioral ecology perspective. *Ethology and Sociobiology, 17,* 247–262.

Miller, E. E. 1997. Could nonshared environmental variation have evolved to assure diversification through randomness? *Evolution and Human Behavior, 18,* 195–221.

Miller, E. K. 2000. The prefrontal cortex and cognitive control. *Nature Reviews Neuroscience, 1,* 59–65.

Miller, G. A., Galanter, E., & Pribram, K. H. 1960. *Plans and the structure of behavior.* New York: Adams-Bannister-Cox.

Miller, G. F. 2000a. *The mating mind: How sexual choice shaped the evolution of human nature.* New York: Doubleday.

Miller, G. F. 2000b. Sexual selection for indicators of intelligence. In G. Bock, J. A. Goode, & K. Webb (Eds.), *The nature of intelligence.* Chichester, U.K.: Wiley.

Miller, G. F. 2001. Aesthetic fitness: How sexual selection shaped artistic virtuosity as a fitness indicator and aesthetic preferences as mate choice criteria. *Bulletin of Psychology and the Arts, 2,* 20–25.

Miller, K. D., Keller, J. B., & Stryker, M. P. 1989. Ocular dominance and column development: Analysis and simulation. *Science, 245,* 605–615.

Miller, K. R. 1999. *Finding Darwin's God: A scientist's search for common ground between God and evolution*. New York: Cliff Street Books.

Minogue, K. 1985. *Alien powers: The pure theory of ideology*. New York: St. Martin's Press.

Minogue, K. 1999. Totalitarianism: Have we seen the last of it? *National Interest, 57,* 35–44.

Minsky, M., & Papert, S. 1988. Epilogue: The new connectionism. In *Perceptrons* (expanded ed.). Cambridge, Mass.: MIT Press.

Mithen, S. J. 1996. *The prehistory of the mind: A search for the origins of art, religion, and science*. London: Thames and Hudson.

Miyashita-Lin, E. M., Hevner, R., Wassarman, K. M., Martinez, S., & Rubenstein, J. L. R. 1999. Early neocortical regionalization in the absence of thalamic innervation. *Science, 285,* 906–909.

Monaghan, E., & Glickman, S. 1992. Hormones and aggressive behavior. In J. Becker, M. Breedlove, & D. Crews (Eds.), *Behavioral endocrinology*. Cambridge, Mass.: MIT Press.

Montagu, A. (Ed.) 1973a. *Man and aggression* (2nd ed.). New York: Oxford University Press.

Montagu, A. 1973b. The new litany of "innate depravity," or original sin revisited. In *Man and aggression*. New York: Oxford University Press.

Moore, G. E. 1903/1996. *Principia ethica*. New York: Cambridge University Press.

Mount, F. 1992. *The subversive family: An alternative history of love and marriage*. New York: Free Press.

Mousseau, T. A., & Roff, D. A. 1987. Natural selection and the heritability of fitness components. *Heredity, 59,* 181–197.

Muravchik, J. 2002. *Heaven on Earth: The rise and fall of socialism*. San Francisco: Encounter Books.

Murdoch, I. 1993. *Metaphysics as a guide to morals*. London: Allen Lane.

Murphy, J. P. M. 1999. Hitler was *not* an atheist. *Free Inquiry,* Spring, 9.

Nagel, T. 1970. *The possibility of altruism*. Princeton, N.J.: Princeton University Press.

Neel, J. V. 1994. *Physician to the gene pool: Genetic lessons and other stories*. New York: Wiley.

Neisser, U. 1967. *Cognitive psychology*. Englewood Cliffs, N.J.: Prentice-Hall.

Neisser, U., Boodoo, G., Bouchard, T. J., Jr., Boykin, A. W., Brody, N., Ceci, S. J., Halpern, D. F., Loehlin, J. C., Perloff, R., Sternberg, R. J., & Urbina, S. 1996. Intelligence: Knowns and unknowns. *American Psychologist, 51,* 77–101.

Nesse, R. M., & Lloyd, A. T. 1992. The evolution of psychodynamic mechanisms. In J. H. Barkow, L. Cosmides, & J. Tooby (Eds.), *The adapted mind: Evolutionary psychology and the generation of culture*. New York: Oxford University Press.

Neville, H. J., & Bavelier, D. 2000. Specificity and plasticity in neurocognitive development in humans. In M. S. Gazzaniga (Ed.), *The new cognitive neurosciences*. Cambridge, Mass.: MIT Press.

Newell, A. 1980. Physical symbol systems. *Cognitive Science, 4,* 135–183.

Newsome, W. T. 2001. Life of faith, life of science. Paper presented at the conference "Science and the Spiritual Quest," Memorial Church, Harvard University, Cambridge, Mass.

Nisbett, R. E., & Cohen, D. 1996. *Culture of honor: The psychology of violence in the South*. New York: HarperCollins.

Nolfi, S., Elman, J. L., & Parisi, D. 1994. Learning and evolution in neural networks. *Adaptive Behavior, 3,* 5–28.

Norenzayan, A., & Atran, S. In press. Cognitive and emotional processes in the cultural transmission of natural and nonnatural beliefs. In M. Schaller & C. Crandall (Eds.), *The psychological foundations of culture*. Mahwah, N.J.: Erlbaum.

Nowak, M. A., May, R. M., & Sigmund, K. 1995. The arithmetic of mutual help. *Scientific American, 272,* 50–55.

Nozick, R. 1974. *Anarchy, state, and utopia*. New York: Basic Books.

Nozick, R. 1981. *Philosophical explanations*. Cambridge, Mass.: Harvard University Press.

Nunney, L. 1998. Are we selfish, are we nice, or are we nice because we are selfish? (Review of E. Sober and D. S. Wilson's "Unto others"). *Science, 281,* 1619–1621.

Orians, G. H. 1998. Human behavioral ecology: 140 years without Darwin is too long. *Bulletin of the Ecological Society of America, 79,* 15–28.

Orians, G. H., & Heerwgen, J. H. 1992. Evolved responses to landscapes. In J. H. Barkow, L. Cosmides, & J. Tooby (Eds.), *The adapted mind: Evolutionary psychology and the generation of culture*. New York: Oxford University Press.

Ortega y Gasset, J. 1932/1985. *The revolt of the masses*. Notre Dame, Ind.: University of Notre Dame Press.

Ortega y Gasset, J. 1935/2001. *Toward a philosophy of history*. Chicago: University of Illinois Press.

Orwell, G. 1949/1983. *1984*. New York: Harcourt Brace Jovanovich.

Padden, C. A., & Perlmutter, D. M. 1987. American Sign Language and the architecture of phonological theory. *Natural Language and Linguistic Theory, 5,* 335–375.

Paglia, C. 1990. *Sexual personae: Art and decadence from Nefertiti to Emily Dickinson*. New Haven, Conn.: Yale University Press.

Paglia, C. 1992. *Sex, art, and American culture*. New York: Vintage.

Panksepp, J., & Panksepp, J. B. 2000. The seven sins of evolutionary psychology. *Evolution and Cognition, 6,* 108–131.

Passmore, J. 1970. *The perfectibility of man*. New York: Scribner.

Patai, D. 1998. *Heterophobia: Sexual harassment and the future of feminism*. New York: Rowman & Littlefield.

Patai, D., & Koertge, N. 1994. *Professing feminism: Cautionary tales from the strange world of women's studies*. New York: Basic Books.

Patai, R., & Patai, J. 1989. *The myth of the Jewish race* (rev. ed.). Detroit: Wayne State University Press.

Patterson, O. 1995. For whom the bell curves. In S. Fraser (Ed.), *The Bell Curve wars: Race, intelligence, and the future of America*. New York: Basic Books.

Patterson, O. 1997. *The ordeal of integration*. Washington, D.C.: Civitas.

Patterson, O. 2000. Taking culture seriously: A framework and an Afro-American illustration. In L. E. Harrison & S. P. Huntington (Eds.), *Culture matters: How values shape human progress*. New York: Basic Books.

Pedersen, N. L., McClearn, G. E., Plomin, R., & Nesselroade, J. R. 1992. Effects of early rearing environment on twin similarity in the last half of the life span. *British Journal of Developmental Psychology, 10,* 255–267.

Pennock, R. T. 2000. *Tower of Babel: The evidence against the new creationism*. Cambridge, Mass.: MIT Press.

Pennock, R. T. (Ed.) 2001. *Intelligent design: Creationism and its critics*. Cambridge, Mass.: MIT Press.

Peretz, I., Gagnon, L., & Bouchard, B. 1998. Music and emotion: Perceptual determinants, immediacy, and isolation after brain damage. *Cognition, 68,* 111–141.

Perloff, M. 1999. In defense of poetry: Put the literature back into literary studies. *Boston Review, 24,* 22–26.

Perry, B. D. 1997. Incubated in terror: Neurobiological factors in the "cycle of violence." In J. D. Osofsky (Ed.), *Children in a violent society*. New York: Guilford Press.

Persico, N., Postlewaite, A., & Silverman, D. 2001. *The effect of adolescent experience on labor market outcomes: The case of height*. Philadelphia: Department of Economics, University of Pennsylvania.

Petitto, L. A., Zatorre, R. J., Gauna, K., Nikelski, E. J., Dostie, D., & Evans, A. C. 2000. Speech-like cerebral activity in profoundly deaf people while processing signed language: Implications for the neural basis of all human language. *Proceedings of the National Academy of Sciences, 97* 13961–13966.

Petrinovich, L. F. 1995. *Human evolution, reproduction, and morality*. New York: Plenum Press.

Petrinovich, L. F., O'Neill, P., & Jorgensen, M. 1993. An empirical study of moral intuitions: Toward an evolutionary ethics. *Journal of Personality and Social Psychology, 64,* 467–478.

Pinker, S. 1979. Formal models of language learning. *Cognition, 7,* 217–283.

Pinker, S. 1984a. *Language learnability and language development* (Reprinted with a new introduction, 1996). Cambridge, Mass.: Harvard University Press.

Pinker, S. 1984b. Visual cognition: An introduction. *Cognition, 18,* 1–63.

Pinker, S. 1989. *Learnability and cognition: The acquisition of argument structure*. Cambridge, Mass.: MIT Press.

Pinker, S. 1990. A theory of graph comprehension. In R. Friedle (Ed.), *Artificial intelligence and the future of testing*. Mahwah, N.J.: Erlbaum.

Pinker, S. 1991. Rules of language. *Science, 253,* 530–535.

Pinker, S. 1994. *The language instinct*. New York: HarperCollins.

Pinker, S. 1996. Language learnability and language development revisited. In *Language learnability and language development*. Cambridge, Mass.: Harvard University Press.

Pinker, S. 1997. *How the mind works*. New York: Norton.

Pinker, S. 1998. Still relevant after all these years (Review of Darwin's "The expression of the emotions in man and animals, 3rd ed."). *Science, 281,* 522–523.

Pinker, S. 1999. *Words and rules: The ingredients of language*. New York: HarperCollins.

Pinker, S. 2001a. Four decades of rules and associations, or whatever happened to the past tense debate? In E. Dupoux (Ed.), *Language, the brain, and cognitive development*. Cambridge, Mass.: MIT Press.

Pinker, S. 2001b. Talk of genetics and vice-versa. *Nature, 413*, 465–466.

Pinker, S., & Mehler, J. (Eds.) 1988. *Connections and symbols*. Cambridge, Mass.: MIT Press.

Pinker, S., & Prince, A. 1988. On language and connectionism: Analysis of a Parallel Distributed Processing model of language acquisition. *Cognition, 28*, 73–193.

Pinker, S., & Prince, A. 1996. The nature of human concepts: Evidence from an unusual source. *Communication and Cognition, 29*, 307–361.

Plamenatz, J. 1963. *Man and society: A critical examination of some important social and political theories from Machiavelli to Marx* (Vol. 2). London: Longman.

Plamenatz, J. 1975. *Karl Marx's philosophy of man*. New York: Oxford University Press.

Plomin, R. 1990. The role of inheritance in behavior. *Science, 248*, 183–248.

Plomin, R. 1991. Continuing commentary: Why children in the same family are so different from one another. *Behavioral and Brain Sciences, 13*, 336–337.

Plomin, R. 1994. *Genetics and experience: The interplay between nature and nurture*. Thousand Oaks, Calif.: Sage.

Plomin, R., & Daniels, D. 1987. Why are children in the same family so different from one another? *Behavioral and Brain Sciences, 10*, 1–60.

Plomin, R., DeFries, J. C., & Fulker, D. W. 1988. *Nature and nurture in infancy and early childhood*. New York: Cambridge University Press.

Plomin, R., DeFries, J. C., McClearn, G. E., & McGuffin, P. 2001. *Behavior genetics* (4th ed.). New York: Worth.

Plomin, R., DeFries, J. C., McClearn, G. E., & Rutter, M. 1997. *Behavioral genetics* (3rd ed.). New York: W. H. Freeman.

Plomin, R., Owen, M. J., & McGuffin, P. 1994. The genetic basis of complex human behaviors. *Science, 264*, 1733–1739.

Polti, G. 1921/1977. *The thirty-six dramatic situations*. Boston: The Writer, Inc.

Pons, T. M., Garraghty, P. E., Ommaya, A. K., Kass, J. H., Taub, E., & Mishkin, M. 1991. Massive cortical reorganization after sensory deafferentation in adult macaques. *Science, 252*, 1857–1860.

Pope, G. G. 2000. *The biological bases of human behavior*. Needham Heights, Mass.: Allyn & Bacon.

Pratto, F., & Bargh, J. A. 1991. Stereotyping based on apparently individuating information: Trait and global components of sex stereotypes under attention overload. *Journal of Experimental Social Psychology, 27*, 26–47.

Preuss, T. 1995. The argument from animals to humans in cognitive neuroscience. In M. S. Gazzaniga (Ed.), *The new cognitive neurosciences*. Cambridge, Mass.: MIT Press.

Preuss, T. M. 2000. What's human about the human brain? In M. S. Gazzaniga (Ed.), *The new cognitive neurosciences*. Cambridge, Mass.: MIT Press.

Preuss, T. M. 2001. The discovery of cerebral diversity: An unwelcome scientific revolution. In D. Falk & K. Gibson (Eds.), *Evolutionary anatomy of the primate cerebral cortex*. New York: Cambridge University Press.

Price, M. E., Cosmides, L., & Tooby, J. 2002. Punitive sentiment as an anti–free rider psychological device. *Evolution and Human Behavior, 23*, 203–231.

Proctor, R. 1999. *The Nazi war on cancer*. Princeton, N.J.: Princeton University Press.

Provine, R. R. 1993. Laughter punctuates speech: Linguistic, social, and gender contexts of laughter. *Ethology, 95*, 291–298.

Putnam, H. 1973. Reductionism and the nature of psychology. *Cognition, 2*, 131–146.

Quartz, S. R., & Sejnowski, T. J. 1997. The neural basis of cognitive development: A constructivist manifesto. *Behavioral and Brain Sciences, 20*, 537–596.

Quine, W. V. O. 1969. Natural kinds. In W. V. O. Quine (Ed.), *Ontological relativity and other essays*. New York: Columbia University Press.

Rachels, J. 1990. *Created from animals: The moral implications of Darwinism*. New York: Oxford University Press.

Raine, A., Lencz, T., Bihrle, S., LaCasse, L., & Colletti, P. 2000. Reduced prefrontal gray matter volume and reduced autonomic activity in antisocial personality disorder. *Archives of General Psychiatry, 57*, 119–127.

Rakic, P. 2000. Setting the stage for cognition: Genesis of the primate cerebral cortex. In M. S. Gazzaniga (Ed.), *The new cognitive neurosciences*. Cambridge, Mass.: MIT Press.

Rakic, P. 2001. Neurocrationism—Making new cortical maps. *Science, 294,* 1011–1012.

Ramachandran, V. S. 1993. Behavioral and magnetoencephalographic correlates of plasticity in the adult human brain. *Proceedings of the National Academy of Sciences, 90,* 10413–10420.

Ramachandran, V. S., & Blakeslee, S. 1998. *Phantoms in the brain: Probing the mysteries of the human mind.* New York: William Morrow.

Ramachandran, V. S., & Hirstein, W. 1999. The science of art. *Journal of Consciousness Studies, 6/7,* 15–41.

Rapin, I. 2001. An 8-year-old boy with autism. *Journal of the American Medical Association, 285,* 1749–1757.

Ravitch, D. 2000. *Left back: A century of failed school reforms.* New York: Simon & Schuster.

Rawls, J. 1976. *A theory of justice.* Cambridge, Mass.: Harvard University Press.

Recanzone, G. H. 2000. Cerebral cortical plasticity: Perception and skill acquisition. In M. S. Gazzaniga (Ed.), *The new cognitive neurosciences*. Cambridge, Mass.: MIT Press.

Redmond, E. 1994. *Tribal and chiefly warfare in South America.* Ann Arbor: University of Michigan Museum.

Reed, T. E., & Jensen, A. R. 1992. Conduction velocity in a brain nerve pathway of normal adults correlates with intelligence level. *Intelligence, 17,* 191–203.

Reeve, H. K. 2000. Review of Sober & Wilson's "Unto others." *Evolution and Human Behavior, 21,* 65–72.

Reiner, W. G. 2000. Cloacal exstrophy. Paper presented at the Lawson Wilkins Pediatric Endocrine Society, Boston.

Reiss, D., Neiderhiser, J. M., Hetherington, E. M., & Plomin, R. 2000. *The relationship code: Deciphering genetic and social influences on adolescent development.* Cambridge, Mass.: Harvard University Press.

Renfrew, J. W. 1997. *Aggression and its causes: A biopsychosocial approach.* New York: Oxford University Press.

Rice, M. 1997. Violent offender research and implications for the criminal justice system. *American Psychologist, 52,* 414–423.

Richards, R. J. 1987. *Darwin and the emergence of evolutionary theories of mind and behavior.* Chicago: University of Chicago Press.

Richardson, L. F. 1960. *Statistics of deadly quarrels.* Pittsburgh: Boxwood Press.

Ridley, M. 1986. *The problems of evolution.* New York: Oxford University Press.

Ridley, M. 1993. *The red queen: Sex and the evolution of human nature.* New York: Macmillan.

Ridley, M. 1997. *The origins of virtue: Human instincts and the evolution of cooperation.* New York: Viking.

Ridley, M. 2000. *Genome: The autobiography of a species in 23 chapters.* New York: HarperCollins.

Roback, J. 1993. Beyond equality. *Georgetown Law Journal, 82,* 121–133.

Rogers, A. R. 1994. Evolution of time preference by natural selection. *American Economic Review, 84,* 460–481.

Roiphe, K. 1993. *The morning after: Sex, fear, and feminism on campus.* Boston: Little, Brown.

Romer, P. 1991. Increasing returns and new developments in the theory of growth. In W. Barnett, B. Cornet, C. d'Aspremont, J. Gabszewicz, & A. Mas-Collel (Eds.), *International Symposium in Economic Theory and Econometrics*. New York: Cambridge University Press.

Romer, P., & Nelson, R. R. 1996. Science, economic growth, and public policy. In B. L. R. Smith & C. E. Barfield (Eds.), *Technology, R&D, and the economy*. Washington, D.C.: Brookings Institution.

Rose, H., & Rose, S. (Eds.) 2000. *Alas, poor Darwin! Arguments against evolutionary psychology.* New York: Harmony Books.

Rose, S. 1978. Pre-Copernican sociobiology? *New Scientist, 80,* 45–46.

Rose, S. 1997. *Lifelines: Biology beyond determinism.* New York: Oxford University Press.

Rose, S., & the Dialectics of Biology Group. 1982. *Against biological determinism.* London: Allison & Busby.

Rosen, S. 1992. War power and the willingness to suffer. In J. A. Vasquez & M. T. Henehan (Eds.), *The scientific study of peace and war: A text reader*. New York: Lexington Books.

Rossen, M., Klima, E. S., Bellugi, U., Bihrle, A., & Jones, W. 1996. Interaction between language and cognition: Evidence from Williams syndrome. In J. H. Beitchman, N. J. Cohen, M. M. Konstantareas, & R. Tannock (Eds.), *Language, learning, and behavior disorders: Developmental, biological, and clinical perspectives*. New York: Cambridge University Press.

Rossiter, C. (Ed.) 1961. *The Federalist Papers*. New York: New American Library.

Rousseau, J.-J. 1755/1986. *The first and second discourses together with the replies to critics and Essay on the origin of languages*. New York: Perennial Library.

Rousseau, J.-J. 1755/1994. *Discourse upon the origin and foundation of inequality among mankind*. New York: Oxford University Press.

Rousseau, J.-J. 1762/1979. *Emile*. New York: Basic Books.

Rowe, D. 1994. *The limits of family influence: Genes, experience, and behavior*. New York: Guilford Press.

Rowe, D. C. 2001. The nurture assumption persists. *American Psychologist, 56,* 168–169.

Rozin, P. 1996. Towards a psychology of food and eating: From motivation to module to model to marker, morality, meaning, and metaphor. *Current Directions in Psychological Science, 5,* 18–24.

Rozin, P. 1997. Moralization. In A. Brandt & P. Rozin (Eds.), *Morality and health*. New York: Routledge.

Rozin, P., & Fallon, A. 1987. A perspective on disgust. *Psychological Review, 94,* 23–41.

Rozin, P., Markwith, M., & Stoess, C. 1997. Moralization and becoming a vegetarian: The transformation of preferences into values and the recruitment of disgust. *Psychological Science, 8,* 67–73.

Rue, L. 1994. *By the grace of guile: The role of deception in natural history and human affairs*. New York: Oxford University Press.

Rumelhart, D. E., & McClelland, J. L. 1986. PDP models and general issues in cognitive science. In D. E. Rumelhart, J. L. McClelland, & the PDP Research Group (Eds.), *Parallel distributed processing: Explorations in the microstructure of cognition* (Vol. 1: *Foundations*). Cambridge, Mass.: MIT Press.

Rumelhart, D. E., McClelland, J. L., & the PDP Research Group. 1986. *Parallel distributed processing: Explorations in the microstructure of cognition* (Vol. 1: *Foundations*). Cambridge, Mass.: MIT Press.

Ruse, M. 1998. *Taking Darwin seriously: A naturalistic approach to philosophy*. Amherst, N.Y.: Prometheus Books.

Ruse, M. 2000. *Can a Darwinian be a Christian? The relationship between science and religion*. New York: Cambridge University Press.

Rushton, J. P. 1996. Race, intelligence, and the brain: The errors and omissions of the "revised" edition of S. J. Gould's "The mismeasure of man." *Personality and Individual Differences, 23,* 169–180.

Rushton, J. P., Fulker, D. W., Neale, M. C., Nias, D. K. B., & Eysenck, H. J. 1986. Altruism and aggression: The heritability of individual differences. *Journal of Personality and Social Psychology, 50,* 1192–1198.

Rutter, M. 1997. Nature-nurture integration: The example of antisocial behavior. *American Psychologist, 52,* 390–398.

Ryle, G. 1949. *The concept of mind*. London: Penguin.

Sadato, N., Pascual-Leone, A., Grafman, J., Ibañez, V., Delber, M.-P., Dold, G., & Hallett, M. 1996. Activation of the primary visual cortex by Braille reading in blind subjects. *Nature, 380,* 526–528.

Sahlins, M. 1976. *The use and abuse of biology: An anthropological critique of sociobiology*. Ann Arbor: University of Michigan Press.

Salmon, C. A. 1998. The evocative nature of kin terminology in political rhetoric. *Politics and the Life Sciences, 17,* 51–57.

Salmon, C. A., & Symons, D. 2001. *Warrior lovers*. New Haven, Conn.: Yale University Press.

Samelson, F. 1982. Intelligence and some of its testers (Review of S. J. Gould's "The mismeasure of man"). *Science, 215,* 656–657.

Saperstein, A. M. 1995. War and chaos. *American Scientist, 83,* 548–557.

Sapolsky, R. M. 1997. *The trouble with testosterone: And other essays on the biology of the human predicament*. New York: Simon & Schuster.

Sayre-McCord, G. 1988. *Essays on moral realism*. Ithaca, N.Y.: Cornell University Press.

Scarr, S., & Carter-Saltzman. 1979. Twin method: Defense of a critical assumption. *Behavior Genetics, 9,* 527–542.

Scarr, S., & Weinberg, R. A. 1981. The transmission of authoritarian attitudes in families: Genetic resemblance in social-political attitudes? In S. Scarr (Ed.), *Race, social class, and individual differences in IQ*. Mahwah, N.J.: Erlbaum.

Scarry, E. 1999. *Dreaming by the book*. New York: Farrar, Straus & Giroux.

Schaller, M., & Crandall, C. (Eds.) In press. *The psychological foundations of culture*. Mahwah, N.J.: Erlbaum.

Schellenberg, E. G., & Trehub, S. E. 1996. Natural musical intervals: Evidence from infant listeners. *Psychological Science, 7*, 272–277.

Schelling, T. 1960. *The strategy of conflict*. Cambridge, Mass.: Harvard University Press.

Schütze, Y. 1987. The good mother: The history of the normative model "mother-love." In P. A. Adler, P. Adler, & N. Mandell (Eds.), *Sociological studies of child development* (Vol. 2). Greenwich, Conn.: JAI Press.

Schwartz, F. N. 1992. *Breaking with tradition: Women and work, the new facts of life*. New York: Warner Books.

Scott, J. C. 1998. *Seeing like a state: How certain schemes to improve the human condition failed*. New Haven, Conn.: Yale University Press.

Searle, J. R. 1995. *The construction of social reality*. New York: Free Press.

Segal, N. 2000. Virtual twins: New findings on within-family environmental influences on intelligence. *Journal of Educational Psychology, 92*, 442–448.

Segerstråle, U. 2000. *Defenders of the truth: The battle for sociobiology and beyond*. New York: Oxford University Press.

Seligman, M. E. P. 1971. Phobias and preparedness. *Behavior Therapy, 2*, 307–320.

Sen, A. 1984. *Poverty and famines: An essay on entitlement and deprivation*. New York: Oxford University Press.

Sen, A. 2000. East and West: The reach of reason. *New York Review of Books*, July 20.

The Seville Statement on Violence. 1990. *American Psychologist, 45*, 1167–1168.

Shalit, W. 1999. *A return to modesty: Discovering the lost virtue*. New York: Free Press.

Sharma, J., Angelucci, A., & Sur, M. 2000. Induction of visual orientation modules in auditory cortex. *Nature, 404*, 841–847.

Sharpe, G. 1994. *William Hill's bizarre bets*. London: Virgin Books.

Shastri, L. 1999. Advances in SHRUTI: A neurally motivated model of relational knowledge representation and rapid inference using temporal synchrony. *Applied Intelligence, 11*, 79–108.

Shastri, L., & Ajjanagadde, V. 1993. From simple associations to systematic reasoning: A connectionist representation of rules, variables, and dynamic bindings using temporal synchrony. *Behavioral and Brain Sciences, 16*, 417–494.

Shatz, A. 1999. The guilty party. *Lingua Franca*, B17–B21.

Shepard, R. N. 1990. *Mind sights: Original visual illusions, ambiguities, and other anomalies*. New York: W. H. Freeman.

Sherif, M. 1966. *Group conflict and cooperation: Their social psychology*. London: Routledge & Kegan Paul.

Shipman, P. 1994. *The evolution of racism*. New York: Simon & Schuster.

Short, P. 1999. *Mao: A life*. New York: Henry Holt.

Shoumatoff, A. 1985. *The mountain of names: A history of the human family*. New York: Simon & Schuster.

Shweder, R. A. 1990. Cultural psychology: What is it? In J. W. Stigler, R. A. Shweder, & G. H. Herdt (Eds.), *Cultural psychology: Essays on comparative human development*. New York: Cambridge University Press.

Shweder, R. A. 1994. "You're not sick, you're just in love": Emotion as an interpretive system. In P. Ekman & R. J. Davidson (Eds.), *The nature of emotion*. New York: Oxford University Press.

Shweder, R. A., Much, N. C., Mahapatra, M., & Park, L. 1997. The "big three" of morality (autonomy, community, and divinity) and the "big three" explanations of suffering. In A. Brandt & P. Rozin (Eds.), *Morality and health*. New York: Routledge.

Siegal, M., Varley, R., & Want, S. C. 2001. Mind over grammar: Reasoning in aphasia and development. *Trends in Cognitive Sciences, 5*, 296–301.

Silverman, I., & Eals, M. 1992. Sex differences in spatial abilities: Evolutionary theory and data. In J. Barkow, L. Cosmides, & J. Tooby (Eds.), *The adapted mind: Evolutionary psychology and the generation of culture*. New York: Oxford University Press.

Simon, J. L. 1996. *The ultimate resource 2*. Princeton, N.J.: Princeton University Press.

Singer, I. B. 1972. *Enemies, a love story*. New York: Farrar, Straus & Giroux.

Singer, P. 1981. *The expanding circle: Ethics and sociobiology*. New York: Farrar, Straus & Giroux.

Singer, P. 1999. *A Darwinian left: Politics, evolution, and cooperation*. New Haven, Conn.: Yale University Press.

Siple, P., & Fischer, S. D. (Eds.) 1990. *Theoretical issues in sign language research*. Chicago: University of Chicago Press.

Skinner, B. F. 1948/1976. *Walden Two*. New York: Macmillan.

Skinner, B. F. 1971. *Beyond freedom and dignity*. New York: Knopf.

Skinner, B. F. 1974. *About behaviorism*. New York: Knopf.

Skuse, D. H., James, R. S., Bishop, D. V. M., Coppin, B., Dalton, P., Aamodt-Leeper, G., Bacarese-Hamilton, M., Cresswell, C., McGurk, R., & Jacobs, P. A. 1997. Evidence from Turner's Syndrome of an imprinted X-linked locus affecting cognitive function. *Nature, 287*, 705–708.

Sloman, S. A. 1996. The empirical case for two systems of reasoning. *Psychological Bulletin, 119*, 3–22.

Slovic, P., Fischof, B., & Lichtenstein, S. 1982. Facts versus fears: Understanding perceived risk. In D. Kahneman, P. Slovic, & A. Tversky (Eds.), *Judgment under uncertainty: Heuristics and biases*. New York: Cambridge University Press.

Smith, A. 1759/1976. *The theory of moral sentiments*. Indianapolis: Liberty Classics.

Smith, A., Jussim, L., & Eccles, J. 1999. Do self-fulfilling prophesies accumulate, dissipate, or remain stable over time? *Journal of Personality and Social Psychology, 77*, 548–565.

Smolensky, P. 1990. Tensor product variable binding and the representation of symbolic structures in connectionist systems. *Artificial Intelligence, 46*, 159–216.

Smolensky, P. 1995. Reply: Constituent structure and explanation in an integrated connectionist/symbolic cognitive architecture. In C. MacDonald & G. MacDonald (Eds.), *Connectionism: Debates on Psychological Explanations* (Vol. 2). Cambridge, Mass.: Blackwell.

Snyderman, M., & Rothman, S. 1988. *The IQ controversy: The media and public policy*. New Brunswick, N.J.: Transaction.

Sommers, C. H. 1994. *Who stole feminism?* New York: Simon & Schuster.

Sommers, C. H. 1998. Why Johnny can't tell right from wrong. *American Outlook*, 45–47.

Sommers, C. H. 2000. *The war against boys: How misguided feminism is harming our young men*. New York: Touchstone Books.

Sougné, J. 1998. Connectionism and the problem of multiple instantiation. *Trends in Cognitive Sciences, 2*, 183–189.

Sowell, T. 1980. *Knowledge and decisions*. New York: Basic Books.

Sowell, T. 1985. *Marxism: Philosophy and economics*. New York: Quill.

Sowell, T. 1987. *A conflict of visions: Ideological origins of political struggles*. New York: Quill.

Sowell, T. 1994. *Race and culture: A world view*. New York: Basic Books.

Sowell, T. 1995a. Ethnicity and IQ. In S. Fraser (Ed.), *The Bell Curve wars: Race, intelligence, and the future of America*. New York: Basic Books.

Sowell, T. 1995b. *The vision of the anointed: Self-congratulation as a basis for social policy*. New York: Basic Books.

Sowell, T. 1996. *Migrations and cultures: A world view*. New York: Basic Books.

Sowell, T. 1998. *Conquests and cultures: An international history*. New York: Basic Books.

Spann, E. K. 1989. *Brotherly tomorrows: Movements for a cooperative society in America, 1820–1920*. New York: Columbia University Press.

Spelke, E. 1995. Initial knowledge: Six suggestions. *Cognition, 50*, 433–447.

Spelke, E. S., Breinlinger, K., Macomber, J., & Jacobson, K. 1992. Origins of knowledge. *Psychological Review, 99*, 605–632.

Sperber, D. 1985. Anthropology and psychology: Towards an epidemiology of representations. *Man, 20*, 73–89.

Sperber, D. 1994. The modularity of thought and the epidemiology of representations. In L. Hirschfeld & S. Gelman (Eds.), *Mapping the mind: Domain specificity in cognition and culture*. New York: Cambridge University Press.

Spiller, R. J. 1988. S. L. A. Marshall and the ratio of fire. *RUSI Journal, 133*.

Sponsel, L. 1996. The natural history of peace: The positive view of human nature and its potential. In T. Gregor (Ed.), *A natural history of peace*. Nashville, Tenn.: Vanderbilt University Press.

Sponsel, L. 1998. Yanomami: An area of conflict and aggression in the Amazon. *Aggressive Behavior, 24*, 97–122.

Staddon, J. R. 1999. On responsibility in science and law. In E. Paul, F. Miller, & J. Paul (Eds.), *Responsibility* (Vol. 16). New York: Cambridge University Press.

Steiner, G. 1967. *Language and silence: Essays on language, literature, and the inhuman.* New Haven, Conn.: Yale University Press.

Steiner, G. 1984. *Antigones: How the Antigone legend has endured in Western literature, art, and thought.* New Haven, Conn.: Yale University Press.

Steiner, W. 2001. *Venus in exile: The rejection of beauty in 20th-century art.* New York: Free Press.

Stevens, M. 2001. Only causation matters: Reply to Ahn et al. *Cognition, 82,* 71–76.

Stevens, P. 2001. Magical thinking in complementary and alternative medicine. *Skeptical Inquirer,* 32–37.

Stevens, W. 1965. *The necessary angel.* New York: Random House.

Stevenson, L., & Haberman, D. L. 1998. *Ten theories of human nature.* New York: Oxford University Press.

Stoolmiller. 2000. Implications of the restricted range of family environments for estimates of heritability and nonshared environment in behavior-genetic adoption studies. *Psychological Bulletin, 125,* 392–407.

Storey, R. 1996. *Mimesis and the human animal.* Evanston, Ill.: Northwestern University Press.

Stromswold, K. 1998. Genetics of spoken language disorders. *Human Biology, 70,* 297–324.

Stromswold, K. 2000. The cognitive neuroscience of language acquisition. In M. S. Gazzaniga (Ed.), *The new cognitive neurosciences.* Cambridge, Mass.: MIT Press.

Stryker, M. P. 1994. Precise development from imprecise rules. *Science, 263,* 1244–1245.

Sulloway, F. J. 1995. Birth order and evolutionary psychology: A meta-analytic overview. *Psychological Inquiry, 6,* 75–80.

Sulloway, F. J. 1996. *Born to rebel: Family conflict and radical genius.* New York: Pantheon.

Sur, M. 1988. Visual plasticity in the auditory pathway: Visual inputs induced into auditory thalamus and cortex illustrate principles of adaptive organization in sensory systems. In M. A. Arbib & S. Amari (Eds.), *Dynamic interactions in neural networks* (Vol. 1: *Models and data*). New York: Springer-Verlag.

Sur, M., Angelucci, A., & Sharma, J. 1999. Rewiring cortex: The role of patterned activity in development and plasticity of neocortical circuits. *Journal of Neurobiology, 41,* 33–43.

Swim, J. K. 1994. Perceived versus meta-analytic effect sizes: An assessment of the accuracy of gender stereotypes. *Journal of Personality and Social Psychology, 66,* 21–36.

Symons, D. 1979. *The evolution of human sexuality.* New York: Oxford University Press.

Symons, D. 1995. Beauty is in the adaptations of the beholder: The evolutionary psychology of human female sexual attractiveness. In P. R. Abramson & S. D. Pinkerton (Eds.), *Sexual nature, sexual culture.* Chicago: University of Chicago Press.

Szathmáry, E., Jordán, F., & Pál, C. 2001. Can genes explain biological complexity? *Science, 292,* 1315–1316.

Tajfel, H. 1981. *Human groups and social categories.* New York: Cambridge University Press.

Talmy, L. 2000. The cognitive culture system. In L. Talmy (Ed.), *Toward a cognitive semantics* (Vol. 2: *Typology and process in concept structuring*). Cambridge, Mass.: MIT Press.

Taylor, J. K. 1992. *Reclaiming the mainstream: Individualist feminism rediscovered.* Buffalo, N.Y.: Prometheus Books.

Taylor, S. E. 1989. *Positive illusions: Creative self-deception and the healthy mind.* New York: Basic Books.

Tesser, A. 1993. The importance of heritability in psychological research: The case of attitudes. *Psychological Review, 100,* 129–142.

Tessier-Lavigne, M., & Goodman, C. S. 1996. The molecular biology of axon guidance. *Science, 274,* 1123–1132.

Tetlock, P. E. 1999. Coping with tradeoffs: Psychological constraints and political implications. In A. Lupia, M. McCubbins, & S. Popkin (Eds.), *Political reasoning and choice.* Berkeley: University of California Press.

Tetlock, P. E., Kristel, O. V., Elson, B., Green, M. C., & Lerner, J. 2000. The psychology of the unthinkable: Taboo tradeoffs, forbidden base rates, and heretical counterfactuals. *Journal of Personality and Social Psychology, 78,* 853–870.

Teuber, M. 1997. Gertrude Stein, William James, and Pablo Picasso's Cubism. In W. G. Bringmann, H. E. Luck, R. Miller, & C. E. Early (Eds.), *A pictorial history of psychology.* Chicago: Quintessence Publishing.

Thaler, R. H. 1994. *The winner's curse.* Princeton, N.J.: Princeton University Press.

Thiessen, D., & Young, R. K. 1994. Investigating sexual coercion. *Society, 31,* 60–63.

Thompson, P. M., Cannon, T. D., Narr, K. L., van Erp, T. G. M., Poutanen, V.-P., Huttunen, M., Lönnqvist, J., Standertskjöld-Nordenstam, C.-G., Kaprio, J., Khaledy, M., Dail, R., Zoumalan, C. I., & Toga, A. W. 2001. Genetic influences on brain structure. *Nature Neuroscience, 4,* 1–6.

Thornhill, R. 1998. Darwinian aesthetics. In C. Crawford & D. L. Krebs (Eds.), *Handbook of evolutionary psychology: Ideas, issues, and applications.* Mahwah, N.J.: Erlbaum.

Thornhill, R., & Palmer, C. T. 2000. *A natural history of rape: Biological bases of sexual coercion.* Cambridge, Mass.: MIT Press.

Thornhill, R., & Palmer, C. T. 2001. Rape and evolution: A reply to our critics (Preface to the paperback ed.), *A natural history of rape: Biological bases of sexual coercion* (paperback ed.). Cambridge, Mass.: MIT Press.

Tierney, P. 2000. *Darkness in El Dorado: How scientists and journalists devastated the Amazon.* New York: Norton.

Tinbergen, N. 1952. Derived activities: Their causation, biological significance, origin, and emancipation during evolution. *Quarterly Review of Biology, 27,* 1–32.

Tomasello, M. 1999. *The cultural origins of human cognition.* Cambridge, Mass.: Harvard University Press.

Tong, R. 1998. *Feminist thought: A more comprehensive introduction* (2nd ed.). Boulder, Colo.: Westview Press.

Tooby, J., & Cosmides, L. 1990. On the universality of human nature and the uniqueness of the individual: The role of genetics and adaptation. *Journal of Personality, 58,* 17–67.

Tooby, J., & Cosmides, L. 1992. Psychological foundations of culture. In J. Barkow, L. Cosmides, & J. Tooby (Eds.), *The adapted mind: Evolutionary psychology and the generation of culture.* New York: Oxford University Press.

Tooby, J., & DeVore, I. 1987. The reconstruction of hominid evolution through strategic modeling. In W. G. Kinzey (Ed.), *The evolution of human behavior: Primate models.* Albany, N.Y.: SUNY Press.

Tooley, M. 1972. Abortion and infanticide. *Philosophy and Public Affairs, 2,* 37–65.

Toussaint-Samat, M. 1992. *History of food.* Cambridge, Mass.: Blackwell.

Tramo, M. J., Loftus, W. C., Thomas, C. E., Green, R. L., Mott, L. A., & Gazzaniga, M. S. 1995. Surface area of human cerebral cortex and its gross morphological subdivisions: In vivo measurements in monozygotic twins suggest differential hemispheric effects of genetic factors. *Journal of Cognitive Neuroscience, 7,* 267–91.

Tribe, L. 1971. Trial by mathematics: Precision and ritual in the legal process. *Harvard Law Review, 84,* 1329–1393.

Trivers, R. 1971. The evolution of reciprocal altruism. *Quarterly Review of Biology, 46,* 35–57.

Trivers, R. 1972. Parental investment and sexual selection. In B. Campbell (Ed.), *Sexual selection and the descent of man.* Chicago: Aldine.

Trivers, R. 1974. Parent-offspring conflict. *American Zoologist, 14,* 249–264.

Trivers, R. 1976. Foreword. In R. Dawkins, *The selfish gene.* New York: Oxford University Press.

Trivers, R. 1981. Sociobiology and politics. In E. White (Ed.), *Sociobiology and human politics.* Lexington, Mass.: D. C. Heath.

Trivers, R. 1985. *Social evolution.* Reading, Mass.: Benjamin/Cummings.

Trivers, R. 1998. As they would do to you: A review of E. Sober & D. S. Wilson's "Unto others." *Skeptic, 6,* 81–83.

Trivers, R., & Newton, H. P. 1982. The crash of Flight 90: Doomed by self-deception? *Science Digest,* 66–68.

Trivers, R. L., & Willard, D. E. 1973. Natural selection of parental ability to vary the sex ratio of offspring. *Science, 179,* 90–91.

Turkheimer, E. 2000. Three laws of behavior genetics and what they mean. *Current Directions in Psychological Science, 5,* 160–164.

Turkheimer, E., & Waldron, M. 2000. Nonshared environment: A theoretical, methodological, and quantitative review. *Psychological Bulletin, 126,* 78–108.

Turner, F. 1985. *Natural classicism: Essays on literature and science.* New York: Paragon.

Turner, F. 1995. *The culture of hope.* New York: Free Press.

Turner, F. 1997. Modernism: Cure or disease? *Critical Review, 11,* 169–180.

Turner, M. 1991. *Reading minds: The study of English in the age of cognitive science.* Princeton, N.J.: Princeton University Press.

Turner, M. 1996. *The literary mind.* New York: Oxford University Press.

Tversky, A., & Kahneman, D. 1973. Availability: A heuristic for judging frequency and probability. *Cognitive Psychology, 4,* 207–232.

Tversky, A., & Kahneman, D. 1974. Judgment under uncertainty: Heuristics and biases. *Science, 185,* 1124–1131.

Twain, M. 1884/1983. *Adventures of Huckleberry Finn.* In D. Voto (Ed.), *The portable Mark Twain.* New York: Penguin.

Valero, H., & Biocca, E. 1965/1996. *Yanoáma: The story of Helena Valero, a girl kidnapped by Amazonian Indians.* New York: Kodansha.

Valian, V. 1998. *Why so slow? The advancement of women.* Cambridge, Mass.: MIT Press.

van den Berghe, P. L. 1981. *The ethnic phenomenon.* Westport, Conn.: Praeger.

Van Essen, D. C., & Deyoe, E. A. 1995. Concurrent processing in the primate visual cortex. In M. S. Gazzaniga (Ed.), *The cognitive neurosciences.* Cambridge, Mass.: MIT Press.

Van Valen, L. 1974. Brain size and intelligence in man. *American Journal of Physical Anthropology, 40,* 417–424.

Vandell, D. L. 2000. Parents, peer groups, and other socializing influences. *Developmental Psychology, 36,* 699–710.

Vasquez, J. A. 1992. The steps to war: Toward a scientific explanation of Correlates of War findings. In J. A. Vasquez & M. T. Henehan (Eds.), *The scientific study of peace and war: A text reader.* New York: Lexington Books.

Veblen, T. 1899/1994. *The theory of the leisure class.* New York: Penguin.

Venable, V. 1945. *Human nature: The Marxian view.* New York: Knopf.

Venter, C., et al. 2001. The sequence of the human genome. *Science, 291,* 1304–1348.

Verhage, M., Maia, A. S., Plomp, J. J., Brussaard, A. B., Heeroma, J. H., Vermeer, H., Toonen, R. F., Hammer, R. E., van der Berg, T. K., Missler, M., Geuze, H. J., & Südhoff, T. C. 2000. Synaptic assembly of the brain in the absence of neurotransmitter secretion. *Science, 287,* 864–869.

Vonnegut, K. 1968/1998. *Welcome to the monkey house.* New York: Doubleday.

Waddington, D. H. 1957. *The strategy of the genes.* London: Allen & Unwin.

Wakefield, J. C. 1992. The concept of mental disorder: On the boundary between biological facts and social values. *American Psychologist, 47,* 373–388.

Walker, L. J. 1984. Sex differences in the development of moral reasoning: A critical review. *Child Development, 55,* 677–691.

Walker, P. L. 2001. A bioarchaeological perspective on the history of violence. *Annual Review of Anthropology, 30,* 573–596.

Walker, R. (Ed.) 1995. *To be real: Telling the truth and changing the face of feminism.* New York: Anchor Books.

Wang, F. A., Nemes, A., Mendelsohn, M., & Axel, R. 1998. Odorant receptors govern the formation of a precise topographic map. *Cell, 93,* 47–60.

Ward, K. 1998. *Religion and human nature.* New York: Oxford University Press.

Warren, M. A. 1984. On the moral and legal status of abortion. In J. Feinberg (Ed.), *The problem of abortion.* Belmont, Calif.: Wadsworth.

Watson, G. 1985. *The idea of liberalism.* London: Macmillan.

Watson, J. B. 1924/1998. *Behaviorism.* New Brunswick, N.J.: Transaction.

Weiskrantz, L. (Ed.) 1988. *Thought without language.* New York: Oxford University Press.

Weizenbaum, J. 1976. *Computer power and human reason.* San Francisco: W. H. Freeman.

White, S. H. 1996. The relationships of developmental psychology to social policy. In E. Zigler, S. L. Kagan, & N. Hall (Eds.), *Children, family, and government: Preparing for the 21st century.* New York: Cambridge University Press.

Whorf, B. L. 1956. *Language, thought, and reality: Selected writings of Benjamin Lee Whorf.* Cambridge, Mass.: MIT Press.

Wilkinson, M. J. In press. The Greek-Turkish-American triangle. In M. Abramovitz (Ed.), *Turkey and the United States.* New York: Century Foundation.

Wilkinson, R. 2000. *Mind the gap: Hierarchies, health, and human evolution.* London: Weidenfeld and Nicholson.

Willerman, L., Schultz, R., Rutledge, J. N., & Bigler, E. D. 1991. In vivo brain size and intelligence. *American Journal of Physical Anthropology, 15,* 223–238.

Williams, G. C. 1966. *Adaptation and natural selection: A critique of some current evolutionary thought.* Princeton, N.J.: Princeton University Press.

Williams, G. C. 1988. Huxley's evolution and ethics in sociobiological perspective. *Zygon: Journal of Religion and Science, 23,* 383–407.

Williams, J. M. 1990. *Style: Toward clarity and grace.* Chicago: University of Chicago Press.

Williams, K., Harkins, S., & Latané, B. 1981. Identifiability as a deterrent to social loafing: Two cheering experiments. *Journal of Personality and Social Psychology, 40,* 303–311.

Wilson, D. S., & Sober, E. 1994. Re-introducing group selection to the human behavior sciences. *Behavioral and Brain Sciences, 17,* 585–608.

Wilson, E. O. 1975/2000. *Sociobiology: The new synthesis* (25th-anniversary ed.). Cambridge, Mass.: Harvard University Press.

Wilson, E. O. 1984. *Biophilia.* Cambridge, Mass.: Harvard University Press.

Wilson, E. O. 1994. *Naturalist.* Washington, D.C.: Island Press.

Wilson, E. O. 1998. *Consilience: The unity of knowledge.* New York: Knopf.

Wilson, J. Q. 1993. *The moral sense.* New York: Free Press.

Wilson, J. Q., & Herrnstein, R. J. 1985. *Crime and human nature.* New York: Simon & Schuster.

Wilson, M., & Daly, M. 1992. The man who mistook his wife for a chattel. In J. H. Barkow, L. Cosmides, & J. Tooby (Eds.), *The adapted mind: Evolutionary psychology and the generation of culture.* New York: Oxford University Press.

Wilson, M., & Daly, M. 1997. Life expectancy, economic inequality, homicide, and reproductive timing in Chicago neighborhoods. *British Medical Journal, 314,* 1271–1274.

Wilson, R. A., & Keil, F. C. 1999. *The MIT Encyclopedia of the Cognitive Sciences.* Cambridge, Mass.: MIT Press.

Witelson, S. F., Kigar, D. L., & Harvey, T. 1999. The exceptional brain of Albert Einstein. *Lancet, 353,* 2149–2153.

Wolfe, T. 1975. *The painted word.* New York: Bantam Books.

Wolfe, T. 1981. *From Bauhaus to our house.* New York: Bantam Books.

Wolfe, T. 2000. Sorry, but your soul just died. In *Hooking up.* New York: Farrar, Straus & Giroux.

Wrangham, R. 1999. Is military incompetence adaptive? *Evolution and Human Behavior, 20,* 3–17.

Wrangham, R. W., & Peterson, D. 1996. *Demonic males: Apes and the origins of human violence.* Boston: Houghton Mifflin.

Wright, F. A., Lemon, W. J., Zhao, W. D., Sears, R., Zhuo, D., Wang, J.-P., Yang, H.-Y., Baer, T., Stredney, D., Spitzner, J., Stutz, A., Krahe, R., & Yuan, B. 2001. A draft annotation and overview of the human genome. *Genome Biology, 2,* 0025.1–0025.18.

Wright, L. 1995. Double mystery. *New Yorker,* August 7, 45–62.

Wright, R. 1994. *The moral animal: Evolutionary psychology and everyday life.* New York: Pantheon.

Wright, R. 2000. *NonZero: The logic of human destiny.* New York: Pantheon.

Yinon, Y., & Dovrat, M. 1987. The reciprocity-arousing potential of the requestor's occupation, its status, and the cost and urgency of the request as determinants of helping behavior. *Journal of Applied Social Psychology, 17,* 429–435.

Young, C. 1999. *Ceasefire! Why women and men must join forces to achieve true equality.* New York: Free Press.

Zahavi, A., & Zahavi, A. 1997. *The handicap principle: A missing piece of Darwin's puzzle.* New York: Oxford University Press.

Zahn-Wexler, C., Radke-Yarrow, M., Wagner, E., & Chapman, M. 1992. Development of concern for others. *Developmental Psychology, 28,* 126–136.

Zentner, M. R., & Kagan, J. 1996. Perception of music by infants. *Nature, 383,* 29.

Zhou, R., & Black, I. B. 2000. Development of neural maps: Molecular mechanisms. In M. S. Gazzaniga (Ed.), *The new cognitive neurosciences.* Cambridge, Mass.: MIT Press.

Zimbardo, P. G., Maslach, C., & Haney, C. 2000. Reflections on the Stanford Prison Experiment: Genesis, transformations, consequences. In T. Blass (Ed.), *Current perspectives on the Milgram paradigm.* Mahwah, N.J.: Erlbaum.

Zimler, J., & Keenan, J. M. 1983. Imagery in the congenitally blind: How visual are visual images? *Journal of Experimental Psychology: Learning, Memory, and Cognition, 9,* 269–282.

INDEX

aristocracy, 5–6, 301–2
Aristotle, 266
Arlo and Janis, 163
Arnhart, Larry, 299
Art (Bell), 413
artificial intelligence, 33–34, 61, 105, 106
arts, 216–17, 400–420
  brain and, 405
  human nature and, 404–20
  modernism and, 409–13, 417–18
  postmodernism and, *see* postmodernism
  prevalence of, 404–5
  psychological roots of, 404–9, 412, 417
  sexual attraction and, 407–8
  three ailing areas of, 403–4
  universal tastes and, 408–9
  visual system and, 405, 412, 417–18
Asimov, Isaac, 133
associationism, 18–19, 21, 62, 79, 81
Astell, Mary, 337–38
Astonishing Hypothesis, The (Crick), 41
Atran, Scott, 230
Austad, Steven, 397
Australia, 68–69, 404
autism, 46, 62

Baby and Child Care (Spock), 20
Baddeley, Alan, 209–10
Bailey, Ronald, 131
Baker, Mark, 38
Bakunin, Mikhail, 295, 331
Bambi, 12
Barash, David, 366
Barry, Dave, 383, 425
Barthes, Roland, 208
Bates, Elizabeth, 35–36
Bauhaus, 418
Bazelon, David, 181
Beatles, the, 402
beauty, 53, 387–88, 405
  denial of, 413–14
Beauvoir, Simone de, 171
Becker, Gary, 357

behavioral genetics, 45–51, 111, 124, 134, 142, 413
  family effects in, 378–87
  mind-matter divide and, 45–51
  three laws of, 372–80, 393
  unique environment in, 380–81
  see also heritability
behaviorism, 19–21, 40, 124, 170, 177
Behavior of Organisms, The (Skinner), 20
Behe, Michael, 130
Bell, Clive, 413
Bell, Quentin, 407, 414
Bell Curve, The (Herrnstein and Murray), viii, 301, 302
Benbow, Camilla, 342, 353, 356
Benedict, Ruth, 25
Benny, Jack, 278
Bentham, Jeremy, 285
Berkeley, George, 22
Berlin, Isaiah, 151, 170, 287
Berra, Yogi, 322
Bethell, Tom, 130
Betzig, Laura, 342
Bever, Tom, 80
Beyond Freedom and Dignity (Skinner), 169
Bible, 2, 128–29
Bierce, Ambrose, 240
Big Chill, The, 264
biology:
  intuitive, 220
  reductionism and, 69–72
  soul concept and, 224–27
biophilia, 405
birth order, 389–90
"Black Male" art exhibit, 217, 411
Black Panthers, 111, 301
Blank Slate, 11, 17
  origin of term, 5
  rise of, 16–17
  see also specific topics
blind people, 94–96
Block, Ned, 11
Boas, Franz, 22, 25, 66–67, 207
Boehm, Christopher, 298

Bogart, Humphrey, 163
bonobos, 45
Borges, Jorge Luis, 59
Bork, Robert, 130
Born to Rebel (Sulloway), 381, 389
Boston Globe, 86, 309, 339, 360, 370, 382, 394–95
Botticelli, Sandro, 408
Bouchard, Thomas, 378, 381
Bourdieu, Pierre, 407–8, 413
bourgeoisie, 128, 152, 157, 158, 410, 416
bowerbirds, 407–8
Bowles, Samuel, 303
Boyd, Brian, 417
Braceras, Jennifer, 353
Braille, 94–96
brain, 21, 41–45, 74, 83–100, 423–24
   anatomy of, 44
   art and, 405, 412
   cognitive neuroscience and, 41–45
   complexity of, 197
   corpus callosum severed in, 43
   damage to, 42–45, 98–100, 265
   development of, 83–100, 227, 386–87, 396–97
   genetics and, 49, 90–94, 98
   hemispheres of, 43, 99
   inhibition and, 44
   plasticity of, 44–45, 74, 83–100
   sex differences in, 347
   visual cortex of, 87–97
   see also neural plasticity
"Brain Is Wider Than the Sky, The" (Dickinson), 423–24
Brain Storm (Dooling), 176
Brando, Marlon, 375
Brasília, 170
Brazelton, T. Berry, 386, 394
Brecht, Bertolt, 170
Breggin, Peter, 314
Breland, Keller, 20
Breland, Marian, 20
Brennan, William, 181
Britain, 16, 68, 71, 144, 296
Broca, Paul, 44
Brooks, Rodney, 61

Brown, Donald, 55, 57, 435–39
Brown, Roger, 205
Brownmiller, Susan, 361–62, 363–64, 365, 368
Bruer, Jon, 386–87
Bryan, William Jennings, 130
Buckley, William F., 130, 262
Buddha, 163
Bueno de Mesquita, Bruce, 319
Bukharin, Nikolai, 156
Burke, Edmund, 287, 289
Buruma, Ian, 280
Bush, George W., 12, 130, 274
Buss, David, 316
Butler, Judith, 415, 416
Byatt, A. S., 419

Calvin and Hobbes, 187
Cambodia, 152, 155, 158
Canada, 16, 311, 331, 333
cannibalism, 306–7, 320
capitalism, 161, 246–47, 290–91, 297, 302–4, 393
capital punishment, 181–82, 331
Carey, Susan, 222
Carnegie, Andrew, 16
Carroll, Joseph, 417
Cashdan, Elizabeth, 342
categorization, 201–7, 228–29
Centers for Disease Control, 312
cerebral palsy, 99
Cézanne, Paul, 409
Chagnon, Napoleon, 115–19, 314, 323, 334, 431
Chamberlain, Neville, 333
Chandigarh, 170
Cheers, 403
Chekhov, Anton, xi
child abuse, 164–65, 308–9
child development:
   chance in, 396
   family effects in, 249, 378–99
   heritability of traits in, 373–78
childrearing, *see* parenting
chimpanzees, 45, 61–62, 89, 134, 143, 316, 367

neural circuitry and, 60
socialization and, 377, 391, 399
stability and change in, 66
Culture of Honor (Nisbett and Cohen), 327

Dahmer, Jeffrey, 263
Daly, Martin, 135, 164–65, 182, 254, 304, 313, 319, 325, 327
Damasio, Antonio, 99–100
Damasio, Hannah, 99–100
Daniels, Denise, 381
Danto, Arthur, 409
Darkness in El Dorado (Tierney), 116
Darrow, Clarence, 130
Darwin, Charles, 2, 15, 28, 30, 51, 66, 132, 151, 186, 254, 285, 305
Darwinian Left, A (Singer), 298, 300
Dawkins, Richard, 53, 112, 113, 114, 191, 241, 318–19
deafness, 95, 391
Death by Government (Rummel), 332
decision making, 40, 42–44, 51, 58, 174–75, 302–3
Declaration of Independence, 145
deconstructionism, 198, 208, 209
   see also postmodernism
Deep Blue, 33–34
Degler, Carl, 17
de Kenessey, Stefania, 417
Delaney Clause, 278
Delay, Tom, 129
democracy, 296–98
Denfeld, Rene, 343
Dennett, Dan, 10, 177, 216
Derrida, Jacques, 208
Derrière Guard, 417
Descartes, René, 8, 9–10, 42, 126, 215
determinism, 112–13, 122, 127, 174–85
deterrence, 180–85, 324–29, 330–32
Devil's Dictionary, The (Bierce), 240
DeVore, Irven, 111, 238
de Waal, Frans, 168, 298
Dialectical Biologist, The (Levins), 126
dialectical biology, 113, 126, 135

Dialogue Concerning the Two Chief World Systems (Galileo), 138
Diamond, Jared, 68–69
Dickeman, Mildred, 342
Dickens, Charles, 291–92
Dickinson, Emily, 423–24
Dictator game, 256, 257
Didion, Joan, 342–43
difference feminism, 342
Discovery Institute, 161
discrimination, 141, 145–49, 201–2, 204–7, 214, 217, 311–12
   age, 148
   sex, 337–39, 341, 351, 354, 355, 357
Disney, Walt, 11
Disraeli, Benjamin, 287
Dissanayake, Ellen, 404–5, 406
Divale, W. T., 57
Dooling, Richard, 176
Dostoevsky, Fyodor, 41, 307
Double-Blind Dictator game, 257
Douglas, William O., 181, 265
Dred Scott decision, 292
drug policies, 331–32
Dryden, John, 6
dualism, 8–9, 10
   see also Ghost in the Machine;
      mind-matter divide; soul
Dunbar, Robin, 298
Durham decision, 184
Durkheim, Emile, 23–24, 25, 108, 156, 284, 286, 427
Dutton, Denis, 404, 406–7, 415, 417
Dworkin, Andrea, 171, 365
Dworkin, Ronald, 288

Eagly, Alice, 309
Easterlin, Nancy, 417
Eastwood, Clint, 219
economics:
   behavioral, 256–58, 302–4
   human nature as seen in, 256, 285–86, 302–3
   intuitive, 221, 233–36, 302–3
education, 222–23, 235–36, 301
   arts and humanities in, 401

Hrdy, Sarah Blaffer, 250, 342
Hubel, David, 97, 108
Huckleberry Finn, The Adventures of,
   (Twain), 428–31
Human Genome Project, xi, 74, 75–78,
   135, 377
humanities, 6, 31, 60, 68, 69–72, 75, 134,
   285, 356, 400–420
Human Universals (Brown), 435–39
Humboldt, Alexander von, 301
Hume, David, 79, 178, 180, 279, 296,
   408
Hummel, John, 80
Hunt, Morton, 128
hunter-gatherer societies, 53, 63, 68,
   233–34, 294, 306–7, 316
Hurtado, Magdalena, 342
Huston, Anjelica, 432
hyperreality, 214

identity politics, 206–7
Ifaluk, 38–39
imagery, *see* psychology, of imagery
images, 213–18
imitation, 60–62, 63–65
immigration, 17, 391
inequality, 141–58, 304
infanticide, 227, 248–49
information, 31–34, 238
Inge, Dean, 287
Inherit the Wind, 130
Inquisition, 137
insanity defense, 183–84
intellectual property, 238
intelligence, 33, 44, 106–7, 238, 375
   denials of, 149–50
   heritability of, 47, 146–47, 150, 297,
      374–75, 376–78
   multiple, 219
intelligence quotient (IQ) tests, viii,
   106–7, 134–35, 145, 146–47, 201,
   301–2, 378
   behavioral genetics and, 373, 379
Intelligent Design, 130, 133
interest, lending at, 234–35
intuitive psychology, *see* theory of mind

Iroquois Federation, 296
Israel, 246, 257, 311
It Takes a Village (Clinton), 394

Jackson, Andrew, 328
James, Oliver, 393
James, William, 19, 56, 203, 417–18
Japan, 310, 369, 408
Japanese-Americans, 312
Japanese language, 37–38, 71
Jefferson, Thomas, 145
Jensen, Arthur, 107
Jespersen, Otto, 14–15, 22
Jesus Christ, 193
Jews, 16, 17, 22, 66, 130, 141, 143, 152,
   154, 201, 204, 217, 235, 251, 274,
   431–32
Jivaro, 117–18
John Paul II, Pope, 130, 186, 224
Johnson, Lyndon, 286, 292, 298, 308, 313
Johnson, Philip, 130
Johnson, Samuel, 30, 137, 142, 418
Jones, Owen, 165, 176, 367, 371
Jong, Erica, 253, 254
Joshua, 137
judicial activism, 291
Jumpers (Stoppard), 321–22
Junger, Sebastian, 258
junk DNA, 78
Just Society, 286

Kagan, Jerome, 394–95
Kahneman, Daniel, 302
Kamin, Leon, 112, 113, 114, 122–23, 126,
   378
Kaminer, Wendy, 342
Kant, Immanuel, 180, 193, 287, 301,
   332–33
Kantor, J. R., 19–20
Karamazov, Dmitri, 85
Kasparov, Garry, 34
Kass, Leon, 130, 133, 274, 339–40
Katz, Lawrence, 97–98
Keegan, John, 333
Keeley, Lawrence, 56, 57
Keil, Frank, 230

mind-matter divide (*continued*)
see also dualism; Ghost in the
Machine; soul
Minogue, Kenneth, 170
Minsky, Marvin, 80
"Misbehavior of Organisms, The"
(Breland and Breland), 20
Mismeasure of Man, The (Gould), 149
M'Naughten rule, 183–84
modernism, 170–71, 409–13, 417–18
Money, John, 349
Montagu, Ashley, 24, 26, 27, 124, 134,
258, 307
Monty Python's Flying Circus, 72
Moore, G. E., 150
moralistic fallacy, 162–63, 178, 313
moralization 275–77
morality, 269–80
basis of, 168–69, 187–90, 192–93, 224,
274–75
cross-cultural differences in, 166–69,
271–75
emotions and, 271–72, 279
religion and, 186–90
science and, 103–4, 138–39
self-deception and, 264–66
universality of, 168–69, 187–88, 193,
271–75
Moral Majority, 276
moral progress, 166–68
Mount, Ferdinand, 246
Muller, Hermann, 153
Murdoch, Iris, 343, 418–19
Murdock, George, 24
Murray, Charles, viii, 302
music, 402, 403, 405, 409, 410, 417
Mutual Assured Destruction, 325
mutualism, 242
Myth of the First Three Years, The
(Bruer), 386–87

Napoleon I, emperor of France, 295
National Center for Science Education,
129
National Endowment for the Arts, 401
National Institute of Mental Health, 312

National Institutes of Health, 314
National Public Radio, 166
National Science Foundation, 359
Native Americans, 6, 12, 22–23, 57,
115–19, 124, 212, 296, 298, 332
Natural Classicism, 417
natural history, intuitive, 220
Natural History of Rape, A (Thornhill
and Palmer), viii, 161, 359–69
naturalistic fallacy, 150, 162–63, 164
natural selection, 28, 50, 51–52, 54, 55,
83, 101, 142, 231, 249
Dawkins on, 318–19
sex ratios and, 343
see also evolution
Navajo language, 37
Nazism, 153–58, 180–81, 272–73
Neel, James, 115–19
Neill, A. S., 222
neologisms, 211–13
nepotism, 245–46, 253, 294
Nesse, Randolph, 264
neural development, 83–100, 227,
386–87, 396–97
neural networks, 21, 42, 78–83, 92
neural plasticity, 44–45, 74, 83–100,
384–87
brain damage and, 98–100
of cortex vs. subcortical structures, 89
developmental biology and, 90–100,
386–87, 396–97
primary sensory cortex and, 87–91,
93–94
neuroscience, 111, 131, 341–42, 386–87
cognitive, 41–45
commercial applications of, 87
and Ghost in the Machine, 42, 44, 129
mind-matter divide and, 41–45
see also brain
Newell, Alan, 105
New Formalism, 417
*New Know-Nothings, The: The Political
Foes of the Scientific Study of
Human Nature* (Hunt), 128
Newman, Barnett, 413
Newton, Huey, 111

Newton, Sir Isaac, 30
New Yorker, 46, 116, 179
New York Review of Books, 109, 262
New York Times, 86, 179, 339, 349, 414
Nietzsche, Friedrich, 131, 139, 207
nihilism, 269
   religion concerns about, 186–90
   secular concerns about, 190–94
Nim Chimpsky, 61
1984 (Orwell), 425–26
Nisbett, Richard, 327, 328
Nixon, Richard M., 180, 236, 298
Noble Savage, 6, 26–27, 31, 118, 119,
    162, 261, 293, 381, 421
   communalism and, 255
   evolution and, 55–56
   feminism and, 339
   neuroscience and, 44
   radical science defense of, 124–26, 134
   rape and, 362
   violence and, 312, 336
nonrival goods, 238
Non Sequitur (comic strip), 179
Not in Our Genes (Kamin), 112, 126
Nozick, Robert, 149, 151
nuclear weapons, 324, 325
number sense, 192, 209–10, 220, 223
Nurture Assumption, The (Harris), viii,
   381, 392
Nussbaum, Martha, 172

Oakshott, Michael, 290
Ofili, Chris, 414
Oklahoma City bombing (1995), 309
olfactory (smell) system, 93
Onion, The, 414, 425
orangutans, 367
Ortega y Gasset, José, 24, 308
Orwell, George, 321, 425–28
Ozick, Cynthia, 343

Paddock, Paul, 236–37
Paddock, William, 236–37
Paglia, Camille, 342, 343, 369–70
Paine, Thomas, 288
Painted Word, The, 414

paleontology, 55, 306
Palmer, Craig, viii, 161, 176, 359–69
Papert, Seymour, 80
Parallel Distributed Processing
   (Rumelhart, McClelland et al.), 21
parenting, viii, ix, x, 164–65, 171–72,
   378–99
   behavioral genetics and, 378–87
   conflicts in, 249–51
   individualized, 387–90
   sex differences in, 252–54, 345, 350, 357
   stepparenting, 164–65
Parsons, Talcott, 285, 286
Pascal, Blaise, 419
Passmore, John, 159
Pasteur, Louis, 154
Patai, Daphne, 342
Paul, Elizabeth, 253
peers, 390–92, 395–96, 399
Percy, Walker, 209
perfectability, 27, 159–73
Perfect Storm, The (Junger), 258
Perry, Bruce, 171
personality, 46–51, 135, 373
   socialization vs., 395
   see also traits
Petitto, Laura, 61, 95
phantom limbs, 98
philosophy, 5–13, 18, 22, 23, 33, 34–35,
    38–39, 64–65, 69–70, 101, 137–39,
    145–48, 150–52, 159, 162, 164,
    166–68, 169–70, 172, 174–78,
    179–85, 187–89, 192–93, 207, 216,
    227–29, 239–40, 274–75, 279–80,
    287–88, 296–98, 318–32, 335–36,
    337–38
Philosophy and Literature, 415
physical fallacy, 234
physics, 30, 137, 239
   intuitive, 220, 223, 239
Picasso, Pablo, 409
Pindar, 287
Piss Christ (Serrano), 414
Plato, 192, 284, 285
Plomin, Robert, 381
Pocahontas, 12

FOR THE BEST IN PAPERBACKS, LOOK FOR THE

In every corner of the world, on every subject under the sun, Penguin represents quality and variety—the very best in publishing today.

For complete information about books available from Penguin—including Penguin Classics, Penguin Compass, and Puffins—and how to order them, write to us at the appropriate address below. Please note that for copyright reasons the selection of books varies from country to country.

In the United States: Please write to *Penguin Group (USA), P.O. Box 12289 Dept. B, Newark, New Jersey 07101-5289* or call 1-800-788-6262.

In the United Kingdom: Please write to *Dept. EP, Penguin Books Ltd, Bath Road, Harmondsworth, West Drayton, Middlesex UB7 0DA.*

In Canada: Please write to *Penguin Books Canada Ltd, 10 Alcorn Avenue, Suite 300, Toronto, Ontario M4V 3B2.*

In Australia: Please write to *Penguin Books Australia Ltd, P.O. Box 257, Ringwood, Victoria 3134.*

In New Zealand: Please write to *Penguin Books (NZ) Ltd, Private Bag 102902, North Shore Mail Centre, Auckland 10.*

In India: Please write to *Penguin Books India Pvt Ltd, 11 Panchsheel Shopping Centre, Panchsheel Park, New Delhi 110 017.*

In the Netherlands: Please write to *Penguin Books Netherlands bv, Postbus 3507, NL-1001 AH Amsterdam.*

In Germany: Please write to *Penguin Books Deutschland GmbH, Metzlerstrasse 26, 60594 Frankfurt am Main.*

In Spain: Please write to *Penguin Books S. A., Bravo Murillo 19, 1° B, 28015 Madrid.*

In Italy: Please write to *Penguin Italia s.r.l., Via Benedetto Croce 2, 20094 Corsico, Milano.*

In France: Please write to *Penguin France, Le Carré Wilson, 62 rue Benjamin Baillaud, 31500 Toulouse.*

In Japan: Please write to *Penguin Books Japan Ltd, Kaneko Building, 2-3-25 Koraku, Bunkyo-Ku, Tokyo 112.*

In South Africa: Please write to *Penguin Books South Africa (Pty) Ltd, Private Bag X14, Parkview, 2122 Johannesburg.*